ELEMENTS OF COMPUTER-AIDED DESIGN AND MANUFACTURING

ELEMENTS OF COMPUTER-AIDED DESIGN AND MANUFACTURING

CAD/CAM

Y. C. PAO
The University of Nebraska, Lincoln

JOHN WILEY & SONS
New York   Chichester   Brisbane   Toronto   Singapore

Library of Congress Cataloging in Publication Data:

Pao, Y. C.
    Elements of computer-aided design and manufacturing,
CAD/CAM.

    Includes indexes.
    1. CAD/CAM systems.   I. Title.
TS155.6.P36   1984   670.42′7   84-5256
ISBN 0-471-88194-5

Printed in the United States of America

10 9 8 7 6 5 4 3 2 1

PREFACE

The original manuscript of this book was prepared for a pilot course "Introduction to Computer-Aided Design (CAD)" offered during the spring semester of the 1982–83 academic year at the University of Nebraska. It was a part of a coordinated effort by the College of Engineering and Technology at the University of Nebraska—Lincoln to bring computer-aided design and manufacturing technology into the college curriculum.

The acquisition of two Tektronix 4054 computer graphics systems makes possible the development of the computer programs, displays, and hardcopies presented in this text. The first 4054 system was acquired in March 1982 by a $22,000 grant from the Fred J. Kelly, II Fund of the University of Nebraska Foundation. It was the result of a teaching and research proposal that I submitted as principal investigator and assisted by Professors R. T. DeLorm, L. Kersten, C. W. Martin, R. N. McDougal, and G. M. Smith. The second 4054 system was later acquired by a fund from the College of Engineering and Technology, the University of Nebraska—Lincoln. I am grateful for all of this support.

Many of the BASIC programs presented herein are the translated versions of FORTRAN programs reported in works that I had previously published. The following graduate assistants and friends have contributed to the preparation of these programs: L. C. Chang, T. A. Huang, W. T. Kao, C. M. Lin, M. N. Maheshwari, G. K. Nagendra, J. Nikkola, K. A. Peterson, R. M. Sedlacek, D. S. S. Shy, A. J. Wang, J. D. Wilson, and S. J. Zitek. Some of them were or are supported by the General Motors CAD/CAM fellowship program. I would like to acknowledge them and the General Motors Corporation for their contributions to this text.

Leon Hill of the Boeing Company, Dr. Han-Chung Wang of the IBM Thomas J. Watson Research Center, and Erik L. Ritman, M.D., Ph.D., of the Mayo Clinic read the manuscript thoroughly and made numerous constructive suggestions for the improvement of this book. I am indeed indebted to these long-time friends for their invaluable contributions.

I would like to thank Professors Donald R. Riley, Gary L. Kinzel and Lawrence L. Durocher of the University of Minnesota, the Ohio State University, and the University of Bridgeport, respectively, and Dr. Kenneth W. Neves of Boeing Computer Services Company for their helpful comments while reviewing the manuscript. These reviewers have provided many constructive suggestions for changes,

v

particularly the addition of the APPLE, IBM, and TRS-80 BASIC versions of some of the developed BASIC programs to facilitate the readers who may not have a high-resolution Tektronix 4054 system but possesses one of the popular microcomputers. Appendix D has therefore been included to partially fulfill such a need and to present some applications of the PLOT-10 software. Many thanks go to Professor David W. Brooks of the Chemistry Department and Professor George R. Schade of the Mechanical Engineering Department, both at the University of Nebraska-Lincoln, for allowing me and my assistants to use their IBM Personal and APPLE microcomputers.

Mrs. Louise Simmons has typed many of the author's publications. Again, a great portion of this text is the result of her expert skill. The pleasure of working with the editorial staff should also be mentioned. They are dedicated, conscientiously hard-working, and most cooperative. It is indeed a happy and rewarding experience. I thank them wholeheartedly, and particularly to Bill Stenquist, Susan Winick, Elyse Rieder, and Cindy Stein. Last, but not least, my wife Rosaline Shao-Ann's patience and understanding during the preparation of this text should be especially mentioned and acknowledged.

Y. C. Pao

CONTENTS

PART TWO. CAD/CAM OF ELEMENTS AND SYSTEMS   205

ELEMENTS OF COMPUTER-AIDED DESIGN AND MANUFACTURING

PART ONE

INTERACTIVE COMPUTER GRAPHICS AND SIMPLE EXAMPLES OF CAD

CHAPTER 1

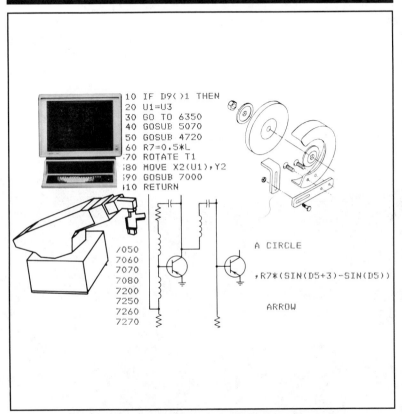

```
10 IF D9<>1 THEN
20 U1=U3
30 GO TO 6350
40 GOSUB 5070
50 GOSUB 4720
60 R7=0.5*L
70 ROTATE T1
80 MOVE X2(U1),Y2
90 GOSUB 7000
10 RETURN
```

```
7050
7060
7070
7080
7200
7250
7260
7270
```

A CIRCLE

,R7*(SIN(D5+3)-SIN(D5))

ARROW

INTRODUCTION

1.1 COMPUTER-AIDED DESIGN AND COMPUTER-AIDED MANUFACTURING (CAD/CAM)

Design and manufacturing of marketable products are the principal concern of the engineering profession. Designers apply imagination and creativity and usually go through the process of making sketches, drawing flowcharts and block diagrams, building test models, and conducting experimental tests. They have to work within cost limitations, select from a finite number of available materials and manufacturing techniques, and overcome other existing restrictions. Through **trial** processes, the designer becomes increasingly familiar with the problem and finally arrives at a satisfactory solution.

In practical situations, various possibilities are open to the designer for meeting the design **specifications**. It often begins with an initial design that meets a number, but not all, of the specifications. **Adjustments** need to be made in order to improve, refine, and if possible **optimize** the design. Hence, a considerable amount of repeated work is involved. Computers are best suited for doing repetitive work at extremely high speed. It is in the adjustment step that the computers play a very important role (Fig. 1-1).

Figure 1-1 Use of interactive graphics device for altering the design of a certain part of a passenger car. (Courtesy of Ford Motor Company.)

TABLE 1-1
Application of CAD/CAM in Industry

Application	No. of Representatives Using the Application[a]
Computer-Aided Design	
1. Designing of machine and structural elements and circuit boards	21
2. Problem solving	17
3. Interactive graphics design	14
Computer-Aided Drafting	
4. Editing and alteration of existing drawings	14
5. Automated drafting	25
Computer-Aided Manufacturing	
6. Scheduling inventory, control processes, personnel records, etc.	20
7. Numerically controlled machines	19
8. Materials handling and monitoring	8
9. Interactive parts nesting	4
10. Nondestructive and other testing	3

Source: Adapted from "Some Common Characteristics in Industrial Applications of CAD/CAM," *Engineering Design Graphics Journal*, American Society of Engineering Education, Fall 1981. Courtesy of Professor R. N. McDougal.
[a] See Table 1-3.

When a design that satisfies all specifications has been chosen, manufacturing of the design may also proceed with the aid of computers. Minicomputers may be used to numerically control the motions of the machine tools. Many other aspects of the manufacturing processes may also be controlled in real time by use of computers.

Hatvany et al.[1] performed a survey of the applications of computer-aided design in 1977. More recently, Professor R. N. McDougal has published the results of a survey of industrial application in the United States of computer-aided design and computer-aided manufacturing, abbreviated hereon as CAD and CAM, respectively. Tables 1-1, 1-2, and 1-3 are presented to show the variety of applications and industries involved in CAD, CADR, and CAM where CADR is commonly used as an abbreviation for computer-aided drafting. According to the November 1981 issue of the *Engineering Times* published by the National Society of Professional Engineers, "the worldwide market for computer-aided design and manufacturing (CAD/CAM) will reach more than $5.8 billion in 1986, up from $900 million in 1980."

In a recent article[2] by S. H. Chasen, the activities of the CAD/CAM field are

[1] T. Hatvany, W. M. Newman, and M. A. Sabin, "World Survey of Computer-Aided Design," *Computer-Aided Design*, Vol. 9, No. 2, 1977.

[2] S. H. Chasen, "Guidelines for Acquiring CAD/CAM Information," *Computers in Mechanical Engineering*, Vol. 1, No. 1, August 1982, pp. 37–42.

TABLE 1-2
Industrial Computer-Aided Applications

Application	No. of Representatives Using the Applications[a]
Computer-Aided Drafting	39
1. Automated drafting	25
2. Editing and alteration of existing drawings	14
Computer-Aided Design	
1. Designing of machine and structural elements and circuit boards	21
2. Problem solving	17
3. Interative graphics design	14
Computer-Aided Manufacturing	
1. Scheduling, inventory, control processes, personnel, records, etc.	20
2. Numerically controlled machines	19
3. Materials handling and monitoring	8
4. Interactive parts nesting	4
5. Nondestructive and other testing procedures	3

Source: Adapted from "Some Common Characteristics in Industrial Applications of CAD/CAM," *Engineering Design Graphics Journal*, American Society of Engineering Education, Fall 1981. Courtesy of Professor R. N. McDougal.
[a] See Table 1-3.

explored in five major categories. The author provides information on (1) publication media, (2) educational services, (3) users, (4) vendors, and (5) consultants. As he asserted that the field is changing at such a rapid rate, the article could not possibly be exhaustive. It did provide good sound advice for potential users of CAD/CAM in acquiring essential knowledge of the field and in selecting the available hardware and software in the market.

1.2 ENGINEERING DRAFTING AND COMPUTER PLOTTING

Designers begin their trial designs by making sketches on paper. As the design gradually develops into final form, carefully scaled drawings are then needed for checking whether or not the sizes of the parts involved will fit properly. Especially for manufacturing, more detailed descriptions of every part will have to be drawn. **Engineering drafting** is a required course in almost all engineering curricula, in which the basic techniques and rules are learned regarding scales, dimensioning, lettering, and drawing orthographic and auxiliary views of the designed objects.

Descriptive geometry plays an essential role in engineering drafting, as it determines the geometric information connected with **three-dimensional designs**. For the reason that all drawings are two-dimensional, projections of three-dimensional objects have to be taken. Equations must be derived for calculation of the angles

TABLE 1-3
Industrial Representatives Using CAD/CAM

Industrial Representative	Use of CAD/CAM[a]
Allis Chalmers	2,4,5,6,7,9
American Plywood Assn.	1,2,6,8
AMOCO (Std. Oil Co. of Ind.)	3,5,6,7
ARMCO Building Systems	1,5,7
Bechtel Power Corp. (Enq. Div.)	2,5
Bell Laboratories	1,5
Black & Veatch	1,3,5,6
Brunswick Co. (Defence Div.)	7
Burlington Northern	1,5
City of Los Angeles (County Engineer)	1,2
Colt Industries (Fairbanks Morse Pump Div.)	1,2,7
Deere & Co. (Manufacturing Eng. Div.)	2,3,4,6,7,8,9,10
Firestone Tire & Rubber Co.	5,6,7
Fisher Controls Co.	1,2,6,7
Ford Tractor Div.	1,2,3,4,5,6
General Dynamics (Electric Boat Div.)	2,3,4,7,9,10
General Motors Corp.	1,2,3,4,5,6,7,8,9,10
General Portland Inc.	6
International Paper Co.	6,7,8
Johns-Manville Sales Corp.	4,5,6
Martin Marietta Aerospace	1,3,4,5,6,7,8
McDonnell Douglas (Aircraft Co.)	3,5,7
McDonnell Douglas (Astronautics Co.)	1,3,4,5,7
Monsanto Co. (Eng. Dept.)	1,5
Phillips Petroleum Co. (Corporate Eng. Div.)	2,3,4,5
Proctor & Gamble	2,3,4,5,6,7
Republic Steel	1,5,7
Sandia Laboratories	1,2,3,4,6,7
Texaco (Computer Services Dept.)	1,3,4,5
The Schemmer Assoc. Inc. (Architectural Dept.)	1,5,6
Trane Co.	1,2,4,5,7,9
Union Oil Co. of California	2,5,6,8
Union Pacific R.R. Co.	5,6,8
United Technologies Research Center	1,6
Valmont Ind. Inc.	2,3
Western Electric	1,2,4,5,6,8
Westinghouse Elec. Corp. (R & D Center)	1,5,7

Source: Adapted from "Some Common Characteristics in Industrial Applications of CAD/CAM," *Engineering Design Graphics Journal*, American Society of Engineering Education, Fall 1981. Courtesy of Professor R. N. McDougal. [a]Refer to Table 1-1 for explanation of application of CAD/CAM

between lines, between a line and a plane, between planes, between a plane and a surface, and for the calculation of the areas, lengths, and other dimensions.

Computers can help engineering drafting in various ways (Fig. 1-2). For its speed and precision, a computer can easily take over the task of freehand lettering

Figure 1-2   Calcomp's high-speed 960 plotter. (Courtesy of California Computer Products, Inc.)

of alphabetic and numeric characters. And if a standard part is often used in designs, the layout of that part may as well be routinely plotted by a computer.

Computer plotting, automated drafting, computer-aided drafting, computer graphics, and many other terms all have been used interchangeably. In a follow-up survey by Professor R. N. McDougal, various applications of computer-aided drafting in education (Table 1-4) and in industry (Table 1-5) have been tallied. As the field is expanding at an extremely rapid pace, it should not be a surprising revelation that in a later publication the lists will have increased to tenfold or even 100-fold.

TABLE 1-4
Use of Computer Graphics in Education

University	Computer Graphics Application	Departments Using Computer Graphics	Level of Student Instruction	Text Authors[a]
Air Force Inst. of Tech.	Solutions	School of Eng.	Undergrad and Grad.	Notes and manuals
Alaska, Univ. of	Sol. and art.[b]	CE., Eng. Sci.	Fresh.	Notes and manuals
Arizona, Univ. of	Solutions	Aero., ME., CE.	Undergrad and Grad.	Newman and Sproull
Bridgeport, Univ. of	Solutions	EE., ME.	Sr. and Grad.	Newman and Sproull
British Columbia, Univ. of	Solutions	CE.	Sr.	Notes and manuals
Calif. State Univ., Long Beach	Solutions	ME. and others	Soph., Jr., Sr., Grad.	Notes and manuals
Calif. Univ. of, Davis	Solutions	ME., CE.	Fresh., Sr., Grad.	Notes and manuals
City Univ. of New York	Solutions	CE.	Undergrad. and Grad.	Notes and manuals
Clemson Univ.	Solutions	Eng. Graphics	Fresh., Jr., Sr.	Daniel Ryan Wolfgang Giloi
Concordia Univ.	Sol. and art.	CE.	Adv. undergrad. and Grad.	Notes and manuals
Connecticut, Univ. of	Solutions	Eng. Graph., ME.	Fresh., Soph., Sr., Grad.	Newman and Sproull
Dayton, Univ. of	Solutions	ME.	Sr., and Grad.	Notes and manuals
Florida Inst. of Tech.	Solutions	Math. Sci. Dep't.	Undergrad. and Grad.	Newman and Sproull Wolfgang Giloi
Georgia Inst. of Tech.	Solutions	C.S., EE., ME.	Sr., Grad.	Newman and Sproull
Harvard Univ.	Sol. and art.	Architecture	Grad.	Rogers and Adams Newman and Sproull, Parslow, Prouse, and Green
Illinois Inst. of Tech.	Solutions	All Eng. Dep'ts.	Jr., Sr., Grad.	Notes and manuals
Illinois, Univ. of, Urbana–Champlain	Solutions	Aero., Met., ME., G.E.	Soph., Jr., Sr.	Notes and manuals
Kansas, Univ. of	Sol. and art.	ME., C.S., CE., Pet. E.	Jr.	Notes and manuals
Louisiana Tech. Univ.	Solutions	ME.	Jr., Sr., Grad.	Notes and manuals
Maryland, Univ. of	Solutions	CE.	Upper level	Notes and manuals
Michigan State Univ.	Solutions	ME.	Soph., Jr., Sr., Grad.	Rogers and Adams
Michigan Tech. Univ.	Solutions	ME., EM.	Fresh., Sr.,	Rogers and Adams
Missouri, Univ. of, Columbia	Solutions	Mech. and Aero. Eng.	Soph., Jr., Sr.	Notes and manuals
Missouri, Univ. of, Rolla	Solutions	Aero., CE., Chem. E., EM., ME.	Soph., Jr.	Notes and manuals

TABLE 1-4 (*Continued*)

University	Computer Graphics Application	Departments Using Computer Graphics	Level of Student Instruction	Text Authors[a]
Nebraska, Univ. of, Lincoln	Sol. and art.	ME., CE., IE., EE.	Fresh., Sr.	DeLorm and Kersten
New Brunswick, Univ. of	Solutions	ME.	Sr.	Notes and manuals
North Carolina State Univ.	Solutions	EE., GE., C.S.	Jr., Sr.,	Notes and manuals
Northwestern Univ.	Solutions	Chem. Eng.	Grad.	Notes and manuals
Oakland Univ.	Solutions	Unknown	Upper level	Notes and manuals
Oklahoma State Univ.	Solutions	School of Tech.	Fresh., Soph., Jr., Sr.	Notes and manuals
Ohio Univ.	Solutions	Eng. Graphics	Fresh.	Notes and manuals
Pennsylvania, Univ. of	Sol. and art.	C. and Urban Eng. C.S.	Soph., Jr., Grad.	Newman and Sproull Rogers and Adams
Purdue Univ.	Sol. and art.	ME.	Upper level, Grad.	Newman and Sproull Rogers and Adams
Rochester, Univ. of	Solutions	EE., ME.	Sr. and Grad.	Notes and manuals
Rose-Hulman Inst. of Tech.	Solutions	Eng. Design	Undergrad.	Notes and manuals
South Dakota State Univ.	Solutions	G.E.	Fresh.	Notes and manuals
Southeastern Mass. Univ.	Sol. and art.	EE., Math.	Fresh., Grad.	Notes and manuals
Stevens Inst. of Tech.	Solutions	CE., Chem., Phys., Econ.	Soph., Jr., Sr.	Notes and manuals
Tennessee Tech. Univ.	Solutions	Eng. Sc. and Mech.	Jr., Grad.	Notes and manuals
Texas, Univ. of, at Austin	Solutions	ME.	Fresh., Soph.	Notes and manuals
Toronto, Univ. of	Solutions	ME., CCED	Undergrad. and Grad.	Notes and manuals
Tuskegee Inst.	Sol. and art.	EE.	Sr., Grad.	W. R. Bennet
U.S. Coast Guard Academy	Solutions	Dep't of Appl. Sc. and Eng.	Jr., Sr.	Notes and manuals
Vermont, Univ. of	Solutions	CE.	Soph., Jr., Sr.	Notes and manuals
Virginia, Univ. of	Solutions	ME., Comp. Sc.	Jr., Sr., Grad.	Notes and manuals
Western Michigan Univ.	Solutions	IE., Eng. Graph.	Sr.	Notes and manuals
Wisconsin, Univ. of, Madison	Sol. and art.	G. E., ME., CE.	Fresh., Soph., Jr., Grad.	Notes and manuals
Wisconsin, Univ. of, Milwaukee	Sol. and art.	Syst. Des. Dep't.	Fresh., Grad.	Notes and manuals

Source: Adapted from "CG as a Design Tool," *Engineering Design Graphics Journal*, American Society of Engineering Education, Winter 1981. Courtesy of Professor R. N. McDougal.
[a] Refer to original article for references.
[b] Solutions and artistic drawings.

TABLE 1-5
Use of Computer Graphics in Industry

Industry	Application	Industry	Application
Bechtel	Computer aid design in nuclear power plant work, Finite element interactive graphics, Generalized contour mapping program, Computer-aided drafting		Graphical representation of information for design Equipment arrangement drawings Plotting
Bendix	Automated drafting in: Schematics PC board layouts LSI design Automated design in: Gear train design Simulation work	Exxon	Preparation and updating of piping diagrams Visual display of structural members Mapping of open pit mines
Bethlehem Steel	Computer-aided manufacturing Computer-aided lofting	Firestone	Equipment and plant layout work
Black & Veatch	Not used	Fisher Controls	Electronic P/C board design Computer-aided drafting Design layout Finite element modeling Seismic charting
Boeing Computer	Computer-aided drafting Computer-aided manufacturing	General Dynamics Convair Division	Computer-aided design Computer-aided drafting
Brown & Root	Computer-aided drafting		
Chrysler	Computer manikin to forecast the reaction of people	General Dynamics Electric Boat Division	Computer-aided drafting Finite element modeling Nondestructive ultrasonic testing Computer-aided manufacturing
Coca Cola	Not used		
Conaco	Design of off-shore structures Line drawings		
Corn Products	Not used	General Dynamics Quincy Shipbuilding Division	Computer-aided drafting Computer-aided manufacturing Optimizing nesting of structural parts
Dow Chemical	Not used		
DuPont	Computer-aided pipe sketches Code diagrams		

TABLE 1-5 (*Continued*)

Industry	Application	Industry	Application
General Motors	Preparation of body surface shapes	Rockwell International Graphic Systems Division	Computer-aided drafting
Georgia Pacific	Not used		
Goodyear	Not used	Sandia Laboratories	Computer-aided design Computer-aided drafting Analysis
Indland Steel	Design and manufacturing of rolls (used to shape steel)		Computer-aided manufacturing
International Harvester	Computer-aided design Computer-aided manufacturing	Standard Oil (Indiana)	Genigraphics Geological/geophysical drawings
Kodak	Computer-aided drafting		Computerized drafting
Kraft	Not used	Suntech Group	Computer-aided drafting
Mobil	Not used	Tennessee Gas Pipeline	Computer-aided drafting Computer-aided analysis and design
Monsanto	Design of tracking device Pollution abatement design studies	Texaco	Computer-aided drafting Structural design
Owens-Illinois	Computer-aided design Computer-aided drafting	Union Oil	Not used
Phillips Petroleum	Automated drafting Computerized image generator	Uniroyal	Computer-aided drafting Computer-aided manufacturing
Ralston Purina	Not used	United Technologies Pratt & Whitney	Computer-aided drafting and design Computer-aided manufacturing
Republic Steel	Automated drafting Automated design Computer-aided manufacturing		
Rockwell International Automotive Operations	Computer-aided design Finite element modeling Detail drafting Manufacturing	Warner Lambert	Not used
		Westinghouse	Computer-aided design

Source: Adapted from "CG as a Design Tool," *Engineering Design Graphics Journal,* American Society of Engineering Education, Winter 1981. Courtesy of Professor R. N. McDougal.

Figure 1-3   Use of computer graphics display for viewing the exterior shape of a car by rotation. (Courtesy of Chrysler Corporation.)

1.3   INTERACTIVE GRAPHICS

The computer can be programmed to implement alteration of a part or all of trial designs and to plot the modified layouts. The plotted drawings are one form of **hardcopies** and the **hardcopy** approach is too slow. Furthermore, the computer plotter may be a super draftsman but it is no substitute for the ingenuity of the human mind. During the trial process of computer-aided design, human intervention is inevitable. The **display** method makes it possible for the user and the computer to conduct a conversation through an interact-and-respond approach (Figs. 1-3, 1-4, and 1-5). Only when one is finally satisfied with a displayed design should a hard copy be produced.

It is thus indispensable in CAD/CAM applications to include extensive discussion on computer graphics hardware and software. The former should cover not only the computer itself but also the interactive graphics devices, whereas the latter explores the programming languages pertaining to interactive operations.

Figure 1-4 Computer graphics and peripheral equipment. (Courtesy of Tektronix, Inc.)

Figure 1-5 Computer interactive graphics operations with a Calcomp's 4000 Vistagraphic system. (Courtesy of California Computer Products, Inc.)

1.4 ROBOTICS[3]

The word *robot* was derived from Czech *robota*, meaning compulsory labor. Nowadays, it is used to indicate the mechanical devices capable of performing human tasks by remote control or by internal programmed control. As the robots never feel tired or become emotional, they can be counted on to do repetitive, hazardous, and tedious works. Most robots are composed of a power supply unit, a mechanical unit to reach, grab, push, and so on, and a control unit (Fig. 1-6*a–c*). It is the control unit that determines the capabilities of a robot. The brain of a robot that sends out the commands for desired motions is the control unit. Therefore, computer programming of the control process can be incorporated into robots. As a result, special robot programming languages have evolved and become a field of intensive research. Figure 1-7 summarizes the development.

Robots are being programmed to be able to learn on their own. Not only will a robot have functional capabilities like human eyes, ears, and mouth so that it can see, hear, and talk (Fig. 1-8), but it will also have an electronic brain to think and write programs of its own. Scientists and engineers are setting goals for this so-called fifth-generation computers to be equipped with artificial intelligence. In order to achieve these goals, both core memory and processing speed must be drastically increased to the range of 100 billion to 1 trillion bits to be accessible within several seconds.

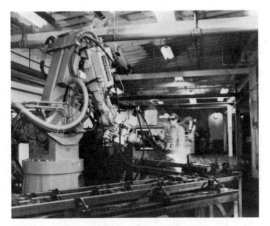

Figure 1-6*a* Robot welder. (Courtesy of Dahlstrom Manufacturing Company, Jamestown, New York.)

---

[3] For an overview of robotics, readers are referred to an article by W. B. Gevarter in *Computers in Mechanical Engineering*, Vol. 1, No. 1, August 1982, pp. 43–49.

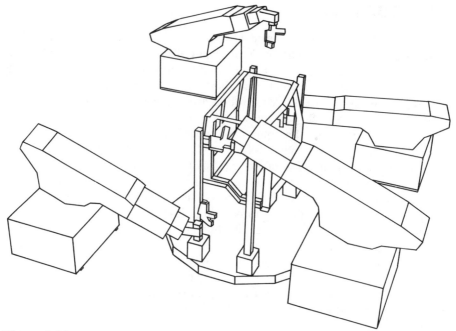

Figure 1-6*b* Four robots are used for a body-cab welding process. (Courtesy of the SAMMIE Group, the University of Nottingham, England.)

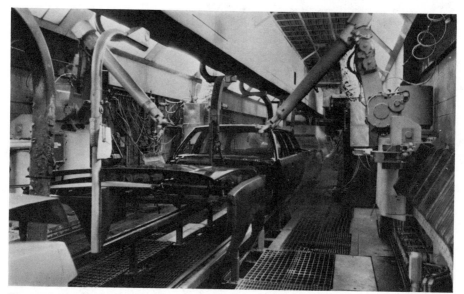

Figure 1-6*c* A pair of hydraulically driven N/C (numerically controlled) painters are spraying the body and front assembly of a station wagon. (Courtesy of General Motors Corporation.)

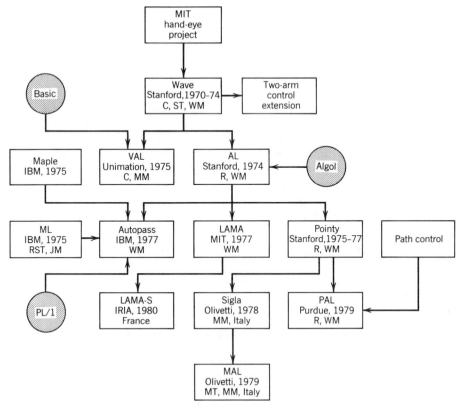

Figure 1-7 Robot programming languages evolution and classification. (Courtesy of Professors D. D. Ardayfio and H. J. Pottinger, University of Missouri—Rolla, and *Mechanical Engineering.*)

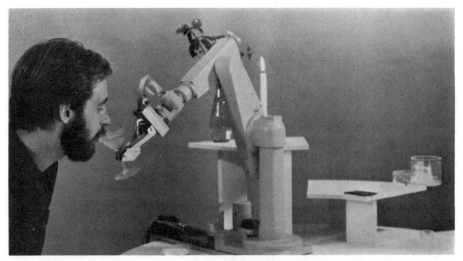

Figure 1-8 Use of robots in aiding the disabled through voice commands. (Courtesy of Professor Larry Leifer, Stanford University.)

1.5 CAD/CAM CURRICULUM DEVELOPMENT

"Computer aided design and manufacturing are revolutionizing industrial design and production in the United States and abroad," said Dean John E. Gibson of the University of Virginia and the 4C Consortium program.[4] This program funded by the National Science Foundation attempts to integrate CAD/CAM into engineering curricula "as quickly and efficiently as possible." The 12 universities participating in this 4C consortium program are Carnegie-Mellon University, Cornell University, Dartmouth College, Duke University, Georgia Institute of Technology, Illinois Institute of Technology, North Carolina State University, Rensselaer Polytechnic Institute, The Johns Hopkins University, University of Florida, University of Virginia, and Washington University in St. Louis.

The 4C program demonstrates that some coordinated efforts are underway besides those varied instructional activities reported in Table 1-1 about computer-aided drafting, design, and manufacturing. There are also reports of established centers, laboratories, and special facilities for interactive computer graphics at the University of Minnesota,[5] Michigan State University,[6] Rensselaer Polytechnic Institute,[7] Ohio State University,[8] Cornell University,[9] and many others.

1.6 OBJECTIVES AND SCOPE OF THIS BOOK

This book is designed to introduce the basic elements that are indispensable in application of interactive computer graphics for automated drafting, design, analysis, control, and manufacturing. The materials covered in this book are kept at a level suitable for the sophomore or junior engineering students, consisting of mostly linear problems and elementary practical examples. The coverage of this book can best be described with aid of a block diagram shown in Fig. 1-9 outlining the procedure of CAD/CAM.

Chapters 2 and 3 discuss the hardware and software required for CAD/CAM studies. Display equipment and command languages are the eyes and nerve system

---

[4] K. W. Hickerson, "Consortium Formed, Universities Exploring Integrating CAD/CAM into Engineer Curricula," *Engineering Times*, National Society of Professional Engineers, October 1981.

[5] F. Kelso and D. R. Riley, "Teaching Computer-Aided Drafting," *Mechanical Engineering*, Vol. 104, No. 10, October 1982, pp. 70–71.

[6] Computer-Aided Design short course offering announcement, summer 1982, the Case Center for Computer-Aided Design, Michigan State University.

[7] S. J. Derby and R. N. Smith, "Interactive Computer Graphics at RPI," *Mechanical Engineering*, Vol. 104, No. 9, September 1982 pp. 24–29.

[8] *Advanced Design Methods Laboratory 1981 Annual Report*, Department of Mechanical Engineering, Ohio State University.

[9] "The Advances of Computer Graphics," *Engineering: Cornell University Quarterly*, Vol. 16, No. 3, Winter 1981–82, Ithaca, New York.

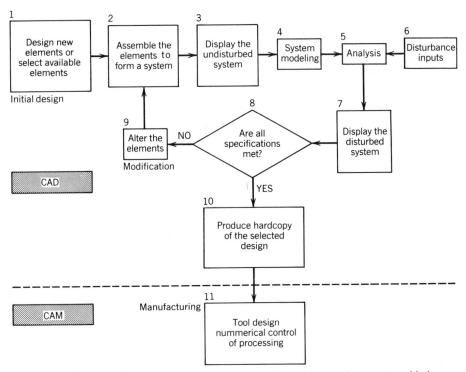

Figure 1-9 Block-diagram outline of computer-aided design and computer-aided manufacturing (CAD/CAM).

of the CAD/CAM processes diagrammed in Fig. 1-9. The capabilities of currently available interactive computer graphics devices are delineated in Chapter 2. As far as programming languages are concerned, BASIC is the main language while FORTRAN and other languages are occasionally used in this book. Since BASIC is the most commonly used language on home and high school microcomputers, it is not formally introduced, but the commands are reviewed throughout this book.

Chapter 4 is devoted to the fundamental concepts of plotting, displaying, and viewing designed objects from different angles. Chapter 5 explains the interaction of user and graphics unit. Various devices capable of implementing the interactive requirements and programming preparations for achieving human intervention are described.

Chapter 6 presents a number of elementary practical examples of CAD by applications of interactive computer graphics. These examples may be termed as CAD of the **elements** before they are assembled as components of a system.

In order to apply CAD in realistic engineering systems, characteristics of the elements that constitute the system must be understood and be modeled for mathematical analysis. Chapter 7 reviews the properties of the commonly used elements, methods of mathematical modeling, and analytical solution of the governing equations that are the Steps 4 and 5 in Fig. 1-9. This chapter lays the groundwork for later discussions of transfer-function, finite-element, analog-computer, and numerical techniques for systems analysis.

Chapter 8 is devoted to systems analysis by the transfer-function method. Computer-aided simplification of the block diagram is particularly important in the deletion and addition of elements of a system during the modification process (Step 9 in Fig. 1-9). It also facilitates consideration of various disturbance inputs affecting the system (Steps 6 and 7 in Fig. 1-9).

The finite-element analyses that lead to the solution of matrix equations are covered in Chapter 9. The analogy between various engineering systems is explained with presentation of fluidity and admittance matrix equations derived from consideration of flow and electric networks. Computer programs and numerical examples are given to elucidate the details involved in applications of the finite-element method.

Numerical solution should be stressed for consistency as long as one uses computers to aid drafting, design, and so on. Various methods for approximate analysis are presented in Chapter 10. Finite difference, iterative solutions, and numerical integration are the main topics covered so as to keep the scope of this book at a junior level; however, appropriate references are given to direct readers to the senior/graduate level sources.

Hybrid computer systems also play an important role in CAD/CAM by utilizing both digital and analog computers. These systems interface the laboratory monitoring equipment with digital computers by use of analog-to-digital and digital-to-analog converters. Such an interfacing arrangement is commonplace in control systems, hence particularly important in CAM. Chapter 11 discusses the applications of analog computers and the analog-to-digital and digital-to-analog converters. It is demonstrated that for certain cases, nonlinear governing equations of a system can be conveniently solved by use of analog computers.

Chapter 12 covers a number of advanced and miscellaneous topics. Parametric and optimum studies leading to best designs are discussed with practical examples. Since CAM involves CAD of manufacturing equipment and control processes already covered in the preceding chapters, emphasis is placed on the topic of numerical control (NC) of the manufacturing process. An example of NC program in APT language for machining of a CAM profile is presented to demonstrate the concept and procedure. It is the intent of the book only to guide the readers to the threshold of CAM; for an in-depth study of CAM, senior/graduate level courses in academia and short courses offered by industry on CAM are more appropriate than this book.

The purpose of this book is to help the readers to have a solid grasp on the fundamental elements of CAD/CAM, to develop the skills of writing module and database programs, and to build abilities in assembling subprograms and making effective use of interactive graphics devices in their CAD/CAM endeavors. Use of market-available CAD/CAM software packages is deemphasized herein. Indeed, these packages will expedite the design process in senior/graduate level courses in the specialized fields. The intended mission of this junior level book is to advocate that the readers should generate their own software when they learn the basic of CAD/CAM.

Professors Juricic and Barr of the University of Texas in Austin systematically categorized CAD/CAM teaching into four levels[10]. With the lowest level 1 being the introduction that requires no programming skill and the highest level 4 being the instruction that employs CAD/CAM for topics of special interest, this book can be considered as an attempt to fulfill the needs of the levels 2 and 3, involving all disciplines of engineering and dedicated to an integrated system approach.

[10] D. Juricic and R. E. Barr, "Graphics and CAD—A Systematic Approach," *Mechanical Engineering*, Vol. 104, No. 9, September 1982, pp. 48–53.

CHAPTER 2

Cray-1 Supercomputer. (Courtesy of Cray Research, Inc.)

HARDWARE
FOR CAD/CAM

2.1 INTRODUCTION

The concept of "computer aided" relies on computer technology to produce better designs and manufacturing processes, faster and with fewer errors than the classical, manual approach. Computer systems applicable for CAD/CAM must therefore be able to assist in one or many aspects of the automated process, which should include drafting, analysis, control, and production. Speed, precision, cost, and availability of specialized hardware and software are among the principal factors in deciding how a particular CAD/CAM task should be pursued.

As far as speed is concerned, the computer technology has advanced to a rate of 100–250 million floating-point operations per second on a Cray-1 supercomputer[1] with a memory of 1,048,576 64-bit words. The use of array processors enables the operation speed to be drastically improved. On the other hand, the cost of a microcomputer of 16K memory may not be as formidable as that of the supercomputer, but one must ask whether its memory size and operation speed are adequate for the CAD/CAM task at hand.

For visual study of designed objects, engineering drawings are nowadays commonly displayed on television-like cathode ray tubes. Small microcomputers may not be adequate in displaying smooth graphs. This leads to the question about the **graphic resolution** of the computer display screen. If the microcomputer lacks sufficient memory to tackle a CAD/CAM problem, can it be linked to a larger computer?

In CAD/CAM studies, a part of the design or process drawing may need to be magnified, viewed from different angles, and changed. A computer graphics unit may therefore require additional attachments to perform the **interactive** operations. These so-called **peripheral devices** are thus necessary for **interfacing** with the graphics terminal.

In this chapter, attempts are made to resolve these questions as they pertain to the currently available hardware and its capabilities that could arise in CAD/CAM applications. Commonly used hardware terms are explained in the course of introducing various prevailing computer graphics equipment—from bits, bytes, and words of computer memory cells to modems for communications between computers via telephone lines, from pixels on cathode ray tube of computer graphics devices to the hand maneuverable joystick for interactive control of the program being executed by the computer. It is a virtually impossible task, however, to be exhaustive on coverage of computer hardware for CAD/CAM in view of such rapid expansion of this field. Hopefully, having acquired the background knowledge presented herein, readers will be able to further their own exploration of this exciting field.

---

[1] R. M. Russel, "The Cray-1 Computer System," *Communications of the ACM*, Vol. 21, 1978, pp. 63–72.

2.2 ADDING MACHINES, CALCULATORS, AND COMPUTERS

Early man counted his possessions by use of **stone**, which in Latin is *calx*. From that, the English word **calculate** was derived. **Abacus** was perhaps the most notable device, among many others, developed as an early computer (Fig. 2-1*a*). As far as the first **mechanical** machines are concerned, these can be traced back to the seventeenth century, when Pascal designed an adding machine. As we all know, a programming language is now named after him. Many mechanical computing machines were built since then; Leibniz, Thomas, Muller, Babbage, Jacquard, and Hollerith all made significant contributions. Generally speaking, these mechanical machines were bulky, heavy, and slow in computation.

Dr. Herman Hollerith, a statistician, developed the idea of cards punched with coded holes and helped the statistical analysis of the 1890 U.S. census data (Fig. 2-1*b*). His firm, Computing-Tabulating-Recording Company, eventually become the International Business Machines Corporation; the 80-column cards are often called **IBM cards**.

In 1944, Howard Aiken of Harvard University completed the Mark I electro-mechanical machine. It was an Automatic Sequence Controlled Calculator. This 50-foot-long and 8-foot-high computer could perform complicated calculations such as sines and cosines. It was used for more than 15 years until the computers built with electronic tubes took over. The first electronic computer was ENIAC (Electronic Numerical Intergrator and Computer), built at the University of Pennsylvania for the U.S. Army Ordanance Department to calculate ballistic tables. It was still a bulky, 30-by-50-foot machine, but it was versatile and fast in computations of 5000 additions or 350 multiplications per second. There were other electronic computers built for specific uses. UNIVAC I and IBM 650, computers using vacuum tubes, were produced in large numbers in the 1950s.

The introduction of the transistor revolutionized computer designs. Transistors enabled smaller, lighter, and more reliable computers to be made. The evolution of computers from mechanical, electromechanical, electronic vacuum tubes, and

Figure 2-1*a*   Abacus, an old computation device.

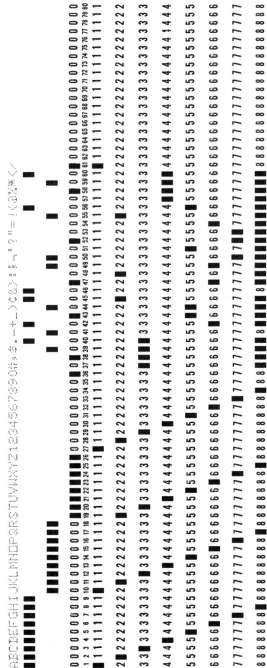

Figure 2-1b Hollerith 80-column card punched with coded holes representing alphanumeric and other characters.

Figure 2-2 Desktop microcomputer. (Courtesy of Apple Computer, Inc.)

then to integrated circuits has created terms of the first-, second-, third-, and fourth-generation computers, respectively. Nowadays, the development of computers with intelligence has already been marked as the fifth generation.

The idea of the integrated circuit was conceived by Jack St. Clair Kilby of Texas Instrument Inc., but the improved process of integrating circuits on microchips was the invention of Robert Noyce of the Intel Corporation. Their work made possible the rapid advance of microelectronic technology.

The microminiaturized computer system, or **microcomputer**, unlike the **mainframe** computers such as IBM 370 or Cray-1, has the memory, control and arithmetic units all in a single integrated-circuit chip (Fig. 2-2). As diagrammed in Fig. 2-3, the **central processing unit**, abbreviated CPU, is where the interactions between the memory unit and control unit, and between control unit and arithmetic unit, take place. CPU requires the support of the peripheral input and output devices.

The control unit of a computer system manages the flow of data into and out of all the other units. It controls where to get the data whether from the internal memory or from peripheral input unit, delivers the data to the arithmetic unit to be operated on such as subtracting or dividing, and controls where the calculated result of the arithmetic unit should be forwarded whether temporarily kept in the internal memory or stored in a peripheral tape or printed.

The memory unit and input/output devices are discussed in ensuing sections.

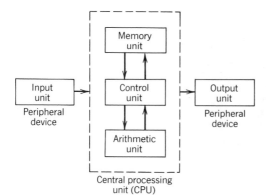

Figure 2-3 Central processing unit and peripheral devices of digital computers.

2.3 BITS, BYTES, AND WORDS

Computer memory storages are composed of basic cell units, called **bits**. *Bit* is the shortened word of *binary digit*. A bit is used to store a binary digit of either 0 or 1. For storing decimal integer numbers greater than 1, one bit is apparently insufficient and multiple bits are therefore necessary. In case of a decimal integer number $(15)_{10}$ to be stored in the computer, it has to be converted into a binary number equal to $(1111)_2$, which requires that 1 be stored in 4 consecutive bits.

Usually, a fixed number of bits is assigned for storing numbers and characters in a computer memory. This fixed-size combination of consecutive bits is called a **word**. The word sizes may differ from one computer series to another. For instance, IBM 360/370 series computers use 32-bit words while CDC 6000–7000 series computers use 60-bit words. Cray-1 computer uses 64-bit words.

For convenience, a word is often subdivided into **bytes**. IBM 360/370 computers have 4-byte words; that means one byte is composed of 8 consecutive bits. CDC 6000–7000 computer words are composed of 10 bytes each; that means one byte is composed of 6 consecutive bits. The location of a word in the computer memory storage is **addressed** by its left-most byte. For example, if a word A is immediately followed by a word B and if the address of A is X, the address of B will be $X + 4$ in a 4-byte computer and $X + 6$ in a 6-byte computer.

A microcomputer of 1K memory actually has 2^{10}, or 1024, bytes. In order to provide memory in compact sizes, integrated-circuit technology has been widely used for miniaturization of computer storage. Figure 2-4 shows the magnified picture of a 64K memory chip having a U.S. coin in the background. The more advanced magnetic bubble memory can store 1 megabits on one chip. Quite often, the computer memory is referred to as a **core**, because it can also be made of extremely small magnetic cores.

The most commonly used memory types are the random access memory (RAM) and read only memory (ROM). ROM stores data, commands, and programs intended to be read only. It is manufactured so that no change can be made by the user. RAM, which can be read and written, is the type presently in use for the main memory of microcomputers. By random it means that one location of the

Figure 2-4 A 64K memory ship. (Courtesy of International Business Machines Corporation.)

memory is as easily accessible as any other. In some mainframe computers, the serial memory, which arranges storages in sequential order, is still in use along with the RAM.

2.4 ARRAY PROCESSOR[2]

In conventional execution of an instruction, the computer first locates the instruction, in the memory, carries out the operations specified by the instruction, which requires the retrieval of data from memory to the central processor to be acted upon, returns the resultant data to memory, and so on. Thus, only one operation can take place at any instant as it occupies the **single** pathway between a single memory and the logic and control units. **Array processors** are built on the concept of **multiple** communication pathways.

Based on this idea of parallel organization, the basic arithmetic operations of addition and multiplication can also be implemented in a **pipeline** manner. That is, the addition and multiplication are divided into suboperations, and the sub-

---

[2] For a more elaborated discussion on array processors and pipeline floating point arithmetic, the article written by H. F. Davis in *Industrial Research*, November 1977, pp. 83–85, should be read.

operations are carried out using multiple adders. For example, a set of numbers *A* are to be added to those of set *B* while *C* are to be added to *D*, and the addition is to be carried out, for simplicity of discussion, in two suboperations. The pipelining operation executes the first suboperation of *A*'s and *B*'s; the adders used for this suboperation will begin the first suboperation of *C*'s and *D*'s while the other adders execute the second suboperation of *A*'s and *B*'s. The latter adders may meanwhile be shared for implementing a suboperation of additions of other sets of numbers.

2.5   INPUT AND OUTPUT DEVICES

Internal memory of the computer requires less access time but is more expensive than the peripheral mass storage devices. Popular peripheral storage devices include cassette tape, disk pack, floppy disk, and magnetic tapes (Fig. 2-5). As output devices, printers, punched cards, hardcopies, and video monitors are the most commonly used.

Tapes and disks both rely on the magnetic recording process. The former stores data sequentially, whereas the latter can store data either sequentially or randomly. The random-access arrangement is possible because the disk is divided

Figure 2-5   Mass storage devices. (Top left and bottom left: Courtesy 3M; top right: Courtesy Wang; bottom right: Elyse Rieder.)

Figure 2-6   (Courtesy of Apple Computer Inc.)

into concentric rings called **tracks**, which are further divided into sectors. Each sector is as easily accessible as any other by the recording head.

Printers make hardcopy of the program listing, input data and the obtained results. There are various types of printers. The hardcopy can be printed by impact or nonimpact, in serial with one hammer or line-by-line with a row of hammers. It can be printed with dot matrix or solid characters. For example, a thermal printer is nonimpact and uses dot matrix for the characters (Fig. 2-6). It uses special papers and the characters are printed by a voltage or heat process.

2.6   BUSES, PORTS, AND INTERRUPTS

For communication between various components of a computer system such as between the central processing unit and the input/output peripheral devices, the data and commands travel through a **bus** structure made of a group of wires or lines. A data bus enables data to be input to or output from the CPU, whereas a control bus identifies the external device. An address bus is also required to specify the memory location for the data storage or retrieval in the CPU or in an external device. When an external device is involved, the channel of the address bus becomes a **port**. A port may be connected to an input/output device such as a cassette tape, through which the computer can read or write data.

Since the computer operates at a faster speed than the input/output devices, it carries on its operations while the user enters requests by slowly typing on the keyboard. **Interrupt** lines are therefore necessary for getting the attention of the computer, as well as for obtaining other information required for the desired action by the CPU that is communicated through a bus device (Fig. 2-7).

Figure 2-7 Interfacing buses: RS-232 and IEEE-488. (Courtesy Inmac.)

For interfacing nearby peripheral input/output device with microcomputers, the RS-232 bus is often used for transmitting data in a byte-by-byte fashion. IEEE-488 bus is also widely employed for networking of several processors to share various peripherals and also for connecting laboratory instruments directly to a computer; it uses 24 or more wires.

Not only should there be an agreement in format and speed between the transmitting and receiving devices, but the message sent through a bus must also be checked for errors possibly resulting from noise or other causes. The "check bits" arrangement often takes up one or more lines of a bus.

2.7 MODEMS

Modem is the abbreviation of *mo*dulator and *dem*odulator, which performs the translation of digital signals into analog signals and the analog signals back to digital. Since the telephone network transmits voices as analog signals of varying voltages over a pair of wires, the modems enable the computers to telecommunicate with each other. The modem sends and receives different tones of binary bits 1 and 0 through the telephone lines in serial, that is, one bit at a time. The capability of a modem is thus measured by its speed in transmitting data.

Baud is the measure of the speed of transmitting data, referred to as bits per second. The telephone lines generally can transmit at 1200 baud, that is, 1200 bits per second. Because of possible interferences such as noise, echo, and so on, the speed has to be reduced to allow the modem to recognize the signals. Unless special telephone lines are used to minimize the interferences, the most common transmitting speed of a modem is 300 baud. It amounts to approximately 30 characters per second as at least 7 bits are required to represent a character according to the ASCII (American Standard Code for Information Interchange), and some extra bits are needed for describing the relationship of adjacent characters.

Many types of acoustic couplers and direct-connect modems of half and full duplex are available on the market (Fig. 2-8). Rubber cups are used for the acoustic-coupler modem whereas the direct-connect type plugs into a standard

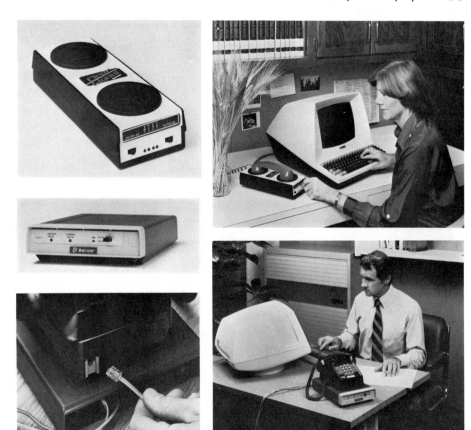

Figure 2-8 Modems: Accoustic couplers and direct-connect, half or full duplex. (Courtesy Inmac.)

telephone jack. The half and full duplex modems handle one-way and two-way data communications, respectively. Two-way communication means transmitting and receiving at the same time, whereas one-way can only either transmit or receive. The concept and developed technology of **multiplexing** permits a number of signals to share a single communication channel and be used by different users at different frequencies. This has resulted in the time-sharing uses of mainframe computers for telecommunications with a large number of remote input/output terminals.

2.8 GRAPHICS DISPLAY

Video monitoring of pictures on a TV-like screen is called a **display**, sometimes called a **softcopy** as opposed to a hardcopy, which is a printed version of the display. The screen actually is the flat front of a cathode ray tube, abbreviated as

CRT (Fig. 2-9a). As sketched in Fig. 2-9b, CRT has a flat end, the inside of which is covered with a phosphor, and at the other end has an electron gun. The deflector plates when charged with different voltages cause the electron gun to hit at different points of the phosphor to emit light for a short period of time. By varying the intensity of the beam of the electrons and voltage on the deflector

Figure 2-9a Assortment of cathode ray tubes. (Courtesy of Westinghouse Electric Corporation.)

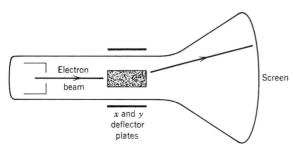

Figure 2-9b Cathode ray tube (CRT).

plates, pictures and graphs can be generated on the screen. Various phosphors can be utilized for emitting different colors; that is how color displays are generated.

There are two prevailing methods of displaying objects on the screen, called vector graphics and raster graphics. The **vector graphics** computer draws the object by use of a series of line segments. The coordinates of the beginning and ending points of these line segments are stored in the computer memory and used through a digital-to-analog converter to control the electron gun of the CRT to light up straight lines on the screen. Because of its vast storage requirement, the vector graphics method is not used in most microcomputers.

Raster graphics is based on the scanning of a screen as a gridwork of points, each of which can be turned on or off individually. By sweeping the electron beam from left to right and from top to bottom, the intensity of the electron beam can be varied by the computer to generate a desired picture on the screen. The quality of the displayed picture by the raster graphics method is thus determined by the number of points that can be turned on or off. This is commonly referred to as the **resolution** of a graphic terminal. The question of how the resolution will affect the quality of a display will be fully addressed in Chapter 4.

There are increasing numbers of computer graphics devices available in the market; AED, Digital, Evans and Sutherland, Lexidata, Megatex, Ramtek, and Tektronix are among the notables. The Tektronix 4054 computer graphics system shown in Fig. 2-10 has been used for development of the CAD/CAM and other programs reported in this book.

2.9   INTERACTIVE DEVICES

Graphics systems generally have an illuminated or blinking cursor on the screen, at which characters can be entered and displayed. It is a "prompting" signal asking for user's action. A keyboard is commonly used for entering the characters or the command. For changing the position of the cursor, such **interaction** of the user with the graphics systems may also be achieved by other input devices such as by use of a joystick, thumbwheels, lightpen, or the stylus of a graphics tablet. These interactive display devices all send input signals, which the graphics systems use to place the cursor at a particular location on the screen and sometimes to display at that location a specific character with designated brightness and color.

The lightpen works on a photodiode concept. As the screen of the display is continuously refreshed by the raster scan process, the point on the screen that is touched by the lightpen, when passed by the sweeping electron beam, will cause a jump in light output and subsequently make it possible to enter the coordinates of that point to the graphics system for implementation of a precoded action.

Joystick and thumbwheels allow the user to adjust the cursor manually in various directions by hand and in the meantime to view its changing position on the screen.

Graphics tablets are made of meshed wires, each carrying a unique signal. By pressing a penlike device, a stylus, on the tablet, each location triggers a uniquely

1. Tektronix exclusive
   Direct View Storage Tube
2. Data tape cartridges
3. Ten user definable keys
4. The full 128 character
   ASCII keyboard
5. Five functions calculator
   key pad
6. Built—in thumbwheels

Figure 2-10   Tektronix 4054 computer graphics system. (Courtesy of Tektronix, Inc.)

coded signal to activate the computer graphics system to implement a specific pre-arranged programmed task. Various types of tablets can be constructed by placing copper wires in gridwork form with each carrying different pulses, by using voltage gradients within the plate, or by acoustic pickup. Applications of the above-mentioned interactive devices are discussed in Chapter 5.

CHAPTER 3

```
700  REM  **  PLOT THE SPECIFIED SECTION  **
710  PAGE
720  W=U2-U1 MAX V2-V1
730  WINDOW U1,U2,V1,V2
740  VIEWPORT 30,30+100*(U2-U1)/W,0,100*(V2-V1)/W
750  GOSUB 850
760  GOSUB 270
770  RETURN
```

```
                    PARTNO  CAMPROFILE
                            CLPRNT
                            CUTTER/1.25
                    SETPT = POINT/0,0,0
                    C1    = CIRCLE/15,10,3.75
                    C2    = CIRCLE/10,7.5,2.5
                    P1    = POINT/15,10,0
                    P2    = POINT/10,7.5,0
                    P3    = POINT/0,10,0
                    L1    = LINE/P3,RIGHT,TANTO,C1
                    L2    = LINE/P3,LEFT,TANTO,C1
                    XYPLN = PLANE/P1,P2,P3
                            FROM/SETPT
                            GO/PAST,L2,TO,XYPLN
                            GORGT/L2,TANTO,C1
                            GOFWD/C1,TANTO,L1
                            GOFWD/L1,TANTO,C2
                            GOFWD/C2,PAST,L2
                            GOLFT/L2,TO,L1
                            GOTO/SETPT
                            FINI
```

SOFTWARE
FOR CAD/CAM

3.1   INTRODUCTION

We have reviewed in the preceding chapters the capabilities and limitations of the available CAD/CAM equipment. How this equipment can be effectively utilized should be the next immediate concern in applications of CAD/CAM. Communications between the user and a CAD/CAM device require proper languages, and rules governing these languages have to be established. The questions that need to be addressed include the transfer of alphabetic, numeric, operational, and other characters into and from the computer graphics terminal through the use of a keyboard or other peripheral devices, the numerical manipulation of numerical data, the processing of alphabetic and other characters, and the visual display, hardcopy, and line printing of the results.

In this chapter, binary, octal, decimal, and hexidecimal number systems are first discussed, and then conversions of decimal integers and real numbers into binary numbers are explained. The reader can thus understand how all numerical data are stored and processed as binary digits of 0's and 1's. In order for the alphabetic, operational, and other special characters to be processed, they must also be converted into binary numbers. The standard codes developed for such required conversions are therefore delineated.

When characters and graphs are displayed on the cathode ray tube of a computer graphics terminal, the user should know about the commands that enable the character sizes to be changed and the smoothness of the graphs to be improved. A section of this chapter is therefore specifically devoted to the topic of "graphic resolution," which should always be of prominent concern in producing accurate displays during CAD exercises (see Section 3.5).

Presently, most of the CAD/CAM computer programs are written in the problem-oriented languages of BASIC, FORTRAN[1], and PASCAL[2]. How the commands coded in such languages are translated into binary digits of 0's and 1's so that the machine can understand, a process called **compilation**, is also explained in this chapter.

3.2   NUMBER SYSTEMS

In Chapter 2 binary and decimal numbers were briefly mentioned. For a firm grasp of the machine manipulations of numeric data, it is necessary to discuss the binary, octal, decimal, and hexidecimal number systems in detail. Table 3-1 summarizes the relationships among these four number systems.

While the computer can process numbers only in the form of a series of binary digits of 0's and 1's, all numeric data in the decimal system using the digits 0 through 9 therefore require speedy conversion into 0's and 1's. For such conversion, the octal number system, which uses the digits 0 through 7, and the hexide-

---

[1] FORTRAN is the abbreviation for *formula translator*.

[2] PASCAL was introduced by Professor Nichlas Wirth of the Engineering University at Zurich, Switzerland, in the late 1960s.

TABLE 3-1
Relationships among Number Systems

Binary (2)	Octal (8)	Decimal (10)	Hexadecimal (16)
0000	0	0	0
0001	1	1	1
0010	2	2	2
0011	3	3	3
0100	4	4	4
0101	5	5	5
0110	6	6	6
0111	7	7	7
1000	10	8	8
1001	11	9	9
1010	12	10	A
1011	13	11	B
1100	14	12	C
1101	15	13	D
1110	16	14	E
1111	17	15	F

cimal number system, which uses the numeric digits 0 through 9 together with the alphabetic characters A through F, can be very conveniently utilized as go-betweens. The hexidecimal number system also condenses the excessively long binary expressions into more compact short forms for the output purpose, particularly in the core-dump of the computer storage contents during debugging of programs.

To explain how the octal and hexidecimal number systems help the conversion of a decimal number into binary form, first let the given decimal integer x be related to its binary equivalent as follows:

$$(x)_{10} = \sum_{i=0}^{n} b_{n-i} 2^{n-i} = b_n 2^n + b_{n-1} 2^{n-1} + \cdots + b_1 2^1 + b_0 2^0 = (b_n b_{n-1} \cdots b_1 b_0)_2$$

where b_{i-1}'s are the remainders obtained during the ith divisions, for $i = 1, 2, \ldots,$ $n + 1$, and can take on a binary digit of either 1 or 0, and n is the number of divisions of the given decimal integer x by 2 required in order to arrive at a final quotient b_n less than 2. As an example, consider the case of $x = 53$; it takes 5 divisions for its conversion into a binary integer. The divisions are

$$
\begin{array}{ccccc}
26 & 13 & 6 & 3 & 1\,(=b_5) \\
2\,\overline{)\,53} & 2\,\overline{)\,26} & 2\,\overline{)\,13} & 2\,\overline{)\,6} & 2\,\overline{)\,3} \\
\underline{4} & \underline{2} & \underline{12} & \underline{6} & \underline{2} \\
13 & 6 & b_2 = 1 & b_3 = 0 & b_4 = 1 \\
\underline{12} & \underline{6} & & & \\
b_0 = 1 & b_1 = 0 & & &
\end{array}
$$

As a result, we obtain

$$(53)_{10} = \sum_{i=0}^{5} b_{5-i} 2^{5-i} = (b_5 b_4 b_3 b_2 b_1 b_0)_2 = (110101)_2$$

If the decimal integer 53 is converted into an octal or hexidecimal number, only one division needs to be carried out. That is,

$$
\begin{array}{c}
6 \\
8 \overline{)\ 53} \\
48 \\
\overline{5}
\end{array}
\qquad \text{or} \qquad
\begin{array}{c}
3 \\
16 \overline{)\ 53} \\
48 \\
\overline{5}
\end{array}
$$

That is to say

$$(53)_{10} = (65)_8 = (35)_{16} = (110101)_2$$

Conversion of octal and hexidecimal numbers into binary numbers can be easily achieved by looking up the 3-bit and 4-bit expansions, respectively, from Table 3-1. Since the octal integer 6 in 3-bit binary is 110 and 5 is 101, so $(65)_8 = (110101)_2$. Likewise, the hexidecimal 3 in 4-bit binary is 0011 and 5 is 0101, so $(35)_{16} = (00110101)_2$.

In a similar manner, the decimal portion of real numbers can also be converted into binary numbers except that the negative powers of 2 have to be used. Table 3-2 lists the relationships between decimal numbers and the negative powers of 2. The conversion process is more involved. The following examples are presented for illustration of the procedure:

$$(0.25)_{10} = (0 \times 2^{-1}) + (1 \times 2^{-2}) = (0.01)_2$$

$$
\begin{aligned}
(0.89)_{10} &= (0.5 + 0.25 + 0.125 + 0.0078125 + 0.00390625 + \cdots) \\
&= (1 \times 2^{-1}) + (1 \times 2^{-2}) + (1 \times 2^{-3}) + (1 \times 2^{-7}) + (1 \times 2^{-8}) + \cdots \\
&= (0.11100011 \cdots)_2
\end{aligned}
$$

TABLE 3-2
Decimal Values for Various Power of 2, 8, and 16

Power of				Power of			
2	8	16	Decimal Value	2	8	16	Decimal Value
−12	−4	−3	0.000 244 140 625	1			2
−11			0.000 488 281 25	2			4
−10			0.000 976 562 5	3	1		8
−9	−3		0.001 953 125	4		1	16
−8		−2	0.003 906 25	5			32
−7			0.007 812 5	6	2		64
−6	−2		0.015 625	7			128
−5			0.031 25	8		2	256
−4		−1	0.062 5	9	3		512
−3	−1		0.125	10			1024
−2			0.25	11			2048
−1			0.5	12	4	3	4096

3.3 INTEGER (FIXED-POINT) AND REAL (FLOATING-POINT) NUMBERS

Computers store and manipulate integers and real numbers differently. In computer jargon, these numbers are called **fixed-point** and **floating-point** numbers, respectively. Assuming that a computer has words composed of 32 bits, the bits of a word will be numbered from 0 at left to 31 at right, as shown in Fig. 3-1. The left-most bit 0 is used as a sign bit. When it contains binary digits 0 and 1, the number represented by bits 1 through 31 will be positive and negative, respectively.

When a word is used for storing an **integer**, a binary digit 1 appearing on bit 1 signals that one unit of 2^{30} is a part of that number. If a binary digit 0 appears right-most, bit 31 of the word, no unit of 2^0 is contributing to the number. So it should be evident that the largest integer that can be stored in a 32-bit word is when all bits 1 through 31 contain a binary digit 1. The decimal value of this integer can be easily calculated to be

$$(11 \cdots 11)_2 = (1 \times 2^{30}) + (1 \times 2^{29}) + \cdots + (1 \times 2^1) + (1 \times 2^0)$$

$$(31 \text{ ones}) \quad = 2^{31} - 1 = (2147483647)_{10}$$

And it should also be easy to generalize that the ranges of integers that can be stored in a *n*-bit word are $-(2^{n-1} - 1)$ to $2^{n-1} - 1$.

In order to deal with numbers exceeding 2^{31} and also decimal numbers, a 32-bit word is divided into three parts, as shown in Fig. 3-2. Bit 0 is again used for the sign of the real number as for an integer, but bits $1-7$ and $8-31$ are assigned to the exponent and fraction portions, respectively, of that real number when it has been converted into normalized exponent notation. To express the number in normalized exponent notation, it is necessary to shift the decimal point of a real number to the left of its first significant figure and to maintain its value by multiplying the altered number with appropriate powers of 10. For example, if the real number is -16.25, then its normalized exponent notation is -0.1625×10^2. It is

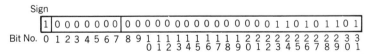

Sign

| 1 | 0 0 0 0 0 0 0 | 0 0 0 0 0 0 0 0 0 0 0 0 0 0 0 1 1 0 1 0 1 1 0 1 |

Bit No. 0 1 2 3 4 5 6 7 8 9 1 1 1 1 1 1 1 1 1 2 2 2 2 2 2 2 2 2 2 3 3
                              0 1 2 3 4 5 6 7 8 9 0 1 2 3 4 5 6 7 8 9 0 1

Figure 3-1 The decimal integer -429 being stored in binary form in a 32-bit word.

Sign   characteristic   Fraction

| 1 | 0 1 0 0 1 0 1 | 0 1 0 0 0 0 0 0 0 0 0 0 0 0 0 0 0 0 0 0 0 0 0 0 |

Bit No. 0 1 2 3 4 5 6 7 8 9 1 1 1 1 1 1 1 1 1 2 2 2 2 2 2 2 2 2 2 3 3
                              0 1 2 3 4 5 6 7 8 9 0 1 2 3 4 5 6 7 8 9 0 1

Figure 3-2

easy to identify the sign, exponent, and fraction of a real number expressed in normalized exponent notation. Obviously, for the number -16.25, it has a minus sign, and its exponent and fraction are 2 (base 10) and 0.1625, respectively.

The exponent defined above, however, is based on the decimal system. In order to express both the fraction and exponent parts of a real number in the binary system, it is more convenient first to convert both the integral and the fraction parts into binary digits and then to normalize it by shifting the position of the decimal point. For example, according to Table 3-1, $(-16.25)_{10} = (-10000.01)_2$. After shifting the decimal point to the left of the first significant figure, the number is normalized to become -0.1000001×2^5. Thus, as a binary number, the sign remains $-$, but the exponent is 5 (base 2) or $(101)_2$, and the fraction is 0.1000001.

As the exponent of a real number is to be kept in bits 1–7, the largest real number that can be stored is thus $2^7 - 1 = 128 - 1 = 127$. This is when the exponent is restricted to the positive numbers. In order to accommodate for small decimal numbers that have negative exponents, bit 1 of a 32-bit word will have to be reserved for keeping the sign of the exponent. Conventionally, this situation is circumvented by adding $2^6 = 64$ to the exponent to generate a **characteristic**, then stored in bits 1–7. In other words, a 32-bit word can, in this way, store floating-point numbers in the range from 2^{-64} (2^{0-64}) to 2^{63} (2^{127-64}). Returning to the number -16.25, the exponent 5 is adjusted to have a characteristic of $5 + 64 = 69 = (0100101)_2$. And the entire number is stored as a 32-bit word in the form shown in Fig. 3-2. Hence, it is evident that one must be very attentive to the difference with numbers such as 1 and 1. because they are stored in core memory following completely different formats.

3.4 PROCESSING OF CHARACTERS

Computers process numeric data as well as characters. The preceding section explained the methods of converting decimal integers and real numbers into binary numbers. The numeric data must be in the form of 0's and 1's in order for the computer to carry out the specified arithmetic operations. For processing of the alphabetic (A–Z), operational ($+$, $-$, /, *, etc.), and special (%, @, #, etc.) characters, conversion of these characters into binary numbers composed of 0's and 1's again needs to be implemented. To that end, various rules have been developed to represent the characters in binary codes. Tables 3-3 and 3-4 are the two most commonly used codes.

Usually, a character is represented by an 8-bit binary code. A 32-bit word can thus store 4 characters. If a byte has 8 bits, then each byte stores one character. It is easy to understand that other than the characters that are well known, various special characters, such as the characters used in the foreign languages shown in Fig. 3-3, can be created and given appropriate binary codes. This is the reason that many graphics devices have developed their own **font** characters.

TABLE 3-3
ASCII (American Standard Codes for Information Interchange)
for Commonly Used Keyboard Characters

Char.	ASCII	Char.	ASCII	Char.	ASCII
A	1000001	a	1100001	Ø (Zero)	0110000
B	1000010	b	1100010	1	0110001
C	1000011	c	1100011	2	0110010
D	1000100	d	1100100	3	0110011
E	1000101	e	1100101	4	0110100
F	1000110	f	1100110	5	0110101
G	1000111	g	1100111	6	0110110
H	1001000	h	1101000	7	0110111
I	1001001	i	1101001	8	0111000
J	1001010	j	1101010	9	0111001
K	1001011	k	1101011	:	0111010
L	1001100	l	1101100	;	0111011
M	1001101	m	1101101	<	0111100
N	1001110	n	1101110	=	0111101
O (oh)	1001111	o	1101111	>	0111110
P	1010000	p	1110000	?	0111111
Q	1010001	q	1110001	@	1000000
R	1010010	r	1110010		
S	1010011	s	1110011	[(Left bracket)	1011011
T	1010100	t	1110100	\ (Reverse slash)	1011100
U	1010101	u	1110101	] (Right bracket)	1011101
V	1010110	v	1110110	↑ (Up arrow)	1011110
W	1010111	w	1110111	__ (Underscore)	1011111
X	1011000	x	1111000	` (Accent grave)	1100000
Y	1011001	y	1111001		
Z	1011010	z	1111010	(An incomplete list)	

TABLE 3-4
EBCDIC (Extended Binary Coded Decimal Interchange Codes)
for Alphanumeric Characters

Char.	EBCDIC	Char.	EBCDIC	Char.	EBCDIC	Char.	EBCDIC
						0	1111 0000
A	1100 0001	J	1101 0001			1	0001
B	0010	K	0010	S	1110 0010	2	0010
C	0011	L	0011	T	0011	3	0011
D	0100	M	0100	U	0100	4	0100
E	0101	N	0101	V	0101	5	0101
F	0110	O	0110	W	0110	6	0110
G	0111	P	0111	X	0111	7	0111
H	1000	Q	1000	Y	1000	8	1000
I	1100 1001	R	1101 1001	Z	1110 1001	9	1111 1001

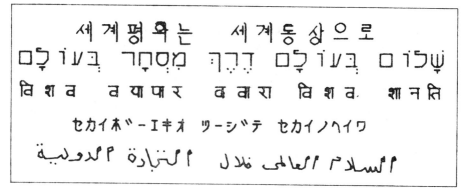

Figure 3-3 Special characters such as the above foreign language characters can be coded and displayed on graphic terminals. (Courtesy of IBM Corporation.)

3.5 DISPLAY OF CHARACTERS AND GRAPHS ON CATHODE RAY TUBE

A normal screen size of the cathode ray tube has a width to height ratio, called **aspect ratio**, of 4 to 3. The smallest unit of the screen that can be turned on or off is called a **pixel**, an abbreviation for picture element. For example, the Tektronix 4054 graphics terminal has 4096 pixels horizontally per each line in the x-direction and 3125 lines in the y-direction. It is thus said that the display device is of 4096 by 3125 **resolution**. Radio Shack TRS-80 Model I has a 128 by 48 resolution. Many microcomputers do not have their own display screens but use a modulator to connect to a standard television receiver so that the program statements and/or graphs can be displayed on the television screen when the microcomputer is in **text** and/or **graphics modes**. The resolution provided by such an arrangement is about 256 by 192.

It is quite easy to comprehend how a graph can be generated on the screen by selectively turning on a series of neighboring pixels to construct the desired curve. This is done when the display unit is in a graphics mode. If an expression composed of alphanumeric characters needs to be displayed, the graphics terminal is then in a text mode. In that case, a **block** of pixels is used for display of one character, as illustrated in Fig. 3-4. Many microcomputers use a block of 5 by 7 matrix of pixels; however, the block size varies from one graphics device to another and may be changed in some graphics devices by commands entered by the user, as will be demonstrated in Chapter 4 when CHARSIZE is introduced.

Not only can a pixel be turned on or off, the color graphics display units also specify a color for each pixel. This requires additional memory to store the on/off and color status of every pixel. Usually the memory for graphics display is separated from the main memory of the computer. The more color selections that can be chosen on a display unit, the more bits required for storing the color codes at every pixel. As a consequence, a smaller amount of memory storage is available for specifying the on/off status and hence a poorer resolution is achieved.

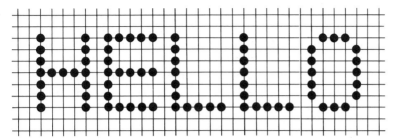

Figure 3-4 Characters are displayed using blocks of pixels.

Conversely, the better the resolution required, the fewer color selections possible on a display unit if it has a fixed-size video memory.

3.6 PROGRAMMING LANGUAGES

As has been explained in Chapter 2, computers are built with memory cells, each of which can store binary digit 0 or 1. In the preceding sections, rules and codes have been introduced so that integers, real numbers, and characters can also be translated into binary digits 0's and 1's. All of this underlines the fact that computers can communicate with users in language of binary numbers, or **machine language** of 0's and 1's. It is obvious that if one chooses to communicate with the computers in machine language, the rules and codes that vary from machine to machine will have to be fully understood and the program written for implementing a certain task will be very involved and detailed.

Standard software is available, however, that enables the users to write their programs in English-like languages such as BASIC, FORTRAN, ALGOL, COBOL, PASCAL, and so on and automatically translate the programs into machine language. This software is called **compilers** and the process is called **compilation**. The standard procedure of compilation consists of first taking the program written in **source** (problem-oriented) **language**, such as BASIC or FORTRAN, then replacing the statements in the **source program** by symbolic codes called **mnemonics** to develop an intermediate **assembler program**, and finally translating the **assembly language** into a machine language, **object** program. For example, $A = B + 1.25$ is a statement in the source program. It will be replaced by a series of steps as in operating a calculator.

Recall B	(Press RCL button and then the memory number in which the value B is stored)
Add	(Press + button)
Enter 1.25	(Press buttons "1", ".", "2", and "5")
Store A	(Press STO button and then the memory number assigned for A, in which the sum is to be stored)

The above illustration is not exactly in precise form of assembly language but facilitates the explanation to users who are familiar with the operation of calculators. As with the characters that can be coded in binary numbers as explained in an earlier article, these assembly commands, Recall, Add, Enter, and so on, can also be translated into binary numbers in accordance with the established rules, such as $(1000)_8$ for CLR (clear), $(1400)_8$ for ADD, $(3000)_8$ for STO, and so on. As can be seen, the symbolic codes are already in octal numbers and only need to be expanded to 3 binary bits for each octal bit.

While compilers relieve programmers from writing the tedious format of assembly and machine languages, they also produce less efficient programs. The reason is that the compilation software is prepared for general cases; for a specific program, more precise coding in either assembly or machine language could certainly be attempted to achieve the optimal results that save both storage and computing time.

In this text, the problem-oriented language BASIC will be used on the Tektronix 4054 Computer Graphics System. Some FORTRAN versions of the developed programs are presented in Appendix D. As microcomputers are nowadays easily accessible at the high schools and at individual homes, most readers of this book should have some exposure to various types of source languages, if not BASIC probably FORTRAN for engineers and applied scientists. Based on this assumption, BASIC language will not be introduced in an orderly fashion, but all statements adopted in this text will be explained when they appear for the first time in a program. To demonstrate how the developed programs written in BASIC language on Tektronix 4054 can be converted in other BASIC languages on prevailing microcomputers, a number of examples are presented in Appendix D for APPLE, IBM, and TRS microcomputers.

CHAPTER 4

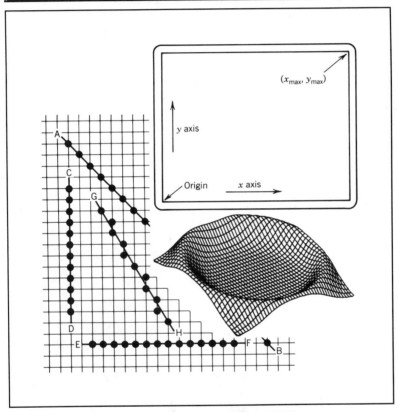

COMPUTER
PLOTTING
AND DISPLAY

4.1   INTRODUCTION

Designers conceive their ideas first mentally, proceed to make sketches, then modify the sketches, and finally decide on one or several choices that are to their satisfaction and also meet the design specifications. Traditionally, the chosen design or designs will have to be precisely drawn by a draftsman using ruler, compass, protractor, and other devices. The modern technique of computer plotting (Fig. 4-1) facilitates precision drafting, especially for repetitive routine drawing of certain standard parts composed of shapes such as a circle, a rectangle, a pair of arrowed lines for dimensioning, a center line, and so on. In this chapter, the fundamental concept of computer plotting is discussed with examples and developed computer programs.

Very thin objects such as a gasket that can be made by cutting from a flat material of thin uniform thickness can be considered two-dimensional; most designed objects are, however, **three-dimensional** as their thickness may vary. In order to present the three-dimensional features of a designed object, it has to be plotted or displayed as a series of projected views taken from different angles, on a two-dimensional paper or display screen. Moreover, many designed parts in CAD/CAM are to be fitted and assembled. Moving an object from one location to another in the trial fitting process is often required. This calls for translation of the displayed view and also alteration of the sizes and dimensions of the object in order to make a successful assembly with the other components.

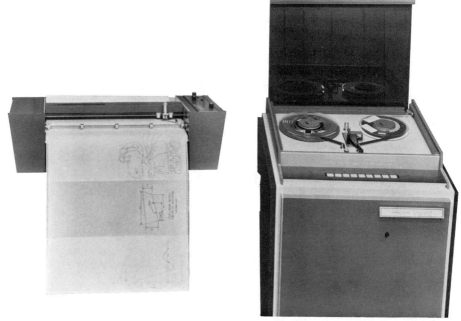

Figure 4-1   Digital computer plotting system. (Courtesy of Houston Instrument.)

Thus, the above-described basic needs in displaying a three-dimensional object involve three geometric transformations, namely, rotation, translation, and scaling. In the ensuing sections, matrix rotation will be used to express the mathematical equations involved in these three types of maneuvers. Illustrative examples and listings of computer programs for displaying the object are provided for demonstration of the rotation, translation, and scaling processes.

In the effective use of a display screen, where an image of an object should be placed requires the proper specification of a VIEWPORT statement. It defines where and how large a two-dimensional area is to be designated for imaging the projected view of an object. And in order to fit into that area, the user's data will have to be properly scaled. Otherwise, only a clipped portion of the whole projected view will appear in the selected viewport. These problems can be resolved by skillful utilization of the WINDOW statement and the clipping technique. Examples and BASIC computer programs are presented in this chapter to elucidate the application of VIEWPORT, WINDOW, and clipping techniques for CAD/CAM.

Perspective viewing, hidden-line and contour plotting, and display also play important roles in presenting three-dimensional surfaces and data in easily comprehensible graphic forms. These techniques are discussed as well in this chapter with examples and computer programs.

4.2 BASIC CONCEPT OF COMPUTER PLOTTING

We press a pen or pencil on paper and move it from one place to another in order to draw a line. If the pen or pencil is lifted and then pressed on a new spot, a new line may be drawn from there. That is the way a sketch is made. These two basic states of pen **touching** and **not touching** the paper combined with appropriate computer-controlled motions of moving the pen or the paper constitute the essential operations of computer plotting.

When it is desired to move, without the plotting pen touching the paper, to a new **destination** of coordinates (X, Y), various plotters may achieve this by use of the commands such as the following:

	Plotter(s)	Language
`CALL PLOT(X, Y, 3)`	CALCOMP,[1] COMPLOT[2]	FORTRAN
`MOVE X, Y`	Tektronix	BASIC

If the pen is to draw a line from the present position to a new destination (X, Y), the respective commands are

	Plotter(s)	Language
`CALL PLOT(X, Y, 2)`	CALCOMP, COMPLOT	FORTRAN
`DRAW X, Y`	Tektronix	BASIC

---

[1] California Computer Products, Inc., Anaheim, California.

[2] Houston Instrument, Bellaire, Texas.

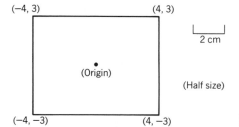

Figure 4-2

The constants 2 and 3 appear as the third argument of the above CALL PLOT statements referring to the **pen down** (touching the paper) and **pen up** (not touching the paper) states, respectively.

PLOT is a **subroutine** name of an independent program made available by the plotter manufacturers for the users to request various plot actions with the CALL statements as demonstrated above. More examples will be given on CALL PLOT applications later.

Since the pen movement is specified by the coordinates X and Y of the destination point, the values of X and Y have no meaning unless one knows where the origin of the coordinate system is and what the units are (centimeter, inch, or others) of the surface (a plot paper, a display screen, or others) on which the plot is to appear. Since the plotter and screen sizes vary from one apparatus to another, this problem will have to be addressed later when the commands WINDOW and VIEWPORT are discussed.

Let us assume that the origin of the coordinate system has been set at the center of a plot paper or a display screen. The following statements will generate a plot of a 8×6 rectangle as shown in Fig. 4-2.

BASIC	FORTRAN
MOVE −4, −3	CALL PLOT(−4., −3.,3)
DRAW −4, 3	CALL PLOT(−4., 3.,2)
DRAW 4, 3	CALL PLOT(4., 3.,2)
DRAW 4, −3	CALL PLOT(4., −3.,2)
DRAW −4, −3	CALL PLOT(−4., −3.,2)

Notice that the first two arguments of CALL PLOT must either be real variables (X and Y) or real constants ($-4.$, $-3.$, and so on). In fact, they can be any expression that yields *real* results.[3]

4.3 SMOOTHNESS OF PLOTTED CURVES

After learning how a straight line can be computer-plotted, we are ready to plot curves. All curves are plotted with a series of connected straight line segments.

[3] T. C. Smith and Y. C. Pao, *Introduction to Digital Computer Plotting*, Gordon and Breach Science Publishers, New York, 1973.

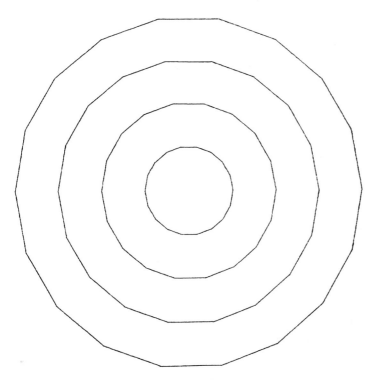

Figure 4-3

Smoothness of the plotted curves is thus determined by the number of line segments and depends on the spacing of the end points of each line segment. The following BASIC Program CIRCLES has generated a plot intended to be four concentric circles of radii equal to 10, 20, 30, and 40 units and with a common center at (90, 60) as shown in Fig. 4-3. All four circles are drawn with 18 straight line segments that resulted from choosing an angle increment of 20 degrees. The plot shows that the innermost circle is barely acceptable as a smooth circle; for the other three, using a finer angle increment, that is, more line segments, is mandatory.

PROGRAM CIRCLES	FORTRAN Version

```
100 SET DEGREES              X1=90
110 X1=90                    Y1=60
120 Y1=60                    C1=20*3.1416/180
130 C1=20                    DO 5 IR=1,4
140 FOR R=40 TO 10 STEP -10  R=50-IR*10
150 MOVE X1+R,Y1             CALL PLOT(X1+R,Y1,3)
160 FOR C=C1 TO 360 STEP C1  DO 3 IC=1,18
170 X=X1+R*COS(C)            C=C1*IC
180 Y=Y1+R*SIN(C)            X=X1+R*COS(C)
190 DRAW X,Y                 Y=Y1+R*SIN(C)
200 NEXT C                   CALL PLOT(X,Y,2)
210 NEXT R                 3 CONTINUE
220 END                    5 CONTINUE
                             END
```

A few explanations are in order as far as programming is concerned. Program CIRCLES uses the previously introduced MOVE and DRAW statements and some other statements that appear for the first time. Statements 170 and 180 employ two so-called **library functions**, namely the trigonometric functions cosine COS and sine SIN both with one argument C. Statement 100 simply declares that the angles to appear hereon in the entire program should be in degrees rather than in radians until they are converted to radians by another SET statement.

Statements 110, 120, 130, 170, and 180 are **arithmetic statements** that have the general form of

$$\text{variable} = \text{arithmetic expression}$$

In BASIC language, **variables** must be named with one or two characters, that is, one capital **alphabetic character** (A through Z) alone or followed by a second **numeric character** (0 through 9); unlike the FORTRAN language, in which variables for some computers may be named with up to eight alphanumeric characters. The equal sign in an arithmetic statement has the meaning of **assignment** or **replacement**. For example, the statement X = X1 + R*COS (C) requires the arithmetic expression X1 + R*COS (C) to be executed and the result to be assigned to or to replace the previous value of X.

R*COS(C) is the translation of R multiplied by cosine of the angle C. The complete **arithmetic operation signs** used in BASIC and FORTRAN consist of the following:

	BASIC	**FORTRAN**
Addition +	+	+
Subtraction −	−	−
Multiplication ×	*	*
Division ÷	/	/
Exponentiation	↑	**

For example, $a + b - c \times d/e^f$ may be translated into A+B−C*D/E ↑ F in BASIC and A+B−C*D/E**F in FORTRAN.

Notice that a BASIC or FORTRAN program has to be ended with an END statement. It should also be noted that the **statement number** or **line number** is assigned to all statements in BASIC programs and they must be in increasing order but not necessarily in equal increments. The increasing-order requirement is necessary in order to accommodate for the **renumbering** request that the user may wish to use for better organization of his or her program. The general form for a renumbering request is

$$\text{REN (RENUMBER)}^4 \; n_1, n_2, n_3$$

which specifies that all of the statements including and following line n_3 are to be renumbered with new line numbers starting with n_1 and be equally incremented by n_2.

---

[4] In fact, only the first three characters need to be used for a BASIC command.

4.4 DRAWING REPEATED PATTERNS

It should be apparent that in Fig. 4-3, the concentric circles are drawn by generating the straight line segments of equal lengths, then rotating and connecting them, and by use of different radii. The Program CIRCLES that draws Fig. 4-3 lists $X1 = 90$, $Y1 = 60$, and $C1 = 20$ indicating that the center of the concentric circles is at (90, 60) and the 18 line segments approximating each circle are to be generated using an angle increment of 20 degrees. The repeated generation and subsequent plotting of the line segments are implemented by the **block of statements** numbered 140 through 210. The variables R and C are to be changed from 40 to 10 in **decrements** of -10, and from 20 to 360 degrees in **increments** of 20 degrees, for the radii and angles, respectively, in plotting the four concentric circles.

The combination of FOR and NEXT statements in BASIC langugage, and the combination of DO and CONTINUE statements in FORTRAN language make it possible to carry out the desired repetitive operations. In Program CIRCLES, the block of statements numbered 140 through 210, beginning with FOR and ending with NEXT R, is to be repeatedly executed with R and C, two **index variables** appearing in the FOR and NEXT statements, changed according to the details specified in the FOR statement. The general form of such **looping** blocks is

```
FOR I = I1 TO I2 STEP I3

: (statements may or may not involve I)

NEXT I
```

The looped computation begins with the index variable I assigned with the value of $I1$ and proceeds to execute the block of statements following the FOR statement. When the NEXT I statement is executed, the index variable I will have the value of $I1$ incremented by the value of $I3$ if $I3 > 0$ and decremented if $I3 < 0$. If $I3$ has a value of 1, STEP $I3$ can be dropped from the FOR statement. The looping is to be continued until I has the value of $I2$ or $I_1 + nI_3$ ($< I2$) where $n = \text{INT}[(I2 - I1)/I3]$, where INT is a library integer function that truncates a real number. For example, $n = 4$ if $I1 = 1$, $I2 = 10$, and $I3 = 2$. That is to say, the looping will be proceeded 5 times for $I = I1 = 1$, then $I = 3, 5, 7, 9$. The quotient of $(I2 - I1)/I3 = (10 - 1)/2 = 4.5$ will be truncated and yield a result of $n = 4$. Notice that in a FOR statement, either variables or constants may be assigned to $I1$, $I2$, and $I3$. Statements 140 and 160 in Program CIRCLES serve as examples.

In fact, Program CIRCLES shows an example of **nested loops**. It contains an outer R loop and an inner C loop. In such a case, the outer loop index variable R is to have an initial value equal to 40 and the inner loop is to be completed for C starting with a value of $C1$, which is 20 degrees, incremented also by the value of $C1$ until reaching a value of 360°. The inner loop is to be completely recycled when R is decremented to 30, 20, and finally 10.

Another example is presented in Fig. 4-4 to demonstrate the use of nested

```
100 D1=8
110 D2=5
120 FOR X=50 TO 110 STEP 10
130 X1=X
140 X2=X1+D1
150 FOR Y=90 TO 30 STEP -10
160 Y1=Y
170 Y2=Y1-D2
180 GOSUB 220
190 NEXT Y
200 NEXT X
210 END
220 MOVE X1,Y1
230 DRAW X1,Y2
240 DRAW X2,Y2
250 DRAW X2,Y1
260 DRAW X1,Y1
270 RETURN

RUN
```

Figure 4-4

loops. A program called R.ARRAY enables the rectangular rows and columns of rectangles to be drawn. This example also introduces the concept of **subprograms**. Statements 220 through 270 constitute an independent program, which is developed for the sole purpose of drawing a rectangle with the coordinates $(X1, Y1)$ and $(X2, Y2)$ as one of its diagonals. This subprogram can be reached via a GOSUB 220 statement, numbered 180 in Program R.ARRAY. When the last statement of the subprogram, 270 RETURN, is implemented, the computer returns to the **calling program** and executes the statement immediately following the GOSUB 220, namely, the statement NEXT Y numbered 190. In FORTRAN language, the GOSUB 220 may be substituted by CALL RECT(X1, Y1, X2, Y2) and the subprogram may be prepared as

```
SUBROUTINE RECT(X1,Y1,X2,Y2)
CALL PLOT(X1,Y1,3)
CALL PLOT(X1,Y2,2)
CALL PLOT(X2,Y2,2)
CALL PLOT(X1,Y2,2)
CALL PLOT(X1,Y1,2)
RETURN
END
```

For later applications in this text, two subroutines written in BASIC language and called CIRCLE and CENTLN have been prepared for drawing circles and center lines. Their listings are given below. Notice that the REM (shortened command of REMARK) statements are nonexecutable and are to be used for labeling purposes.

```
100 REM * SUBROUTINE CIRCLE
110 REM * DRAW A CIRCLE OF RADIUS R1 CENTERED AT (C1.C2)
120 MOVE C1+R1.C2
130 FOR J=5 TO 360 STEP 5
140 DRAW C1+R1*COS(J),C2+R1*SIN(J)
150 NEXT J
160 RETURN
170 REM * SUBRUTINE CENTLN
180 REM * DRAW CENTER LINES OF A CIRCLE OF RADIUS R1
190 MOVE -1.25*R1+C1,C2
200 DRAW -0.1*R1+C1,C2
210 MOVE -0.05*R1+C1,C2
220 DRAW 0.05*R1+C1,C2
230 MOVE 0.1*R1+C1,C2
240 DRAW 1.25*R1+C1,C2
250 MOVE C1,-1.25*R1+C2
260 DRAW C1,-0.1*R1+C2
270 MOVE C1,-0.05*R1+C2
280 DRAW C1.0.05*R1+C2
290 MOVE C1.0.1*R1+C2
300 DRAW C1,1.25*R1+C2
310 RETURN
```

The RUN command that appears after the listing of the Program R. ARRAY is included to show that it needs to be entered from the keyboard of the graphic display device in order to execute the program and produce the display of Fig. 4-4.

4.5 DISPLAY AND RESOLUTION

A design can be plotted on paper as well as be displayed on the cathode ray tube or, as it is commonly called, the screen of a computer display terminal. Smoothness of a curve when it is displayed on the screen depends not only on the size and number of line segments used to approximate the curve but also on the resolution of the display unit. As has been explained in Chapter 3, the resolution of a display screen is the matrix points that can be lighted up. The screen is like a gridwork, and the intercepts of the horizontal and vertical lines, called dots, grids, nodes, or pixels, are usable points. For example, the Radio Shack TRS-80 Color Computer has 256 × 192 (horizontal × vertical) pixels, whereas the Tektronix 4054 Display Terminal has 4096 × 3125 pixels. It is up to the user to effectively utilize these pixels on the screen to be lighted to draw diagrams, graphs, and characters.

As shown in Fig. 4-5a, no matter how many line segments one uses to approximate a circular arc, its displayed image is composed of the usable points on the screen. The resolution of the graphics unit dictates the appearance of the curve. Even for a straight line, its image on a low-resolution screen will be comprised of a series of lighted points, as demonstrated by the line GH in Fig. 4-5b. Only special cases of horizontal lines, vertical lines, and those lines with ±45° slopes could have "straight-line" appearances on the screen, such as the lines EF, CD, and AB illustrated in Fig. 4-5b.

Hence the user of a particular graphics terminal should bear in mind the resolution of the equipment available at hand when he or she decides on how many line segments ought to be used to approximate a curve. Otherwise, unnecessary

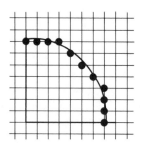

Figure 4-5a

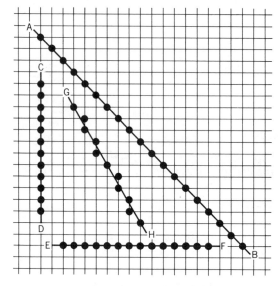

Figure 4-5b

computations for generating the fine line segments only take up the time that could be better utilized for other tasks, for the reason that the equipment cannot display at such a high resolution anyway.

In Appendix D, the graphics resolution of APPLE, IBM, and TRS microcomputers is further discussed. How the developed BASIC programs in this text should be converted so that they can be implemented on these microcomputers is also explained with a number of examples.

4.6   GEAR TOOTH PROFILES

Having discussed the software and hardware of CAD/CAM and the basic concepts of computer plotting and graphics displays, we can see that a practical application of this knowledge will be a helpful guide in studying other topics. Let us examine how the gear tooth profiles can be drawn by use of computer graphics.

Cycloidal shapes, which minimize frictional resistance, are considered the best possible profile for gear teeth—a finding dating back as early as 1495 to Leonardo da Vinci. A **cycloid** is the locus generated by a point on the circumference of a circle that rolls without slipping on a straight line. **Epicycloid** and **hypocycloid** are

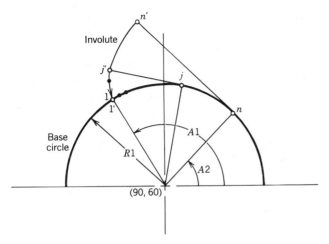

Figure 4-6

loci generated when the circle is rolled on the outside and inside of another circle, respectively.

In practical applications of circular gearing, it is commonly required that the ratio of angular velocities of the driving and driven gears be a constant. **Involute** curves are the most popular shape adopted for the gear profile to satisfy this requirement. It is constructed by unwinding a taut cord from around a **base circle**. Figure 4-6 demonstrates how an involute is formed step by step. First, a chosen arc $1n$ is divided into equal increments with radial lines from the center of a base circle. Tangents are then drawn at these partitioning points. The arc lengths 11, 12, . . . , $1n$ are calculated and used to mark new points $1', 2', . . . , n'$ on the respective tangent lines $11', 22', . . . , nn'$. The curve joining the points $1', 2', . . . , n'$ is the involute.

Figure 4-7 is presented to illustrate how a gear tooth is plotted using a sub-

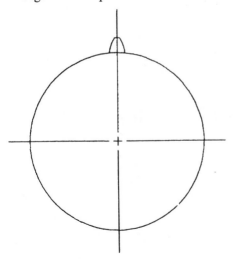

Figure 4-7

routine INVOLU developed by application of the involute curves. A small program G. TOOTH that calls the subroutine INVOLU has produced this plot.

In subroutine INVOLU, one-dimensional subscripted array variables $X2$ and $Y2$ are used. For declaration of the use of a subscripted variable, a DIMENSION statement should be placed in the program preceding the statement, in which the array variable appears for the first time, and should specify how many elements are in the array. For example, Statement 200 declares that there will be 22 elements in both arrays $X2$ and $Y2$. When a particular element of an array is involved in a statement, the subscript of the array has to be specifically defined with a constant or a variable or an expression such as $X2(5)$ or $X2(J)$ or $X2(23 - J)$. One should be cautioned that a subscripted variable with a zero or negative subscript is invalid on most computers. Two-dimensional subscripted array variables for handling matrices will be discussed later.

In the main program, an input statement READ is used. It has the general form of the command word READ followed by a list of variables separated by commas. Statement 120 illustrates the reading of five variables, $C1$, $C2$, $R1$, $A1$, and $A2$. DATA statement numbered 110 in the program provides the values of these variables. There could be many READ statements in a program and as many DATA statements should be provided.

PROGRAM G. TOOTH

```
100 SET DEGREES
110 DATA 96,60,20,95,60
120 READ C1,C2,R2,A1,A2
130 GOSUB 150
140 END
150 REM * SUBROUTINE INVOLU - PLOTS GEAR TOOTH INVOLUTE *
160 REM    GEAR CENTER AT (C1,C2), BASE RADIUS R1,
170 REM    INITIAL AND ENDING ANGLES ARE A1 AND A2
180 N=11
190 A3=(A1-A2)/10
200 DIM X2(22),Y2(22)
210 MOVE C1+R1,C2
220 FOR J=5 TO 360 STEP 5
230 DRAW C1+R1*COS(J),C2+R1*SIN(J)
240 NEXT J
250 FOR J=1 TO N
260 A4=A1-(J-1)*A3
270 L=R1*0.017453*(J-1)*A3
280 X2(J)=R1*COS(A4)+C1-L*COS(A4-90)
290 Y2(J)=R1*SIN(A4)+C2-L*SIN(A4-90)
300 X2(23-J)=-X2(J)+2*C1
310 Y2(23-J)=Y2(J)
320 NEXT J
330 MOVE -1.25*T1+C,C2
340 DRAW -0.1*R1+C1,C2
350 MOVE -0.05*R1+C1,C2
360 DRAW 0.05*R1+C1,C2
370 MOVE 0.1*R1+C1,C2
380 DRAW 1.25*R1+C1,C2
390 MOVE C1,-1.25*R1+C2
400 DRAW C1,-0.1*R1+C2
```

```
410 MOVE C1,-0.05*R1+C2
420 DRAW C1.0.05*R1+C2
430 MOVE C1.0.1*R1+C2
440 DRAW C1,1.5*R1+C2
450 MOVE X2(1).Y2(1)
460 DRAW X2,Y2
470 RETURN
```

4.7 ROTATION

In Fig. 4-7, we observe that in order to completely draw all teeth of a gear the profile of the drawn tooth could be rotated and then relocated. This leads to the discussion of **rotation** of an object.

Suppose that the square $ABCD$ shown in Fig. 4-8 is to be displayed as $A'B'C'D'$ by rotating it about the z axis by an angle θ_z. To that end, the coordinates of A', B', C', and D' will have to be calculated in terms of the coordinates of A, B, C, and D, and also θ_z. Noting that the rotation is about the origin 0, the lengths $0A$ and $0A'$ are thus equal. Also based on the relationships between rectangular and polar coordinates, we can obtain

$$x_{A'} = 0A' \cos \theta_{A'} = 0A \cos (\theta_A + \theta_z)$$
$$= 0A \cos \theta_A \cos \theta_z - 0A \sin \theta_A \sin \theta_z = x_A \cos \theta_z - y_A \sin \theta_z$$

and

$$y_{A'} = 0A' \sin \theta_{A'} = 0A \sin (\theta_A + \theta_z)$$
$$= 0A \cos \theta_A \sin \theta_z + 0A \sin \theta_A \cos \theta_z = x_A \sin \theta_z + y_A \cos \theta_z$$

In matrix notation, the above equations can be written as

$$\begin{bmatrix} x_{A'} \\ y_{A'} \end{bmatrix} = \begin{bmatrix} \cos \theta_z & -\sin \theta_z \\ \sin \theta_z & \cos \theta_z \end{bmatrix} \begin{bmatrix} x_A \\ y_A \end{bmatrix} \tag{a}$$

Image that the square $ABCD$ is not on the xy plane but is floating parallel to and at a constant distance z_A above the xy plane. The rotation of the square $ABCD$ about the z axis will not change the z coordinates of its four vertices. That is to say, $z_{A'} = z_A$ and the matrix equation (a) can be expanded to the form

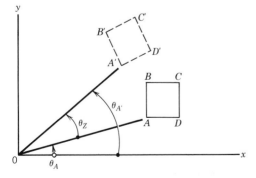

Figure 4-8 Rotation in x–y plane.

$$\begin{bmatrix} x_{A'} \\ y_{A'} \\ z_{A'} \end{bmatrix} = \begin{bmatrix} \cos\theta_z & -\sin\theta_z & 0 \\ \sin\theta_z & \cos\theta_z & 0 \\ 0 & 0 & 1 \end{bmatrix} \begin{bmatrix} x_A \\ y_A \\ z_A \end{bmatrix}$$

(b)

Similarly, if the rotation is about the y axis by an angle θ_y, the coordinates of the point A' can be calculated by use of the matrix equation

$$\begin{bmatrix} x_{A'} \\ y_{A'} \\ z_{A'} \end{bmatrix} = \begin{bmatrix} \cos\theta_y & 0 & \sin\theta_y \\ 0 & 1 & 0 \\ -\sin\theta_y & 0 & \cos\theta_y \end{bmatrix} \begin{bmatrix} x_A \\ y_A \\ z_A \end{bmatrix}$$

(c)

And after a rotation about the x axis by an angle θ_x, the matrix equation is

$$\begin{bmatrix} x_{A'} \\ y_{A'} \\ z_{A'} \end{bmatrix} = \begin{bmatrix} 1 & 0 & 0 \\ 0 & \cos\theta_x & -\sin\theta_x \\ 0 & \sin\theta_x & \cos\theta_x \end{bmatrix} \begin{bmatrix} x_A \\ y_A \\ z_A \end{bmatrix}$$

(d)

For convenience in discussion, Eqs. (b), (c), and (d) are abbreviated as

$$\mathbf{P}(A') = \mathbf{R}(\theta_z)\mathbf{P}(A)$$

(e)

$$\mathbf{P}(A') = \mathbf{R}(\theta_y)\mathbf{P}(A)$$

(f)

$$\mathbf{P}(A') = \mathbf{R}(\theta_x)\mathbf{P}(A)$$

(g)

A BASIC language subroutine Z.ROTATE is listed below for transformation of any data point (x_1, y_1) on the display screen into new coordinates (x_2, y_2) based on Eq. (a) when the point undergoes a **counterclockwise** rotation of C3 degrees about the z axis, which is outward normal to the screen. A Program GEAR making use of this subroutine is also presented. In the program, the arrays T1 and T2 containing the coordinates of the points for drawing the gear tooth profile are continuously transformed by different angles of rotation C3 in accordance with the loop composed of statements 200–270. Program GEAR also shows the application of the subroutines CIRCLE, CENTLN, INVOLU, and Z.ROTATE. Figure 4-9 is the resulting plot when the center of the gear base circle (C1, C2) is located

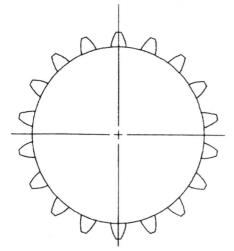

Figure 4-9

```
100 REM * PROGRAM GEAR *
110 SET DEGREES
120 DIM Q1(22),Q2(22)
130 DATA 96,60,20,95,60
140 READ C1,C2,R1,A1,A2
150 GOSUB 450
160 GOSUB 510
170 GOSUB 320
180 MOVE T1(1),T2(1)
190 DRAW T1,T2
200 FOR C3=20 TO 340 STEP 20
210 FOR I=1 TO 22
220 X1=T1(I)-C1
230 Y1=T2(I)-C2
240 GOSUB 650
250 Q1(I)=X2+C1
260 Q2(I)=Y2+C2
270 NEXT I
280 MOVE Q1(1),Q2(1)
290 DRAW Q1,Q2
300 NEXT C3
310 END
320 REM * SUBROUTINE INVOLU - PLOTS GEAR TOOTH INVOLUTE *
330 N=11
340 A3=(A1-A2)/10
350 DIM T1(22),T2(22)
360 FOR J=1 TO N
370 A4=A1-(J-1)*A3
380 L=R1*0.017453*(J-1)*A3
390 T1(J)=R1*COS(A4)+C1-L*COS(A4-90)
400 T2(J)=R1*SIN(A4)+C2-L*SIN(A4-90)
410 T1(23-J)=-T1(J)+2*C1
420 T2(23-J)=T2(J)
430 NEXT J
440 RETURN
450 REM * SUBROUTINE CIRCLE
460 MOVE C1+R1,C2
470 FOR J=5 TO 360 STEP 5
480 DRAW C1+R1*COS(J),C2+R1*SIN(J)
490 NEXT J
500 RETURN
510 REM * SUBROUTINE CENTLN
520 MOVE -1.25*R1+C1,C2
530 DRAW -0.1*R1+C1,C2
540 MOVE -0.05*R1+C1,C2
550 DRAW 0.05*R1+C1,C2
560 MOVE 0.1*R1+C1,C2
570 DRAW 1.25*R1+C1,C2
580 MOVE C1,-1.25*R1+C2
590 DRAW C1,-0.1*R1+C2
600 MOVE C1,-0.05*R1+C2
610 DRAW C1,0.05*R1+C2
620 MOVE C1,0.1*R1+C2
630 DRAW C1,1.5*R1+C2
640 RETURN
650 REM * SUBROUTINE Z.ROTATE
660 X2=COS(C3)*X1-SIN(C3)*Y1
670 Y2=SIN(C3)*X1+COS(C3)*Y1
680 RETURN
```

Figure 4-9a (*continued*)

at (96, 60), both in graphic display units (GDU), the radius $R1$ is equal to 20 GDU, and the beginning and ending angles $A1$ and $A2$ are equal to 95° and 60°, respectively. Statements 130 and 140 specify these data.

SUBROUTINE Z. ROTATE

```
100 REM *SUBROUTINE Z.ROTATE - TRANSFORMS (X1,Y1) INTO (X2,Y2)
110 REM    BY ROTATING ABOUT Z-AXIS BY C3 DEGREES.
120 REM  Z IS OUTWARD, NORMAL TO THE SCREEN.
130 REM  C3 IS POSITIVE IF ROTATION IS COUNTERCLOCKWISE.
140 X2=COS(C3)*X1-SIN(C3)*Y1
150 Y2=SIN(C3)*X1+COS(C3)*Y1
160 RETURN
```

4.8 TRANSLATION

An object displayed on a screen can also be moved to a new position by translation. As illustrated in Fig. 4-10, the square $ABCD$ moves to a new position $A'B'C'D'$ when the required translations in x, y, and z directions are specified. Suppose that x_T, y_T, and z_T (not shown in Fig. 4-10) are the specified translations in x, y, and z directions, respectively. It should be apparent that the coordinates of the vertices of the square at the new position can be obtained by vector addition, such as

$$\begin{bmatrix} x_{A'} \\ y_{A'} \\ z_{A'} \end{bmatrix} = \begin{bmatrix} x_A \\ y_A \\ z_A \end{bmatrix} + \begin{bmatrix} x_T \\ y_T \\ z_T \end{bmatrix} \tag{h}$$

We shall abbreviate Eq. (h) as

$$\mathbf{P}(A') = \mathbf{P}(A) + \mathbf{V}_T \tag{i}$$

And it should be clear that if the square $ABCD$ undergoes a simple translation, the translative vector, $\mathbf{V}_T$, should be applied for all four vertices, A, B, C, and D to arrive at the new positions A', B', C', and D', respectively.

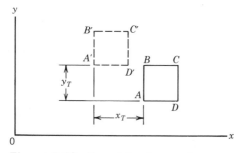

Figure 4-10   Translation in x–y plane.

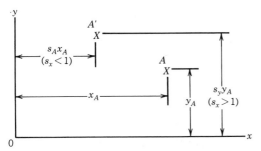

Figure 4-11 Scaling in $x-y$ plane.

4.9 SCALING

Often it is necessary to compress or stretch an object being displayed. To that end, a scaling procedure is called for. As illustrated in Fig. 4-11, the coordinates of a point A, (x_A, y_A, z_A), can be scaled by specifying the scaling factors (s_x, s_y, s_z) so that the new position of A, A', will have the coordinates $(s_x x_A, s_y y_A, s_z z_A)$. Notice that the z coordinate of the points A and A' are not shown in Fig. 4-11. In matrix notation, the scaling for the point A can be written as

$$\begin{bmatrix} x_{A'} \\ y_{A'} \\ z_{A'} \end{bmatrix} = \begin{bmatrix} s_x & 0 & 0 \\ 0 & s_y & 0 \\ 0 & 0 & s_z \end{bmatrix} \begin{bmatrix} x_A \\ y_A \\ z_A \end{bmatrix} \tag{j}$$

And the abbreviated equation is

$$\mathbf{P}(A') = \mathbf{S}\mathbf{P}(A) \tag{k}$$

where $\mathbf{S}$ is a diagonal matrix. For the special case of homogeneous scaling, $s_x = s_y = s_z = s$, Eq. (k) can be simplified as

$$\mathbf{P}(A') = s\mathbf{P}(A) \tag{l}$$

4.10 DISPLAY PROGRAM AND EXAMPLES

Program ROTATE has been developed to demonstrate the basic applications of the transformation of scaling, translation, and rotation. A brick of dimensions $2 \times 3 \times 4$ is used as an object for displaying. The front, right, and top surfaces of the brick are labeled with "F," "R," and "T," respectively. The origin of the coordinate system is located at the left lower corner of the front surface. The x and y axes are on the front surface directed to the right and upward along the sides of 4 and 2 units long, respectively.

Figures 4-12 and 4-13 are generated with the coordinates of the vertices of the brick specified by lines 170–250. No translation and scaling are to be carried out by ROTATE as specified by lines 290 and 300. The reason for not deliberating the scaling and translation processes here is that later in this chapter, the

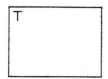

Figure 4-12   Top (T), front (F), and right (R) views of a 2 × 3 × 4 brick displayed by use of Program ROTATE.

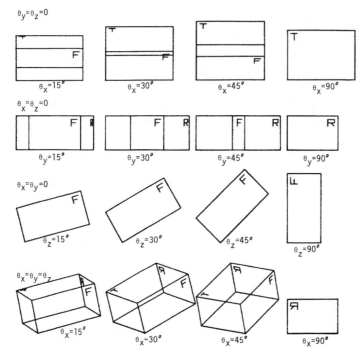

Figure 4-13   Various rotated views of a 2 × 3 × 4 brick generated by Program ROTATE with DATA changes as listed below.

```
270 VIEWPORT 50,130,20,100
280 REMARK   C-ROTATION: S-SCALE: T-TRANSLATION
290 READ S1,S2,S3,T1,T2,T3
300 DATA 1,1,1,0,0,0
310 S=28
315 W=6
320 FOR VO=75 TO 0 STEP -25
330 FOR HO=0 TO 75 STEP 25
340 READ C1,C2,C3
350 DATA 15,0,0
360 VIEWPORT HO,HO+S,VO,VO+5
370 GOSUB 560
```

Continued on next page.

```
380 GOSUB 720        (TRANSF & Scaling)
400 WINDOW X1,X1+W,Y1,Y1+W
410 GOSUB 840          (Plotting)
420 DATA 30,0,0
430 DATA 45,0,0
440 DATA 90,0,0
450 DATA 0,-15,0
460 DATA 0,-30,0
470 DATA 0,-45,0
480 DATA 0,-90,0
490 DATA 0,0,15
500 DATA 0,0,30
510 DATA 0,0,45
520 DATA 0,0,90
522 DATA 15,15,15
524 DATA 30,30,30
526 DATA 45,45,45
528 DATA 90,90,90
530 NEXT HO
540 NEXT VO
550 END
```

VIEWPORT and WINDOW statements will be introduced; they allow effective implementation of scaling and translation in displaying three-dimensional objects. The front, top, and right views of the brick are plotted in Fig. 4-12. They are the results when the viewing eye is at infinity and on the z, y, and x-axes, respectively. These so-called orthographic views will be further discussed in a later section on the perspective views. Figure 4-13 presents a variety of views of the brick resulting from different angles of rotation about the three coordinate axes. The labels "*F*," "*T*," and "*R*" on the surfaces of the brick are very helpful in depicting the distinction between these rotated views.

The subroutine TRANSF, which performs the required scaling, translation, and rotation, has been prepared in both BASIC and FORTRAN languages. There are several statements in the Program ROTATE and subroutine TRANSF that need to be explained.

Subroutine TRANSF

1. FORTRAN Version

```
        SUBROUTINE TRANSF(XI,YI,ZI,CX,CY,CZ,TX,TY,TZ,SX,SY,SZ,N,
     *                 XO,YO,ZO)
C
C       INPUT N POINTS WITH COORDINATES (XI,YI,ZI)
C       TO BE TRANSFORMED TO (XO,YO,ZO) WHEN ROTATION
C       ANGLES CX,CY,CZ, TRANSLATIONS TX,TY,TZ, AND
C       SCALING SX,SY,SZ ARE SPECIFIED.
C
        DIMENSION XI(N),YI(N),ZI(N),XO(N),YO(N),ZO(N)
        A=COS(CZ)*COS(CY)
        B=SIN(CZ)*COS(CY)
        C=-SIN(CY)
        D=-SIN(CZ)*COS(CX)+COS(CZ)*SIN(CY)*SIN(CX)
```

```
E=COS(CZ)*COS(CX)+SIN(CZ)*SIN(CY)*SIN(CX)
F=COS(CY)*SIN(CX)
G=SIN(CZ)*SIN(CX)+COS(CZ)*SIN(CY)*COS(CX)
H=-COS(CZ)*SIN(CX)+SIN(CZ)*SIN(CY)*COS(CX)
P=COS(CY)*COS(CX)
DO 2 I=1,N
XO(I)=A*SX*XI(I)+D*SX*YI(I)+G*SX*ZI(I)+TX
YO(I)=B*SY*XI(I)+E*SY*YI(I)+H*SY*ZI(I)+TY
ZO(I)=C*SZ*XI(I)+F*SZ*YI(I)+P*SZ*ZI(I)+TZ
2 CONTINUE
RETURN
END
```

2. BASIC Version

Subroutine TRANSF is listed as lines 480 through 750 in Program ROTATE

Program ROTATE

The statements 170, 200, and 230 are examples of how an entire one-dimensional subscripted array can be read by mentioning its name alone. An INI(INITIAL) statement initializes all controls, variables, and many by-default arrangements of the graphic display unit. Specific details will be given when the occasion arises. The PAGE statement clears the screen and makes it ready for display or entering of new statements.

The logical IF statements also appear in the program. They have the general form of

$$n_1 \text{ IF (a numeric or logical expression) THEN } n_2$$
$$n_3 \text{ (next statement)}$$

It tests the true or false condition of the numeric or logical expression and transfers to the line n_2 or n_3, respectively. The following are some valid examples:

```
100 IF A THEN 150
110 IF B*C + 1 THEN 200
120 IF J < >2 THEN 175
```

In line 120, $J < >2$ is a logical expression asking, "Is J not equal to 2, true or false?" $< >$ is a relational operator meaning "either greater or less than" or in other words, "not equal to." Many relational operators are self-explanatory but they will nonetheless be explained when they appear in a program for the first time in this text.

PROGRAM ROTATE

```
100 INIT
110 PAGE
120 SET DEGREES
130 REMARK    8 PTS FOR BOX, 16 PTS FOR LETTERS "F","T","R"
140 N=24
150 DIM XO(N),YO(N),ZO(N),X(N),Y(N),Z(N),X5(N),Y5(N),Z5(N)
160 REMARK  FIRST 8 FOR BOX,THEN 16 FOR LETTERS
170 READ XO
```

```
180 DATA 0,4,4,0,0,4,4,0
190 DATA 3.4,3.4,3.4,3.8,3.7,0.2,0.4,0.6,0.4,4,4,4,4,4,4,4,4
200 READ Y0
210 DATA 0,0,2,2,0,0,2,2
220 DATA 1.4,1.6,1.8,1.8,1.6,2,2,2,2,1.4,1.6,1.8,1.8,1.6,1.6,1.4
230 READ Z0
240 DATA 0,0,0,0,3,3,3,3
250 DATA 3,3,3,3,3,0.2,0.2,0.2,0.6,0.6,0.6,0.6,0.2,0.2,0.4,0.2
260 REM : C-ROTATION; S-SCALING; T-TRANSLATION
270 FOR L=1 TO 3
280 READ S1,S2,S3,T1,T2,T3,C1,C2,C3
290 DATA 1,1,1,0,0,0,0,0,0
300 READ H0,V0,S
310 DATA 50,50,20
320 VIEWPORT H0,H0+S,V0,V0+S
330 REM: SUBS- TRANSFER & SCALING
340 GOSUB 470
350 GOSUB 630
360 IF L>1 THEN 380
370 W=X2-X1 MAX Y2-Y1
380 WINDOW X1,X1+W,Y1,Y1+W
390 REM : SUB - PLOTTING
400 GOSUB 750
410 DATA 1,1,1,0,0,0,90,0,0
420 DATA 50,70,20
430 DATA 1,1,1,0,0,0,0,-90,0
440 DATA 80,50,20
450 NEXT L
460 END
470 REM : SUBROUTINE TRANF - TRANSLATE THEN ROTATE & SCALE
480 A=COS(C3)*COS(C2)
490 B=SIN(C3)*COS(C2)
500 C=-SIN(C2)
510 D=-SIN(C3)*COS(C1)+COS(C3)*SIN(C2)*SIN(C1)
520 E=COS(C3)*COS(C1)+SIN(C2)*SIN(C3)*SIN(C1)
530 F=COS(C2)*SIN(C1)
540 G=SIN(C3)*SIN(C1)+COS(C3)*SIN(C2)*COS(C1)
550 H=-COS(C3)*SIN(C1)+SIN(C3)*SIN(C2)*COS(C1)
560 P=COS(C2)*COS(C1)
570 FOR I=1 TO N
580 X(I)=A*S1*(X0(I)+T1)+D*S1*(Y0(I)+T2)+G*S1*(Z0(I)+T3)
590 Y(I)=B*S2*(X0(I)+T1)+E*S2*(Y0(I)+T2)+H*S2*(Z0(I)+T3)
600 Z(I)=C*S3*(X0(I)+T1)+F*S3*(Y0(I)+T2)+P*S3*(Z0(I)+T3)
610 NEXT I
620 RETURN
630 REM SCLAING OF DATA
640 X1=X(1)
650 X2=X(1)
660 Y1=Y(1)
670 Y2=Y(1)
680 FOR I=2 TO N
690 X1=X1 MIN X(I)
700 X2=X2 MAX X(I)
710 Y1=Y1 MIN Y(I)
720 Y2=Y2 MAX Y(I)
730 NEXT I
740 RETURN
750 REMARK  COMMENCE PLOTTING
760 MOVE X(5),Y(5)
770 DRAW X(6),Y(6)
780 DRAW X(2),Y(2)
```

```
790  DRAW  X(1),Y(1)
800  DRAW  X(5),Y(5)
810  DRAW  X(8),Y(8)
820  DRAW  X(7),Y(7)
830  DRAW  X(3),Y(3)
840  DRAW  X(4),Y(4)
850  DRAW  X(8),Y(8)
860  MOVE  X(6),Y(6)
870  DRAW  X(7),Y(7)
880  MOVE  X(3),Y(3)
890  DRAW  X(2),Y(2)
900  MOVE  X(1),Y(1)
910  DRAW  X(4),Y(4)
920  REMARK DRAW "F"
930  MOVE  X(9),Y(9)
940  DRAW  X(11),Y(11)
950  DRAW  X(12),Y(12)
960  MOVE  X(10),Y(10)
970  DRAW  X(13),Y(13)
980  REMARK DRAW "T"
990  MOVE  X(14),Y(14)
1000 DRAW  X(16),Y(16)
1010 MOVE  X(17),Y(17)
1020 DRAW  X(15),Y(15)
1030 REMARK DRAW "R"
1040 MOVE  X(18),Y(18)
1050 DRAW  X(20),Y(20)
1060 DRAW  X(21),Y(21)
1070 DRAW  X(22),Y(22)
1080 DRAW  X(19),Y(19)
1090 MOVE  X(23),Y(23)
1100 DRAW  X(24),Y(24)
1110 RETURN
```

4.11 VIEWPORT

A display screen usually has a two-dimensional, finite-sized region for drawing graphs, diagrams, and characters. As illustrated in Fig. 4-14, this region can be defined as $0 \leq x \leq x_{max}$ and $0 \leq y \leq y_{max}$. In most cases, the origin is located at the lower left corner and the coordinates (x, y) are expressed in **graphic display units** (GDU).

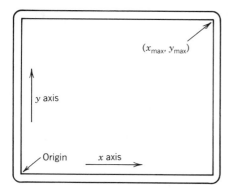

(x_{max}, y_{max})

y axis

Origin   x axis

Figure 4-14

The graphic display units for x_{max} and y_{max} in a Tektronix 4054 system are 130 and 100, respectively.

Not necessarily the entire screen needs to be utilized. The command VIEWPORT enables the entire available region or a part of it to be specified for displaying use. The general form is

$$\text{VIEWPORT } X1, X2, Y1, Y2$$

where $0 \leq X1 < X2 \leq x_{max}$ and $0 \leq Y1 < Y2 \leq y_{max}$. When a display unit is properly initialized, the entire screen will be assigned by default. That is, automatically $X1$, $X2$, $Y1$, and $Y2$ are set equal to 0, x_{max}, 0, and y_{max}, respectively, for the VIEWPORT statement.

Microcomputers, such as APPLE II, IBM Personal, and Radio Shack TRS-80, have different setups of viewports. First of all, the coordinate system has an origin located at the upper left corner of the display screen and a y-axis pointing downward. Furthermore, the maximum x and y values allowed for displays on the screen depend on the version of the microcomputer and also depend on the supporting graphics software.

In Appendix D, the problem of how to run the computer programs, developed in this text for Tektronix 4054 system, on microcomputers is addressed. Especially, the transformation equations needed for the different $x-y$ coordinate systems are derived and illustrations of their applications to a number of the developed programs are presented. Prior to these discussions, it is necessary, however, to explain the relationship between the user's data points used for the display and the mapped points on a display screen. To that end, we need to introduce the WINDOW statement.

4.12 WINDOW

Once a viewport has been specified, either by default or by a user's statement, that particular region in graphical display units is available for display use. For a general case, the user's data may not be in graphical display units. A conversion or scaling procedure must therefore be carried out before the data can be adequately displayed in the chosen viewport. A command WINDOW maps the user's data domain onto the viewport. It has the general form

$$\text{WINDOW } H1, H2, V1, V2$$

where $H1$, $H2$, $V1$, and $V2$ are the limiting values that the user wishes to map to $X1$, $X2$, $Y1$, and $Y2$ in the VIEWPORT statement, respectively.

Figure 4-15 shows a number of examples on use of the VIEWPORT and WINDOW statements. Statements 250–320 draw an area enclosed by four vertices (x_i, y_i) for $i = 1, 2, 3, 4$. Statement 120 specifies a 80×80 (GDU) VIEWPORT at the upper right corner of the screen. A square of 2×2 in user's units symmetric in both x and y axes is chosen for plotting, which is defined with the statements 130–150. The first WINDOW statement, 160, requests that the entire VIEWPORT

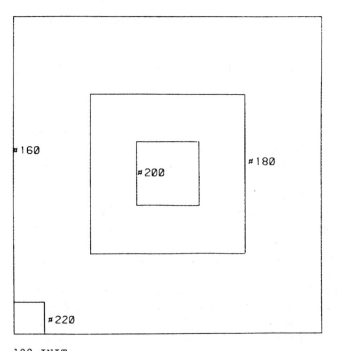

```
100 INIT
110 DIM X(4),Y(4)
120 VIEWPORT 50,130,20,100
130 READ X,Y
140 DATA -1,-1,1,1
150 DATA -1,1,1,-1
160 WINDOW -1,1,-1,1
170 GOSUB 250
180 WINDOW -2,2,-2,2
190 GOSUB 250
200 WINDOW -5,5,-5,5
210 GOSUB 250
220 WINDOW 0,10,0,10
230 GOSUB 250
240 END
250 MOVE X(1),Y(1)
260 FOR I=2 TO 5
270 J=I
280 IF J<5 THEN 300
290 J=1
300 DRAW X(J),Y(J)
310 NEXT I
320 RETURN
```

RUN   **Figure 4-15**

be used to plot the square. The second and third WINDOW statements, 180 and 200, result in scaling the square to 1/2 and 1/5 sizes, respectively.

The fourth WINDOW statement, 220, is of special interest because it demonstrates the concept of **clipping**. The values 0, 10, 0, and 10 (user's unit) in the WINDOW statement are to correspond, respectively, to 50, 130, 20, and 100

```
100 WINDOW 0,8,0,5
110 DATA 0,0,8,5
120 READ X1,Y1,X2,Y2
130 FOR H=1 TO 7
140 C1=50+(H-1)*10
150 FOR V=1 TO 7
160 C2=90-(V-1)*10
170 VIEWPORT C1,C1+X2,C2-Y2,C2
180 GOSUB 220
190 NEXT V
200 NEXT H
210 END
220 MOVE X1,Y1
230 DRAW X1,Y2
240 DRAW X2,Y2
250 DRAW X2,Y1
260 DRAW X1,Y1
270 RETURN

RUN
```

Figure 4-16

(GDU) in the VIEWPORT statement. Except for its upper right corner, (1, 1), and quadrant, most of the square falls outside of the viewport. Because the WINDOW statement specifies that only points with positive x between 0 and 10 and positive y between 0 and 10 are visible, 3/4 of the square has thus been **clipped** away.

Figure 4-16 is presented to illustrate that the shape of an object, in this case a rectangle described by the statements 110–120, can be moved about on the screen by changing the VIEWPORT statement. This is the same display presented in Chapter 3 when the commands MOVE, DRAW, FOR, and NEXT were used. Here, how the same plot can be obtained by applying the WINDOW and VIEWPORT commands now is demonstrated.

Figure 4-17 is another example in which both VIEWPORT and WINDOW statements have to be adjusted in order to plot the same concentric circles previously shown in Fig. 4-3.

A subroutine MAX.MIN that determines the bounds of x and y coordinate values of the displayed points has been developed and will be in frequent use for windowing and other applications throughout this text. Its listing is presented below.

```
100 REM* SUB. MAX,MIN - FINDS THE MAXIMA AND MINIMA OF X AND Y ARRAYS
110 REM  OF LENGTH N1 AND STORES THEM IN X1,Y1 (MIN) & X2,Y2 (MAX).
120 X1=X(1)
130 X2=X(1)
140 Y1=Y(1)
150 Y2=Y(1)
160 FOR I1=2 TO N1
170 IF X(I1)>X1 THEN 190
180 X1=X(I1)
190 IF X(I1)<X2 THEN 210
```

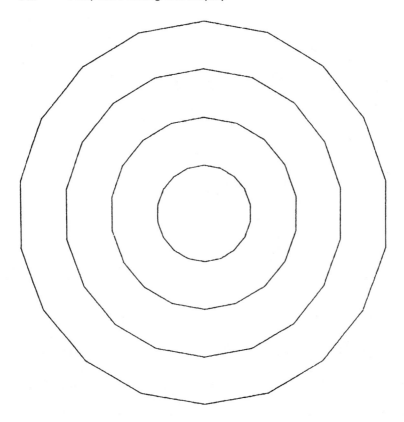

```
100 INIT
110 SET DEGREES
120 X1=90
130 Y1=60
140 C1=20
150 FOR S=1 TO 4
160 R=40-(S-1)*10
170 VIEWPORT 90-R,90+R,60-R,60+R
180 WINDOW 90-R,90+R,60-R,60+R
190 GOSUB 220
200 NEXT S
210 END
220 MOVE X1+R,Y1
230 FOR C=C1 TO 360 STEP C1
240 X=X1+R*COS(C)
250 Y=Y1+R*SIN(C)
260 DRAW X,Y
270 NEXT C
280 RETURN

RUN
```

Figure 4-17

```
200 X2=X(I1)
210 IF Y(I1)>Y1 THEN 230
220 Y1=Y(I1)
230 IF Y(I1)<2 THEN 250
240 Y2=Y(I1)
250 NEXT I1
260 RETURN
```

In Appendix D, a subroutine called VIEW.WIN is presented that replaces the VIEWPORT and WINDOW statements so that BASIC programs can be run on the microcomputers when these two statements are not available.

4.13   CLIPPING

To further demonstrate the concept of clipping, Fig. 4-18 and its plotting program are presented. The program first plots a viewport of 80×80 GDU with statements 120–160. The subroutine consisting of statements 390–440 draws a rectangle when the coordinates of the endpoints $(X1, Y1)$ and $(X2, Y2)$ of one of its diagonals are specified. The object to be clipped is a circle of radius equal to 40 and centered

```
100 INIT
110 SET DEGREES
120 DATA 50,20,130,100
130 READ X1,Y1,X2,Y2
140 VIEWPORT 50,130,20,100
150 WINDOW 50,130,20 100
160 GOSUB 390
170 DATA 0,0,10,40
180 READ X1,Y1,C1,R
190 WINDOW -40,40,-40,40
200 GOSUB 320
210 REM * CLIPPING *********
220 DATA 10,30,20,40
230 READ X1,X2,Y1,Y2
240 GOSUB 390
250 VIEWPORT 75,105,45,75
260 WINDOW 10,30,20,40
270 GOSUB 390
280 DATA 0,0,10,40
290 READ X1,Y1,C1,R
300 GOSUB 320
310 END
320 MOVE X1+R,Y1
330 FOR C=C1 TO 360 STEP C1
340 X=X1+R*COS(C)
350 Y=Y1+R*SIN(C)
360 DRAW X,Y
370 NEXT C
380 RETURN
390 MOVE X1,Y1
400 DRAW X1,Y2
410 DRAW X2,Y2
420 DRAW X2,Y1
430 DRAW X1,Y1
440 RETURN

RUN
```

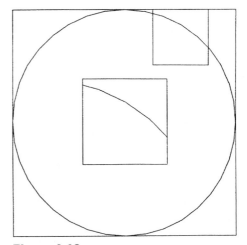

Figure 4-18

at (0, 0). The circle is drawn with the aid of the subroutine defined by statements 320–380. Statements 170–200 draw the complete circle.

The clipping request is implemented by statements 220–240 and illustrated by the small square on the upper right corner. To enlarge the clipped region and to plot it at the center of the original viewport, statement 250 is included to define the new VIEWPORT, whereas statement 260 clips the desired portion of the circular arc.

Figure 4-19 shows an application of the clipping technique in viewing the parti-

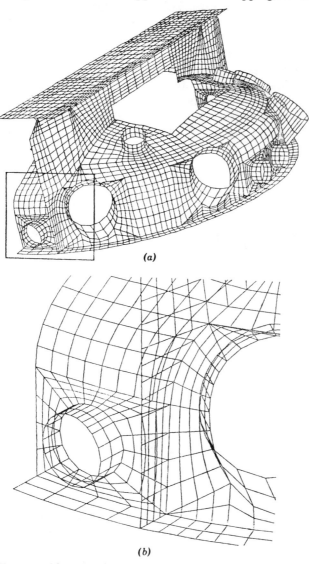

(a)

(b)

Figure 4-19 (a) Define a clipped substructure and (b) plot the clipped region after enlargement and rotation. (Courtesy of PDA Engineering, Santa Ana, California.)

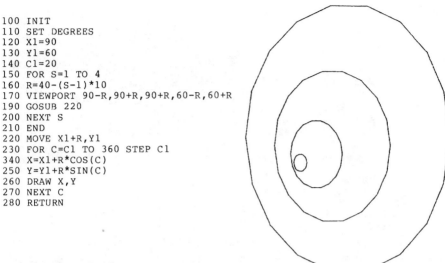

```
100 INIT
110 SET DEGREES
120 X1=90
130 Y1=60
140 C1=20
150 FOR S=1 TO 4
160 R=40-(S-1)*10
170 VIEWPORT 90-R,90+R,90+R,60-R,60+R
190 GOSUB 220
200 NEXT S
210 END
220 MOVE X1+R,Y1
230 FOR C=C1 TO 360 STEP C1
340 X=X1+R*COS(C)
250 Y=Y1+R*SIN(C)
260 DRAW X,Y
270 NEXT C
280 RETURN
```

Figure 4-20

tioning pattern of a selected substructure in finite element analysis that is to be introduced in Chapter 9.

EXERCISE

The program and plot presented in Fig. 4-20 show four distorted circles of radii equal to 40, 30, 20, and 10. Aside from the inappropriate shapes resulting from approximating the circumferences of these circles by the connected linear segments, which can easily be refined by use of finer angle increments, they appear more like ellipses than circles. It is due to improper use of the VIEWPORT statement without the aid of an accompanying WINDOW statement to declare how the data should be scaled.

Explain why the circles appear as ellipses and add a WINDOW statement to correct the outcome. *Hint:* Without a WINDOW statement, the display device will, by default, place the origin of the user's coordinate system at the left lower corner of the viewport.

4.14   PERSPECTIVE PROJECTION OF THREE-DIMENSIONAL OBJECTS

Photographic pictures, such as Fig. 4-21, showing three-dimensional structures are themselves two-dimensional. The display screen is also two-dimensional. Three-dimensional objects to be displayed are **projected** onto the screen. Figure 4-22 illustrates schematically the concept of **perspective projection**. The image of a three-dimensional object displayed on a screen is determined by the relative positions of the object, the screen, and the observer's eye. If the screen is placed between the eye and the object, the image is smaller than the object; if the object is placed between the eye and the screen, the image is larger than the object.

Figure 4-21   Washington, D.C., Metro. (Courtesy of CRS Group, Inc.)

Let us consider the simple case where a global three-dimensional coordinate system XYZ is chosen in such a way that the eye, often called the **viewpoint**, is located on the Z axis. To further simplify the discussion, a two-dimensional coordinate system xy is chosen on the image plane. The x and y axes are parallel to the X and Y axes, respectively, resulting in the Z axis, often called the **optical axis**, being perpendicular to the image plane. This type of consideration for perspective views is classified as the **central** projection.

In order to draw the accurate image of an object on the screen, the three-dimensional coordinates (X, Y, Z) of a large number of points on the object must be transformed into two-dimensional coordinates (x, y) on the screen. For consideration of the central projection of an object, let P be a generic point of the object and p be its projection on the screen. Based on the relationships of the similar right triangles shown in Fig. 4-22, the following transformation equation can be obtained

$$x = 0p' = (QP')(E0)/(EQ)$$
$$= X(Z_E - Z_0)/(Z_E - Z) \tag{a}$$
$$y = Y(Z_E - Z_0)/(Z_E - Z)$$

where Z_E and Z_0 are the distances of the viewpoint and the image plane, respectively, to the origin of the global coordinate system.

The limiting case of $Z_E \to \infty$, that is, when the viewpoint is at infinity, reduces Eqs. (a) to

$$x = X \qquad \text{and} \qquad y = Y \tag{b}$$

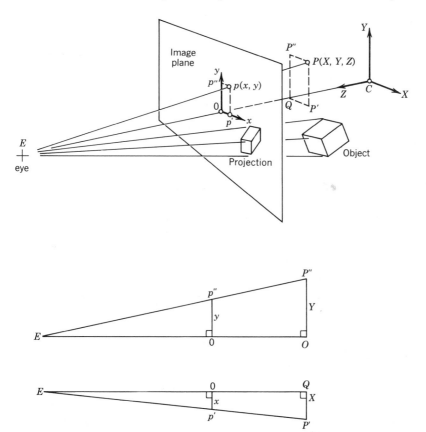

Figure 4-22

It is the so-called **orthographic** projection when all points on the object having the same X and Y coordinates but different Z coordinates will be projected onto the same point on the screen.

Figure 4-23 is presented to demonstrate different perspective projections by use of the previously mentioned brick. Program PERSPECT has been employed to produce these sample cases. It should be clear that Fig. 4-12, presented earlier, is an example of orthographic projection when the viewing eye is at infinity, that is, $Z_E = \infty$. Line 800 HOME in Program PERSPECT causes the cursor to be positioned at the upper left corner of the screen.

PROGRAM PERSPECT

```
100 REM * PROGRAM PERSPECT *
110 INIT
120 PAGE
130 N=24
140 DIM XO(N),YO(N),ZO(N),X(N),Y(N),Z(N)
150 REM * FIRST 8 FOR BOX,THEN 16 FOR LETTERS F, T, AND R.
160 READ XO
```

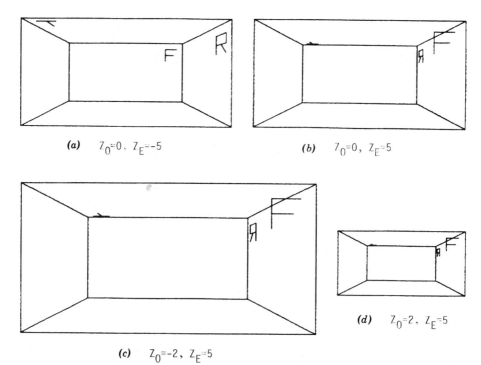

(a) $Z_0=0$, $Z_E=-5$

(b) $Z_0=0$, $Z_E=5$

(c) $Z_0=-2$, $Z_E=5$

(d) $Z_0=2$, $Z_E=5$

Figure 4-23 Origin of the global $X-Y-Z$ coordinate system is at the centroid of the brick and $X-Y$ plane bisects the brick parallel to the front (F) side of the brick. Both (a) and (b) have the screen $X-Y$ plane placed coincident with the $X-Y$ plane, but (a) is viewed from the back side and (b) from the front side. Both (c) and (d) view the brick from front of the brick, but (c) has the screen placed behind the brick and (d) in front of the brick.

```
170 DATA 0,4,4,0,0,4,4,0
180 DATA 3.4,3.4,3.4,3.8,3.7,0.2,0.4,0.6,0.4,4,4,4,4,4,4,4
190 READ YO
200 DATA 0,0,2,2,0,0,2,2
210 DATA 1.4,1.6,1.8,1.8,1.6,2,2,2,2,1.4,1.6,1.8,1.8,1.6,1.6,1.4
220 READ ZO
230 DATA 0,0,0,0,3,3,3,3
240 DATA 3,3,3,3,3,0.2,0.2,0.2,0.6,0.6,0.6,0.6,0.2,0.2,0.4,0.2
250 REM * TRANSLATE SO THAT X-Y PLANE BISECTS"
260 REM      THE R AND T PLANES OF THE BRICK.
270 READ T1,T2,T3
280 DATA -2,-1,-1.5
290 FOR I=1 TO N
300 X(I)=X0(I)+T1
310 Y(I)=Y0(I)+T2
320 Z(I)=Z0(I)+T3
330 NEXT I
340 REM * PERSPECTIVE VIEWING *
350 PRINT "INPUT AXIAL COORDINATES OF SCREEN AND EYE ";
360 INPUT Z1,Z2
370 FOR I=1 TO N
380 LET X(I)=X(I)*(Z2-Z1)/(Z2-Z(I))
```

```
390 LET Y(I)=Y(I)*(Z2-Z1)/(Z2-Z(I))
400 NEXT I
410 REM * COMMENCE PLOTTING
420 WINDOW -5,5,-5,5
430 VIEWPORT 10,90,10,90
440 REMARK DRAW X-Y PROJECTION OF 3-D BOX
450 MOVE X(5),Y(5)
460 DRAW X(6),Y(6)
470 DRAW X(2),Y(2)
480 DRAW X(1),Y(1)
490 DRAW X(5),Y(5)
500 DRAW X(8),Y(8)
510 DRAW X(7),Y(7)
520 DRAW X(3),Y(3)
530 DRAW X(4),Y(4)
540 DRAW X(8),Y(8)
550 MOVE X(6),Y(6)
560 DRAW X(7),Y(7)
570 MOVE X(3),Y(3)
580 DRAW X(2),Y(2)
590 MOVE X(1),Y(1)
600 DRAW X(4),Y(4)
610 REMARK DRAW "F"
620 MOVE X(9),Y(9)
630 DRAW X(11),Y(11)
640 DRAW X(12),Y(12)
650 MOVE X(10),Y(10)
660 DRAW X(13),Y(13)
670 REMARK DRAW "T"
680 MOVE X(14),Y(14)
690 DRAW X(16),Y(16)
700 MOVE X(17),Y(17)
710 DRAW X(15),Y(15)
720 REMARK DRAW "R"
730 MOVE X(18),Y(18)
740 DRAW X(20),Y(20)
750 DRAW X(21),Y(21)
760 DRAW X(22),Y(22)
770 DRAW X(19),Y(19)
780 MOVE X(23),Y(23)
790 DRAW X(24),Y(24)
800 HOME
810 END
```

4.15   HIDDEN-LINE ALGORITHM

The rotated and perspective projections of the $2 \times 3 \times 4$ brick previously shown in Figs. 4-13 and 4-23, respectively, provide to a certain degree the depth information of the object. The depth description of a displayed object can be further enhanced by removal of the **hidden lines** as well as by application of the **shading** technique. In order to maintain the scope of this text at an introductory level, only the hidden-line technique will be discussed here. Figures 4-24 and 4-25 exemplify how application of the hidden-line technique can greatly improve the presentation of the perspective views of three-dimensional objects. To be more specific, the goal of applying the hidden-line technique, taking the case of the rotated brick,

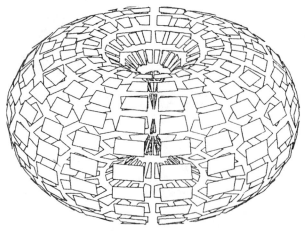

Figure 4-24 "Black hole" drawn with hidden-line
technique (Courtesy of PDA Engineering, Santa Ana,
California.)

is to remove the invisible lines in Fig. 4-26a so that the resulting plot will appear
as Fig. 4-26b.

Lerman[5] had presented a paper in which the method for removal of the hidden
lines was explained in very plain terms easily understandable to beginners, as
opposed to most other mathematical presentations on the hidden-line algorithms.
He suggested that all surfaces of a three-dimensional object be approximated by a
number of flat surfaces. Each flat surface will have straight-line boundaries called
edges. The endpoints of an edge are called vertices, such as the corner points 1
through 8 for the brick shown in Fig. 4-27. To remove the hidden lines, the vertices
of a three-demensional object are first projected onto a display screen. Whether
or not a point is visible will depend on whether it is within the area enclosed by
the edges connecting the projected vertices. For simplicity of discussion, let the
plane containing points 2, 3, 8, and 5 of the brick shown in Fig. 4-27 coincide
with the paper. Notice that points 4 and 6 are outside of the boundary edges
2-3-8-5-2, whereas points 1 and 7 are inside, as illustrated in Fig. 4-27. It should
also be recognized that the edges 2-5 and 3-8, which do not really exist, are drawn
using imaginary lines. The exterior points 4 and 6 are apparently visible, whereas
the interior points 1 and 7 need further tests. Since the edges 2-6, 3-4, 4-8, and
5-6 are outside of the boundary 2-3-8-5-2, these lines should be drawn and the
boundary should be **updated** to become 2-3-4-8-5-6-2.

Returning to interior points 1 and 7, as one looks down at the paper containing
the updated boundary of the vertices, point 1 is on the same side as the eyes in
front of the paper while point 7 is behind the paper. Point 1 is therefore visible,

---

[5] H. N. Lerman, "A Planar Solution to 3-Dimensional Plotting," *Software Age*, July 1970,
pp. 16–19.

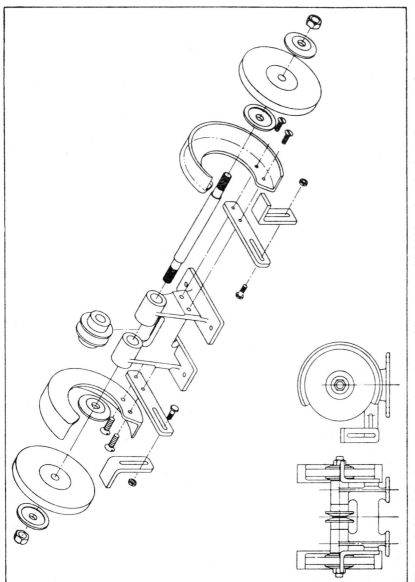

Figure 4-25 Effective representation of assembling parts with hidden-line plotting. (Courtesy of IBM Corporation.)

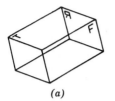 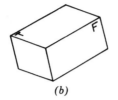

(a)   *(b)*

Figure 4-26 *(a)* Rotated view of the $2 \times 3 \times 4$ brick for $\theta_x = \theta_y = \theta_z = 30°$ previously shown in Fig. 4-13 and *(b)* with the hidden lines removed.

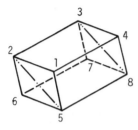

Figure 4-27   Procedure used for removal of hidden lines.

and point 7 is invisible in this case; Lerman has pointed out that if there is a cut-out or a window, then the point may also be visible. Based on the described procedure, it can then be determined that edges 3-7, 6-7, and 2-8 shown in Fig. 4-26*a* are invisible. These invisible edges can either be removed or drawn with dashed lines as shown in Figs. 4-26*b* and 4-27, respectively.

In a similar manner, a three-dimensional surface $z(x, y)$ defined over a rectangular domain $x_1 \leq x \leq x_2$ and $y_1 \leq y \leq y_2$ can also be plotted with the hidden-line technique. First, the $z(x, y_1)$ curve on the $y = y_1$ plane over the (x_1, x_2) interval is considered as entirely visible and plotted. Then, the next curve $z(x, y_j)$ is approximated as a series of linear segments and each segment is examined to see whether it is inside or outside of the boundary of the first curve. As Fig. 4-28 shows, the boundary is continuously updated and the series of $z(x, y_j)$ curves for y_j values in range of y_1 to y_2 are incorporated consecutively. Figure 4-28 was plotted by a BASIC Program PLOT3D developed following the above-described hidden-line algorithm.

There are many new commands in Program PLOT3D that need to be explained. The two-dimensional subscripted array variables are introduced and the library functions INT, ABS, EXP, LGT, and SQR are used in the program. These functions perform the tasks of truncation, removal of the sign, raising the base e (natural logarithm base) and base 10, and taking the square root, respectively, of an arithmetic expression. In many cases, the arithmetic expression could consist of a single constant, or a single variable alone.

A CHA (CHARSIZE) statement specifies the size of characters ranging from 4, the largest, to 3, to 2, then to 1, the smallest available on a graphic display device such as Tektronix 4054, which will automatically reset to CHA 4 by default if it is initialized. Smaller size of characters allows more lines and more characters per line on the screen.

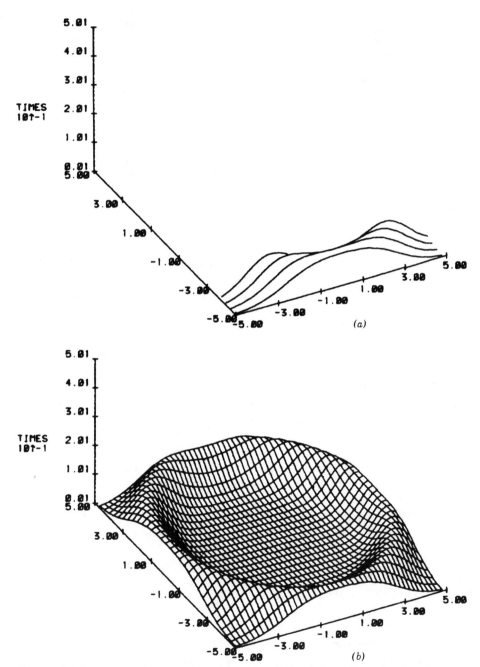

Figure 4-28   Screen printout and resulting plot of $v(x, y)$ three-dimensional surface using the hidden-line technique, by application of Program PLOT3D. (*a*) Interrupted plot to show the continuous updating of boundary lines in application of the hidden-line technique. (*b*) Completed plot.

The PRI (PRINT) statements output the value of a variable or a constant or a string of characters on the display screen or to a peripheral device. The following are example PRINT statements:

```
150 PRINT
160 PRINT 1.25
170 PRINT A
180 PRINT I*J-0.357
190 PRINT C$
200 PRINT "PRINT ABCDE"
210 PRINT A,B;C$
220 PRINT USING 300: A,C$
300 IMAGE 5D.2D,5X,20A
```

Lines 150, 160, 170, and 180 print a blank line, a constant, and the value of a variable and an arithmetic expression, respectively. Lines 190 and 200 print the characters in a string variable and specified characters, respectively. The comma used in line 210 separating two variables causes the output data to be printed in prearranged separate fields called automatic TAB,[6] whereas a semicolon suppresses this arrangement and demands only one character space between the two data. Lines 150 through 210 are examples of nonformated print statements. Line 220 illustrates the use of a formated print statement, in which the value of the variable A and the characters in the string variable C$ are to be printed according to line 300. The IMAGE specifies the particular formats to be used, namely 5 and 2 digits for the integral and decimal parts of the numerical value of A (5D.2D) and after 5 blank spaces (5X), followed by the first 20 characters of C$ (20A). There are other format codes that will be introduced later.

In Program PLOT3D, line 1110 demonstrates the use of a flexible field of format of FD.2D. The "F" is called a **field modifier** requesting as many digits as necessary for printing the integral part of a decimal number. Underlined characters J and H are used in the PRINT statements, such as in lines 120, 180, and 1440. They enable the cursor of the display device to be moved about on the screen. J and H demand that the cursor be moved down one line and back one space, respectively. For example, line 1440 specifies that after printing out the five characters "TIMES," the cursor should be moved back five spaces (following the command of HHHHH) and then be moved down one line (following the command J) so that the characters "10↑" will be lined up with "TIMES." The outcome can be seen in Fig. 4-28b as the characters label the z axis representing the function $v(x, y)$.

The INPUT statements input numeric data and characters from the keyboard of a graphics display unit or from a specified peripheral unit. The following examples illustrate various uses of the INPUT statement:

---

[6] A TABed field in a Tektronix 4054 graphics system is 18 characters.

```
100 DIM E(10)
150 INPUT A
160 INPUT B$
170 INPUT C, D$, E, F, G$
180 INPUT @33 : H, I, J, K, L$
```

Lines 150, 160, and 170 demonstrate how numeric data, a string of characters, and a combination of them, respectively, can be entered from the keyboard. Notice that the mentioning of the variable E requires 10 data be entered because E has been dimensioned as an array with 10 elements. Line 180 is an example of the data being entered from a peripheral unit designated as 33, which on a Tektronix 4054 graphics system is assigned to the magnetic tape.

Lines 630 and 640 in Program PLOT3D show applications of the logical operators AND, which can be used to connect two logical expressions to ask the question, "Are *both* conditions specified by the two logical expressions true or false?" The arithmetic operators MAX and MIN have also been used in the program, lines 470 and 480. They perform the comparison of the numeric results of the arithmetic expressions involved and then keep the largest and smallest values, respectively, of them. The general forms involving MAX or MIN sandwiched between arithmetic expressions are

```
100 A = B MAX C*D-5 MAX 3 MAX ...
200 X(I) = 1.25 MIN G*EXP(4 + H/X(5)) MIN ...
```

A GO TO n statement unconditionally transfers the next program step to line n. Lines 650, 690, and 740 in Program PLOT3D are examples of unconditional GO TO statements. As will be used later on, there are also **computed** GO TO statements that allow multiple choices of transfer. The general form of a computed GO TO statement is

$$n \text{ GO TO V OF } n_1, n_2, n_3, \ldots, n_k$$

The next statement, to which the program is to continue, will be line $n_1, n_2, \ldots,$ or n_k, depending on whether the value of the variable V is equal to 1, 2, ..., or k, respectively.

Line 880 in Program PLOT3D instructs that the x and y data both be scaled to 20%. This scaling command has the general form of

SCALE arithmetic expression 1, arithmetic expression 2

The numeric results of the two arithmetic expressions are to be used as the horizontal and vertical scale factors.

In application of the Program PLOT3D, the user may choose a number of options by responding to the input questions as demonstrated in Fig. 4-28. The options include (1) whether or not the crosshatching, that is, drawing the z curves on $x = $ constant planes as shown in Fig. 4-28b, is desired; (2) whether there should be no annotation or annotation of the x and y axes with their respective true values or with the point numbers; and (3) whether to include various debugging printout requests.

```
* PROGRAM PLOT3D *

    HIDDEN-LINE PLOT OF V(X,Y) OVER A RECTANGULAR REGION
    (XMIN,XMAX) BY (YMIN,YMAX).
    THERE ARE N1 POINTS ALONG X AND N2 CURVES ALONG Y
    DIRECTIONS, RESPECTIVELY.

DO YOU WANT TO SEE A SAMPLE PLOT, Y/N? : Y

PLOT V=(X↑2+Y↑2)↑.5*EXP(-2*SIN((X↑2+Y↑2)↑.5))*2
```

NUMBER OF POINTS ALONG X DIERCTION FOR ALL CURVES = 45

XMIN = -5 XMAX = 5

NUMBER OF CURVES FOR EQUALLY INCREMENTED Y VALUES = 30

YMIN= -5 YMAX = 5

NCROSS= 2 (1/2=Y/N FOR CROSSHATCHING.)

NZNORM= 5 (UNITS USED ALONG Z-AXIS.)

IAXIS= 0 (0/1=ANNOTATE WITH TRUE VALUE/NUMBER
          2=NO ANNOTATION.)

```
100 PAGE
110 PRINT "* PROGRAM PLOT3D *"
120 PRINT "    HIDDEN-LINE PLOT OF V(X,Y) OVER A RECTANGULAR REGION"
130 PRINT "    (XMIN,XMAX) BY (YMIN,YMAX),"
140 PRINT "    THERE ARE N1 POINTS ALONG X AND N2 CURVES ALONG Y"
150 PRINT "    DIRECTIONS, RESPECTIVELY. "
160 INIT
170 VIEWPORT 40,130,5,95
180 PRINT "_DO YOU WANT TO SEE A SAMPLE PLOT, Y/N? : ";
190 INPUT A$
200 IF A$="N" THEN 3470
210 GOSUB 3190
220 CHARSIZE 2
230 PRINT "_NUMBER OF POINTS ALONG X DIERCTION FOR ALL CURVES = ";N1
240 PRINT "_XMIN = ";X7(1);"   XMAX = ";X7(2)
250 PRINT "_NUMBER OF CURVES FOR EQUALLY INCREMENTED Y VALUES = ";N2
260 PRINT "_YMIN= ";Y7(1);"   YMAX = ";Y7(2)
270 PRINT "_NCROSS= ";I1;" (1/2=Y/N FOR CROSSHATCHING.)"
280 PRINT "_NZNORM= ";I2;" (UNITS USED ALONG Z-AXIS.)"
290 PRINT "_IAXIS= ";I3;" (0/1=ANNOTATE WITH TRUE VALUE/NUMBER"
300 PRINT "                2=NO ANNOTATION.)"
310 I0=1
320 J0=2
330 L0=2
340 I4=1
350 N4=1
360 T1=0.26794919
370 N0=INT(2*N1/3/N2/M5)
380 IF N0=0 THEN 3170
390 N3=N0+1
400 M0=N0
410 N1=N1+1-N0
420 L=1
430 Z1=V(1,1)
440 Z2=V(1,1)
450 FOR K1=2 TO N2
460 FOR K2=2 TO N1
470 Z1=Z1 MIN V(K1,K2)
480 Z2=Z2 MAX V(K1,K2)
```

```
490 NEXT K2
500 NEXT K1
510 M=0
520 N=0
530 IF ABS(Z2)=0 THEN 550
540 M=INT(LGT(Z2))
550 IF ABS(Z1)=0 THEN 570
560 N=INT(LGT(Z1))
570 IF ABS(Z2)=>1 THEN 590
580 M=M-1
590 IF ABS(Z1)=>1 THEN 610
600 N=N-1
610 E1=10^(-N+2)
620 E2=10^(-M+2)
630 IF Z1=>0 AND Z2=>0 THEN 790
640 IF Z1<=0 AND Z2<=0 THEN 660
650 GO TO 700
660 Z1=0.01*INT(Z1*E1-1)
670 Z2=0.01*INT(Z2*E1)
680 E0=-N
690 GO TO 890
700 IF M<N THEN 750
710 Z1=0.01*INT(Z1*E2-1)
720 Z2=0.01*INT(Z2*E2+1)
730 E0=-M
740 GO TO 780
750 Z1=0.01*INT(Z1*E1-1)
760 Z2=0.01*INT(Z2*E1+1)
770 E0=-N
780 GO TO 890
790 Z1=0.01*INT(Z1*E2)
800 Z2=0.01*INT(Z2*E2+1)
810 E0=-M
820 X0=7.5*M5/N1
830 Y0=5/N2
840 Z0=I2/(Z2-Z1)
850 CHARSIZE 2
860 MOVE 70,15
870 W=0.1
880 SCALE 0.2,0.2
890 MOVE 0,0
900 IF I3=2 THEN 1450
910 X1=X7(1)
920 Y1=Y7(1)
930 X2=(X7(2)-X7(1))/(5*M5)
940 Y2=(Y7(2)-Y7(1))/5
950 IF I3=0 THEN 1000
960 X2=N1/(5*M5)
970 Y2=N2/5
980 X1=0
990 Y1=0
1000 REMARK

1010 J=5*M5+1
1020 FOR K=1 TO J
1030 X4=X1+(K-1)*X2
1040 X6=(K-1)*1.5
1050 DRAW X6,X6*T1
1060 MOVE X6,X6*T1+W
1070 DRAW X6,X6*T1-W
1080 MOVE X6-1*W,X6*T1-5*W
```

```
1090 IF I3=1 THEN 1130
1100 PRINT USING 1110:X4
1110 IMAGE FD.2D
1120 GO TO 1150
1130 PRINT USING 1140:X4
1140 IMAGE FD
1150 MOVE X6,X6*T1
1160 NEXT K
1170 MOVE 0,0
1180 FOR K=1 TO 6
1190 Y4=Y1+(K-1)*Y2
1200 Y6=K-1
1210 DRAW -Y6,Y6
1220 MOVE -Y6,Y6+W
1230 DRAW -Y6,Y6-W
1240 MOVE -Y6-10*W,Y6-3*W
1250 IF I3=1 THEN 1280
1260 PRINT USING 1110:Y4
1270 GO TO 1290
1280 PRINT USING 1140:Y4
1290 MOVE -Y6,Y6
1300 NEXT K
1310 J=I2+1
1320 FOR K=1 TO J
1330 Z4=(K-1)/I0+Z1
1340 Z6=Y4+K-1
1350 DRAW -Y6,Z6
1360 MOVE -Y6+W,Z6
1370 DRAW -Y6-W,Z6
1380 MOVE -Y6-10*W,Z6
1390 PRINT USING 1110:Z4
1400 MOVE -Y6,Z6
1410 NEXT K
1420 IF E0=0 THEN 1450
1430 MOVE -7.7,4.55+0.5*I2
1440 PRINT "TIMES          10^";E0
1450 X5=10^E0
1460 T4=T1*X0
1470 FOR K2=1 TO N2
1480 FOR K1=1 TO N1
1490 V(K2,K1)=(V(K2,K1)*X5-Z1)*X0+Y0*(K2-1)+(K1-1)*T4
1500 NEXT K1
1510 NEXT K2
1520 IF I0=1 THEN 1640
1530 T=N2
1540 N2=N1
1550 N1=T
1560 J0=1
1570 L0=N0+1
1580 IF I1=2 THEN 1600
1590 L0=N2
1600 I4=N0
1610 L=L+1
1620 M0=1
1630 N3=2
1640 X3=X0*N4*I4
1650 FOR J=1 TO N1
1660 I(1)=1
1670 I(2)=J
1680 X(J)=(I(J0)-1+N0*(1-I(I0)))*X0
1690 Y(J)=V(I(I0),I(J0))
```

```
1700 IF IO<>1 AND I1<>2 THEN 1740
1710 O(J)=Y(J)
1720 U(J)=O(J)
1730 GO TO 1760
1740 O(J)=S(J,1)
1750 U(J)=S(J,2)
1760 NEXT J
1770 S(1,1)=O(N1-1)
1780 S(1,2)=U(N1-1)
1790 S1=Y(1)
1800 L3=N1
1810 MOVE X(1),Y(1)
1820 FOR K=2 TO L3
1830 DRAW X(K),Y(K)
1840 NEXT K
1850 L4=0
1860 M4=0
1870 FOR K=LO TO N2 STEP I4
1880 I(1)=K
1890 I(2)=1
1900 S2=V(I(IO),I(JO))
1910 S3=(S2-S1)/MO
1920 S1=S2
1930 N5=N3+N1
1940 FOR K1=N3 TO N1
1950 L2=N5-K1
1960 O(L2)=O(L2-MO)
1970 U(L2)=U(L2-MO)
1980 NEXT K1
1990 VO=0
2000 FOR K1=1 TO MO

2010 I(2)=K1
2020 T2=VO
2030 VO=V(I(IO),I(JO))
2040 X(K1)=(I(JO)-1+NO*(1-I(IO)))*XO
2050 Y(K1)=VO
2060 O(K1)=VO
2070 U(K1)=S2-S3*(K1-1)
2080 IF U(K1)<=VO THEN 2100
2090 O(K1)=U(K1)
2100 IF VO=>U(K1) THEN 2120
2110 U(K1)=VO
2120 NEXT K1
2130 L3=MO
2140 N6=MO
2150 N7=1
2160 T2=VO
2170 J=N3
2180 REMARK
2190 J2=J
2200 I(2)=J
2210 T3=T2
2220 T2=VO
2230 VO=V(I(IO),I(JO))
2240 IF VO=>O(J) THEN 2680
2250 IF VO<=U(J) THEN 2430
2260 IF N6=1 THEN 2360
2270 IF I1=2 AND VO<V(I(IO)-1,I(JO)-I4) THEN 2970
2280 B(1,1)=(VO-T2)/X3
2290 B(2,1)=(O(J)-T6)/X3
```

```
2300 B(1,2)=V0-B(1,1)*(X(N6-1)+X3)
2310 B(2,2)=O(J)-B(2,1)*(X(N6-1)+X3)
2320 L3=L3+1
2330 X(L3)=(B(2,2)-B(1,2))/(B(1,1)-B(2,1))
2340 Y(L3)=B(1,1)*X(L3)+B(1,2)
2350 GO TO 2970
2360 IF N7=1 THEN 2910
2370 IF I1=2 AND V0>V(I(I0)-1,I(J0)-I4) THEN 2970
2380 B(1,1)=(V0-T2)/X3
2390 B(2,1)=(U(J)-T5)/X3
2400 B(1,2)=V0-B(1,1)*(X(N7-1)+X3)
2410 B(2,2)=U(J)-B(2,1)*(X(N7-1)+X3)
2420 GO TO 2320
2430 IF N6=1 THEN 2470
2440 IF J=N3 THEN 2470
2450 M4=1
2460 GO TO 2970
2470 T5=U(J)
2480 U(J)=V0
2490 IF J<>N3 THEN 2510
2500 N7=N3
2510 X(N7)=(I(J0)-1+N0*(1-I(I0)))*X0
2520 Y(N7)=V0
2530 IF N7>1 THEN 2630
2540 B(1,1)=(V0-T2)/X3
2550 B(2,1)=(T5-U(J-1))/X3
2560 B(1,2)=V0-B(1,1)*X(N7-1)+B(1,2)
2570 B(2,2)=T5-B(2,1)*X(N7)
2580 N7=N7+1
2590 X(N7)=X(N7-1)
2600 X(N7-1)=(B(2,2)-B(1,2))/(B(1,1)-B(2,1))
2610 Y(N7-1)=B(1,1)*X(N7-1)+B(1,2)
2620 Y(N7)=V0
2630 L3=N7
2640 N6=1
2650 N7=N7+1
2660 M4=0
2670 GO TO 2910
2680 IF N7=1 THEN 2710
2690 L4=1
2700 GO TO 2970
2710 IF J<>N3 THEN 2730
2720 N6=N3
2730 T6=O(J)
2740 O(J)=V0
2750 X(N6)=(I(J0)-1+N0*(1-I(I0)))*X0
2760 Y(N6)=V0
2770 IF N6>1 THEN 2870
2780 B(1,1)=(T6-O(J-1))/X3
2790 B(2,1)=(V0-T2)/X3
2800 B(1,2)=T6-B(1,1)*X(N6)
2810 B(2,2)=V0-B(2,1)*X(N6)
2820 N6=N6+1
2830 X(N6)=X(N6-1)
2840 X(N6-1)=(B(2,2)-B(1,2))/(B(1,1)-B(2,1))
2850 Y(N6-1)=B(1,1)*X(N6-1)+B(1,2)
2860 Y(N6)=V0
2870 L3=N6
2880 N7=1
2890 N6=N6+1
2900 L4=0
2910 REMARK
2920 J=J+1
```

```
2930 IF J<=N1 THEN 2180
2940 IF N6>1 THEN 2970
2950 IF N7>1 THEN 2970
2960 GO TO 3080
2970 REMARK
2980 MOVE X(1),Y(1)
2990 FOR K3=1 TO L3
3000 DRAW X(K3),Y(K3)

3010 NEXT K3
3020 L=L+1
3030 IF L4=1 THEN 2710
3040 N6=1
3050 IF M4=1 THEN 2470
3060 N7=1
3070 IF J2<N1 THEN 2910
3080 IF I1=2 THEN 3110
3090 S(K,1)=O(N1-1)
3100 S(K,2)=U(N1-1)
3110 NEXT K
3120 IF IO=2 THEN 3160
3130 N4=-1
3140 IO=2
3150 GO TO 1520
3160 END
3170 PRINT "EXECUTION TERMINATED**********NCNT (NO) IS ZERO"
3180 END
3190 REM * RUN A SAMPLE CASE *
3200 N1=45
3210 N2=30
3220 M5=1
3230 I1=2
3240 I2=5
3250 I3=0
3260 PRINT "_PLOT V=(X^2+Y^2)^.5*EXP(-2*SIN((X^2+Y^2)^.5))*2"
3270 DIM X(602),Y(602),O(600),U(600),S(100,2)
3280 DIM V(N2,N1)
3290 DIM B(2,2),I(2),X7(2),Y7(2)
3300 X7(1)=-5
3310 X7(2)=5
3320 Y7(1)=-5
3330 Y7(2)=5
3340 X8=(X7(2)-X7(1))/N1
3350 Y8=(Y7(2)-Y7(1))/N2
3360 Y0=Y7(1)
3370 FOR K2=1 TO N2
3380 X0=X7(1)
3390 FOR K3=1 TO N1
3400 R=SQR(X0^2+Y0^2)
3410 V(K2,K3)=R*EXP(-2*SIN(R))*2
3420 X0=X7(1)+X8*K3
3430 NEXT K3
3440 Y0=Y7(1)+Y8*K2
3450 NEXT K2
3460 RETURN
3470 PRINT "VALUE IS A NCURV BY NPTS ARRAY (#Y BY #X) WHICH IS PLOTTED."
3480 PRINT "_THE RATIO OF NPTS TO NCURV MUST BE AN INTEGER MULTIPLE OF"
3490 PRINT "1.5*MULT.    THIS IS IMPORTANT!  DISTORTION WILL OTHERWISE"
3500 PRINT "RESULT."
3510 PRINT "_ENTER THE NPTS ";
3520 INPUT N1
3530 PRINT "ENTER THE NCURV ";
3540 INPUT N2
```

```
3550 PRINT "ENTER MULT ";
3560 INPUT M5
3570 PRINT "ENTER NCROSS ";
3580 INPUT I1
3590 PRINT "ENTER NZNORM ";
3600 INPUT I2
3610 PRINT "ENTER IAXES ";
3620 INPUT I3
3630 DIM X(602),Y(602),O(600),U(600),S(100,2)
3640 DIM V(N2,N1)
3650 DIM B(2,2),I(2),X7(2),Y7(2)
3660 PRINT "ENTER XMIN AND XMAX ";
3670 INPUT X7(1),X7(2)
3680 PRINT "ENTER YMIN AND YMAX ";
3690 INPUT Y7(1),Y7(2)
3700 PRINT "ENTER THE VALUE ARRAY. THIS WILL BE A ";N2;" BY ";N1;
3710 PRINT " ARRAY.   ENTER IN ROW-MAJOR FORM"
3720 INPUT V
3730 PAGE
3740 PRINT "THE ARRAY VALUE WILL NOW BE PRINTED OUT ROW BY ROW FOR"
3750 PRINT "YOUR INSPECTION.  THE OTHER INPUT DATA WILL BE PRINTED"
3760 PRINT "OUT LATER."
3770 FOR K1=1 TO N2
3780 FOR K2=1 TO N1
3790 PRINT V(K1,K2);
3800 NEXT K2
3810 PRINT
3820 NEXT K1
3830 GO TO 180
```

4.16 PROGRAM CONTOUR

Another way to graphically represent $z(x, y)$ is to plot the z value at a point by use of an appropriate character. A so-called **contour map** will result from such an approach, as illustrated by Fig. 4-29. This section discusses how a contour plot is generated.

A general need in contour plotting is to know the function $z(x, y)$ values at the vertices of a three-node triangular element, as sketched in Fig. 4-30. The reason for discussing a basic **triangular** region is that other types of regions can be subdivided into a number of triangular regions. For covering the entire triangular area with contour lines based on the z values specified at the vertices, it is necessary to interpolate the z values throughout the triangle. Instead of drawing contour lines, **interpolation** of z values at selected increments in x and y directions according to the available character sizes on a display device and use of alphanumeric or font characters will be demonstrated.

The interpolation of z values in a triangular region begins by finding the bounds of the region in x direction. They are determined by the equations, referring to Fig. 4-30,

$$x_{min} = \min(x_i, x_j, x_k) \qquad x_{max} = \max(x_i, x_j, x_k)$$

In order to cover the entire triangular region, a movable line $x = x_m$ sweeps from the boundary $x = x_{min}$ to the other boundary $x = x_{max}$. For each selected x_m value, the bounds in y direction, called y_u and y_l in Fig. 4-30, must be found. Recognizing

NUMBER OF NODES= 12
ISOTHERMAL CONTOUR PLOT, Z IN DEGREES, F.
NUMBER OF ELEMENTS= 12

XMIN= 5

XMAX= 9

YMIN= 0

YMAX= 4

ZMIN= 0
   (CHARACTER 0)

ZMAX= 100
   (CHARACTER 9)

```
9999999999999999999999999999999999999999999999999999
9999999999999999999999999999999999999999999999999999
9999999999999999999999999999999999999999999999999999
9999999999998888888888888888888888888888888899999999
9999999999998888888888888888888888888888888888999999
9999999999998888888888888888888888888888888899999999
9999999999998888888888888888888888888888888899999999
9999999999998888888888888888888888888888888899999999
9999999999998888888888888888877777788888888888888899
9999999999998888888888888888777777777777778888888888
9999999999998888888888888887777777777777777777788888
9999999999998888888888888887777777777777777777777777
9999999999998888888888887777777666666666666666677777777
9999999999998888888887777777766666666666666666666666667
9999999999998888888877777777666666666666666666666666666
9999999999988888887777777666666666665555555555555566666
99999998888888777777766666665555555555555555555555555555
999998888888777777766666666555555555555555555555555555
999888888887777776666666655555555555555555555555555555
88888777777766666666555555554444444444444444444444444
87777777666666665555554444444444444444444444444444444
777666666655555554444444444444444444444444444444444
66666655555554444444333333333333333333333333333333333
66555555544444443333333333333333333333333333333333333
55554444443333333333333333333333333333333333333333333
44444443333333222222222222222222222222222222222222222
44333333322222222222222222222222222222222222222222222
33333222222222222222222222222222222222222222222222222
32222221111111111111111111111111111111111111111111111
22211111111111111111111111111111111111111111111111111
111111111111111111111111111111111111111111111111111111
1000000000000000000000000000000000000000000000000000000
00000000000000000000000000000000000000000000000000000
```

Figure 4-29   A sample 10-level contour plot of the temperature distribution of a flat plate, and screen printout generated by application of Program CONTOUR.

that they are the intercepts of the line $x = x_m$ with two of the three sides of the triangle ijk, one of the sides must be excluded. It is done by testing to see whether x_m is outside of the interval (x_p, x_q) where p, $q = i$, j, k. For example, if x_m is outside of (x_k, x_j) as shown in Fig. 4-30, then the side jk is not involved in finding y_u and y_l for $x = x_m$. Once x_m passes x_k in Fig. 4-30, x_m will then be outside of (x_i, x_k) and, consequently, the side ik will not be involved.

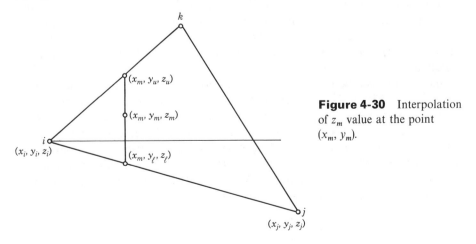

Figure 4-30   Interpolation of z_m value at the point (x_m, y_m).

If y_u and y_l are to be interpolated from the lines connecting nodes i and k and nodes i and j, respectively, as shown in Fig. 4-30, then the equations to be used are

$$y_u = y_i + (y_k - y_i)(x_m - x_i)/(x_k - x_i)$$
$$y_l = y_i + (y_j - y_i)(x_m - x_i)/(x_j - x_i)$$

If $x_k = x_i$, or $x_j = x_i$, then y_u and y_l should take up the maximum and minimum y values, respectively, of the two points, of which their values are equal. Similar interpolation formulas should be used to find z_u and z_l at (x_m, y_u) and (x_m, y_l), respectively, namely by replacing all the y's in the above two equations with their respective z's.

Knowing the z bounds, (z_l, z_u) on the line $x = x_m$, the z value at a moving point $y = y_m$ for $y_l \leq y_m \leq y_u$ can then be interpolated using the formula

$$z(x_m, y_m) = z_m = z_l + (y_m - y_l)(z_u - z_l)/(y_u - y_l)$$

Thus, by varying x_m between x_{min} and x_{max} and for each x_m varying y_m between y_l and y_u, the z values can be completely interpolated to cover the entire triangular region ijk by choosing appropriate increments Δx and Δy. It has been found that the Tektronix 4054 graphics system can be best utilized if the increments are chosen in accordance with the sizes of characters employed for plotting the contours as follows:

Character Size	Δx	Δy
1	0.94	1.5
2	1.00	1.6
3	1.56	2.5
4	1.75	2.8

Notice that both Δx and Δy values listed above are in GDU.

A BASIC Program CONTOUR has been developed with the described interpolation scheme. Figure 4-29 shows a typical application of CONTOUR for the case of plotting the temperature distribution of a plate. Further applications of Program CONTOUR can be found throughout this text. For instance, in Chapter 9 the stress distribution in a structure determined by the finite element method is plotted.

Since z value can be interpolated at every point within the three-node domain as sketched in Fig. 4-30, lines can therefore be drawn to connect all of the points having same z value. They are called the **contour lines**. Figure 4-31 shows a contour-line plot on three-dimensional surfaces with the hidden-line technique.

On BASIC programing, line 730 in Program CONTOUR shows an application of the repeated field operator ( ). 3(10D. 4D) means that the format inside the parentheses is to be repeated three times. Line 310 demonstrates the use of a computed GO TO statement discussed earlier.

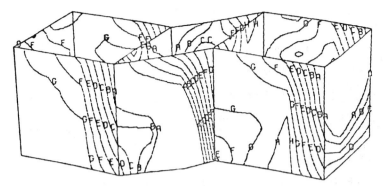

Figure 4-31   Contour plotting with hidden-surface technique.
(Courtesy of PDA Engineering, Santa Ana, California.)

PROGRAM CONTOUR

```
100 PAGE
110 PRINT "* PROGRAM CONTOUR *"
120 INIT
130 VIEWPORT 40,130,5,95
140 PRINT "_WANT TO RUN A SAMPLE OR YOUR OWN PROBLEM? (1 OR 2) : ";
150 INPUT Z9
160 PRINT "_INPUT THE CHARACTER SIZE 1,2,3, OR, 4 FOR PLOTTING. : ";
170 INPUT C
180 CHARSIZE 2
190 IF C<3 THEN 220
200 H=2.5+(C-3)*0.3
210 GO TO 230
220 H=1.5+(C-1)*0.1
230 H1=H*0.625
240 W1=1
250 REM GOSUB FOR INPUT, SCALING, AND PLOT
260 GOSUB 300
270 GOSUB 850
280 GOSUB 2030
290 END
300 REMARK   SUBROUTINE INPUT
310 GO TO Z9 OF 320,510
320 T$="ISOTHERMAL CONTOUR PLOT, Z IN DEGREES, F."
330 N1=12
340 N2=12
350 DIM X(N1),Y(N1),N(3,N2),S(N1)
360 FOR I2=1 TO N2
370 FOR L1=1 TO 3
380 READ N(L1,I2)
390 NEXT L1
400 NEXT I2
410 DATA 1,6,5,1,2,6,2,3,6,3,9,6
420 DATA 3,4,9,4,11,9,5,6,7,7,6,8
430 DATA 8,6,10,6,9,10,10,9,12,9,11,12
440 READ X
450 DATA 0,0,0,0,5,5,9,9,5,9,5,9
460 READ Y
470 DATA 9,7,4,0,9,7,9,7,4,4,0,0
480 READ S
```

```
490 DATA 100,100,100,0,100,84.4,100,100,53.5,52,0,0
500 GO TO 690
510 REMARK
520 PRINT "_INPUT A TITLE DESCRIBING THE PROBLEM AND UNIT FOR Z VALUES.
530 INPUT T$
540 PRINT "_ENTER NUMBER OF NODES : ";
550 INPUT N1
560 PRINT "_ENTER NUMBER OF ELEMENTS : ";
570 INPUT N2
580 DIM X(N1),Y(N1),N(3,N2),S(N1)
590 PRINT "_ENTER THE X,Y AND Z VALUES FOR EACH NODE AS CALLED FOR"
600 FOR I1=1 TO N1
610 PRINT "AT NODE ";I1;"  ";
620 INPUT X(I1),Y(I1),S(I1)
630 NEXT I1
640 PRINT "_ENTER THREE NODES OF EACH ELEMENT AS CALLED FOR"
650 FOR I2=1 TO N2
660 PRINT "FOR ELEMENT ";I2;" THE NODES ARE   ";
670 INPUT N(1,I2),N(2,I2),N(3,I2)
680 NEXT I2
690 IF W1=0 THEN 840
700 PRINT "  NODE              X            Y            Z"
710 FOR I1=1 TO N1
720 PRINT USING 730:I1,X(I1),Y(I1),S(I1)
730 IMAGE 4D,3(10D.4D)
740 NEXT I1
750 PRINT "_WHEN READY TO CONTINUE, PRESS ANY KEY : ";
760 INPUT K$
770 PRINT "ELEMENT       NODE NUMBERS"
780 FOR I2=1 TO N2
790 PRINT USING 800:I2,N(1,I2),N(2,I2),N(3,I2)
800 IMAGE 4D,3(6D)
810 NEXT I2
820 PRINT "_WHEN READY TO CONTINUE, PRESS ANY KEY : ";
830 INPUT K$
840 RETURN
850 REMARK   SUBROUTINE SCALE
860 X1=X(1)
870 X2=X(1)
880 Y1=Y(1)
890 Y2=Y(1)
900 S1=S(1)
910 S2=S(1)
920 FOR I1=2 TO N1
930 IF X(I1)>X1 THEN 950
940 X1=X(I1)
950 IF X(I1)<X2 THEN 970
960 X2=X(I1)
970 IF Y(I1)>Y1 THEN 990
980 Y1=Y(I1)
990 IF Y(I1)<Y2 THEN 1010
1000 Y2=Y(I1)

1010 IF S(I1)>S1 THEN 1030
1020 S1=S(I1)
1030 IF S(I1)<S2 THEN 1050
1040 S2=S(I1)
1050 NEXT I1
1060 S4=S2-S1
1070 W=X2-X1 MAX Y2-Y1
1080 FOR I1=1 TO N1
1090 S(I1)=(S(I1)-S1)/S4*9
1100 S3=S1
```

```
1110 S4=S2
1120 NEXT I1
1130 RETURN
1140 REM * SUBROUTINE CONTOUR *
1150 REM      PLOT 10-LEVEL S VALUES OF THE I2ND ELEMENT
1160 REM          WITH 0-9 USING CHAR SIZE C.
1170 REM       X, Y AND S FOR N(1-3,I2) SHOULD BE KNOWN,
1180 REM       H1 AND H SHOULD BE DEFINED IN CALLING PROGRAM
1190 REM          ACCORDIN G TO C.
1200 J1=N(1,I2)
1210 J2=N(2,I2)
1220 J3=N(3,I2)
1230 X1=X(J3) MIN X(J1) MIN X(J2)
1240 X2=X(J3) MAX X(J1) MAX X(J2)
1250 Y1=Y(J3) MIN Y(J1) MIN Y(J2)
1260 Y2=Y(J3) MAX Y(J1) MAX Y(J2)
1270 S1=S(J3) MIN S(J1) MIN S(J2)
1280 S2=S(J3) MAX S(J1) MAX S(J2)
1290 M1=INT(X1/H1+0.5)
1300 M2=INT(X2/H1+0.5)
1310 FOR I=M1 TO M2
1320 X0=I*H1
1330 Z1=Y2
1340 Z2=Y1
1350 T1=S2
1360 T2=S1
1370 IF ABS(X0-X(J1))+ABS(X0-X(J2))>ABS(X(J2)-X(J1)) THEN 1520
1380 IF X(J2)=X(J1) THEN 1480
1390 Z=Y(J1)+(Y(J2)-Y(J1))*(X0-X(J1))/(X(J2)-X(J1))
1400 T=S(J1)+(S(J2)-S(J1))*(X0-X(J1))/(X(J2)-X(J1))
1410 IF Z<Y1 THEN 1440
1420 Z1=Z1 MIN Z
1430 T1=T1 MIN T
1440 IF Z>Y2 THEN 1520
1450 Z2=Z2 MAX Z
1460 T2=T2 MAX T
1470 GO TO 1520
1480 Z1=Z1 MIN Y(J2) MIN Y(J1)
1490 T1=T1 MIN S(J2) MIN S(J1)
1500 Z2=Z2 MAX Y(J2) MAX Y(J1)
1510 T2=T2 MAX S(J2) MAX S(J1)
1520 IF ABS(X0-X(J2))+ABS(X0-X(J3))>ABS(X(J2)-X(J3)) THEN 1670
1530 IF X(J3)=X(J2) THEN 1630
1540 Z=Y(J2)+(Y(J3)-Y(J2))*(X0-X(J2))/(X(J3)-X(J2))
1550 T=S(J2)+(S(J3)-S(J2))*(X0-X(J2))/(X(J3)-X(J2))
1560 IF Z<Y1 THEN 1590
1570 Z1=Z1 MIN Z
1580 T1=T1 MIN T
1590 IF Z>Y2 THEN 1670
1600 Z2=Z2 MAX Z
1610 T2=T2 MAX T
1620 GO TO 1670
1630 Z1=Z1 MIN Y(J3) MIN Y(J2)
1640 T1=T1 MIN S(J3) MIN S(J2)
1650 Z2=Z2 MAX Y(J3) MAX Y(J2)
1660 T2=T2 MAX S(J3) MAX S(J2)
1670 IF ABS(X0-X(J1))+ABS(X0-X(J3))>ABS(X(J1)-X(J3)) THEN 1820
1680 IF X(J1)=X(J3) THEN 1780
1690 Z=Y(J3)+(Y(J1)-Y(J3))*(X0-X(J3))/(X(J1)-X(J3))
1700 T=S(J3)+(S(J1)-S(J3))*(X0-X(J3))/(X(J1)-X(J3))
1710 IF Z<Y1 THEN 1740
```

```
1720 Z1=Z1 MIN Z
1730 T1=T1 MIN T
1740 IF Z>Y2 THEN 1820
1750 Z2=Z2 MAX Z
1760 T2=T2 MAX T
1770 GO TO 1820
1780 Z1=Z1 MIN Y(J3) MIN Y(J1)
1790 T1=T1 MIN S(J3) MIN S(J1)
1800 Z2=Z2 MAX Y(J3) MAX Y(J1)
1810 T2=T2 MAX S(J3) MAX S(J1)
1820 L1=INT(Z1/H+0.5)
1830 L2=INT(Z2/H+0.5)
1840 FOR J=L1 TO L2
1850 Y0=J*H
1860 IF Y0>Z2 OR Y0<Z1 THEN 2000
1870 IF Z2>Z1 THEN 1900
1880 S0=T1
1890 GO TO 1910
1900 S0=T1+(Y0-Z1)/(Z2-Z1)*(T2-T1)
1910 IF S0>9 THEN 1960
1920 IF S0<0 THEN 1940
1930 GO TO 1970
1940 S0=0
1950 GO TO 1970
1960 S0=9
1970 MOVE X0-0.5*H1,Y0-0.65*H
1980 PRINT USING 1990:S0
1990 IMAGE 1D
2000 NEXT J

2010 NEXT I
2020 RETURN
2030 REMARK    SUBROUTINE WINDOWING
2040 PAGE
2050 CHARSIZE C
2060 H=H/90*W
2070 H1=H1/90*W
2080 WINDOW X1,X1+W*1.05,Y1,Y1+W*1.05
2090 FOR I2=1 TO N2
2100 GOSUB 1140
2110 NEXT I2
2120 VIEWPORT 0,130,0,100
2130 WINDOW 0,130,0,100
2140 MOVE 40,95
2150 PRINT T$
2160 HOME
2170 CHARSIZE 2
2180 PRINT "NUMBER OF NODES= ";N1
2190 PRINT " NUMBER OF ELEMENTS= ";N2
2200 PRINT " XMIN= ";X1
2210 PRINT " XMAX= ";X2
2220 PRINT " YMIN= ";Y1
2230 PRINT " YMAX= ";Y2
2240 PRINT " ZMIN= ";S3
2250 PRINT "  (CHARACTER 0)"
2260 PRINT " ZMAX= ";S4
2270 PRINT "  (CHARACTER 9)"
2280 RETURN
```

```
* PROGRAM CONTOUR *

WANT TO RUN A SAMPLE OR YOUR OWN PROBLEM? (1 OR 2) : 1

INPUT THE CHARACTER SIZE 1,2,3, OR, 4 FOR PLOTTING. : 4
    NODE          X              Y              Z
     1         0.0000         9.0000       100.0000
     2         0.0000         7.0000       100.0000
     3         0.0000         4.0000       100.0000
     4         0.0000         0.0000         0.0000
     5         5.0000         9.0000       100.0000
     6         5.0000         7.0000        84.4000
     7         9.0000         9.0000       100.0000
     8         9.0000         7.0000       100.0000
     9         5.0000         4.0000        53.5000
    10         9.0000         4.0000        52.0000
    11         5.0000         0.0000         0.0000
    12         9.0000         0.0000         0.0000

WHEN READY TO CONTINUE, PRESS ANY KEY : c
ELEMENT        NODE NUMBERS
     1          1    6    5
     2          1    2    6
     3          2    3    6
     4          3    9    6
     5          3    4    9
     6          4   11    9
     7          5    6    7
     8          7    6    8
     9          8    6   10
    10          6    9   10
    11         10    9   12
    12          9   11   12

WHEN READY TO CONTINUE, PRESS ANY KEY : c
```

EXERCISE

Program CONTOUR is prepared for plotting numeric characters 0 through 9 for description of the contours over a triangular region. It limits the contour levels to 10 because there are only 10 available numeric characters. The capability of this program can be extended by use of any alphabetic, numeric, and font characters available on a graphics display device. Rather than printing the contour values scaled to 10 levels of 0 through 9, the contour values can be scaled to any desired number of levels N, and any N characters can be selected and stored in a string variable V$. Depending on the contour value J at a certain point in the triangular region, which may vary from 1 to N, the Jth character of V$ can then be requested and be printed at that point.

A command called SEG may be conveniently utilized here. It serves for segmentation of a string and has the general form of SEG(V\$, M, N). In application, B\$ = SEG (V\$, 5, 3) for example, result in storing the sixth through eighth characters of V\$ as the first through third characters of B\$. In general, SEG(V\$, M, N) means the segmentation of the $M + 1$st through $M + N$th characters, or in other words, the N characters after the Mth character, from V\$.

As an exercise, extend Program CONTOUR to 20 levels by use of the alphabetic characters A through T. These characters should first be stored in V\$ and are to be used for printing the contour values already scaled to 1 through 20, respectively. To print a scaled contour value equal to 5 at a certain point in the triangular region, the character E should then be selected. This can be easily accomplished by use of SEG(V\$, 4, 1). Following this procedure for all 20 levels, replot the isothermal contours of Fig. 4-29 with 20 levels using the characters A through T.

CHAPTER 5

INTERACTIVE
GRAPHICS

(Courtesy Digital Equipment Corp.)

5.1 INTRODUCTION

In the previous chapters, the CAD/CAM hardware and software have been discussed so that readers can have a general background on how the trial designs and processes can be automated by the effective use of modern computer equipment and by developing the necessary computer programs. They have also been shown how the computed results, diagrams, and graphs are printed or displayed on the peripheral output devices. In particular, when a graphics display terminal is used, the user can visually examine the results on the screen. It is practical and desirable during the visual inspection that the user be allowed to alter the values of the parameters involved in the CAD process and to look at different alternatives.

Figure 5-1 Interactive graphics devices: joystick (left panel) and user-definable keys and thumbwheels (1 and 2 shown in the right panel, respectively.) (Courtesy of Tektronix, Inc.)

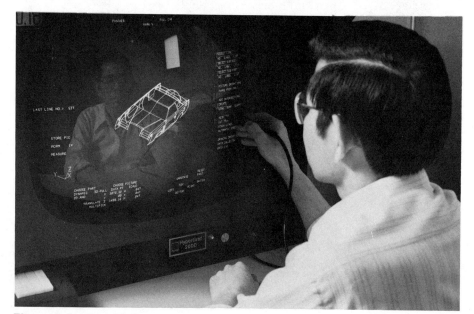

Figure 5-2 Use of lightpen in interactive CAD. (Courtesy of Ford Motor Company.)

Such spontaneous interaction is most needed in implementing the "computer-aided" designs at the speed of machines.

How to effectively utilize the various available devices by preparation of BASIC programs for interaction with these devices and how to produce rapid changes of the graphics in CAD are the topics of this chapter. Joysticks, thumbwheels, definable keys (Fig. 5-1), lightpen (Fig. 5-2), and tablet (Fig. 5-3) are the prevailing devices employed for interactive graphics applications. They will be used for **positioning** of the cursor on the display screen or **pointing** to a portion of a drawing being shown on the screen where a change is to be made.

Special commands in BASIC language are required in using these interactive graphics devices in conjunction with a program being run on the graphics display terminal. Examples of the applications of these commands will be given in a

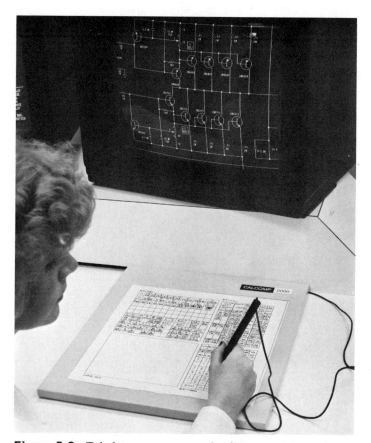

Figure 5-3 Tabulate preprogrammed software modules and use stylus pointing to a specific module on the tablet to request that its application in the program be executed on the display unit.
(Calomp's 200 Series Digitizer. Courtesy of California Computer Products, Inc.)

number of BASIC programs written for demonstration of using thumbwheels, definable keys, and tablet with a Tektronix 4054 computer graphics display system.

In Appendix D, some of the developed BASIC programs have been converted to FORTRAN language as well as in other BASIC versions suited to be run on APPLE, IBM, and TRS microcomputers. The readers should attempt to convert the other programs as exercises in learning the techniques introduced in this text.

5.2 THUMBWHEELS, LIGHTPEN, AND JOYSTICK

The two thumbwheels on the Tektronix 4054 graphics display system shown in Fig. 5-1 can be used as an interactive device in conjunction with a statement in the form of

POINTER Variable 1, Variable 2, string variable

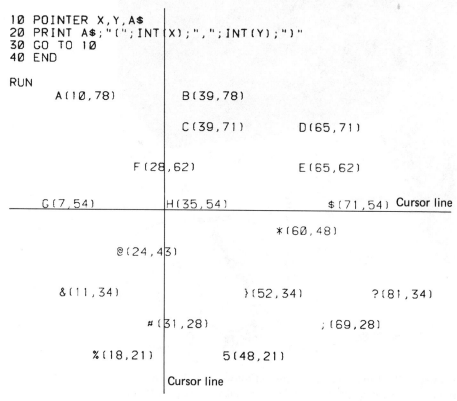

```
10 POINTER X,Y,A$
20 PRINT A$;"(";INT(X);",";INT(Y);")"
30 GO TO 10
40 END

RUN
        A(10,78)            B(39,78)

                            C(39,71)        D(65,71)

                F(28,62)                E(65,62)

    G(7,54)             H(35,54)                $(71,54) Cursor line
                                *(60,48)
            @(24,43)

    &(11,34)                    }(52,34)        ?(81,34)
                    #(31,28)            ;(69,28)

        %(18,21)            5(48,21)

                    Cursor line
```

Figure 5-4 Sample printout of character and values of variables at the location of cursor.

When the above statement is executed, two **cursor lines**, one horizontal and one vertical, will appear on the screen. The two thumbwheels are to be turned to move these two lines vertically and horizontally until the intercept of these two lines is located at a point where the user wants it to be. The user can then type in a string of characters and press the RETURN key. The consequence is that the coordinates of the intercept in GDU become the values of Variable 1 and Variable 2, and the string of characters entered from the keyboard becomes the content of the string variable listed in the POINTER statement. Figure 5-4 illustrates not only the use of a POINTER statement but also the combined use of POINTER and PRINT statements to print characters and numbers at the intercept of the cursor lines. In producing Fig. 5-4, only one character (A–H, 5, %, etc.) has been entered from the keyboard for A$ every time line 10 was executed. Program DEMO. TW is also presented to further demonstrate the results of using PRINT statements together with POINTER statements in a program.

Lightpens and joysticks can also be employed in place of thumbwheels for pointing to a precise location where the cursor is to be positioned on the screen.

PROGRAM DEMO. TW

```
100 INIT
110 DIM X(10),Y(10)
120 FOR I=1 TO 10
130 POINTER X(I),Y(I),M$
140 PRINT M$
150 MOVE 0,55-4*I
160 PRINT X(I),Y(I),M$
170 NEXT I
180 END
```

```
        ABC
          D
          E
          T
           23
            $
            }
```

RUN

71.936	93.056	A
73.856	93.056	B
75.776	93.056	C
75.776	90.752	D
75.776	88.576	E
75.776	86.784	1
77.824	84.608	2
79.488	84.608	3
81.152	81.792	$
81.152	78.848	}

5.3 AN EXAMPLE OF INTERACTIVE CLIPPING

In Chapter 4 the clipping of a certain portion of a drawing has been discussed and demonstrated with an example shown in Fig. 4-18. The purpose of clipping is to focus attention on the selected region by enlarging that portion of the drawing for further scrutiny. Figure 4-18 and the accompanying program are, however, presented for a fixed location and dimensions of the clipped area specified by lines 210–300. These lines need to be changed if a different clipping is desired.

Thumbwheels can be used effectively for clipping. Program CLIP. TW is presented to exemplify the interactive clipping at the command of the user through

```
'* PROGRAM CLIP.TW *'

'_WANT A SAMPLE CLIPPING OR USE THUMBWHEELS ? ENTER 0/1 : ';

'_USE THUMBWHEELS TO SPECIFY THE LEFT LOWER CORNER OF'
'    THE CLIPPING WINDOW.   PRESS `RETURN' KEY.'

'_USE THUMBWHEELS TO SPECIFY THE RIGHT UPPER CORNER OF'
'    THE CLIPPING WINDOW.   PRESS `RETURN' KEY.'
```

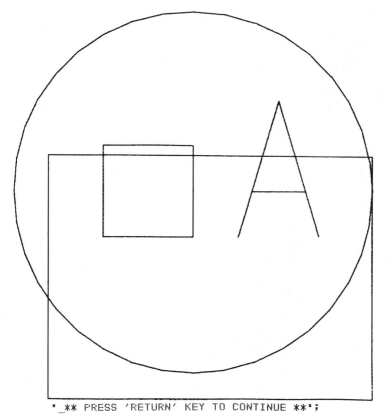

```
'_** PRESS 'RETURN' KEY TO CONTINUE **';
```

Figure 5-5   Use thumbwheels to specify the clipped area.

VIEWPORT FOR THE CLIPPED"
  REGION IS 100 GDU FOR"
  THE WIDER OF THE TWO"
  DIMENSIONS SPECIFIED."
_** CLIPPING END **"

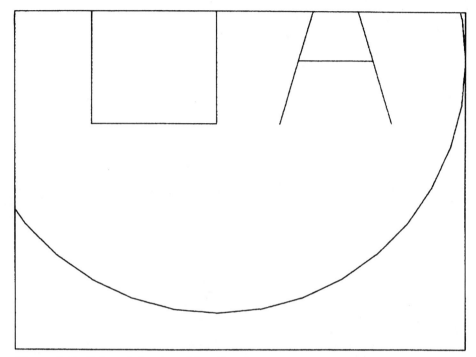

Figure 5-6 Enlarged drawing of the clipped area.

control of the two thumbwheels on the Tektronix 4054 graphics display system.
Figure 5-5 is an example of specifying the clipped area on the screen, and Fig. 5-6
is the resulting enlarged portion of the drawing inside the clipped area.

PROGRAM CLIP.TW

```
100 PAGE
110 INIT
120 CHARSIZE 2
130 PRINT "* PROGRAM CLIP.TW *"
140 SET DEGREES
150 PRINT "_WANT A SAMPLE CLIPPING OR USE THUMBWHEELS ? ENTER 0/1 : ";
160 INPUT Z
170 GOSUB 270
180 GOSUB 480
190 GOSUB 700
200 HOME
210 PRINT "VIEWPORT FOR THE CLIPPED"
220 PRINT "   REGION IS 100 GDU FOR"
230 PRINT "   THE WIDER OF THE TWO"
240 PRINT "   DIMENSIONS SPECIFIED."
```

```
250 PRINT "_** CLIPPING END **"
260 END
270 REM **  PLOT A CIRCLE  **
280 X1=90
290 Y1=50
300 R=40
310 MOVE X1+R,Y1
320 FOR C1=10 TO 360 STEP 10
330 DRAW X1+COS(C1)*R,Y1+SIN(C1)*R
340 NEXT C1
350 REM *DRAW A SQUARE*
360 MOVE 90,60
370 DRAW 90,40
380 DRAW 70,40
390 DRAW 70,60
400 DRAW 90,60
410 REM *DRAW 'A'*
420 MOVE 100,40
430 DRAW 109,70
440 DRAW 118,40
450 MOVE 103,50
460 DRAW 115,50
470 RETURN
480 REM **  CLIPPING  **
490 IF Z=0 THEN 590
500 MOVE 0,92
510 PRINT "_USE THUMBWHEELS TO SPECIFY THE LEFT LOWER CORNER OF"
520 PRINT "_    THE CLIPPING WINDOW.  PRESS 'RETURN' KEY."
530 POINTER U1,V1,C$
540 MOVE 0,10
550 PRINT "_USE THUMBWHEELS TO SPECIFY THE RIGHT UPPER CORNER OF"
560 PRINT "_    THE CLIPPING WINDOW.  PRESS 'RETURN' KEY."
570 POINTER U2,V2,C$
580 GO TO 660
590 U1=45
600 V1=50
610 U2=75
620 V2=75
630 PRINT
640 MOVE 0,92
650 PRI "THE RECTANGLE ABOVE THE **  ** MESSAGE IS THE CLIPPED REGIO
660 GOSUB 780
670 PRINT "_** PRESS 'RETURN' KEY TO CONTINUE **";
680 INPUT A$
690 RETURN
700 REM  **  PLOT THE SPECIFIED SECTION  **
710 PAGE
720 W=U2-U1 MAX V2-V1
730 WINDOW U1,U2,V1,V2
740 VIEWPORT 30,30+100*(U2-U1)/W,0,100*(V2-V1)/W
750 GOSUB 850
760 GOSUB 270
770 RETURN
780 REM ** PLOT THE RECTANGLE **
790 MOVE U1,V1
800 DRAW U1,V2
810 DRAW U2,V2
820 DRAW U2,V1
830 DRAW U1,V1
840 RETURN
850 REM *DRAW THE CLIPPED REGION*
```

```
860 MOVE U1,V1
870 DRAW U1,V2
880 DRAW U2,V2
890 DRAW U2,V1
900 DRAW U1,V1
910 RETURN
```

EXERCISE

Use thumbwheels to enter the coordinates of two chosen points P_1 and P_2 on the screen. Then use thumbwheels again to enter the coordinates of a third point P_3. By preparing a program to draw lines connecting points P_1 and P_3, and points P_2 and P_3, changing the coordinates of P_3 enables a **rubber-band** effect to be demonstrated on the screen.

5.4 GRAPHICS TABLET

Often there is a need to copy a drawing onto the screen of a graphics display unit so that it can be modified as long as the data for displaying the drawing are already transferred into the core memory of the unit. This can be done by placing the drawing on a device called **graphics tablet**, which means a flat surface. By pressing the writing pen called **stylus** or the **four-button cursor**, either one connected to the tablet as shown in Fig. 5-7, on a point of the drawing, the coordinates X and Y together with a status character Z$ will be generated by the tablet and transferred into the core memory of the display unit. It is in response to a statement being executed by the unit in the form of

$$INPUT\ @8: X, Y, Z\$$$

Unit number 8 in the above statement is designated to the graphics tablet. The status Z$ transmits a code number describing how the coordinates X and Y are generated and the status of the switches on the tablet controller. Table 5-1 lists the ASCII status byte values used for a Tektronix 4956 graphics tablet.

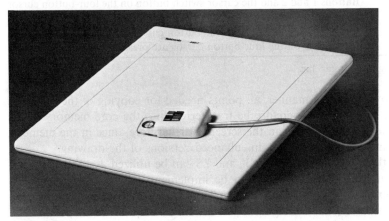

Figure 5-7 Graphics tablet devices. (Courtesy of Tektronix, Inc.)

TABLE 5-1a
Normal Status Byte Values

ASCII Character Status Byte	Meaning
0	The pen (or cursor) is in proximity
1	The pen (or cursor) switch is activated
2	Button 3 on the optional four-button cursor is pressed
4	Button 2 on the optional four-button cursor is pressed
8	Button 1 on the optional four-button cursor is pressed

Source: Courtesy of Tektronix, Inc.

TABLE 5-1b
Special Four-button Cursor Status Byte Values

ASCII Character Status Byte	Meaning
3	Button 3 and the cursor switch button on the four-button cursor are pressed
5	Button 2 and the cursor switch button on the four-button cursor are pressed
6	Buttons 2 and 3 on the four-button cursor are pressed
7	Buttons 2 and 3 and the cursor switch button on the four-button cursor are pressed
9	Button 1 and the cursor switch button on the four-button cursor are pressed
:	Buttons 1 and 3 on the four-button cursor are pressed
;	Buttons 1 and 3 and the cursor switch button on the four-button cursor are pressed
<	Buttons 1 and 2 on the four-button cursor are pressed
=	Buttons 1 and 2 and the cursor switch button on the four-button cursor are pressed
>	Buttons 1, 2, and 3 on the four-button cursor are pressed
?	All buttons on the four-button cursor are pressed

Source: Courtesy of Tektronix, Inc.

In the above-described manner, all points needed for copying of the lines and curves of the drawing can therefore be transferred into the core memory of the display unit for reproducing it on the screen. Furthermore, once in the memory, these data can be modified to obtain enhanced versions of the drawing.

Also, the three input variables X, Y, and Z\$ can be utilized as control parameters for requesting particular tasks to be implemented by a graphics display device. For example, Fig. 5-3 illustrates that the tablet is divided into a number of rectangular blocks, each designated for entering a request for a specific task. By pressing the stylus at any point inside of a block, the X and Y values of that

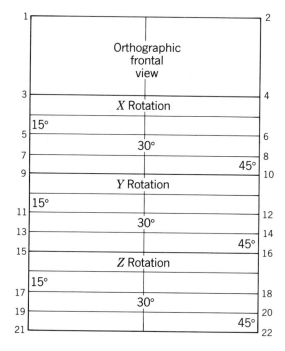

Figure 5-8   Rotation tablet for the 2 × 3 × 4 brick or any other objects constructed in the program.

point transferred into the display unit can then be used to determine which subroutine of the display program should be branched to implement the particular task associated with that block.

Program BRICK.R is presented to demonstrate how the 2 × 3 × 4 brick previously discussed in Chapter 4 can be rotated in an interactive way by use of a tablet. Figure 5-8 is a layout sheet that is to be placed on the tablet so that when the stylus is pressed on a point inside of a particular block, a desired rotation of the brick can be realized and displayed on the screen. The program will instruct the user first to enter the coordinates of the four corner points 1, 2, 21, and 22 shown in Fig. 5-8 by pressing the stylus on them one after another in that order. Automatically, the coordinates of points 3–20 will be generated by the program to define the blocks to be designated for various rotation requests as specified in Fig. 5-8. The user will next be given instructions to proceed placing requests by pressing the stylus to the appropriate blocks one at a time.

For example, once the first step mentioned above is done, the user can press the stylus at any point within the area 1-2-4-3 and see the "F" view of Fig. 4-12. And if one wishes to see the "T" view of Fig. 4-12, the stylus should press the area 7-8-10-9 twice. The user can easily exercise different sequences of rotation (Fig. 5-9), such as those shown in Fig. 4-13, by properly maneuvering the stylus on Fig. 5-8. It should be pointed out that the request sent in by the stylus transforms the *last* displayed view stored in the core memory. The user has to point the stylus to the area 1-2-4-3 in order to return to the unrotated geometry of the brick, namely, the front "F" view.

```
PROGRAM BRICK.R *

USE STYLUS TO PRESS THE FOUR CORNERS OF
  THE *ROTATION* INSTRUCTION TABLE.
     POINT      X        Y
       1       3.60    11.75
       2       9.63    11.71
      21       3.65     3.82
      22       0.67     3.82

COORDINATES OF THE 22 NODES.

       1       3.60    11.75
       2       9.63    11.71
       3       3.61     9.76
       4       9.64     9.74
       5       3.62     9.77
       6       9.64     8.75
       7       3.62     8.28
       8       9.53     8.26
       9       3.62     7.70
      10       9.55     7.76
      11       3.63     7.29
      12       9.65     7.27
      13       3.63     6.30
      14       9.65     6.28
      15       3.64     5.80
      16       9.66     5.79
      17       3.64     5.31
      18       9.66     5.30
      19       3.65     4.32
      20       9.55     4.31
      21       3.65     3.82
      22       9.57     3.82

PRESS *RETURN* KEY TO CONTINUE

ENTER A DISPLAY REQUEST BY PRESSING STYLUS
  INSIDE THE BLOCK OF SELECTED ROTATION.
```

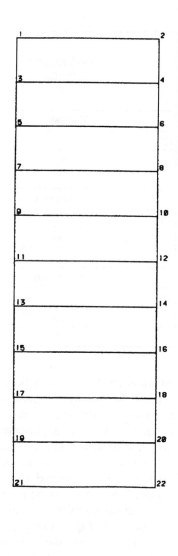

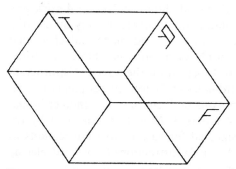

Figure 5-9 Screen displays and messages generated by Program BRICK.R.

PROGRAM BRICK.R

```
100 RINIT
110 PAGE
120 INIT
130 SET DEGREES
140 CHARSIZE 2
150 PRINT "* PROGRAM BRICK.R *"
160 GOSUB 780
```

```
170 GOSUB 230
180 GOSUB 350
190 P5=0
200 P1=0
210 J5=0
220 GO TO 960
230 REM *LABEL THE NODES*
240 J1=1
250 MOVE 56.5,100
260 A3=10
270 FOR J1=1 TO 11
280 FOR J2=1 TO 2
290 PRINT J2+2*(J1-1)
300 RMOVE 30,0
310 NEXT J2
320 RMOVE -30*2,-9.5
330 NEXT J1
340 RETURN
350 REM *ENTER THE COORDINATES OF FOUR CORNERS*
360 DIM H1(22),H2(22)
370 MOVE 0,95
380 PRINT "USE STYLUS TO PRESS THE FOUR CORNERS OF"
390 PRINT "  THE 'ROTATION' INSTRUCTION TABLE."
400 PRINT "_  POINT     X       Y  "
410 PRINT
420 FOR K1=1 TO 2
430 INPUT @8:H1(K1),H2(K1),C$
440 PRINT USING 500:K1,H1(K1),H2(K1)
450 NEXT K1
460 FOR K1=21 TO 22 STEP 1
470 INPUT @8:H1(K1),H2(K1),C$
480 PRINT USING 500:K1,H1(K1),H2(K1)
490 NEXT K1
500 IMAGE  2X,3D,2X,4D,2D,2X,4D,2D
510 H5=H1(1)-H1(21)
520 H6=H1(2)-H1(22)
530 G5=H2(1)-H2(21)
540 G6=H2(2)-H2(22)
550 H1(3)=H5*0.75+H1(21)
560 H1(4)=H6*0.75+H1(22)
570 H2(3)=G5*0.75+H2(21)
580 H2(4)=G6*0.75+H2(22)
590 O6=2
600 FOR K1=5 TO 19 STEP 2
610 H1(K1)=H1(3)-H5*0.75/12*O6
620 H1(K1+1)=H1(4)-H6*0.75/12*O6
630 H2(K1)=H2(3)-G5*0.75/12*O6
640 H2(K1+1)=H2(4)-G6*0.75/12*O6
650 IF K1<>11 AND K1<>17 THEN 680
660 O6=O6+2
670 GO TO 690
680 O6=O6+1
690 NEXT K1
700 PRINT "_COORDINATES OF THE 22 NODES :_"
710 FOR K1=1 TO 22
720 PRINT USING 500:K1,H1(K1),H2(K1)
730 NEXT K1
740 PRINT "_PRESS 'RETURN' KEY TO CONTINUE";
750 INPUT A$
760 PAGE
770 RETURN
```

```
780 REM *PLOT THE 'ROTATION' INSTRUCTION TABLE*
790 Q7=10
800 Q6=10
810 Q9=1
820 CHARSIZE 1
830 S5=57
840 S6=100
850 FOR J6=1 TO Q6
860 MOVE S5,S6
870 RDRAW 30,0
880 RDRAW 0,-9.5
890 RDRAW -30,0
900 RDRAW 0,9.5
910 S5=S5+30
920 S6=S6-9.5
930 S5=S5-Q9*30
940 NEXT J6
950 RETURN
960 REM *MAKE REQUEST FOR A SPECIFIC ROTATION*
970 CHARSIZE 2
980 P5=P5+1
990 IF P5=1 THEN 1010
1000 MOVE X1-0.6*W,Y1+1.5*W

1010 PRINT " ENTER A DISPLAY REQUEST BY PRESSING STYLUS "
1020 PRINT "  INSIDE THE BLOCK OF SELECTED ROTATION."
1030 INPUT @8:H3,H4,C$
1040 IF H3<H1(2) AND H3>H1(3) AND H4<H2(2) AND H4>H2(3) THEN 1150
1050 IF H3<H1(4) AND H3>H1(3) AND H4<H2(4) AND H4>H2(5) THEN 1220
1060 IF H3<H1(6) AND H3>H1(5) AND H4<H2(6) AND H4>H2(7) THEN 1270
1070 IF H3<H1(8) AND H3>H1(7) AND H4<H2(8) AND H4>H2(9) THEN 1320
1080 IF H3<H1(10) AND H3>H1(9) AND H4<H2(10) AND H4>H2(11) THEN 1370
1090 IF H3<H1(12) AND H3>H1(11) AND H4<H2(12) AND H4>H2(13) THEN 142
1100 IF H3<H1(14) AND H3>H1(13) AND H4<H2(14) AND H4>H2(15) THEN 147
1110 IF H3<H1(16) AND H3>H1(15) AND H4<H2(16) AND H4>H2(17) THEN 152
1120 IF H3<H1(18) AND H3>H1(17) AND H4<H2(18) AND H4>H2(19) THEN 157
1130 IF H3<H1(20) AND H3>H1(19) AND H4<H2(20) AND H4>H2(21) THEN 162
1140 END
1150 REM **  FRONT VIEW  **
1160 C1=0
1170 C2=0
1180 C3=0
1190 P1=P1+1
1200 J5=1
1210 GO TO 1660
1220 REM **  15 DEGREE X-ROTATE  **
1230 C1=15
1240 C2=0
1250 C3=0
1260 GO TO 1660
1270 REM **  30 DEGREE X-ROTATE  **
1280 C1=30
1290 C2=0
1300 C3=0
1310 GO TO 1660
1320 REM **  45 DEGREE X-ROTATE  **
1330 C1=45
1340 C2=0
1350 C3=0
1360 GO TO 1660
1370 REM **  15 DEGREE Y-ROTATE  **
```

```
1380 C1=0
1390 C2=15
1400 C3=0
1410 GO TO 1660
1420 REM **  30 DEGREE Y-ROTATE   **
1430 C1=0
1440 C2=30
1450 C3=0
1460 GO TO 1660
1470 REM **  45 DEGREE Y-ROTATE   **
1480 C1=0
1490 C2=45
1500 C3=0
1510 GO TO 1660
1520 REM **  15 DEGREE Z-ROTATE   **
1530 C1=0
1540 C2=0
1550 C3=15
1560 GO TO 1660
1570 REM **  30 DEGREE Z-ROTATE   **
1580 C1=0
1590 C2=0
1600 C3=30
1610 GO TO 1660
1620 REM **  45 DEGREE Z-ROTATE   **
1630 C1=0
1640 C2=0
1650 C3=45
1660 PAGE
1670 REMARK     8 PTS FOR BOX, 16 PTS FOR LETTERS "F","T","R"
1680 N=24
1690 DIM X0(N),Y0(N),Z0(N),X(N),Y(N),Z(N),X5(N),Y5(N),Z5(N)
1700 REMARK  FIRST 8 FOR BOX,THEN 16 FOR LETTERS
1710 IF P5<>1 THEN 1880
1720 READ X0
1730 DATA 0,4,4,0,0,4,4,0
1740 DATA 3.4,3.4,3.4,3.8,3.7,0.2,0.4,0.6,0.4,4,4,4,4,4,4,4
1750 READ Y0
1760 DATA 0,0,2,2,0,0,2,2
1770 DATA 1.4,1.6,1.8,1.8,1.6,2,2,2,1.4,1.6,1.8,1.8,1.6,1.6,1.4
1780 READ Z0
1790 DATA 0,0,0,0,3,3,3,3
1800 DATA 3,3,3,3,3,0.2,0.2,0.2,0.6,0.6,0.6,0.6,0.2,0.2,0.4,0.2
1810 READ S1,S2,S3,T1,T2,T3
1820 DATA 1,1,1,0,0,0
1830 FOR K1=1 TO N
1840 X5(K1)=X0(K1)
1850 Y5(K1)=Y0(K1)
1860 Z5(K1)=Z0(K1)
1870 NEXT K1
1880 IF P5=1 THEN 2010
1890 IF J5=1 THEN 1960
1900 FOR K1=1 TO N
1910 X0(K1)=X(K1)
1920 Y0(K1)=Y(K1)
1930 Z0(K1)=Z(K1)
1940 NEXT K1
1950 GO TO 2010
1960 FOR K1=1 TO N
1970 X0(K1)=X5(K1)
1980 Y0(K1)=Y5(K1)
```

```
1990 Z0(K1)=Z5(K1)
2000 NEXT K1

2010 J5=0
2020 GOSUB 2090
2030 GOSUB 2250
2040 VIEWPORT 0,100,0,100
2050 WINDOW X1-0.6*W,X1+1.5*W,Y1-0.6*W,Y1+1.5*W
2060 GOSUB 2380
2070 GO TO 970
2080 REM  ** MAIN PROGRAM ENDS  **
2090 REM C1, C2, C3 ARE ROTATION ABOUT X-, Y- AND Z-AXES, RESPECTIVI
2100 A=COS(C3)*COS(C2)
2110 B=SIN(C3)*COS(C2)
2120 C=-SIN(C2)
2130 D=-SIN(C3)*COS(C1)+COS(C3)*SIN(C2)*SIN(C1)
2140 E=COS(C3)*COS(C1)+SIN(C2)*SIN(C3)*SIN(C1)
2150 F=COS(C2)*SIN(C1)
2160 G=SIN(C3)*SIN(C1)+COS(C3)*SIN(C2)*COS(C1)
2170 H=-COS(C3)*SIN(C1)+SIN(C3)*SIN(C2)*COS(C1)
2180 P=COS(C2)*COS(C1)
2190 FOR I=1 TO N
2200 X(I)=A*S1*(X0(I)+T1)+D*S1*(Y0(I)+T2)+G*S1*(Z0(I)+T3)
2210 Y(I)=B*S2*(X0(I)+T1)+E*S2*(Y0(I)+T2)+H*S2*(Z0(I)+T3)
2220 Z(I)=C*S3*(X0(I)+T1)+F*S3*(Y0(I)+T2)+P*S3*(Z0(I)+T3)
2230 NEXT I
2240 RETURN
2250 REM SCLAING OF DATA
2260 X1=X(1)
2270 X2=X(1)
2280 Y1=Y(1)
2290 Y2=Y(1)
2300 FOR I=2 TO N
2310 X1=X1 MIN X(I)
2320 X2=X2 MAX X(I)
2330 Y1=Y1 MIN Y(I)
2340 Y2=Y2 MAX Y(I)
2350 W=X2-X1 MAX Y2-Y1
2360 NEXT I
2370 RETURN
2380 REMARK  COMMENCE PLOTTING
2390 MOVE X(5),Y(5)
2400 DRAW X(6),Y(6)
2410 DRAW X(2),Y(2)
2420 DRAW X(1),Y(1)
2430 DRAW X(5),Y(5)
2440 DRAW X(8),Y(8)
2450 DRAW X(7),Y(7)
2460 DRAW X(3),Y(3)
2470 DRAW X(4),Y(4)
2480 DRAW X(8),Y(8)
2490 MOVE X(6),Y(6)
2500 DRAW X(7),Y(7)
2510 MOVE X(3),Y(3)
2520 DRAW X(2),Y(2)
2530 MOVE X(1),Y(1)
2540 DRAW X(4),Y(4)
2550 REMARK DRAW "F"
2560 MOVE X(9),Y(9)
2570 DRAW X(11),Y(11)
2580 DRAW X(12),Y(12)
```

```
2590 MOVE X(10),Y(10)
2600 DRAW X(13),Y(13)
2610 REMARK DRAW "T"
2620 MOVE X(14),Y(14)
2630 DRAW X(16),Y(16)
2640 MOVE X(17),Y(17)
2650 DRAW X(15),Y(15)
2660 REMARK DRAW "R"
2670 MOVE X(18),Y(18)
2680 DRAW X(20),Y(20)
2690 DRAW X(21),Y(21)
2700 DRAW X(22),Y(22)
2710 DRAW X(19),Y(19)
2720 MOVE X(23),Y(23)
2730 DRAW X(24),Y(24)
2740 RETURN
```

EXERCISE

The Program PIE listed below displays the partitioning of a circular area into
a specified number of slices according to the input percentages. Modify the program so that it can be requested interactively by use of the stylus and tablet.
Figure 5-10 is a sample plot of Program PIE. Also modify the program to accommodate for adoption of the characters chosen by the user, rather than using
the fixed set of characters as arranged in the program. For example, the seven

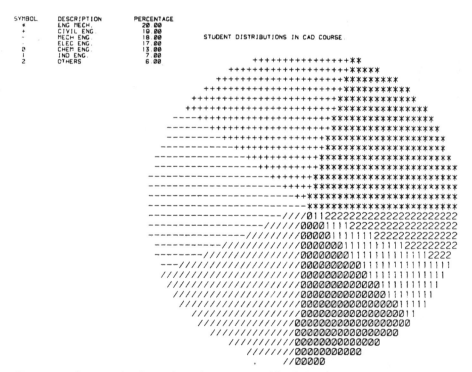

Figure 5-10 Sample plot and captions generated by Program PIE.PLOT.

slices could be labeled with the characters *E*, *C*, *M*, *L*, *I*, *H*, and 0. In Program PIE, line 2440 stores the number of characters in B$ into $H5(L7)$ by use of the LEN function. Line 2310 returns the ASCII character equivalent of the decimal value of $B7(I2)$ by use of the CHR function.

(Program messages on the display screen)

```
¥ PROGRAM PIE PLOT ¥

WANT TO RUN A SAMPLE OR YOUR OWN PROBLEM? (1 OR 2) : 1

INPUT THE CHARACTER SIZE 1,2,3, OR, 4 FOR PLOTTING. : 2

   NODE          X              Y
    1         80.0000        55.0000
    2         95.0000        55.0000
    3         84.6353        60.2658
    4         68.4423        64.5614
    5         66.4276        48.6133
    6         79.0581        40.0296
    7         90.2682        44.0655
    8         93.9466        49.4781

WHEN READY TO CONTINUE, PRESS 'RETURN' KEY .

ELEMENT      NODE  NUMBERS     PERCENTAGE   NOTE
   1          1    2    3        20.00      ENG MECH.
   2          1    3    4        10.00      CIVIL ENG.
   3          1    4    5        18.00      MECH ENG.
   4          1    5    6        17.00      ELEC ENG.
   5          1    6    7        13.00      CHEM ENG.
   6          1    7    8         7.00      IND ENG.
   7          1    8    2         6.00      OTHERS

WHEN READY TO CONTINUE, PRESS 'RETURN' KEY :
```

PROGRAM PIE . PLOT

```
100 PAGE
110 CHARSIZE 2
120 PRINT "* PROGRAM PIE PLOT *"
130 INIT
140 VIEWPORT 40,130,5,95
150 PRINT "_WANT TO RUN A SAMPLE OR YOUR OWN PROBLEM? (1 OR 2) : ";
160 INPUT Z9
170 PRINT "_INPUT THE CHARACTER SIZE 1,2,3, OR, 4 FOR PLOTTING. : ";
180 INPUT C
190 CHARSIZE 2
200 IF C<3 THEN 230
210 H=2.5+(C-3)*0.3
220 GO TO 240
230 H=1.5+(C-1)*0.1
240 H1=H*0.625
250 W1=1
260 REM GOSUB FOR INPUT AND PLOT.
270 GOSUB 310
280 GOSUB 1540
290 GOSUB 2250
300 END
310 REMARK  SUBROUTINE INPUT
320 GO TO Z9 OF 330,400
330 T$="STUDENT DISTRIBUTIONS IN CAD COURSE.         "
340 N1=8
350 N2=7
360 DIM X(N1),Y(N1),N(3,N2),D7(N1),P7(N2)
370 READ P7
380 DATA 20,19,18,17,13,7,6
```

```
390 GO TO 480
400 PRINT "_ENTER A TITLE FOR THE PIE DIAGRAM"
410 INPUT T$
420 PRINT "_ENTER NUMBER OF ELEMENTS : ";
430 INPUT N2
440 N1=N2+1
450 DIM X(N1),Y(N1),N(3,N2),D7(N1),P7(N2)
460 PRINT "_INPUT THE PERCENTAGES : "
470 INPUT P7
480 GOSUB 1950
490 GOSUB 2360
500 FOR I2=1 TO N2
510 N(1,I2)=1
520 N(2,I2)=I2+1
530 IF I2=N2 THEN 560
540 N(3,I2)=I2+2
550 GO TO 570
560 N(3,I2)=2
570 NEXT I2
580 IF W1=0 THEN 770
590 PRINT "_   NODE            X              Y "
600 FOR I1=1 TO N1
610 PRINT USING 620:I1,X(I1),Y(I1)
620 IMAGE 4D,2(3X,10D.4D)
630 NEXT I1
640 PRINT "_WHEN READY TO CONTINUE, PRESS 'RETURN' KEY : ";
650 INPUT K$
660 PRI "_ELEMENT       NODE  NUMBERS        PERCENTAGE        NOTE          "
670 FOR I2=1 TO N2
680 IF I2=1 THEN 710
690 C$=SEG(E$,H6(I2)-H5(I2)+1,H5(I2))
700 GO TO 720
710 C$=SEG(E$,H6(I2)-H5(I2),H5(I2))
720 PRINT USING 730:I2,N(1,I2),N(2,I2),N(3,I2),P7(I2),C$
730 IMAGE 4D,4X,3(6D),5X,6D.2D,8X,25A
740 NEXT I2
750 PRINT "_WHEN READY TO CONTINUE, PRESS 'RETURN' KEY : ";
760 INPUT K$
770 RETURN

780 REM * SUBROUTINE CONTOUR *
790 J1=N(1,I2)
800 J2=N(2,I2)
810 J3=N(3,I2)
820 IF P9=7 THEN 880
830 IF D7(I2)<=180 THEN 880
840 IF H7<>1 THEN 870
850 GOSUB 2120
860 GO TO 880
870 GOSUB 2050
880 X1=X(J3) MIN X(J1) MIN X(J2)
890 X2=X(J3) MAX X(J1) MAX X(J2)
900 Y1=Y(J3) MIN Y(J1) MIN Y(J2)
910 Y2=Y(J3) MAX Y(J1) MAX Y(J2)
920 M1=INT(X1/H1+0.5)
930 M2=INT(X2/H1+0.5)
940 FOR I=M1 TO M2
950 X0=I*H1
960 Z1=Y2
970 Z2=Y1
980 IF ABS(X0-X(J1))+ABS(X0-X(J2))>ABS(X(J2)-X(J1)) THEN 1080
```

```
990 IF X(J2)=X(J1) THEN 1060
1000 Z=Y(J1)+(Y(J2)-Y(J1))*(X0-X(J1))/(X(J2)-X(J1))
1010 IF Z<Y1 THEN 1030
1020 Z1=Z1 MIN Z
1030 IF Z>Y2 THEN 1080
1040 Z2=Z2 MAX Z
1050 GO TO 1080
1060 Z1=Z1 MIN Y(J2) MIN Y(J1)
1070 Z2=Z2 MAX Y(J2) MAX Y(J1)
1080 IF ABS(X0-X(J2))+ABS(X0-X(J3))>ABS(X(J2)-X(J3)) THEN 1300
1090 IF X(J3)=X(J2) THEN 1280
1100 IF D7(I2)>180 THEN 1170
1110 IF X0-80=-R7 THEN 1150
1120 K7=(X0-80)^2
1130 Z=(R7^2-K7)^0.5+55
1140 GO TO 1230
1150 Z=55
1160 GO TO 1230
1170 IF X0-80=-R7 THEN 1220
1180 Z=-((R7^2-(X0-80)^2)^0.5)+55
1190 IF H7<>1 THEN 1230
1200 Z=ABS(Z-55)+55
1210 GO TO 1230
1220 Z=55
1230 REM
1240 Z1=Z1 MIN Z
1250 REM   IF Z>Y2 THEN 1450
1260 Z2=Z2 MAX Z
1270 GO TO 1300
1280 Z1=Z1 MIN Y(J3) MIN Y(J2)
1290 Z2=Z2 MAX Y(J3) MAX Y(J2)
1300 IF ABS(X0-X(J1))+ABS(X0-X(J3))>ABS(X(J1)-X(J3)) THEN 1400
1310 IF X(J1)=X(J3) THEN 1380
1320 Z=Y(J3)+(Y(J1)-Y(J3))*(X0-X(J3))/(X(J1)-X(J3))
1330 IF Z<Y1 THEN 1350
1340 Z1=Z1 MIN Z
1350 IF Z>Y2 THEN 1400
1360 Z2=Z2 MAX Z
1370 GO TO 1400
1380 Z1=Z1 MIN Y(J3) MIN Y(J1)
1390 Z2=Z2 MAX Y(J3) MAX Y(J1)
1400 L1=INT(Z1/H+0.5)
1410 L2=INT(Z2/H+0.5)
1420 FOR J=L1 TO L2
1430 Y0=J*H
1440 IF Y0>Z2 OR Y0<Z1 THEN 1490
1450 REM
1460 MOVE X0-0.5*H1,Y0-0.65*H
1470 PRINT USING 1480:A$
1480 IMAGE 1A
1490 NEXT J
1500 NEXT I
1510 IF H7<>2 THEN 1530
1520 GOSUB 2200
1530 RETURN
1540 REMARK    SUBROUTINE WINDOWING
1550 PAGE
1560 CHARSIZE C
1570 H=H/3
1580 H1=H1/3
1590 WINDOW 65,65+30*1.05,40,40+30*1.05
```

```
1600 P9=0
1610 M7=41
1620 FOR I2=1 TO N2
1630 H7=0
1640 L7=M7+I2
1650 IF L7<>44 AND L7<>46 AND L7<>34 AND L7<>58 THEN 1680
1660 L7=L7+1
1670 M7=M7+1
1680 A$=CHR(L7)
1690 DIM B7(50)
1700 B7(I2)=L7
1710 GOSUB 780
1720 IF H7<>1 THEN 1740
1730 GOSUB 780
1740 NEXT I2
1750 VIEWPORT 0,130,0,100
1760 WINDOW 0,130,0,100
1770 MOVE 55,95
1780 CHARSIZE 2
1790 PRINT "_";T$
1800 HOME
1810 CHARSIZE C
1820 RETURN
1830 PAGE
1840 REM ** THIS SUB PLOTS A CIRCLE  **
1850 R7=15
1860 X(1)=80
1870 X(2)=80+R7
1880 Y(2)=55
1890 Y(1)=55
1900 FOR I1=1 TO N1-2
1910 X(I1+2)=R7*COS(D7(I1))+80
1920 Y(I1+2)=R7*SIN(D7(I1))+55
1930 NEXT I1
1940 RETURN
1950 REM * THIS SUB FINDS THE NODAL COORDINATES *
1960 S7=0
1970 SET DEGREES
1980 FOR I1=1 TO N2
1990 D7(I1)=P7(I1)/100*360
2000 S7=D7(I1)+S7

2010 D7(I1)=S7
2020 NEXT I1
2030 GOSUB 1840
2040 RETURN
2050 REM ** **
2060 F7=X(J3)
2070 F8=Y(J3)
2080 X(J3)=80+COS(180)*R7
2090 Y(J3)=55+SIN(180)*R7
2100 H7=1
2110 RETURN
2120 X(J3)=F7
2130 Y(J3)=F8
2140 F7=X(J2)
2150 F8=Y(J2)
2160 X(J2)=65
2170 Y(J2)=55
2180 H7=2
2190 RETURN
```

```
2200 X(J2)=F7
2210 Y(J2)=F8
2220 H7=0
2230 P9=7
2240 RETURN
2250 REM ** COMMENT **
2260 HOME
2270 CHARSIZE 2
2280 PRINT "    SYMBOL       DESCRIPTION             PERCENTAGE   "
2290 FOR I2=1 TO N2
2300 C$=SEG(E$,H6(I2)-H5(I2)+1,H5(I2))
2310 D$=CHR(B7(I2))
2320 PRINT USING 2340:D$,C$,P7(I2)
2330 NEXT I2
2340 IMAGE 5X,2A,7X,15A,3X,5D.2D
2350 RETURN
2360 REM  **  THIS SUB FOR DESCRIPTION  **
2370 DIM H5(N2),H6(N2),E$(300)
2380 E$=" "
2390 H6(1)=0
2400 IF Z9=1 THEN 2520
2410 FOR L7=1 TO N2
2420 PRINT "_INPUT THE DESCRIPTION FOR ELEMENT ";L7
2430 INPUT B$
2440 H5(L7)=LEN(B$)
2450 IF L7=1 THEN 2480
2460 H6(L7)=H5(L7)+H6(L7-1)
2470 GO TO 2490
2480 H6(L7)=H5(L7)
2490 E$=E$&B$
2500 NEXT L7
2510 GO TO 2650
2520 REM  **  SAMPLE DESCRIPTION  **
2530 E$="ENG MECH,CIVIL ENG,MECH ENG,ELEC ENG,CHEM ENG,IND ENG,OTHERS"
2540 H5(1)=9
2550 H5(2)=10
2560 H5(3)=9
2570 H5(4)=9
2580 H5(5)=9
2590 H5(6)=8
2600 H5(7)=6
2610 H6(1)=9
2620 FOR L7=2 TO 7
2630 H6(L7)=H6(L7-1)+H5(L7)
2640 NEXT L7
2650 RETURN
```

5.5 DEFINABLE KEYS

As shown in Fig. 5-2, the Tektronix 4054 graphics system has a set of 10 keys that could be utilized to enter requests for 20 different specific tasks prearranged in a BASIC program. The use of the SHIFT key enables the upper-designated keys numbered 11 through 20 to be entered (Fig. 5-11). These 20 definable keys are to be activated with a SET KEY statement in the program and they can be unset with a SET NOKEY statement. When the graphics system is powered up or

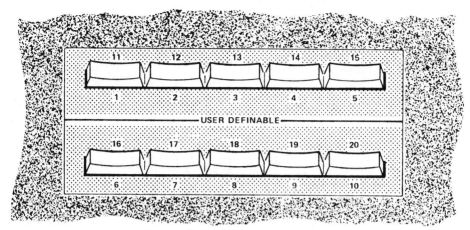

Figure 5-11   User-definable keys of the Tektronix 4054 graphics terminal. (Courtesy of Tektronix, Inc.)

an INIT statement is executed in the program, the system will automatically be set at a NOKEY condition until a SET KEY statement is executed.

While a BASIC program is executing, pressing the user-definable key i causes the current instruction, say line X, to be completed and then GO TO line $4i$ as diagrammed in Fig. 5-12. Lines $4i$ through $4i + 3$ should be arranged so that a specific, user-definable function will be implemented. After completion of the user-definable function, the program is to return to the line following line X to resume normal program execution.

Program E. MODULE has been developed to demonstrate the use of definable keys for drawing the basic electric elements commonly encountered in the circuit designs. The elements, shown in Fig. 5-13, can be requested by pressing the definable keys. As the key table accompanying Fig. 5-13 shows, each definable key when pressed is to perform a specific function. For example, pressing key 1 stops the execution of program E. MODULE, pressing key 8 requests a resistor to be drawn, and pressing key 19 stores all drawn elements on a data file that at present is set to 19 on the cassette tape.

Line 9495 KILL 19 is a statement that erases the file 19 on the cassette tape to make it ready for storage of new data. Line 9500 FIND 19 is a statement that instructs the tape to be positioned at the very beginning of file 19 so that data can be entered sequentially.

Lines 4 through 89 in program E. MODULE demonstrate how the definable keys are to be set up so that the desired functions can be coded as subroutines in the program. Line 155 is an example of the use of the SET KEY statement. Line 3 CALL "WAIT", 1000 instructs the program to halt execution for as long as 1000 seconds and then resume the next statement or until an interrupt condition occurs such as when another definable key is pressed.

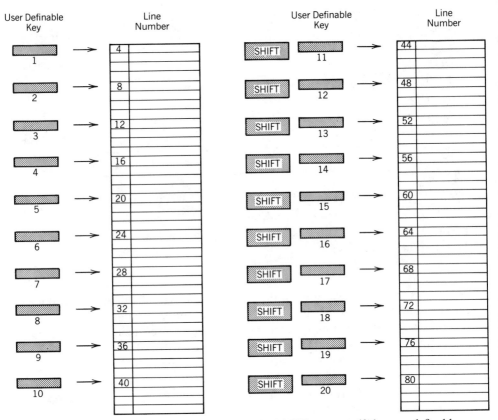

Figure 5-12 Beginning line numbers to use in BASIC program if the user-definable keys 1 through 20 are to be adopted for interactive graphics applications. (Courtesy of Tektronix, Inc.)

KEY TABLE

1 ------- STOP		11 ------- RECTANGLE
2 ------- CIRCLE		12 ------- REVISE
3 ------- SWITCH		13 ------- BIPOLAR
4 ------- BATTERY		14 ------- SEMICIRCLE
5 ------- LAMP		15 ------- DASH
6 ------- CAPACITOR		16 ------- OSCILLATOR
7 ------- GROUND		17 ------- DIODE
8 ------- RESISTOR		18 ------- ARROW
9 ------- INDUCTOR		19 ------- STORE
10 ----- LINE		20 ------- RETRIEVE

PROGRAM E. MODULE

Figure 5-13

STEP	KEY	BEGINNING X	Y	ENDING X	Y
1	2	54.3	75.3	66.7	75.3
2	3	66.7	59.6	54.1	59.6
3	4	54.1	48.8	66.7	48.8
4	5	66.7	37.6	66.7	33.4
5	6	66.7	21.9	54.1	21.9
6	7	85.4	75.4	73.6	75.3
7	8	85.4	59.8	73.5	59.8
8	9	73.5	48.9	85.4	48.9
9	10	85.4	35.6	73.3	35.6
10	11	73.3	20.4	85.4	23.8
11	13	93.1	75.4	106.2	75.4
12	14	99.7	56.6	99.7	63.2
13	15	92.9	48.8	106.2	48.9
14	16	92.8	35.7	106.2	35.7
15	17	92.7	22.0	106.2	22.0
16	18	113.9	70.9	125.1	88.0

DESIGN ENDS

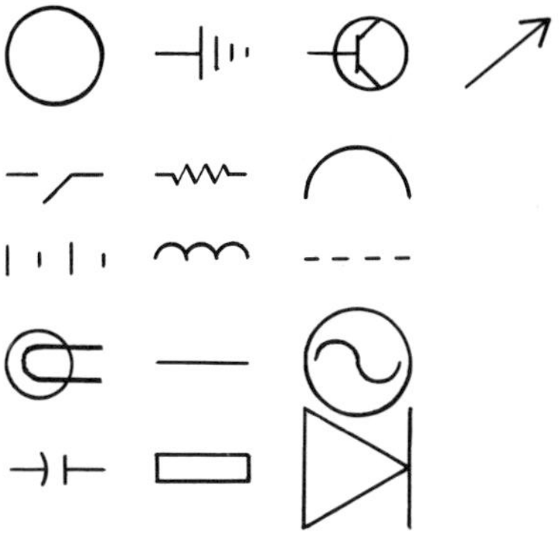

PROBLEM E. MODULE

```
1 REM * PROGRAM E.MODULE *
2 GO TO 150
3 CALL "WAIT",1000
4 GOSUB 3750
8 GOSUB 6250
9 GO TO 3
12 GOSUB 6000
13 GO TO 3
16 GOSUB 2620
17 GO TO 3
20 GOSUB 8000
21 GO TO 3
24 GOSUB 370
25 GO TO 3
28 GOSUB 1280
29 GO TO 3
32 GOSUB 1570
33 GO TO 3
36 GOSUB 2000
37 GO TO 3
40 GOSUB 2470
41 GO TO 3
44 GOSUB 3880
45 GO TO 3
48 GOSUB 4100
49 GO TO 3
52 GOSUB 8700
```

```
53 GO TO 3
56 GOSUB 7500
57 GO TO 3
60 GOSUB 6600
61 GO TO 3
64 GOSUB 3280
65 GO TO 3
68 GOSUB 8500
69 GO TO 3
72 GOSUB 7200
73 GO TO 3
76 GOSUB 9490
77 GO TO 3
80 GOSUB 9700
89 GO TO 3
150 CHARSIZE 1
151 RINIT
152 PAGE
153 INIT
154 SET DEGREES
155 SET KEY
160 GOSUB 200
162 GO TO 318
200 PRINT "* PROGRAM E.MODULE *"
201 PRINT "_KEY TABLE_"
203 PRINT "`1'--------STOP        `11'---------RECTANGLE"
205 PRINT "`2'--------CIRCLE      `12'---------REVISE"
210 PRINT "`3'--------SWITCH      `13'---------BIPOLAR"
220 PRINT "`4'--------BATTERY     `14'---------SEMICIRCLE"
230 PRINT "`5'--------LAMP        `15'---------DASH"
240 PRINT "`6'--------CAPACITOR   `16'---------OSCILLATOR"
250 PRINT "`7'--------GROUND      `17'---------DIODE"
260 PRINT "`8'--------RESISTOR    `18'---------ARROW"
270 PRINT "`9'--------INDUCTOR    `19'---------STORE"
280 PRINT "`10'-------LINE        `20'---------RETRIEVE"
317 RETURN
318 P5=0
319 U7=0
320 U1=0
330 D9=0
340 DIM P9(100),X1(100),Y1(100),X2(100),Y2(100)
350 GOSUB 4930
360 GO TO 3
370 REM * THIS SUBROUTINE PLOTS A CAPACITOR   *
440 CHARSIZE 1
450 J=6
460 IF D9(>1 THEN 490
470 U1=U3
480 GO TO 500
490 GOSUB 5070
500 GOSUB 4720
510 ROTATE T1
520 MOVE X1(U1),Y1(U1)
530 RDRAW L*0.4,0
540 RMOVE 0,-L*0.15
550 RDRAW 0,0.3*L
560 RMOVE 0.2*L,-0.15*L
570 RDRAW 0.4*L,0
580 RMOVE -0.1*L,0
590 RMOVE 0.3*L*COS(151),0.3*L*SIN(151)
600 FOR T2=151 TO 209 STEP 2
```

```
610 X3=0.3*L*(-COS(T2)+COS(T2+2))
620 Y3=0.3*L*(-SIN(T2)+SIN(T2+2))
630 RDRAW X3,Y3
640 NEXT T2
710 RETURN

1280 REM * THIS SUBROUTINE PLOTS A GROUND *
1290 CHARSIZE 1
1350 J=7
1370 IF D9<>1 THEN 1400
1380 U1=U3
1390 GO TO 1410
1400 GOSUB 5070
1410 REM! PAGE
1420 GOSUB 4720
1430 ROTATE T1
1440 MOVE X1(U1),Y1(U1)
1450 RMOVE 0,-L/(12*3^0.5)
1460 RDRAW 0,L/(6*3^0.5)
1470 FOR J7=1 TO 3
1480 MOVE X1(U1),Y1(U1)
1490 RMOVE J7*L/6,-J7*L/(6*3^0.5)
1500 RDRAW 0,J7*L/(3*3^0.5)
1510 NEXT J7
1520 MOVE X2(U1),Y2(U1)
1530 RDRAW -L/2,0
1550 RETURN
1570 REM * THIS SUBROUTINE PLOTS A RESISTOR *
1580 CHARSIZE 4
1630 J=8
1640 CHARSIZE 1
1660 IF D9<>1 THEN 1690
1670 U1=U3
1680 GO TO 1700
1690 GOSUB 5070
1700 L=((X1(U1)-X2(U1))^2+(Y1(U1)-Y2(U1))^2)^0.5
1710 IF X1(U1)>X2(U1) AND Y1(U1)=Y2(U1) THEN 1790
1720 IF X1(U1)=X2(U1) AND Y1(U1)>Y2(U1) THEN 1810
1730 IF X1(U1)>X2(U1) AND Y1(U1)>Y2(U1) THEN 1830
1740 IF Y1(U1)=>Y2(U1) THEN 1770
1750 T1=ACS((X2(U1)-X1(U1))/L)
1760 GO TO 1850
1770 T1=ASN((Y2(U1)-Y1(U1))/L)
1780 GO TO 1850
1790 T1=180
1800 GO TO 1850
1810 T1=-90
1820 GO TO 1850
1830 T1=ATN((Y2(U1)-Y1(U1))/(X2(U1)-X1(U1)))
1840 T1=T1+180
1850 MOVE X1(U1),Y1(U1)
1860 ROTATE T1
1870 RDRAW L/5,0
1880 RDRAW L/20,-L/10
1890 RDRAW L/10,L/5
1900 RDRAW L/10,-L/5
1910 RDRAW L/10,L/5
1920 RDRAW L/10,-L/5
1930 RDRAW L/10,L/5
1940 RDRAW L/20,-L/10
1950 RDRAW L/5,0
```

```
1980 RETURN
2000 REM * THIS SUBROUTINE PLOTS AN INDUCTOR *

2060 J=9
2070 IF D9<>1 THEN 2100
2080 U1=U3
2090 GO TO 2120
2100 GOSUB 5070
2120 GOSUB 4720
2130 MOVE X1(U1),Y1(U1)
2140 ROTATE T1
2150 RMOVE L/3,0
2160 ROTATE T1+90
2170 R7=L/6
2180 GOSUB 7610
2190 ROTATE T1
2200 MOVE X1(U1),Y1(U1)
2210 RMOVE L/3*2,0
2220 ROTATE T1+90
2230 GOSUB 7610
2240 MOVE X2(U1),Y2(U1)
2250 GOSUB 7610
2450 RETURN
2470 REM * THIS SUBROUTINE PLOTS A LINE *
2520 J=10
2530 IF D9<>1 THEN 2560
2540 U1=U3
2550 GO TO 2570
2560 GOSUB 5070
2570 MOVE X1(U1),Y1(U1)
2580 DRAW X2(U1),Y2(U1)
2600 RETURN
2620 REM * THIS SUBROUTINE PLOTS A BATTERY *
2670 J=4
2680 GOSUB 9000
2690 MOVE X1(U1),Y1(U1)
2700 ROTATE T1
2710 RMOVE 0,-0.3*L
2720 RDRAW 0,0.6*L
2730 RMOVE L/3,-0.3*L
2740 RMOVE 0,-0.075*L
2750 RDRAW 0,0.15*L
2760 RMOVE L/3,-0.075*L-0.3*L
2770 RDRAW 0,0.6*L
2780 MOVE X2(U1),Y2(U1)
2790 RMOVE 0,-0.075*L
2800 RDRAW 0,0.15*L

3260 RETURN
3280 REM * THIS SUBROUTINE PLOTS AN OSCILLATOR *
3330 J=16
3340 REM! PAGE
3350 CHARSIZE 1
3370 IF D9<>1 THEN 3400
3380 U1=U3
3390 GO TO 3410
3400 GOSUB 5070
3410 L=((X1(U1)-X2(U1))^2+(Y1(U1)-Y2(U1))^2)^0.5
3420 IF X1(U1)>X2(U1) AND Y1(U1)=Y2(U1) THEN 3500
3430 IF X1(U1)=X2(U1) AND Y1(U1)>Y2(U1) THEN 3520
3440 IF X1(U1)>X2(U1) AND Y1(U1)>Y2(U1) THEN 3540
```

```
3450 IF Y1(U1)=>Y2(U1) THEN 3480
3460 T1=ACS((X2(U1)-X1(U1))/L)
3470 GO TO 3560
3480 T1=ASN((Y2(U1)-Y1(U1))/L)
3490 GO TO 3560
3500 T1=180
3510 GO TO 3560
3520 T1=-90
3530 GO TO 3560
3540 T1=ATN((Y2(U1)-Y1(U1))/(X2(U1)-X1(U1)))
3550 T1=T1+180
3560 MOVE X1(U1),Y1(U1)
3565 ROTATE T1
3590 RMOVE 1*L,0
3595 R7=L*0.5
3600 GOSUB 7000
3650 MOVE X1(U1),Y1(U1)
3655 RMOVE 0.5*L,0
3660 ROTATE 90+T1
3671 R7=4/20*L
3673 GOSUB 7610
3678 MOVE X2(U1),Y2(U1)
3679 ROTATE T1
3680 RMOVE -L*0.5,0
3685 ROTATE 270+T1
3690 R7=4*L/20
3700 GOSUB 7610
3730 RETURN
3750 REM * PROGRAM ENDS *
3800 G9=8
3810 RINIT
3820 GOSUB 4530
3830 REM INIT
3850 MOVE 0,10
3860 PRINT "**  DESIGN ENDS  **"
3870 END
3880 REM * THIS SUBROUTINE PLOTS A RECTANGLE *
3930 J=11
3940 GOSUB 9000

4020 MOVE X1(U1),Y1(U1)
4030 DRAW X2(U1),Y1(U1)
4040 DRAW X2(U1),Y2(U1)
4050 DRAW X1(U1),Y2(U1)
4060 DRAW X1(U1),Y1(U1)
4080 RETURN
4100 REM * THIS SUBROUTINE REVISES THE LAST DRAWN ELEMENT *
4150 RINIT
4160 MOVE 0,80
4180 MOVE 0,20
4200 U2=U1
4210 PRINT "INPUT THE NEW KEY NUMBER USING "
4211 PRINT " NUMERIC KEYS ; NOT DEFINABLE KEY!!!"
4220 INPUT P9(U2)
4230 PRINT "ADJUST THE THUMBWHEELS TWICE!!!"
4240 PRINT " TO INPUT THE BEGINNING AND"
4250 PRINT "  ENDING POINTS OF THE ELEMENT"
4260 POINTER X1(U2),Y1(U2),C$
4265 POINTER X2(U2),Y2(U2),C$
4280 D9=1
4290 PAGE
4300 FOR U3=1 TO U1
```

```
4310 GOS P9(U3) OF 3750,6250,6000,2620,8000,370,1280,1570,2000,2470,3880
4315 GOSUB P9(U3)-11 OF 4100,8700,7500,6600,3280,8500,7200
4320 NEXT U3
4330 D9=0
4340 MOVE 0,98.3
4342 GOSUB 200
4468 MOVE 0,70
4470 ROPEN 5
4480 PRINT "!!!!!!!!!!!!!!!!!!!!!!!!!!!!!!!!!!!!!!!!!!!!!"
4490 PRINT "!!  ENTER ANOTHER NEW KEY#          !!"
4500 PRINT "!!!!!!!!!!!!!!!!!!!!!!!!!!!!!!!!!!!!!!!!!!!!!"
4510 RCLOSE
4520 GO TO 3
4530 REM * THIS SUBROUTINE PRINTS OUT THE ELEMENT COORDIANTES *
4580 MOVE 0,70
4590 PRINT "STEP# KEY#  BEGINNING        ENDING "
4600 PRINT "               X      Y       X      Y"
4610 FOR I5=1 TO U1
4620 PRINT USING 4630:I5,P9(I5),X1(I5),Y1(I5),X2(I5),Y2(I5)
4630 IMAGE 2D,4X,2D,2X,4(3D,1D,2X)
4640 NEXT I5
4710 RETURN
4720 REM * THIS SUB FINDS THE LENGTH AND ROTATION ANGLE *
4770 L=((X1(U1)-X2(U1))^2+(Y1(U1)-Y2(U1))^2)^0.5
4780 IF X1(U1)>X2(U1) AND Y1(U1)=Y2(U1) THEN 4860
4790 IF X1(U1)=X2(U1) AND Y1(U1)>Y2(U1) THEN 4880
4800 IF X1(U1)>X2(U1) AND Y1(U1)>Y2(U1) THEN 4900
4810 IF Y1(U1)=>Y2(U1) THEN 4840
4820 T1=ACS((X2(U1)-X1(U1))/L)
4830 GO TO 4920
4840 T1=ASN((Y2(U1)-Y1(U1))/L)
4850 GO TO 4920
4860 T1=180
4870 GO TO 4920
4880 T1=-90
4890 GO TO 4920
4900 T1=ATN((Y2(U1)-Y1(U1))/(X2(U1)-X1(U1)))
4910 T1=T1+180
4920 RETURN
4930 REM * THIS SUBROUTINE INSTRUCTS THE USER *
4990 ROPEN 7
5000 IF U1<>0 THEN 5060
5010 MOVE 0,75
5020 PRINT "******************************************"
5030 PRINT "**          ENTER FIRST KEY#            **"
5040 PRINT "******************************************"
5050 RCLOSE
5060 RETURN
5070 REM * INPUT THE BEGINNING AND ENDING POINTS *
5140 P5=P5+1
5145 U1=P5+U7
5150 P9(U1)=J
5160 RINIT
5170 MOVE 0,65
5180 ROPEN 1
5190 PRINT "KEY# CHOSEN: ";J
5200 PRINT "ADJUST THE THUMBWHEELS"
5210 PRINT " TO INPUT THE BEGINNING"
5220 PRINT " POINT WHEN READY PRESS"
5230 PRINT " 'RETURN' "
```

```
5240 RCLOSE
5250 IF J=7 THEN 5280
5260 POINTER X1(U1),Y1(U1),C$
5270 GO TO 5290
5280 POINTER X2(U1),Y2(U1),C$
5290 MOVE 0,53
5300 PRINT " "
5310 ROPEN 2
5320 PRINT "COORDINATES OF THE BEGINNING POINT"
5325 IF J=7 THEN 5343
5330 PRINT USING 5390;"X=   ",X1(U1)
5340 PRINT USING 5390;"Y=   ",Y1(U1)
5341 GO TO 5350
5343 PRINT USING 5390;"X=   ",X2(U1)
5345 PRINT USING 5390;"Y=   ",Y2(U1)
5350 PRINT "ADJUST THE THUMBWHEELS"
5360 PRINT " TO INPUT THE ENDING"
5370 PRINT " POINT WHEN READY PRESS"
5380 PRINT " 'RETURN' "
5390 IMAGE 4A,5T,3D.1D
5400 RCLOSE
5410 IF J=7 THEN 5440
5420 POINTER X2(U1),Y2(U1),C$
5430 GO TO 5450
5440 POINTER X1(U1),Y1(U1),C$
5450 MOVE 0,36.5
5460 PRINT " "
5470 ROPEN 3
5480 PRINT "COORDINATES OF THE ENDING POINT"
5485 IF J=7 THEN 5503
5490 PRINT USING 5390;"X=   ",X2(U1)
5500 PRINT USING 5390;"Y=   ",Y2(U1)
5501 GO TO 5510
5503 PRINT USING 5390;"X=   ",X1(U1)
5505 PRINT USING 5390;"Y=   ",Y1(U1)
5510 RCLOSE
5520 ROPEN 8
5530 MOVE 0,25
5540 PRINT "!!!!!!!!!!!!!!!!!!!!!!!!!!!!!!!!!!!!!"
5550 PRINT "!!!  ENTER ANOTHER NEW KEY#  !!!"
5560 PRINT "!!!!!!!!!!!!!!!!!!!!!!!!!!!!!!!!!!!!!"
5570 RCLOSE
5580 RETURN
5590 REM ********************************************************************
5600 REM **           U1 IS COUNTER  ;   P9 STORE KEY #              **
5610 REM **          D9 IS INDENTIFICATION FOR WHICH CALLING         **
5620 REM **   G9 IS THE IDENTIFIER TO STORE THE DATA OR NOT          **
5630 REM **          R7 IS THE RADIUS ;                              **
6000 REM * THIS SUBROUTINE PLOTS A SWITCH *

6050 J=3
6060 IF D9()1 THEN 6100
6070 U1=U3
6080 GO TO 6110
6100 GOSUB 5070
6110 GOSUB 4720
6120 MOVE X1(U1),Y1(U1)
6130 ROTATE T1
6140 RDRAW L/3,0
6150 RDRAW L/3.5,L/3.5
6160 MOVE X2(U1),Y2(U1)
```

```
6170 RDRAW -L/3,0
6190 RETURN
6250 REM * THIS SUBROUTINE PLOTS A CIRCLE *
6300 J=2
6310 IF D9<>1 THEN 6340
6320 U1=U3
6330 GO TO 6350
6340 GOSUB 5070
6350 GOSUB 4720
6360 R7=0.5*L
6370 ROTATE T1
6380 MOVE X2(U1),Y2(U1)
6390 GOSUB 7000
6410 RETURN
6600 REM * THIS SUBROUTINE PLOTS A DASH LINE *
6650 J=15
6660 GOSUB 9000
6670 MOVE X1(U1),Y1(U1)
6680 ROTATE T1
6690 RDRAW L/7,0
6700 RMOVE L/7,0
6710 RDRAW L/7,0
6720 RMOVE L/7,0
6730 RDRAW L/7,0
6740 RMOVE L/7,0
6750 DRAW X2(U1),Y2(U1)
6790 RETURN
7000 REM * THIS SUBROUTINE PLOTS A CIRCLE *

7050 FOR D5=0 TO 357 STEP 3
7060 RDRAW R7*(COS(D5+3)-COS(D5)),R7*(SIN(D5+3)-SIN(D5))
7070 NEXT D5
7080 RETURN
7200 REM * THIS SUBROUTINE PLOTS A ARROW *
7250 J=18
7260 GOSUB 9000
7270 MOVE X1(U1),Y1(U1)
7279 L=6
7280 DRAW X2(U1),Y2(U1)
7290 ROTATE T1
7295 IF L<-6 THEN 7300
7300 RMOVE -L/4,L/16
7310 DRAW X2(U1),Y2(U1)
7320 RMOVE -L/4,-L/16
7330 DRAW X2(U1),Y2(U1)
7350 RETURN
7500 REM * THIS SUBROUTINE PLOTS A SEMICIRCLE *
7550 J=14
7560 GOSUB 9000
7570 ROTATE T1
7580 MOVE X1(U1),Y1(U1)
7590 R7=L
7600 RMOVE 0,-L
7605 GOSUB 7610
7606 RETURN
7610 FOR D5=-90 TO 87 STEP 3
7620 RDRAW R7*(COS(D5+3)-COS(D5)),R7*(SIN(D5+3)-SIN(D5))
7630 NEXT D5
7660 RETURN
8000 REM * THIS SUBROUTINE PLOTS A LAMP *
```

```
8050 J=5
8060 GOSUB 9000
8065 ROTATE T1-90
8070 MOVE X1(U1),Y1(U1)
8080 RDRAW 2*L,0
8115 R7=L/2
8120 FOR D5=-90 TO 88 STEP 2
8130 RDRAW R7*(COS(D5+2)-COS(D5)),R7*(SIN(D5+2)-SIN(D5))
8135 NEXT D5
8140 DRAW X2(U1),Y2(U1)
8150 R7=L
8160 RMOVE 3*L,-0.5*L
8170 GOSUB 7000
8190 RETURN
8500 REM * THIS SUBROUTINE PLOTS A DIODE *
8550 J=17
8560 GOSUB 9000
8570 MOVE X1(U1),Y1(U1)
8580 ROTATE T1
8590 RDRAW 0,-L/3^0.5
8600 DRAW X2(U1),Y2(U1)
8610 RDRAW -L,L/3^0.5
8620 DRAW X1(U1),Y1(U1)
8630 MOVE X2(U1),Y2(U1)
8640 RMOVE 0,-L/3^0.5
8645 RDRAW 0,2*L/3^0.5
8660 RETURN
8700 REM * THIS SUBROUTINE PLOTS A BIPOLAR *
8750 J=13
8760 GOSUB 9000
8770 ROTATE T1
8780 MOVE X1(U1),Y1(U1)
8782 RDRAW 0.5*L,0
8784 RMOVE 0,-L*0.18
8786 RDRAW 0,0.07*L
8787 RDRAW 0.23*L,-0.23*L
8788 RMOVE 0.23*-L,0.23*L
8790 RDRAW 0,0.29*L
8792 RMOVE 0,-0.07*L
8795 RDRAW 0.23*L,0.23*L
8800 MOVE X2(U1),Y2(U1)
8805 R7=0.35*L
8810 GOSUB 7000
8840 RETURN
9000 REM * THIS SUBROUTINE CHECKS THE CALLING ROUTINE *

9050 IF D9<>1 THEN 9100
9060 U1=U3
9070 GO TO 9110
9100 GOSUB 5070
9110 GOSUB 4720
9120 RETURN
9490 REM  * THIS SUB STORES THE DATA *
9492 RINIT
9495 KILL 19
9500 FIND 19
9510 WRITE @33:U1
9520 FOR I5=1 TO U1
9530 WRITE @33:X1(I5),Y1(I5),X2(I5),Y2(I5),P9(I5)
```

```
9545 MOVE 0,60
9550 ROPEN 7
9560 PRINT "!!!!!!!!!!!!!!!!!!!!!!!!!!!!!!!!!!!!!!!!!!!!!!!!!!!!!"
9562 PRINT "!!!      DATA HAS BEEN STORED ON FILE 19      !!!"
9564 PRINT "!!!              ENTER ANOTHOR NEW KEY#        !!!"
9565 PRINT "!!!!!!!!!!!!!!!!!!!!!!!!!!!!!!!!!!!!!!!!!!!!!!!!!!!!!"
9570 RCLOSE
9600 RETURN
9700 REM * THIS SUB RETRIEVES THE DATA FROM TAPE *
9705 RINIT
9710 FIND 19
9720 READ @33:U7
9725 FOR I5=1 TO U7
9730 READ @33:X1(I5),Y1(I5),X2(I5),Y2(I5),P9(I5)
9740 NEXT I5
9750 D9=1
9760 FOR U3=1 TO U7
9770 GOS P9(U3) OF 3750,6250,6000,2620,8000,370,1280,1570,2000,2470,388
9780 GOSUB P9(U3)-11 OF 4100,8700,7500,6600,3280,8500,7200
9790 NEXT U3
9800 D9=0
9805 MOVE 0,60
9810 ROPEN 8
9820 PRINT "!!!!!!!!!!!!!!!!!!!!!!!!!!!!!!!!!!!!!!!!!!!!!!!!!!!!"
9823 PRINT "!!!         ENTER ANOTHER NEW KEY#       !!!"
9825 PRINT "!!!!!!!!!!!!!!!!!!!!!!!!!!!!!!!!!!!!!!!!!!!!!!!!!!!!"
9827 RCLOSE
9830 RETURN
```

EXERCISE

Prepare a BASIC program by use of definable keys and tablet to demonstrate the rubber-band effect on screen. Assign a definable key so that when it is pressed, one can change the coordinates of points P_1 and P_2 in drawing lines P_1P_3 and P_2P_3. The coordinates of points P_1, P_2, and P_3 are to be entered by use of the stylus of the tablet.

5.6 MODULES

Program E. MODULE is an example of the development of modules for frequent uses in CAD. In this particular instance, the modules are for drawing the electric elements: resistor, capacitor, inductor, diode, and so on, as listed in the key table of Fig. 5-13. In CAD of circuits, the elements are to be drawn at specified locations and their sizes should be compatible with the neighboring elements. Hence, when these modules are coded as subroutines, they ought to provide such flexibility.

Program E. MODULE provides that each element is requested by pressing a designated definable key making **flashing** messages appear on the screen asking the user to adjust the thumbwheels to enter two desired locations of the cursor. Between these two specified points, the requested element will be drawn. As can be expected, the size of the drawn element will be proportional to the distance between these two points. Quite often, during the course of drawing a complete circuit, one element may not be connected accurately to the neighboring elements and therefore needs to be redrawn. The definable key 12 has been assigned to implement the task of "REVISE" and in order to redraw the complete circuit again, "STORE" and "RETRIEVE" of the drawn elements become necessary. Definable keys 19 and 20, respectively, are thus assigned for such purpose.

Figure 5-14 is presented to illustrate the use of Program E. MODULE for drawing the circuit of a UHF amplifier. There are 81 elements involved, for which the coordinates of the endpoints entered by maneuvering the thumbwheels are as printed. These coordinates will be automatically listed when the definable key 1 is pressed to terminate the program.

In Program E. MODULE, the commands ROTATE, RMOVE, and RDRAW are used in many places. The RMOVE and RDRAW statements have the general forms

RMOVE numeric expression 1, numeric expression 2

RDRAW numeric expression 1, numeric expression 2

They instruct that the cursor be moved to the point without drawing a line, or drawing a line, respectively, from the present position. The destination point has coordinates in **user** data values calculated or specified by the two numeric expressions following the commands RMOVE or RDRAW. RMOVE and RDRAW differ from MOVE and DRAW in that the former pair uses X and Y coordinates relative to the present position of the cursor, whereas the latter pair refers to the origin defined in the program.

ROTATE is to be followed by a rotation angle measured in the current trigonometric units specified by the SET DEGREE or SET RADIAN statement. A ROTATE A1 statement placed in front of a set of RMOVE and RDRAW statements causes the drawing to be rotated by A1 degrees or radians.

Lines 4820, 4840, and 4900 of Program E. MODULE uses the library functions ACS (arc cosine), ASN, (arc sine) and ATN (arc tangent), respectively. Lines 9510 and 9530 write the data onto the peripheral device unit 33, which is the cassette tape, by use of the WRITE command. (READ retrieves the data.)

Since the elements are drawn and added on to the circuit one by one on the screen of the graphics display system, the messages for instructing the user to adjust the thumbwheels cannot be "fixed" but are "flashing" on the screen; otherwise the PAGE key has to be pressed to clear the message and it will erase the circuit drawing as well. The next section discusses how the "flashing" displays can be arranged to appear on the screen and can be erased without affecting the "fixed" display on the screen.

5.7 DYNAMIC GRAPHICS AND ANIMATION

The need often arises in CAD for part of the display on the screen of a graphics terminal to be changed. This calls for keeping the unchanged portion of the display on the screen and redrawing the part that must be changed. These two requirements are referred to as the **store** and **refresh** modes, respectively. In the store mode, images are drawn on the screen that stay until the PAGE key is pressed whereas, in the refresh mode images are drawn on the screen with intensity strong enough to be visible but not stored. For example, the screen cursor that flashes

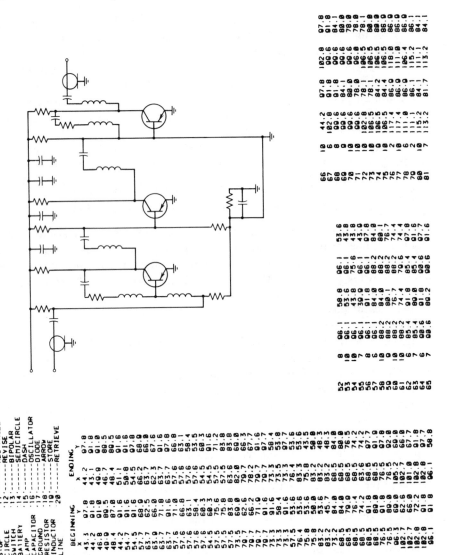

Figure 5-14 UHF amplifier schematic drawn by use of Program E. MODULE.

or blinks is displayed in the refresh mode. Because the refreshed image is not stored, it has to be continually redrawn to remain visible. If the shape and/or position of an image is changed smoothly on the screen, this refresh mode could thus be effectively utilized for **animation** of motions. The visual authenticity of the animated motion depends largely on the **refresh rate** being employed.

Tektronix 4054 computer graphics system can be extended to include an Option 30 Dynamic Graphics Package to provide the refresh mode applications. It will have a refresh rate of 37.5 Hz (cycles per second) and 30K bytes of random-access memory (RAM) for keeping the data in continual redrawing of the refreshed images.

To blink a message on the screen, a refresh file has to be created. Lines 4468–4510 in Program E.MODULE are commands that specify the location where the blinking message is to appear, that a file 5 is to be open on the refresh RAM, what the message is, and finally that file 5 is to be closed. First of all, line 151 RINIT results in erasing, or initializing, all refresh RAM. The ROPEN command creates a new file, and the RCLOSE closes it. Lines 5070–5580 are included in Program E.MODULE for blinking the instructional messages to the user to adjust the thumbwheels so that the location where a requested element is to be drawn can be entered. It further manifests how the refresh RAM can be effectively utilized to display blinking messages.

Program R.SQUARE and Fig. 5-15 are presented to show the use of dynamic graphics for rotation of a square of size 10×10 GDU. Figure 5-15 is the result of requesting the program to display the rotated square in fixed mode. The user may request the refreshed mode and view an animation of a tumbling square across the screen from left to right. Besides the RMOVE, RDRAW, ROTATE, ROPEN, and RCLOSE dynamic graphics commands in Program R.SQUARE, the statements

```
VISIBILITY 1,0
VISIBILITY 1,1
```

are used to instruct, respectively, that first the display to be stored in the refresh RAM file 1 should *not* be visible, when the square is drawn and next that the *completed* display already stored in file 1 is to be made visible on the screen. The first number after the VIS command is the file number in the refresh RAM, and the second number specifies whether the drawing should be invisible if it is equal to zero or visible if it is not equal to zero. In their places, two numeric expressions, in general cases, are allowed after the VIS command.

EXERCISE

Examine the effect on animation of the rotating square by (1) deleting the VIS 1,0 statement, (2) deleting the VIS 1,1 statement, and (3) deleting both VIS statements of Program R.SQUARE. Also, the rotation trajectory may be altered by use of different values for $X1$, increment of X (horizontal translation), and $Z1$, increment of rotation. Experiment with various combinations to study the animated motions.

```
100 PAGE
110 CHARSIZE 4
120 PRINT "* PROGRAM R.SQUARE *"
130 PRINT "_ DEMONSTRATES DYNAMIC GRAPHICS"
140 PRINT "_   OF A ROTATING SQUARE."
150 PRINT "_ INPUT 1/0 FOR FIXED/REFRESHED MODE : ";
160 INPUT M
170 SET DEGREES
180 X=10
190 X1=0.5
200 Y=50
210 Z=5
220 Z1=5
230 IF M=1 THEN 260
240 ROPEN 1
250 VISIBILITY 1,0
260 MOVE X,Y
270 RDRAW 10,0
280 RDRAW 0,10
290 RDRAW -10,0
300 RDRAW 0,-10
310 IF M=1 THEN 340
320 RCLOSE
330 VISIBILITY 1,1
340 X=X+X1
350 ROTATE Z
360 Z=Z+Z1
370 IF Z=720 THEN 390
380 GO TO 230
390 RINIT
400 MOVE 0,10
410 END
```

```
* PROGRAM R.SQUARE *

  DEMONSTRATES DYNAMIC GRAPHICS
     OF A ROTATING SQUARE.

  INPUT 1/0 FOR FIXED/REFRESHED MODE : 1
```

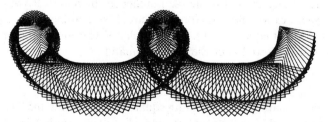

Figure 5-15

CHAPTER 6

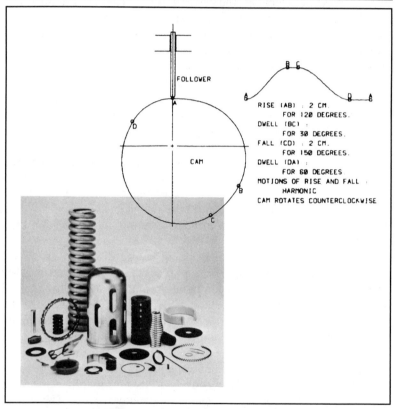

FOLLOWER

CAM

RISE (AB) : 2 CM.
        FOR 120 DEGREES.
DWELL (BC) :
        FOR 30 DEGREES.
FALL (CD) : 2 CM.
        FOR 150 DEGREES.
DWELL (DA) :
        FOR 60 DEGREES
MOTIONS OF RISE AND FALL :
        HARMONIC
CAM ROTATES COUNTERCLOCKWISE

SIMPLE EXAMPLES
OF COMPUTER-AIDED
DRAFTING, DESIGN,
AND ANALYSIS

6.1   INTRODUCTION

In this chapter, interactive graphics techniques are employed for a number of simple examples to illustrate the varied applications of automated drafting in trial designs. Inevitably, equations are involved for deciding on the geometric configuration of a design before it can be displayed for visual inspection, and mathematical analysis needs to be done to assess the design. The detailed explanation of mathematical modeling and analysis of engineering systems are covered in Part Two on CAD/CAM of elements and systems. In this chapter a number of simple examples that do not need elaborate analysis are presented.

Pairing of gears and design of cam profiles are examples of the transmission of rotation and the transformation of rotation to translation. Design of coil springs involves the relationship between force and deformation of elastic bodies. To investigate the strength of selected machine and structural members when they are subjected to loadings, calculations of the cross-sectional area, the location of its centroid, and the moments of inertia are often required. BASIC programs for these studies have been developed; their applications will be discussed.

The Mohr's Circle technique not only helps to **graphically** determine the strongest and weakest directions along which a designed machine or structural member can be subjected to external loads, it can also be applied for calculation of the eigenvalues. A BASIC program has been made available. The lowest resonance frequency of a designed structure, the lowest buckling load of a slender machine member, and the maximum principal stress in a cross section can all be investigated on a graphics display terminal by use of this program.

The basic need of interactive dimensioning is also discussed in this chapter. How to copy a drawing by placing it on a graphics tablet and then using the stylus to transfer all information into the computer graphics unit for display is demonstrated as well. The concept of developing the modules, such as E. MODULE in Chapter 5, for the constant drafting needs of the often used engineering components is further elaborated in discussion of the P. MODULE program prepared for automated layout of pipelines in design of chemical processes.

Kinematics of a crank-slider mechanism is included as an example to demonstrate how the visual display can be effectively utilized for graphical determination of the velocity and acceleration of machine or structural parts. The Wheatstone bridge program, which serves well as an introduction to electric circuit designs, is used as a closing example.

Without explaining the characteristics of the constituent elements of engineering systems, it is indeed awkward in citing various CAD examples to just mention how the designs should be done. However, exposure to these practical cases makes readers aware of the importance of a thorough comprehension of modeling of elements and of design and analysis of systems in CAD/CAM applications. It points to the need of going into the materials covered in Part Two in order to really be involved in full-fledged CAD/CAM studies.

6.2 PAIRING OF GEARS

Gears are used for transmitting torques. When they are paired, reversed rotations and speed changes can be achieved. Applications of gear trains are commonplace, ranging from the smallest in wristwatches to the largest in telescope control systems. In Chapter 4 the plotting/display of the gear tooth was delineated. Here, the problem of several gears coming in contact to form a gear train is to be discussed.

The distance between the centers of the two gears in contact is called **center distance**, C. The contact circle is called the **pitch circle**. If the diameters of the pitch circles of the paired gears are D_1 and D_2, then it is clear that the pitch diameters and the center distance of the two gears are related by the equation

$$C = (D_1 + D_2)/2 \qquad\qquad\text{(a)}$$

The number of teeth per unit of pitch diameter is called **diametral pitch**, or **pitch** p, The pitch can be related to the total number of teeth N as

$$p = N/D \qquad\qquad\text{(b)}$$

In order that the gears are properly meshed, they all must have the same pitch. This condition leads to the equations relating the angular velocities, ω's, N's, and D's of two paired gears. The equations are

$$\omega_1/\omega_2 = D_2/D_1 = N_2/N_1 \qquad\qquad\text{(c)}$$

The ω–D relationship is derived from the fact that the linear velocity, V, at the contact point should be the same for both gears and $V = \omega_1 D_1/2 = \omega_2 D_2/2$. Also, at the contact point on the pitch circle, the angle ϕ of the common tangent to the tooth profiles of both gears measured from the line perpendicular to the line of centers is called the **pressure angle**, which can be written in terms of D and the diameter of the base circle D_b as

$$\phi = \cos^{-1}(D_b/D) \qquad\qquad\text{(d)}$$

Two most commonly used pressure angles are $14.5°$ and $20°$. Once ϕ is decided, the base diameter D_b can then be calculated with Eq. (d) and consequently the involute or cycloidal tooth profile of the gear can be constructed.

Depending on the maximum allowable stress σ_a of the gear material, the maximum horsepower that can be transmitted by a pair of gears can be computed with the equation

$$\text{(H.P.)}_{max} = FV/33{,}000 \qquad\qquad\text{(e)}$$

where the maximum force F allowed at the pitch contact is

$$F = \sigma_a t f \pi/p \qquad\qquad\text{(f)}$$

In Eq. (f), t is the thickness (**face width**) of the gear and f is a so-called form factor, which is selected from Table 6-1 according to the number of teeth N and the pressure angle ϕ.

TABLE 6-1
Form Factor *f*

Number of Teeth N	14.5° Involute	20° Full-depth Involute	20° Stub Involute
12	0.067	0.078	0.099
13	0.071	0.083	0.103
14	0.075	0.088	0.108
15	0.078	0.092	0.111
16	0.081	0.094	0.115
17	0.084	0.096	0.117
18	0.086	0.098	0.120
19	0.088	0.100	0.123
20	0.090	0.102	0.125
21	0.092	0.104	0.127
23	0.094	0.106	0.130
25	0.097	0.108	0.133
27	0.099	0.111	0.136
30	0.101	0.114	0.139
34	0.104	0.118	0.142
38	0.106	0.122	0.145
43	0.108	0.126	0.147
50	0.110	0.130	0.151
60	0.113	0.134	0.154
75	0.115	0.138	0.158
100	0.117	0.142	0.161
150	0.119	0.146	0.165
300	0.122	0.150	0.170
∞ (Rack)	0.124	0.154	0.175

Program GEAR.PAIR has been prepared for the trial designs of gears. Data in Table 6-1 are built in so that the user-specified N and ϕ values enable the proper form factor f value to be automatically selected for calculation of the maximum horsepower. Figure 6-1 is a simple application of Program GEAR.PAIR.

PROGRAM GEAR.PAIR

```
100 REM * PROGRAM GEAR.PAIR *
110 INIT
120 CHARSIZE 2
130 SET DEGREES
140 DIM N(6),D(6),F1(24,4)
150 PAGE
160 PRINT "* PROGRAM GEAR.PAIR *"
170 PRINT "__THIS PROGRAM CALCULATES THE"
180 PRINT " MAXIMUM HORSEPOWER AND DISPLAY"
190 PRINT " OF A PAIR OF GEARS."
200 PRINT "_GIVEN DATA SHOULD INCLUDE THE GEAR"
210 PRINT " PITCH, CENTER DISTANCE AND RPM'S"
220 PRINT " OF THE TWO GEARS PAIRED."
230 PRINT "__THIS PROGRAM IS RESTRICTED TO"
```

PROGRAM GEAR. PAIR

THIS PROGRAM CALCULATES THE
  MAXIMUM HORSEPOWER AND DISPLAY
  OF A PAIR OF GEARS.

GIVEN DATA SHOULD INCLUDE THE GEAR
  PITCH, CENTER DISTANCE AND RPM'S
  OF THE TWO GEARS PAIRED.

THIS PROGRAM IS RESTRICTED TO
  THE FOLLOWING 3 GEAR TYPES:

  1 : 4.5 DEG. FULL-DEPTH INVOLUTE.
  2   20   DEG. FULL-DEPTH INVOLUTE.
  3   20   DEG. STUB INVOLUTE.

ALL DIMENSIONS ARE IN INCHES GEAR.

CENTER DISTANCE = 7.5
SHAFT NO. 1 RPM = 500
SHAFT NO. 2 RPM = 250
GEAR PITCH = 3
GEAR TYPE NO. = 1
ALLOWABLE STRESS FOR
  GEAR PAIR. IN PSI = 12000

MAXIMUM RECOMMENDED GEAR WIDTH = 4.19
DESIRED GEAR WIDTH = 1.5

MAXIMUM HORSEPOWER = 13.9467024618

TO DISPLAY GEARS. ENTER D. OR.
  TO RERUN PROGRAM. ENTER R: N

FOR GEAR NO. 1:
  PITCH DIA. = 5
  NO. OF TEETH = 15

FOR GEAR NO. 2:
  PITCH DIA. = 10
  NO. OF TEETH = 30

COMPUTING! DO NOT INTERRUPT!!

* END OF DESIGN *

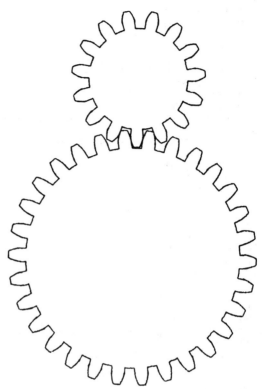

Figure 6-1   Screen menu and resulting display
of a sample application of Program GEAR. PAIR.

```
240 PRINT " THE FOLLOWING 3 GEAR TYPES :_"
250 PRINT " 1. 14.5 DEG. FULL-DEPTH INVOLUTE."
260 PRINT " 2. 20   DEG. FULL-DEPTH INVOLUTE."
270 PRINT " 3. 20   DEG. STUB INVOLUTE GEAR."
280 PRINT "_ALL DIMENSIONS ARE IN INCHES."
290 PRINT "_CENTER DISTANCE = ";
300 INPUT C
310 PRINT "SHAFT NO. 1 RPM = ";
320 INPUT G2
330 PRINT "SHAFT NO. 2 RPM = ";
340 INPUT G1
350 PRINT "GEAR PITCH = ";
360 INPUT P
370 N(2)=2*P*C*G2/(G1+G2)
380 H=INT(N(2))
390 REM  IF N(2)-H=0 THEN 540
400 GO TO 480
410 PRINT "_GEAR PITCH GIVEN WILL NOT WORK FOR THE CENTER"
420 PRINT "DISTANCE AND SHAFT RPM'S GIVEN."
430 PRINT "_WOULD YOU LIKE TO TRY AGAIN WITH DIFFERENT"
440 PRINT "DATA VALUES. ENTER 1 FOR YES, 2 FOR NO."
450 INPUT A
```

```
460 IF A=1 THEN 150
470 IF A)1 THEN 1700
480 PRINT 'GEAR TYPE NO. = ';
490 INPUT T
500 PRINT 'ALLOWABLE STRESS FOR'
510 PRINT ' GEAR PAIR, IN PSI = ';
520 INPUT S1
530 PRINT
540 B1=4*PI/P
550 PRINT USING 560:'MAXIMUM RECOMMENDED GEAR WIDTH = ';B1
560 IMAGE 33A,2D.2D
570 PRINT 'DESIRED GEAR WIDTH = ';
580 INPUT B
590 IF T()1 THEN 630
600 A5=14.5
610 T1=1/P
620 GO TO 690
630 IF T()2 THEN 670
640 A5=20
650 T1=1/P
660 GO TO 690
670 A5=20
680 T1=0.8/P
690 N(1)=G1/G2*N(2)
700 D(1)=N(1)/P
710 D(2)=N(2)/P
720 GOSUB 1690
730 PRINT '__COMPUTING!  DO NOT INTERRUPT!!'
740 C2=9
750 VIEWPORT 40,130,10,100
760 WINDOW 0,18,0,18
770 FOR F=1 TO 2
780 N1=N(F)
790 D1=D(F)
800 GO TO F OF 810,840
810 C1=D(1)/2+1/P+1
820 A8=-90
830 GO TO 860
840 C1=C1+C
850 A8=90-180/N(F)
860 GOSUB 920
870 NEXT F
880 INIT
890 MOVE 0,0
900 PRINT '* END OF DESIGN *'
910 END
920 REM * SUBROUTINE DRAWS SPUR GEAR.
930 REM   SUBROUTINE REQUIRES VALUES FOR P, N1, AND A5 WHERE
940 REM   P=PITCH, N1=NO. OF GEARTEETH, A5=PRESSURE ANGLE.
950 R2=D1/2
960 R1=R2*COS(A5)
970 R3=R2-(1.01+A5/100)/P
980 REM   LOCATING START OF FIRST GEARTOOTH ON BASE CIRCLE IN DEGREES
990 REM   FROM X-AXIS
1000 A6=R2*SIN(A5)/(R1*0.017453)-A5

1010 A1=90+360/(N1*4)+A6
1020 A3=1
1030 DIM X1(100),Y1(100),X2(100),Y2(100),X3(100),X4(100),Y4(100)
1040 DIM X5(100),Y5(100),X6(100),Y6(100),X7(100),Y7(100)
1050 X1=0
```

```
1060 Y1=0
1070 X2=0
1080 Y2=0
1090 X3=0
1100 X4=0
1110 Y4=0
1120 X5=0
1130 Y5=0
1140 X6=0
1150 X7=0
1160 Y7=0
1170 FOR J=1 TO 100
1180 A4=A1-J*A3
1190 X1(J)=R1*COS(A4)
1200 Y1(J)=R1*SIN(A4)
1210 L2=0.017453*J*A3*R1
1220 X2(J)=X1(J)-L2*COS(A4-90)
1230 Y2(J)=Y1(J)-L2*SIN(A4-90)
1240 X3(J)=-X2(J)
1250 X1(J)=X1(J)+C1
1260 Y1(J)=Y1(J)+C2
1270 X2(J)=X2(J)+C1
1280 Y2(J)=Y2(J)+C2
1290 X3(J)=X3(J)+C1
1300 IF Y2(J)>R2+T1+C2 THEN 1320
1310 NEXT J
1320 Q=360/(N1*2)+2*A6
1330 Q=INT(Q)
1340 FOR K=1 TO Q
1350 X7(K)=R3*COS(A1+K-1)+C1
1360 Y7(K)=R3*SIN(A1+K-1)+C2
1370 NEXT K
1380 X7(Q+1)=R1*COS(A1+Q-2*A6)+C1
1390 Y7(Q+1)=R1*SIN(A1+Q-2*A6)+C2
1400 FOR K=1 TO N1
1410 C4=360/N1
1420 C3=K*C4+A8
1430 FOR M1=1 TO Q+1
1440 X4(M1)=COS(C3)*(X7(M1)-C1)-SIN(C3)*(Y7(M1)-C2)+C1
1450 Y4(M1)=SIN(C3)*(X7(M1)-C1)+COS(C3)*(Y7(M1)-C2)+C2
1460 NEXT M1
1470 FOR M=1 TO J
1480 X5(M)=COS(C3)*(X2(M)-C1)-SIN(C3)*(Y2(M)-C2)+C1
1490 Y5(M)=SIN(C3)*(X2(M)-C1)+COS(C3)*(Y2(M)-C2)+C2
1500 X6(M)=COS(C3)*(X3(M)-C1)-SIN(C3)*(Y2(M)-C2)+C1
1510 Y6(M)=SIN(C3)*(X3(M)-C1)+COS(C3)*(Y2(M)-C2)+C2
1520 NEXT M
1530 MOVE X5(1),Y5(1)
1540 FOR L=1 TO Q+1
1550 DRAW X4(L),Y4(L)
1560 NEXT L
1570 MOVE X5(1),Y5(1)
1580 FOR L=1 TO J
1590 DRAW X5(L),Y5(L)
1600 NEXT L
1610 MOVE X6(1),Y6(1)
1620 FOR L=1 TO J
1630 DRAW X6(L),Y6(L)
1640 NEXT L
1650 MOVE X5(J),Y5(J)
1660 DRAW X6(J),Y6(J)
1670 NEXT K
```

```
1680 RETURN
1690 REM   * SUBROUTINE FOR HORSEPOWER CALCULATION *
1700 READ F1
1710 V=PI*(D(1)/12)*G2
1720 IF V=>2000 THEN 1750
1730 S=S1*(600/(600+V))
1740 GO TO 1840
1750 IF V>4000 THEN 1780
1760 S=S1*(1200/(1200+V))
1770 GO TO 1840
1780 PRINT "_PITCH LINE VELOCITY EXCEEDS PROGRAM LIMIT. IF YOU WISH T(
1790 PRINT "TRY AGAIN WITH A DIFFERENT SET OF RPM'S OR CENTER DISTANCI
1800 PRINT "ENTER Y FOR YES, ENTER N FOR NO AND TO END PROGRAM."
1810 INPUT A$
1820 IF A$="Y" THEN 110
1830 GO TO 1675
1840 IF N(1)>N(2) THEN 1870
1850 N3=N(1)
1860 GO TO 1880
1870 N3=N(2)
1880 J=T+1
1890 FOR I=1 TO 24
1900 IF F1(I,1)=>N3 THEN 1940
1910 NEXT I
1920 F2=F1(I,J)
1930 GO TO 1950
1940 F2=F1(I,J)-(F1(I,1)-N3)/(F1(I,1)-F1(I-1,1))*(F1(I,J)-F1(I-1,J))
1950 W=S*B*F2*PI/P
1960 H1=W*V/33000
1970 PRINT "_MAXIMUM HORSEPOWER = ";H1
1980 PRINT "_TO DISPLAY GEARS, ENTER D, OR,"
1990 PRINT " TO RERUN PROGRAM, ENTER R : ";
2000 INPUT A$

2010 IF A$="D" THEN 2040
2020 IF A$="R" THEN 110
2030 GO TO 1675
2040 PRINT "_FOR GEAR NO. 1 :"
2050 PRINT " PITCH DIA. = ";D(1)
2060 PRINT " NO. OF TEETH = ";N(1)
2070 PRINT "_FOR GEAR NO. 2 :"
2080 PRINT " PITCH DIA. = ";D(2)
2090 PRINT " NO. OF TEETH = ";N(2)
2100 GO TO 730                      .
2110 DATA 12,0.067,0.078,0.099
2120 DATA 13,0.071,0.083,0.103
2130 DATA 14,0.075,0.088,0.108
2140 DATA 15,0.078,0.092,0.111
2150 DATA 16,0.081,0.094,0.115
2160 DATA 17,0.084,0.096,0.117
2170 DATA 18,0.086,0.098,0.12
2180 DATA 19,0.088,0.1,0.123
2190 DATA 20,0.09,0.102,0.125
2200 DATA 21,0.092,0.104,0.127
2210 DATA 23,0.094,0.106,0.13
2220 DATA 25,0.097,0.108,0.133
2230 DATA 27,0.099,0.111,0.136
2240 DATA 30,0.101,0.114,0.139
2250 DATA 34,0.104,0.118,0.142
2260 DATA 38,0.106,0.122,0.145
2270 DATA 43,0.108,0.126,0.147
```

```
2280 DATA 50,0.11,0.13,0.151
2290 DATA 60,0.113,0.134,0.154
2300 DATA 75,0.115,0.138,0.158
2310 DATA 100,0.117,0.142,0.161
2320 DATA 150,0.119,0.146,0.165
2330 DATA 300,0.122,0.15,0.17
2340 DATA 400,0.124,0.154,0.175
```

EXERCISE

Refine Program GEAR.PAIR by incorporating subroutine CENTLN so that the center lines can be drawn for the two gears based on the specified center distance. For further refinement, draw the shaft and keyway for both gears. Experiment with various applications of Program GEAR.PAIR for different combinations of RPM ratio and center distance, and with different gear tooth profiles.

6.3 DESIGN OF CAM PROFILES

Cams are commonly used for transforming one type of motion into another. For instance, the automobile shaft aligned with cams transforms rotation into the translation that controls the opening and closing of the intake and exhaust valves of the cylinders (Fig. 6-2). Depending on the desired motions of the follower, various cam profiles need to be designed. Uniform, parabolic, harmonic, and cycloidal motions of the follower are often considered for providing different time

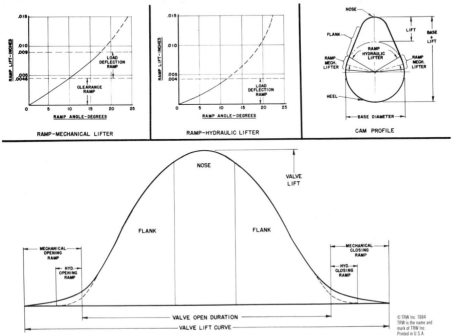

Figure 6-2 Cam Nomenclature. (Courtesy of TRW, Inc.)

rates of change in displacement, velocity, and acceleration. The time rate of change of acceleration is called **jerk**. How to avoid infinite acceleration and jerk is a principal concern in choosing a particular motion of the follower.

Let s, v, and a be the displacement, velocity, and acceleration of the follower, respectively. And let θ and ω be the angular displacement and velocity of the cam shaft, respectively. If the shaft is rotating at a constant angular velocity ω_c, the aim in design of a cam profile is to determine the relationship $r(\theta)$ for one complete revolution of the cam, that is, for $0 \le \theta \le 360°$. Usually, the requirement for the follower's motion, $s(\theta)$, is prescribed in terms of the durations of **rise**, **fall**, and **dwell**, and the type of rise and fall motions desired. This is illustrated by the following example. Suppose that the follower is to rise by 2 cm with harmonic motion in 120° of cam rotation and stay there (dwell) for 30° of cam rotation; it is then to fall 2 cm also with harmonic motion in 150° of cam rotation and stay there for 60° of cam rotation. Figure 6-3 plots $s(\theta)$ of the follower's displacements for a complete revolution of the cam. The desired harmonic motion can be described as

Rise: $s = (H/v)\{1 - \cos[(\theta - \theta_{ro})\pi/(\theta_{r1} - \theta_{ro})]\}$ for $\theta_{ro} \le \theta \le \theta_{r1}$

Dwell: $s = H$ for $\theta_{r1} \le \theta \le \theta_{fo}$

Fall: $s = (H/v)\{1 + \cos[(\theta - \theta_{fo})\pi/\theta_{f1} - \theta_{fo})]\}$ for $\theta_{fo} \le \theta \le \theta_{f1}$

Dwell: $s = 0$ for $\theta_{f1} \le \theta \le \theta_{ro} + 360°$

where H is the maximum amount of rise and the subscripts $r, f, 0$, and 1 denote rise, fall, beginning, and ending, respectively. For the displacement $s(\theta)$ in Fig. 6-3, $H = 2$ cm, $\theta_{ro} = 0°$, $\theta_{r1} = 120°$, $\theta_{fo} = 150°$, and $\theta_{f1} = 300°$.

Let the coordinate of the axis of the cam shaft be (x_c, y_c) and r_0 be the value of $r(\theta)$ when $\theta = 0°$. It can then be easily shown that

$$r(\theta) = r_0 + s(\theta)$$

and the coordinates of point P on the cam profile for an arbitrary θ are

$$x_p = x_c + r \sin\theta \quad \text{and} \quad y_p = y_c + r \cos\theta$$

Figure 6-3 is a sample application of Program CAM developed for design of cam profiles. This program could be modified for consideration of other types of follower motions besides the presently arranged harmonic motion. Such modifications are left to the readers as exercises.

PROGRAM CAM

```
100 INIT
110 CHARSIZE 2
120 SET DEGREES
130 PAGE
140 PRINT "* PROGRAM CAM *"
210 PRINT "_WANT TO SEE A SAMPLE DESIGN? Y/N : ";
220 INPUT Z$
230 IF Z$="Y" THEN 430
```

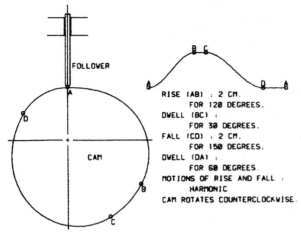

RISE (AB) : 2 CM.
FOR 120 DEGREES.
DWELL (BC) :
FOR 30 DEGREES.
FALL (CD) : 2 CM.
FOR 150 DEGREES.
DWELL (DA) :
FOR 60 DEGREES.
MOTIONS OF RISE AND FALL :
HARMONIC
CAM ROTATES COUNTERCLOCKWISE.

* PROGRAM CAM *

WANT TO SEE SAMPLE DESIGN? Y/N. Y

COMPUTING! DO NOT INTERRUPT!

WANT TO PRINT CAM PROFILE? Y/N:

THETA	R	THETA	R
0	3.00000	5	3.00856
10	3.03407	15	3.07612
20	3.13307	25	3.20665
30	3.29289	35	3.39124
40	3.50000	45	3.61732
50	3.74118	55	3.06947
60	4.90000	65	4.13053
70	4.25882	75	4.38268
80	4.50000	85	4.68876
90	4.79711	95	4.79335
100	4.86603	105	4.92388
110	4.96593	115	4.99144
120	5.00000	125	5.00000
130	5.00000	135	5.00000
140	5.00000	145	5.00000
150	5.00000	155	4.99452
160	4.97815	165	4.95106
170	4.91355	175	4.86693
180	4.88992	185	4.74314
190	4.66013	195	4.58779
200	4.50000	205	4.48674
210	4.30902	215	4.20791
220	4.18453	225	4.00000
230	3.89547	235	3.79289
240	3.69098	245	3.59326
250	3.50000	255	3.41221
260	3.33087	265	3.25686
270	3.19098	275	3.13397
280	3.08645	285	3.04894
290	3.92185	295	3.00548
300	3.00000	305	3.00000
310	3.00000	315	3.00000
320	3.00000	325	3.00000
330	3.00000	335	3.00000
340	3.00000	345	3.00000
350	3.00000	355	3.00000

DESIGN ENDS!!

Figure 6-3   CAD of cam profile.

```
240 PRINT "_INPUT THE FOLLOWING INFORMATION :_"
260 PRINT "  ALL LENGTHS SHOULD BE IN CM."
270 PRINT "  ALL ANGLES SHOULD BE IN DEGREES."
290 PRINT "_COORDINATES OF CAM AXIS=? : ";
300 INPUT C1,C2
310 PRINT "_RISE=FALL=? : ";
320 INPUT H
330 PRINT "_BEGINNING AND ENDING ANGLES OF RISE=? : ";
340 INPUT Z0,Z1
350 PRINT "_ENDING ANGLE OF DWELL AFTER RISE=? : ";
360 INPUT Z2
370 PRINT "_ENDING ANGLE OF FALL=? : ";
380 INPUT Z3
390 PRINT "RADIUS OF CAM BASE CIRCLE=?"
400 INPUT R0
410 PRINT "INCREMENT OF CAM ANGLE=?"
420 INPUT I1
430 REM *A SAMPLE DESIGN*
450 PRINT "_COMPUTING! DO NOT INTERRUPT"
460 READ C1,C2,H,Z0,Z1,Z2,Z3,R0,I1
470 DATA 0,0,2,0,120,150,300,3,5
480 N=360/I1+1
490 DIM P1(N),P2(N),S1(N),S2(N),S3(N)
500 K=0
510 X1=C1
520 X2=C1
530 Y1=C2
540 Y2=C2
550 FOR C=Z0 TO 360+Z0 STEP I1
560 IF C=>Z0 AND C<=Z1 THEN 610
570 IF C=>Z1 AND C<=Z2 THEN 630
580 IF C=>Z2 AND C<=Z3 THEN 650
590 S=0
600 GO TO 660
610 S=H/2*(1-COS(180*(C-Z0)/(Z1-Z0)))
620 GO TO 660
630 S=H
640 GO TO 660
650 S=H/2*(1+COS(180*(C-Z2)/(Z3-Z2)))
660 K=K+1
670 S1(K)=C
680 S2(K)=S
690 R=R0+S
700 S3(K)=R
710 P1(K)=C1+R*SIN(C)
720 IF P1(K)<X1 THEN 750
730 IF P1(K)>X2 THEN 770
740 GO TO 780
750 X1=P1(K)
760 GO TO 780
770 X2=P1(K)
780 P2(K)=C2+R*COS(C)
790 IF P2(K)<Y1 THEN 820
800 IF P2(K)>Y2 THEN 840
810 GO TO 850
820 Y1=P2(K)
830 GO TO 850
840 Y2=P2(K)
850 NEXT C
860 D1=X2-X1
870 D2=Y2-Y1
```

```
880 D=D1
890 IF D2<D1 THEN 910
900 D=D2
910 GOSUB 1090
920 WINDOW 0,130,0,100
930 MOVE 0,10
940 PRINT
950 PRINT "WANT TO PRINT CAM PROFILE? Y/N : "
960 INPUT Z$
970 IF Z$="N" THEN 1070
980 PAGE
990 PRINT "THETA       R       THETA        R    "
1000 FOR L9=1 TO N-1 STEP 2

1010 T1=(L9-1)*I1
1020 T2=L9*I1
1030 PRINT USING 1040:T1,S3(L9),T2,S3(L9+1)
1040 IMAGE 2(5D,5D.5D)
1050 NEXT L9
1070 PRINT "_DESIGN ENDS!!"
1080 END
1090 REM *PLOTTING SUBROUTINE*
1100 WINDOW X1-0.5,X1-0.5+2.5*D,Y1-0.5,Y1-0.5+2.5*D
1110 VIEWPORT 51,130,21,100
1120 REM    "* DRAW CAM PROFILE *"
1130 MOVE P1(1),P2(1)
1140 FOR K=2 TO N
1150 DRAW P1(K),P2(K)
1160 NEXT K
1170 REM    "* DRAW RISE & FALL MOTIONS OF FOLLOWER *"
1180 S=D1/360
1190 MOVE S1(1)+X2,R+S2(1)
1200 FOR K=2 TO N
1210 DRAW X2+S1(K)*S,R+S2(K)
1220 NEXT K
1230 REM    "* DRAW CENTER LINE *"
1240 MOVE X1-0.2,C2
1250 DRAW C1-0.3,C2
1260 MOVE C1-0.1,C2
1270 DRAW C1+0.1,C2
1280 MOVE C1+0.3,C2
1290 DRAW X2+0.2,C2
1300 MOVE C1,2*H+R+0.7
1310 DRAW C1,C2+0.3
1320 MOVE C1,C2+0.1
1330 DRAW C1,C2-0.1
1340 MOVE C1,C2-0.3
1350 DRAW C1,Y1-0.2
1360 REM    "* DRAW FOLLOWER *"
1370 MOVE C1,R
1380 T1=0.2*COS(45)
1390 DRAW C1-T1,R+T1
1400 T2=R+2*H+0.2
1410 DRAW C1-T1,T2
1420 DRAW C1+T1,T2
1430 DRAW C1+T1,R+T1
1440 DRAW C1,R
1450 MOVE C1+T1+1.05,R+H*1.5
1460 DRAW C1+T1+0.05,R+H*1.5
1470 DRAW C1+T1+0.05,R+2*H
1480 DRAW C1+T1+1.05,R+2*H
```

```
1490 MOVE C1-T1-1.05,R+H*1.5
1500 DRAW C1-T1-0.05,R+H*1.5
1510 DRAW C1-T1-0.05,R+2*H
1520 DRAW C1-T1-1.05,R+2*H
1530 REM    "* DRAW CAPTIONS *"
1540 T1=X2+0.1*D
1550 T2=T1+0.2*D
1560 MOVE T1,C2+0.8*R
1570 PRINT "RISE (AB) : ";H;" CM."
1580 MOVE T2,C2+0.6*R
1590 PRINT "FOR ";Z1-Z0;" DEGREES."
1600 MOVE T1,C2+0.4*R
1610 PRINT "DWELL (BC) : "
1620 MOVE T2,C2+0.2*R
1630 PRINT "FOR ";Z2-Z1;" DEGREES."
1640 MOVE T1,C2
1650 PRINT "FALL (CD) : ";H;" CM."
1660 MOVE T2,C2-0.2*R
1670 PRINT "FOR ";Z3-Z2;" DEGREES."
1680 MOVE T1,C2-0.4*R
1690 PRINT "DWELL (DA) :"
1700 MOVE C1+0.4*R,C2-0.4*R
1710 PRINT "CAM"
1720 MOVE T2,C2-0.6*R
1730 PRINT "FOR ";360-Z3;" DEGREES."
1740 MOVE T1,C2-0.8*R
1750 PRINT "MOTIONS OF RISE AND FALL :"
1760 MOVE T2,C2-R
1770 PRINT "HARMONIC"
1780 MOVE T1,C2-1.2*R
1790 PRINT "CAM ROTATES COUNTERCLOCKWISE."
1800 REM *LABEL A,B,C,D ON CAM AND FOLLOWER*
1810 MOVE S1(1)+X2-0.1,R+S2(1)-0.18
1820 PRINT "o"
1830 RMOVE 0,0.2
1840 PRINT "A"
1850 T1=(Z1-Z0)/I1+1
1860 MOVE S1(1)+X2+S1(T1)*S-0.1,R+S2(T1)-0.18
1870 PRINT "o"
1880 RMOVE 0,0.2
1890 PRINT "B"
1900 T1=Z2/I1+1
1910 MOVE S1(1)+X2+S1(T1)*S-0.1,R+S2(T1)-0.18
1920 PRINT "o"
1930 RMOVE 0,0.2
1940 PRINT "C"
1950 T1=Z3/I1+1
1960 MOVE S1(1)+X2+S1(T1)*S-0.1,R+S2(T1)-0.18
1970 PRINT "o"
1980 RMOVE 0,0.2
1990 PRINT "D"
2000 MOVE S1(1)+X2+S1(N)*S-0.1,R+S2(N)-0.18

2010 PRINT "o"
2020 RMOVE 0,0.2
2030 PRINT "A"
2040 MOVE P1(1)-0.1,P2(1)-0.15
2050 PRINT "o"
2060 RMOVE 0.15,-0.35
2070 PRINT "A"
```

```
2080 T1=(Z1-Z0)/I1+1
2090 MOVE P1(T1)-0.1,P2(T1)-0.15
2100 PRINT "o"
2110 RMOVE 0.15,-0.35
2120 PRINT "B"
2130 T1=Z2/I1+1
2140 MOVE P1(T1)-0.1,P2(T1)-0.15
2150 PRINT "o"
2160 RMOVE 0.15,-0.35
2170 PRINT "C"
2180 T1=Z3/I1+1
2190 MOVE P1(T1)-0.1,P2(T1)-0.15
2200 PRINT "o"
2210 RMOVE 0.15,-0.35
2220 PRINT "D"
2230 MOVE C1+0.1*R,R+0.5*H
2240 PRINT "FOLLOWER"
2250 MOVE 0,R+H*2
2260 PRINT
2270 RETURN
```

6.4   DESIGN OF COIL SPRINGS

In this section the design and analysis of coil springs (Fig. 6-4) are discussed. As will become clear in the ensuing paragraphs, design of coil springs is an ideal subject for using interactive graphics, modules, and database. Also, it leads to the need to discuss the Program A.C.I. in the next section.

The spring constant k that relates the applied force F to the resulting displacement δ by the equation $F = k\delta$ can be calculated by the equation[1]

$$k = Gd^4/64R^3n \tag{a}$$

where G is the shear modulus of elasticity, d is the wire diameter, R is the mean radius, and n is the number of active coils in the spring. The mean radius is calculated by subtracting the wire radius from the outside radius of the spring. A coil is inactive once it touches the next coil.

The shear stress induced by the application of an axial force F can be calculated with the equation

$$\tau = 16FRK_w/\pi d^3 \tag{b}$$

where K_w is called Wahl's correction factor,[2] which can be expressed in terms of the spring index, $c = 2R/d$, as

$$K_w = \frac{4c - 1}{4c - 4} + \frac{0.615}{c} \tag{c}$$

When a helical coil is used as a torsion spring, the consequences of applying

---

[1]S. Timoshenko and D. H. Young, *Elements of Strength of Materials*, Van Nostrand Reinhold Company, New York, 1968, pp. 77–85.

[2]A. M. Wahl, "Stress in Heavy Closely Coiled Helical Springs," *Transactions of the American Society of Mechanical Engineers*, Paper No. APM-51-17, Vol. 51, 1929.

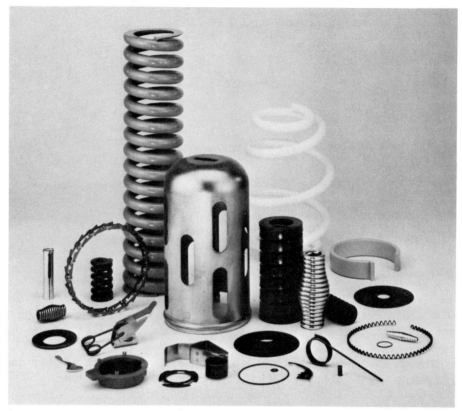

Figure 6-4   An assortment of springs. (Courtesy of Associated Spring, Barnes Group Inc.)

a couple M at its ends are that the spring will be twisted by an angle θ and direct stress σ will be induced. The formulas for computation of θ and average σ derived from the beam equations are

$$\theta = Ml/EI \qquad \text{(d)}$$

$$\sigma = Mc'/I \qquad \text{(e)}$$

where l is the length of the coiled wire equal to $2\pi Rn$, E is the modulus of elasticity, and I is the moment of inertia about the helix axis of the cross section of the wire equal to $\pi d^4/64$ for a round wire and $bh^3/12$ for a rectangular wire of width b and height h. In Eq. (e) c' is equal to $d/2$ for a round wire and $h/2$ for a rectangular wire.

For springs made of heavy wire, stress σ varies significantly across the wire diameter. At the inner edge, the stress can be expressed in terms of the spring index, $c = 2R/d$, as

$$\sigma = \frac{Mc'}{I}\left(\frac{4c-1}{4c-4}\right) \tag{f}$$

In case of a rectangular wire of bxh, the stress at the inner edge is

$$\sigma = \frac{6M}{bh^2}\left[\frac{6(R/h)-1}{6(R/h)-3}\right] \tag{g}$$

EXERCISE

If a helical spring has 12 active coils and is made of 6-mm steel wire by winding around on a 24-mm-diameter mandrel, and another helical spring has 18 active coils and is made of 8-mm steel wire by winding around on a 32-mm-diameter mandrel, what is the equivalent spring constant when these two springs are attached in series? And what is the largest force that may be applied to this two-spring system without exceeding a shear stress of 480 MN/m²? Write a BASIC program for design of coil springs.

Answer: $F = 82\ \text{GN/m}^2$.

6.5 ANALYSIS OF CROSS SECTIONS—AREA, CENTROID, AND MOMENTS OF INERTIA

As illustrated in the preceding section on the design of springs, the spring can be made by winding thin wire around a cylinder. The cross-sectional shape of the wire plays an important role in computation of the spring constant, stresses induced in the wire, and angle of twist when the spring is axially or torsionally loaded. In structural and machine designs and analyses, determination of the area and the location of the centroid of a specified cross-sectional shape and the subsequent computation of the moments of inertia and the radius of gyration are frequently needed. A general program that provides such basic computations is therefore an indispensable tool in CAD. The following paragraphs describe the development of a BASIC Program A. C. I. fulfilling such needs. Also, applications of this program to a number of cross-sectional shapes are presented.

Figure 6-5 depicts that for the general case, a given two-dimensional cross-sectional shape is to be defined by the coordinates (x_i, y_i) of a number of vertices,

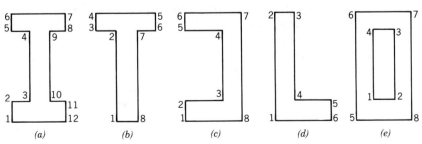

Figure 6-5 Commonly used cross-sectional shapes of structural members.

$i = 1, 2, \ldots, n$. The boundaries are assumed to be comprised of a series of **linear** edges connecting these vertices. Adjacent edges should not have equal slope, since this prevents three vertices from being picked from the same straight edge. It should also be clear that a curved boundary has to be approximated by a series of linear edges. Furthermore, the vertices should be numbered in increasing order by traversing around the boundary in a **clockwise** direction. This sign convention is necessary, as will become evident later, for consideration of both positive and negative areas.

The coordinates (x_i, y_i) of the vertices are to be specified by referring to a conveniently chosen coordinate system so that the cross section lies in the first quadrant. The location of the centroid of the cross section calculated by a developed procedure can then also be referred to the same coordinate system. Likewise, so can the moments of inertia and the radii of gyration.

The cross-sectional area is to be computed by summing the areas under every linear edge of the cross section—that is, to calculate the areas between the edges and the x-axis. Mathematically, the area can be expressed in terms of the coordinates of the n vertices of the cross section as

$$A = \tfrac{1}{2}(x_2 - x_1)(y_1 + y_2) + \tfrac{1}{2}(x_3 - x_2)(y_2 + y_3) + \cdots + \tfrac{1}{2}(x_{n+1} - x_n)(y_n + y_{n+1})$$

$$= \frac{1}{2} \sum_{i=1}^{n} (x_{i+1} - x_i)(y_i + y_{i+1}) \tag{a}$$

where

$$x_{n+1} = x_1 \qquad \text{and} \qquad y_{n+1} = y_1 \tag{b,c}$$

It is evident that the sign of $(x_{i+1} - x_i)$ in Eq. (a) will automatically decide whether a positive or negative contribution will be made to the computation of the net cross-sectional area. And it is precisely for the reason of this particular mathematical formulation based on Eq. (a) that the "clockwise" numbering of the vertices is mandatory. Furthermore, this arrangement facilitates the consideration of multiply connected regions, that is, when the cross sections have holes. In that case, all cutouts need only be defined with vertices that are numbered in increasing order but traversed in a **counterclockwise** direction around the cutout boundary. For instance, the vertices for the cross section shown in Fig. 6-5e 5 through 8 are numbered around the outside boundary in a clockwise sense, whereas vertices 1 through 4 are numbered around the inner boundary in a counterclockwise sense to indicate that there is a rectangular hole.

The coordinates of the centroid (x_c, y_c) of a given cross section are to be determined following the same scheme as for finding the cross-sectional area, that is, by consideration of each linear edge of the cross section. Figure 6-6 shows that for a typical edge $P_i P_{i+1}$, the centroidal coordinates $(\bar{x}_i, \bar{y}_i)$ of the combined areas A_1 (triangular) and A_2 (rectangular) are to be determined by the equations

$$\bar{x}_i = (A_1\bar{x}_1 + A_2\bar{x}_2)/(A_1 + A_2) \tag{d}$$

$$\bar{y}_i = (A_1\bar{y}_1 + A_2\bar{y}_2)/(A_1 + A_2) \tag{e}$$

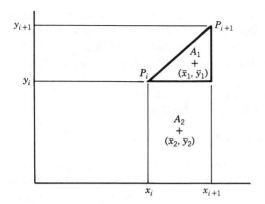

Figure 6-6

where $(\bar{x}_1, \bar{y}_1)$ and $(\bar{x}_2, \bar{y}_2)$ are the centroidal coordinates of the areas A_1 and A_2, respectively. It is easy to show that the terms appearing on the right-hand sides of Eqs. (d) and (e) can be expressed in terms of (x_i, y_i) and (x_{i+1}, y_{i+1}) as follows:

$$\bar{x}_1 = x_i + 2(x_{i+1} - x_i)/3 \tag{f}$$

$$\bar{y}_1 = y_i + (y_{i+1} - y_i)/3 \tag{g}$$

$$\bar{x}_2 = x_i + (x_{i+1} - x_i)/2 \tag{h}$$

$$\bar{y}_2 = y_i/2 \tag{i}$$

$$A_1 = (x_{i+1} - x_i)(y_{i+1} - y_i)/2 \tag{j}$$

$$A_2 = (x_{i+1} - x_i) y_i \tag{k}$$

There are special cases when the line $P_i P_{i+1}$ is horizontal or vertical, which causes A_1 or both A_1 and A_2 to be equal to zero. If $A_1 = A_2 = 0$, Eqs. (d) and (e) then need to be modified to become

$$\bar{x}_i = x_i \tag{l}$$

$$\bar{y}_i = (y_i + y_{i+1})/2 \tag{m}$$

Let the sum of A_1 and A_2 for each linear edge $P_i P_{i+1}$ be designated as A_i. The formulas for finding the centroidal coordinates of composite areas can be extended to the given cross section so that

$$x_c = \left(\sum_{i=1}^{n} \bar{x}_i A_i \right) \bigg/ A \tag{n}$$

$$y_c = \left(\sum_{i=1}^{n} \bar{y}_i A_i \right) \bigg/ A \tag{o}$$

where

$$A = \sum_{i=1}^{n} A_i \tag{p}$$

Equation (p) may as well replace Eq. (a), as the former is used for many applications whereas the latter is solely for the specific purpose of finding the cross-sectional area. Nevertheless, Eq. (a) has helped in explaining the approach of using vertices and proper numbering sequences for simply or multiply connected boundaries of the cross section.

The moments of inertia of a cross-sectional area A about the axes of a coordinate system, again referring to Fig. 6-3, are defined as

$$I_x = \int_A y^2 \, dA \qquad I_y = \int_A x^2 \, dA \qquad (q,r)$$

When the axes happen to pass the centroid of the cross section, the moments of inertia are conventionally denoted as $\bar{I}_x$ and $\bar{I}_y$. If the axes are d_x and d_y distances parallel away from centroid, a so-called parallel-axis theorem[3] can then be applied to make the adjustments. The theorem provides the following equations:

$$I_x = \bar{I}_x + d_x^2 A \qquad (s)$$

$$I_y = \bar{I}_y + d_y^2 A \qquad (t)$$

For simply shaped areas, the formulas for the moments of inertia are readily known. For example, a triangular area of base b and height h parallel to x and y axes, respectively, has $\bar{I}_x = bh^3/36$ and $\bar{I}_y = b^3h/36$, and a rectangular area of b by h, $\bar{I}_x = bh^3/12$ and $\bar{I}_y = b^3h/12$. Based on these available data, the areas A_1 and A_2 for a typical edge P_iP_{i+1} shown in Fig. 6-6 will contribute to the total cross-sectional moments of inertia

$$\begin{aligned} I_{x,i} &= (I_{x,i})_{A_1} + (I_{x,i})_{A_2} \\ &= [(x_{i+1} - x_i)(y_{i+1} - y_i)^3/36 + A_1(\bar{y}_1 - \bar{y}_i)^2] \\ &\quad + [(x_{i+1} - x_i)(y_{i+1} - y_i)^3/12 + A_2(\bar{y}_2 - \bar{y}_i)^2] \end{aligned} \qquad (u)$$

$$\begin{aligned} I_{y,1} &= (I_{y,i})_{A_1} + (I_{y,i})_{A_2} \\ &= [(x_{i+1} - x_i)^3(y_{i+1} - y_i)/36 + A_1(\bar{x}_1 - \bar{x}_i)^2] \\ &\quad + [(x_{i+1} - x_i)^3(y_{i+1} - y_i)/12 + A_2(\bar{x}_2 - \bar{x}_i)^2] \end{aligned} \qquad (v)$$

The total moments of inertia are to be obtained by summing the contributions from all boundary edges of the cross section. That is,

$$I_x = \sum_{i=1}^{n} I_{x,i} \qquad (w)$$

$$I_y = \sum_{i=1}^{n} I_{y,i} \qquad (x)$$

The radii of gyration in x and y directions are to be determined by use of the moments of inertia of mass $I_{m,x}$ and $I_{m,y}$. For a uniform mass, $I_{m,x} = I_x$ and $I_{m,y} = I_y$. The equations for calculation of the radii of gyration are

---

[3] J. L. Meriam, *Dynamics*, John Wiley & Sons, New York, 1978, pp. 467–486.

$$r_x = (I_{m,x}/A)^{1/2} \tag{y}$$

$$r_y = (I_{m,y}/A)^{1/2} \tag{z}$$

They are often required in dynamic analysis of solid bodies.

Figure 6-7 is an example application of Program A. C. I. The users of this program should be able to expand it to cover the calculation of the product of inertia as well, which is usually denoted as I_{xy} and has a similar expression as I_x except the integrand y^2 in Eq. (q) has to be replaced by xy. The parallel-axis theorem for I_{xy} is

$$I_{xy} = \bar{I}_{xy} + d_x d_y A$$

The notation used in the above expression was explained earlier, when Eqs. (s) and (t) were introduced. The extended Program A. C. I including computation of I_{xy} will be needed in a later exercise.

Lines 410 and 420 in Program A. C. I show applications of the AXIS command. The four parameters associated with the AXIS command can generally be described by numeric expressions specifying the tic intervals in x direction and y direction, as well as the coordinates of the location where the origin is to be placed, all in user's data units. $\underline{K}$ control character is used in line 1060 PRINT statement for the purpose of moving the cursor up by one line to achieve better alignment.

PROGRAM A. C. I.

```
100 INIT
110 CHARSIZE 2
120 PAGE
130 PRINT "PROGRAM A.C.I"
140 PRINT "   CALCULATES CROSS-SECTIONAL AREA,"
150 PRINT "  DETERMINES LOCATION OF CENTROID"
160 PRINT "  AND COMPUTE MOMENTS OF INERTIA."
170 PRINT "_INPUT : COORDINATES OF VERTICES "
180 PRINT "          DESCRIBING THE CROSS-SECTIONAL"
190 PRINT "          SHAPE, MUST ALL BE POSITIVE."
200 PRINT "_WANT TO SEE A DEMONSTRATION, Y/N? : ";
210 INPUT A$
220 IF A$="Y" THEN 1260
230 PRINT "_ENTER THE NUMBER OF VERTICES : ";
240 INPUT N1
250 GO TO 1120
260 PRINT "_WANT TO LIST THE COORDINATES,Y/N? : ";
270 INPUT P$
280 IF P$="N" THEN 360
290 PRINT "_NO. OF VERTICES = ";N1
300 FOR K=1 TO N1
310 PRINT USING 320:X(K),Y(K)
320 IMAGE FD.5D,7X,FD.5D
330 NEXT K
340 PRINT "_PUSH KEY `Y' TO CONTINUE ! ";
350 INPUT A$
360 VIEWPORT 50,130,20,100
370 MOVE 0,0
380 SCALE S/80,S/80
```

PROGRAM A.C.I

  CALCULATES CROSS-SECTIONAL AREA.
  DETERMINES LOCATION OF CENTROID
  AND COMPUTE MOMENTS OF INERTIA.

INPUT COORDINATES OF VERTICES
        DESCRIBING THE CROSS-SECTIONAL
        SHAPE. MUST ALL BE POSITIVE.

WANT TO SEE A DEMONSTRATION. Y/N?: Y

WANT TO LIST THE COORDINATES. Y/N?: Y

NO. OF VERTICES = 8

3.00000	1.00000
3.00000	4.00000
1.00000	4.00000
1.00000	5.00000
6.00000	5.00000
6.00000	4.00000
4.00000	4.00000
4.00000	1.00000

PUSH KEY 'Y' TO CONTINUE!Y

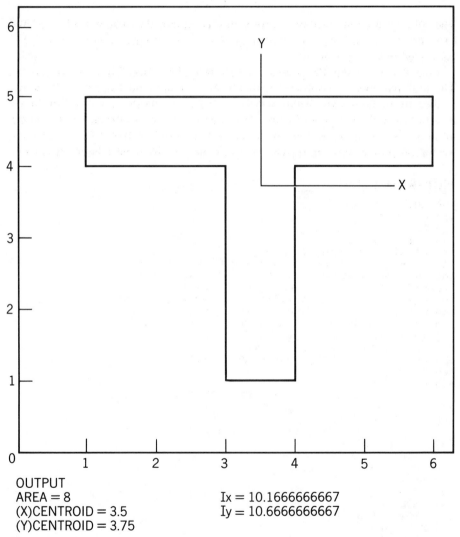

OUTPUT
AREA = 8
(X)CENTROID = 3.5
(Y)CENTROID = 3.75

$I_x = 10.1666666667$
$I_y = 10.6666666667$

Figure 6-7 Analysis of a "T" cross section.

```
390 M=INT(LGT(S))
400 M=10^M
410 AXIS M,M,0,0
420 AXIS 0,0,0.999*S,0.999*S
430 FOR K=0 TO S STEP M
440 MOVE -0.03*S,K-0.01*S
450 PRINT K
460 MOVE K-0.02*S,-0.03*S
470 PRINT K
480 NEXT K
490 MOVE X(N1),Y(N1)
500 DRAW X,Y
510 A=0
520 X0=0
530 Y0=0
540 I1=0
550 I2=0
560 I3=0
570 FOR K=1 TO N1
580 K1=K+1
590 IF K<N1 THEN 610
600 K1=1
610 H=Y(K1)-Y(K)
620 B=X(K1)-X(K)
630 X1=X(K)+2*B/3
640 Y1=Y(K)+H/3
650 X2=X(K)+B/2
660 Y2=Y(K)/2
670 A1=B*H/2
680 A2=2*Y2*B
690 A(K)=A1+A2
700 IF B=0 THEN 750
710 X0(K)=(X1*A1+X2*A2)/A(K)
720 Y0(K)=(Y1*A1+Y2*A2)/A(K)
730 I1(K)=B*H^3/36+A1*(Y1-Y0(K))^2+B*(2*Y2)^3/12+A2*(Y2-Y0(K))^2
740 I2(K)=H*B^3/36+A1*(X1-X0(K))^2+2*Y2*B^3/12+A2*(X2-X0(K))^2
750 NEXT K
760 C1=0
770 C2=0
780 FOR K=1 TO J1
790 C1=C1+E(K)*X0(0)
800 C2=C2+E(K)*Y0(K)
810 NEXT K
820 A0=SUM(E)
830 C1=C1/A0
840 C2=C2/A0
850 GO TO 1050
860 WINDOW 0,130,0,100
870 VIEWPORT 50,130,-80,20
880 MOVE 0,95
890 PRINT "OUTPUT :"
900 MOVE 0,93
910 PRINT "AREA = ";A0
920 MOVE 0,91.5
930 PRINT "(X)CENTROID = ";C1
940 MOVE 0,90
950 PRINT "(Y)CENTROID = ";C2
960 FOR K=1 TO N1
970 I1(K)=I1(K)+E(K)*(Y0(K)-C2)^2
980 I2(K)=I2(K)+E(K)*(X0(K)-C1)^2
```

```
990 NEXT K
1000 MOVE 52,93
1010 PRINT "Ix = ";SUM(I1)
1020 MOVE 52,91.5
1030 PRINT "Iy = ";SUM(I2)
1040 RETURN
1050 MOVE C1+0.3*W,C2
1060 PRINT "_X__";
1070 DRAW C1,C2
1080 DRAW C1,C2+0.3*W
1090 PRINT "Y"
1100 GOSUB 860
1110 END
1120 DIM X(N1),Y(N1),E(N1),X0(N1),Y0(N1)
1130 DIM I1(N1),I2(N1)
1140 PRINT "_ENTER COORDINATES : X1,Y1/X2,Y2/...etc"
1150 PRINT "   (ENTER TWO NUMBERS THEN PUSH 'RETURN')"
1160 PRINT "FOR POSITIVE AREA, ENTER VERTICES IN"
1170 PRINT "   CLOCKWISE ORDER."
1180 FOR L1=1 TO N1
1190 INPUT X(L1),Y(L1)
1200 NEXT L1
1210 GO TO 1340
1220 INPUT X
1230 PRINT "ENTER THE ";N1;" Y VALUES ";
1240 INPUT Y
1250 GO TO 1340
1260 N1=8
1270 DIM X(N1),Y(N1),E(N1),X0(N1),Y0(N1)
1280 DIM I1(N1),I2(N1)
1290 READ X
1300 DATA 3,3,1,1,6,6,4,4
1310 READ Y
1320 DATA 1,4,4,5,5,4,4,1
1330 P$="Y"
1340 REM * SCALING *
1350 X2=X(1)
1360 Y2=Y(1)
1370 FOR K=1 TO N1
1380 X2=X2 MAX X(K)
1390 Y2=Y2 MAX Y(K)
1400 NEXT K
1410 IF X2>Y2 THEN 1440
1420 W=1.05*Y2
1430 GO TO 260
1440 W=1.05*X2
1450 GO TO 260
```

6.6 MOHR'S CIRCLE—EIGENVALUE PROBLEMS

In analysis of stresses in structural members with cross-sectional shapes such as those shown in Fig. 6-5, the direction at which a bending moment is applied affects the maximum and minimum stresses in the member. For the cross-sectional shapes that are symmetric with respect to their own centroidal axes (Fig. 6-5a, b, and e) the locations at which the maximum stress occurs are easy to determine. For the cross-sectional shapes that are not symmetric with respect to their own centroidal axes, however (Fig. 6-5c and d), the maximum and minimum stresses and their locations have to be determined by finding the principal moments of inertia I_u and I_v from the following equations:

$$I_{u,v} = \tfrac{1}{2}(I_x + I_y) \pm \tfrac{1}{2}(I_x - I_y) \cos 2\alpha \mp I_{xy} \sin 2\alpha \qquad (a)$$

where α is the angle between the principal axis u and the x axis measured in a counterclockwise direction from the x axis. α is to be found so that the product of inertia I_{uv} is equal to zero with respect to the principal axes u and v.

Graphically, I_u, I_v, and α can be determined by application of the Mohr's Circle method.[4] This method is best known for its application in determining the principal stresses and planes of two-dimensional stresses. As illustrated in Fig. 6-8, the determination of the principal stresses σ_A and σ_B, and the direction of the principal plane α ($=26.57°$ shown) are based on the specified stress conditions of $\sigma_x = \sigma_V = -100$, $\sigma_y = \sigma_H = -700$, and $\tau_{xy} = \tau_{VH} = 400$, of the same units (say, KN/m^2). σ_A, σ_B, and α satisfy Eq. (a) when I_x, I_y, and I_{xy} are replaced by σ_x, σ_y, and τ_{xy}, and u and v are substituted by A and B. The Mohr's Circle is based on representing the stress conditions on each plane by a point on the σ–τ plane, on which σ and τ are taken as horizontal and vertical axes, respectively. The shear stress τ is considered positive if it causes the element to rotate clockwise. For example, the $\tau_{VH} = 400$ on the V plane causes counterclockwise rotation, and $\tau_{HV} = 400$ on the H plane causes clockwise rotation of the element, so they are negative and positive, respectively, for locating the points V and H on the Mohr's Circle.

By use of proper scales for the σ and τ axes, the points V and H can be located on the σ–τ plane. The line VH is then drawn, intercepting the σ axis at C. C is to be used as the center and either CV or CH as radius to draw the Mohr's Circle. The stress values at the extreme points A and B are the desired principal stresses. Since the V and H planes are actually $90°$ apart but points V and H are $180°$ apart on the Mohr's Circle, all angles measured from the circle should be halved to obtain actual solutions. From Fig. 6-8 one can see that the angle VCA measures $53.14°$. This indicates that the principal plane at which maximum tensile stress $\sigma_A = 100$ occurs is ($\alpha=$) $26.57°$ measured in counterclockwise direction from the V plane. There are other valuable results that are directly measurable from the Mohr's Circle. For example, the maximum shear stress is 500 and the accompanying normal stress is a compressive stress of 400 at the plane D or E, which is $45°$ away from the principal planes.

Besides the above-mentioned applications of the Mohr's Circle method for determination of the maximum and minimum stresses and moments of inertia, the method can also be employed for graphical solution of the eigenvalues of a symmetric matrix and for vibration and buckling analyses. Program MOHR'S.C has been prepared for general application of the Mohr's Circle. Figure 6-8 is a sample display of a stress problem.

---

[4] Y. C. Pao, "A General Program for Computer Plotting of Mohr's Circle," *Computers & Structures*, Vol. 2, 1972, pp. 625–635.

* PROGRAM DIM. NING *

INPUT NORMAL STRESSES SX AND SY. SHEAR STRESS SXY:
-120.-700.400

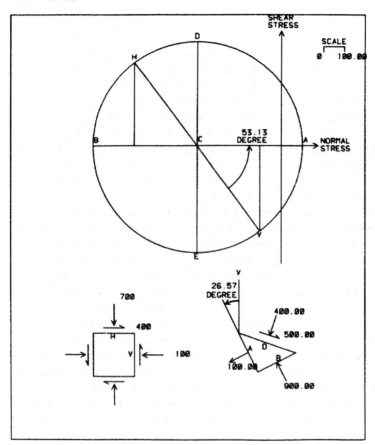

Figure 6-8

PROGRAM MOHR'S.C

```
100 PAGE
110 CHARSIZE 2
120 PRINT "* PROGRAM MOHR'S.C *"
130 INIT
140 SET DEGREES
150 GO TO 240
160 REMARK SUB FOR DRAW ARROW AT Z DEGREES
170 IF J=1 THEN 210
180 RDRAW 0.14*COS(180+Z+26.57),0.14*SIN(180+Z+26.57)
190 RMOVE -0.14*COS(180+Z+26.57),-0.14*SIN(180+Z+26.57)
200 IF J=2 THEN 230
210 RDRAW 0.14*COS(180+Z-26.57),0.14*SIN(180+Z-26.57)
220 RMOVE -0.14*COS(180+Z-26.57),-0.14*SIN(180+Z-26.57)
230 RETURN
240 PRINT "_INPUT NORMAL STRESSES SX AND SY, SHEAR STRESS SXY : "
250 INPUT S1,S2,S3
260 WINDOW 0,10,0,10
270 VIEWPORT 10,100,10,100
280 PAGE
290 MOVE 0,0
300 DRAW 0,10
310 DRAW 8.5,10
320 DRAW 8.5,0
330 DRAW 0,0
340 J=3
350 MOVE 2,1.5
360 RDRAW 1,0
370 RDRAW 0,1
380 RDRAW -1,0
390 RDRAW 0,-1
400 MOVE 2.86,1.93
410 PRINT "V"
420 MOVE 2.43,2.33
430 PRINT "H"
440 A1=ABS(S1)
450 A2=ABS(S2)
460 A3=ABS(S3)
470 S=SGN(S3)
480 IF S1=0 THEN 700
490 IF S1>0 THEN 570
500 Z=180
510 MOVE 3.1875,2
520 GOSUB 160
530 Z=0
540 MOVE 1.8125,2
550 GOSUB 160
560 GO TO 630
570 Z=0
580 MOVE 3.6875,2
590 GOSUB 160
600 Z=180
610 MOVE 1.3125,2
620 GOSUB 160
630 MOVE 3.1875,2
640 RDRAW 0.5,0
650 MOVE 1.8125,2
660 RDRAW -0.5,0
670 MOVE 3.7475,1.93
680 PRINT USING 690:A1
```

```
690  IMAGE 5D
700  IF S2=0 THEN 910
710  IF S2>0 THEN 790
720  Z=270
730  MOVE 2.5,2.6875
740  GOSUB 160
750  Z=90
760  MOVE 2.5,1.3125
770  GOSUB 160
780  GO TO 850
790  Z=90
800  MOVE 2.5,3.1875
810  GOSUB 160
820  Z=270
830  MOVE 2.5,0.8125
840  GOSUB 160
850  MOVE 2.5,1.3125
860  RDRAW 0,-0.5
870  MOVE 2.5,2.6875
880  RDRAW 0,0.5
890  MOVE 2.43,3.2475
900  PRINT USING 690:A2
910  IF S3=0 THEN 1200
920  IF S3>0 THEN 1020
930  MOVE 2.75,1.375
940  RDRAW -0.125,-0.0625
950  MOVE 2.25,2.625
960  RDRAW 0.125,0.0625
970  MOVE 1.875,2.25
980  RDRAW -0.0625,-0.125
990  MOVE 3.125,1.75
1000 RDRAW 0.0625,0.125

1010 GO TO 1100
1020 MOVE 2.25,1.375
1030 RDRAW 0.125,-0.0625
1040 MOVE 2.75,2.625
1050 RDRAW -0.125,0.0625
1060 MOVE 1.875,1.75
1070 RDRAW -0.0625,0.125
1080 MOVE 3.125,2.25
1090 RDRAW 0.0625,-0.125
1100 MOVE 2.25,1.375
1110 RDRAW 0.5,0
1120 MOVE 2.25,2.625
1130 RDRAW 0.5,0
1140 MOVE 1.875,1.75
1150 RDRAW 0,0.5
1160 MOVE 3.125,1.75
1170 RDRAW 0,0.5
1180 MOVE 2.81,2.56
1190 PRINT USING 690:A3
1200 C=0.5*(S1+S2)
1210 R=SQR(0.25*(S1-S2)^2+S3^2)
1220 MOVE 7.5,9.125
1230 RDRAW 0,0.125
1240 RDRAW 0.5,0
```

```
1250 RDRAW 0,-0.125
1260 M=R-C
1270 IF C-R<=0 THEN 1290
1280 M=0
1290 IF ABS(C+R)>ABS(C-R) THEN 1350
1300 IF ABS(C-R)>2*R THEN 1330
1310 L=2*R
1320 GO TO 1390
1330 L=ABS(C-R)
1340 GO TO 1390
1350 IF ABS(C+R)>2*R THEN 1380
1360 L=2*R
1370 GO TO 1390
1380 L=ABS(C+R)
1390 F=5/L
1400 MOVE 7.36,8.925
1410 PRINT "0"
1420 MOVE 7.47,9.2825
1430 PRINT "SCALE"
1440 MOVE 7.6,8.925
1450 PRINT USING 1460:0.5/F
1460 IMAGE 5D.2D
1470 M7=M*F
1480 C7=C*F
1490 R7=R*F
1500 S4=S1*F
1510 S5=S2*F
1520 S6=S3*F
1530 O1=2+M7
1540 O2=6.875
1550 MOVE O1,O2
1560 I=SGN(S1-S2)
1570 P=0.5*ASN(A3/R)
1580 R8=I*S
1590 FOR A=0 TO 360
1600 DRAW O1+C7+R7*COS(A),O2+R7*SIN(A)
1610 NEXT A
1620 MOVE O1+S4,O2
1630 RDRAW 0,-S6
1640 RDRAW S5-S4,2*S6
1650 RDRAW 0,-S6
1660 MOVE O1,O2-1.1*R7
1670 RDRAW 0,2.2*R7
1680 Z=90
1690 MOVE O1,O2+1.1*R7
1700 GOSUB 160
1710 RMOVE -0.3,0.22
1720 PRINT "SHEAR_____"
1730 RMOVE 0,-0.18
1740 PRINT "STRESS"
1750 IF C+R>0 THEN 1780
1760 L=0
1770 GO TO 1790
1780 L=C7+R7
1790 E=L+0.375
1800 MOVE O1,O2
1810 RDRAW E,0
1820 Z=0
1830 MOVE O1+E,O2
1840 GOSUB 160
1850 RMOVE 0.075,0.02
```

```
1860 PRINT "NORMAL_____"
1870 RMOVE 0,-0.18
1880 PRINT "STRESS"
1890 MOVE O1+C7,O2+R7
1900 RDRAW 0,-2*R7
1910 MOVE O1,O2
1920 RDRAW -M7,0
1930 MOVE O1+(S4+C7)/2,O2-S6/2
1940 D=P/45
1950 FOR K=1 TO 90
1960 A=2*P-K*D
1970 DRAW O1+C7+0.5*R7*I*COS(A),O2-0.5*R7*S*SIN(A)
1980 NEXT K
1990 Z=90+(S-1)*90
2000 MOVE O1+C7+R7/2*I,O2
2010 GOSUB 160
2020 RMOVE -0.28,0.2*S+0.02
2030 PRINT USING 2040;2*P
2040 IMAGE 3D.2D,"_____"
2050 RMOVE 0,-0.18
2060 PRINT "DEGREE"
2070 MOVE O1+S4-0.05,O2-S6-(0.07+0.11*S)
2080 PRINT "V"
2090 MOVE O1+S5-0.05,O2+S6-(0.07-0.11*S)
2100 PRINT "H"
2110 MOVE O1+C7+R7*I+(0.04-0.09*(1-I)),O2+0.04
2120 PRINT "A"
2130 MOVE O1+C7-R7*I+(0.04-0.09*(1-I)),O2+0.04
2140 PRINT "B"
2150 MOVE O1+C7-0.05,O2+(R7+(0.04+0.07*(1-S)))*S
2160 PRINT "D"
2170 MOVE O1+C7-0.05,O2+(-R7-(0.04+0.07*(1+S)))*S
2180 PRINT "E"
2190 MOVE O1+C7+(0.04-0.09*(1-R8)),O2+0.06
2200 PRINT "C"
2210 O5=2+4-R8/2
2220 O6=1.5+1
2230 MOVE O5,O6
2240 C8=COS(P)
2250 S8=SIN(P)
2260 X1=0.5*R8*S8
2270 Y1=-0.5*C8
2280 X2=X1+Y1*R8
2290 Y2=Y1-0.5*S8
2300 DRAW O5+2*X1,O6+2*Y1
2310 C9=COS(P+45)
2320 S9=SIN(P+45)
2330 X=R8*SQR(2)*S9
2340 Y=-SQR(2)*C9
2350 DRAW O5+X,O6+Y
2360 DRAW O5,O6
2370 DRAW O5,O6+1.25
2380 MOVE O5,O6
2390 RDRAW -1.75*X1,-1.75*Y1
2400 X3=X/2+X1
2410 Y3=Y/2+Y1
2420 X4=X3+X1
2430 Y4=Y3+Y1
2440 MOVE O5+X1-0.05*(1-R8),O6+Y1
2450 PRINT "A"
2460 MOVE O5,O6+0.75
2470 MOVE O5+X3-0.05*(1-R8),O6+Y3
2480 PRINT "B"
```

```
2490 MOVE 05+X/2-0.05*(1+R8),06+Y/2-0.18
2500 PRINT "D"
2510 MOVE 05,06+0.75
2520 FOR K=1 TO 90
2530 A=K*D/2
2540 DRAW 05-R8*0.75*SIN(A),06+0.75*COS(A)
2550 NEXT K
2560 Z=90+S*(P+90)
2570 GOSUB 160
2580 MOVE 05-0.35-R8*0.4,06+1
2590 PRINT USING 2600;P
2600 IMAGE 3D.2D
2610 RMOVE 0,-0.21
2620 PRINT "DEGREE"
2630 MOVE 05-0.05,06+1.3125
2640 PRINT "V"
2650 IF C+R*I=0 THEN 2790
2660 MOVE 05+X1,06+Y1
2670 DRAW 05+X2,06+Y2
2680 MOVE 05+X2-0.28,06+Y2-0.2
2690 PRINT USING 2700;ABS(C+R*I)
2700 IMAGE 5D.2D
2710 IF C+R*I>0 THEN 2760
2720 Z=90+S*(P-90)
2730 MOVE 05+X1,06+Y1
2740 GOSUB 160
2750 GO TO 2790
2760 Z=90+S*(P+90)
2770 MOVE 05+X2,06+Y2
2780 GOSUB 160
2790 IF C-R*I=0 THEN 2920
2800 MOVE 05+X3,06+Y3
2810 DRAW 05+X4,06+Y4
2820 MOVE 05+X4-0.28,06+Y4-0.2
2830 PRINT USING 2700;ABS(C-R*I)
2840 IF C-R*I>0 THEN 2890
2850 Z=90+S*P
2860 MOVE 05+X3,06+Y3
2870 GOSUB 160
2880 GO TO 2920
2890 Z=270+S*P
2900 MOVE 05+X4,06+Y4
2910 GOSUB 160
2920 IF C=0 THEN 3090
2930 X5=X/2+R8*0.1875*C9
2940 Y5=Y/2+0.1875*S9
2950 X6=X/2+R8*0.6875*C9
2960 Y6=Y/2+0.6875*S9
2970 MOVE 05+X5,06+Y5
2980 DRAW 05+X6,06+Y6
2990 MOVE 05+X6-0.28,06+Y6-0.02
3000 PRINT USING 2700;ABS(C)

3010 IF C>0 THEN 3060
3020 Z=270+S*(P-45)
3030 MOVE 05+X5,06+Y5
3040 GOSUB 160
3050 GO TO 3090
3060 Z=90+S*(P-45)
```

```
3070 MOVE 05+X6,06+Y6
3080 GOSUB 160
3090 Z=90+S*(P-135)
3100 MOVE 05+X5-0.25*COS(Z),06+Y5-0.25*SIN(Z)
3110 RDRAW 0.5*COS(Z),0.5*SIN(Z)
3120 J=(3-S)/2
3130 GOSUB 160
3140 IF S<0 THEN 3180
3150 PRINT USING 3160;R
3160 IMAGE 4D.2D
3170 GO TO 3200
3180 PRINT USING 3190;R
3190 IMAGE "_____",4D.2D
3200 END
```

EXERCISE

Apply Programs A.C.I and MOHR'S.C for finding the principal axes of the cross section shown in Figure 6-9.

Answer: $I_u = 11.264 \times 10^{-5} \text{ m}^4$
$I_v = 0.896 \times 10^{-5} \text{ m}^4$
$\alpha = 10.1°$

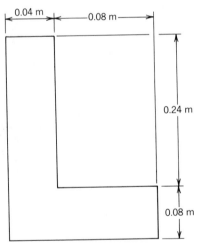

Figure 6-9

6.7   INTERACTIVE DIMENSIONING

One of the fundamental needs in drafting is dimensioning of the designed geometric configurations. For example, the cross sections of commonly adopted structural members shown in Fig. 6-5 may have to be dimensioned. Figure 6-9 shows that the dimensioning of a given geometry involves drawing the lines and arrows and labeling with numerical values. More involved dimensionings can be seen in

Fig. 6-8, where the directions and values of the stresses need to be accurately and clearly labeled. In fact, all the dimensioning algorithms are incorporated in Program MOHR'S.C.

As an illustration of interactive dimensioning, assume that a rectangular cross section needs to be dimensioned by labeling its width and height. Figure 6-10 is the result of interactive dimensioning using Program DIM.NING, for which the definable keys and thumbwheels are employed.

Program DIM.NING is evolved from Program E.MODULE by keeping the modules controlled by keys 1, 10, 12, 18, 19, and 20 for "STOP," drawing "LINE," "REVISE," drawing "ARROW," and "STORE" and "RETRIEVE," respectively. Since dimensioning has the basic need of drawing lines and arrows as well as labeling proper characters at specified locations, key 9 has been selected for control of directing to the module for **labeling**. As a result of this added module, the "REVISE," "STORE," and "RETRIEVE" modules need to be accordingly modified.

From the listing of Program DIM.NING, the readers have the opportunity of reviewing the applications of SEG and LEN commands for maneuvering the

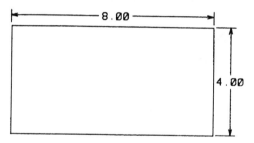

```
* PROGRAM DIM.NING *
  KEY TABLE
   *1* ------- STOP          *12* ------- REVISE
 * NINE * ------- LABELING    *18* ------- ARROW
  *10* ------- LINE           *19* ------- STORE
                              *20* ------- RETRIEVE
INPUT THE LABEL CHARACTER SIZE: 4
```

STEP	KEY	BEGINNING		ENDING	
		X	Y	X	Y
1	10	53.5	77.3	104.3	77.3
2	10	104.3	77.3	104.3	50.0
3	10	104.3	50.8	53.4	50.0
4	10	53.4	50.8	53.4	77.4
5	9	105.2	62.7	4.00	
6	9	76.3	78.5	9.00	
7	10	84.8	89.0	104.2	88.0
8	10	75.3	89.9	53.2	88.0
9	10	53.2	81.3	53.2	78.1
10	10	104.3	78.1	104.3	81.4
11	10	188.7	65.9	188.7	77.4
12	10	108.7	62.3	100.7	50.0
13	10	105.0	59.9	110.0	50.0
14	10	110.0	77.4	104.0	77.4

```
** DESIGN ENDS **
```

Figure 6-10 Interactive dimensioning of a rectangle by use of Program DIM.NING.

character strings. The statement S$=S$&C$ is an example of using the **string concatenation operator** & to join together two strings. If S$ and C$ are "4" and ". 25 CM," respectively, before the statement is executed, the statement will change the content of S$ from "4" to "4.25 CM."

At present, the lines, arrows, and labels are drawn separately. Each element needs to be entered by adjusting the thumbwheels twice together with entering of the label character from the keyboard. As an exercise, readers may want to simplify this procedure by requiring fewer interactive steps. Modify Program DIM.NING to input only the locations of the arrow heads and the starting location of the label, which will automatically generate the additional lines needed in dimensioning. Apply the modified program for dimensioning of the rectangle shown in Fig. 6-10, which is plotted by use of Program DIM.NING with 14 steps, and list the number of steps for comparison of the two versions.

Line 2097 of Program DIM.NING uses a REP function, which replaces a portion of a string by another string. It has the general form

$$A\$=REP(B\$,N,L)$$

B$ is to replace a portion of A$ starting with the Nth character of A$ and there are L characters involved.

PROGRAM DIM.NING

```
1 REM * PROGRAM DIM.NING *
2 GO TO 150
3 CALL "WAIT",1000
4 GOSUB 3750
36 GOSUB 2000
37 GO TO 3
40 GOSUB 2470
41 GO TO 3
48 GOSUB 4100
49 GO TO 3
72 GOSUB 7200
73 GO TO 3
76 GOSUB 9490
77 GO TO 3
80 GOSUB 9700
89 GO TO 3
150 CHARSIZE 1
151 RINIT
152 PAGE
153 INIT
154 SET DEGREES
155 SET KEY
156 DIM H6(50),S$(500)
157 S$='
158 L1=0
160 GOSUB 200
162 MOVE 0,78
164 PRINT "INPUT THE LABEL CHARACTER SIZE : ";
166 INPUT C9
180 GO TO 318
200 PRINT "* PROGRAM DIM.NING *"
201 PRINT "_KEY TABLE_"
```

```
203 PRINT "`1'--------STOP                                      "
205 PRINT "                        `12'--------REVISE"
260 PRINT "                        `18'--------ARROW"
270 PRINT "`9'-------LABELLING `19'--------STORE"
280 PRINT "`10'------LINE          `20'--------RETRIEVE"
317 RETURN
318 P5=0
319 U7=0
320 U1=0
330 D9=0
340 DIM P9(100),X1(100),Y1(100),X2(100),Y2(100)
350 GOSUB 4930
360 GO TO 3

2000 REM *THIS SUB. DOES THE LABELLING*

2010 J=9
2030 IF D9(>1 THEN 2060
2040 U1=U3
2050 GO TO 2070
2060 GOSUB 5070
2062 X1(U1)=X2(U1)
2064 Y1(U1)=Y2(U1)
2070 MOVE X2(U1),Y2(U1)
2071 CHARSIZE C9
2072 IF D9(>1 THEN 2083
2073 IF T=1 THEN 2080
2074 B=H6(T-1)+1
2075 E=H6(T)-H6(T-1)
2076 C$=SEG(S$,B,E)
2077 PRINT C$
2078 T=T+1
2079 RETURN
2080 B=1
2081 E=H6(1)
2082 GO TO 2076
2083 INPUT C$
2085 IF L1=0 THEN 2088
2086 H6(L1+1)=H6(L1)+LEN(C$)
2087 GO TO 2089
2088 H6(1)=LEN(C$)
2089 L1=L1+1
2090 IF L1=1 THEN 2095
2091 B=H6(L1-1)+1
2092 GO TO 2097
2095 B=1
2097 S$=REP(C$,B,LEN(C$))
2100 CHARSIZE 1
2200 GOSUB 5520
2450 RETURN
2470 REM * THIS SUBROUTINE PLOTS A LINE *
2520 J=10
2530 IF D9(>1 THEN 2560
2540 U1=U3
2550 GO TO 2570
2560 GOSUB 5070
2570 MOVE X1(U1),Y1(U1)
2580 DRAW X2(U1),Y2(U1)
2600 RETURN

3750 REM * PROGRAM ENDS *
3800 G9=8
```

```
3810 RINIT
3820 GOSUB 4530
3830 REM INIT
3850 MOVE 0,10
3860 PRINT "**  DESIGN ENDS  **"
3870 END

4100 REM * THIS SUBROUTINE REVISES THE LAST DRAWN ELEMENT *
4150 RINIT
4160 MOVE 0,80
4180 MOVE 0,20
4200 U2=U1
4210 PRINT "INPUT THE NEW KEY NUMBER USING "
4211 PRINT "  NUMERIC KEYS ; NOT DEFINABLE KEY!!!"
4220 INPUT P9(U2)
4221 IF P9(U2)=9 THEN 4270
4230 PRINT "ADJUST THE THUMBWHEELS TWICE!!!"
4240 PRINT "  TO INPUT THE BEGINNING AND"
4250 PRINT "  ENDING POINTS OF THE ELEMENT"
4260 POINTER X1(U2),Y1(U2),C$
4265 POINTER X2(U2),Y2(U2),C$
4268 GO TO 4470
4270 L1=L1-1
4271 D9=0
4272 P5=P5-1
4372 GOSUB 2000
4470 PAGE
4472 GOSUB 200
4475 U7=U2
4482 GOSUB 9750
4520 RETURN
4530 REM * THIS SUBROUTINE PRINTS OUT THE ELEMENTS DRAWN. *
4580 MOVE 0,70
4590 PRINT "STEP# KEY#  BEGINNING       ENDING "
4600 PRINT "            X     Y      X      Y"
4602 T=1
4610 FOR I5=1 TO U1
4612 IF P9(I5)=9 THEN 4632
4620 PRINT USING 4630:I5,P9(I5),X1(I5),Y1(I5),X2(I5),Y2(I5)
4630 IMAGE 2D,4X,2D,2X,4(3D,1D,2X)
4631 GO TO 4680
4632 IF T=1 THEN 4637
4633 B=H6(T-1)+1
4634 E=H6(T)-H6(T-1)
4635 C$=SEG(S$,B,E)
4636 GO TO 4640
4637 B=1
4638 E=H6(1)
4639 GO TO 4635
4640 PRINT USING 4642:I5,P9(I5),X2(I5),Y2(I5),C$
4641 T=T+1
4642 IMAGE 2D,4X,2D,2X,2(3D,1D,2X),20A
4680 NEXT I5
4710 RETURN
4720 REM * THIS SUB FINDS THE LENGTH AND ROTATION ANGLE *
4770 L=((X1(U1)-X2(U1))^2+(Y1(U1)-Y2(U1))^2)^0.5
4780 IF X1(U1)>X2(U1) AND Y1(U1)=Y2(U1) THEN 4860
4790 IF X1(U1)=X2(U1) AND Y1(U1)>Y2(U1) THEN 4880
4800 IF X1(U1)>X2(U1) AND Y1(U1)>Y2(U1) THEN 4900
4810 IF Y1(U1)=>Y2(U1) THEN 4840
4820 T1=ACS((X2(U1)-X1(U1))/L)
4830 GO TO 4920
```

```
4840 T1=ASN((Y2(U1)-Y1(U1))/L)
4850 GO TO 4920
4860 T1=180
4870 GO TO 4920
4880 T1=-90
4890 GO TO 4920
4900 T1=ATN((Y2(U1)-Y1(U1))/(X2(U1)-X1(U1)))
4910 T1=T1+180
4920 RETURN
4930 REM * THIS SUBROUTINE INSTRUCTS THE USER *
4990 ROPEN 7
5000 IF U1<>0 THEN 5060

5010 MOVE 0,75
5020 PRINT "*****************************************"
5030 PRINT "**          ENTER FIRST KEY#          **"
5040 PRINT "*****************************************"
5050 RCLOSE
5060 RETURN
5070 REM * INPUT THE BEGINNING AND ENDING POINTS *
5140 P5=P5+1
5145 U1=P5+U7
5150 P9(U1)=J
5160 RINIT
5170 MOVE 0,65
5180 ROPEN 1
5190 PRINT "KEY# CHOSEN: ";J
5200 PRINT "ADJUST THE THUMBWHEELS"
5210 PRINT " TO INPUT THE BEGINNING"
5220 PRINT " POINT WHEN READY PRESS"
5230 PRINT " 'RETURN' "
5240 RCLOSE
5250 IF J=9 THEN 5280
5260 POINTER X1(U1),Y1(U1),C$
5270 GO TO 5290
5280 POINTER X2(U1),Y2(U1),C$
5290 MOVE 0,53
5300 PRINT " "
5310 ROPEN 2
5320 PRINT "COORDINATES OF THE BEGINNING POINT"
5325 IF J=9 THEN 5343
5330 PRINT USING 5390:"X=  ",X1(U1)
5340 PRINT USING 5390:"Y=  ",Y1(U1)
5341 GO TO 5350
5343 PRINT USING 5390:"X=  ",X2(U1)
5345 PRINT USING 5390:"Y=  ",Y2(U1)
5346 IF J=9 THEN 5580
5350 PRINT "ADJUST THE THUMBWHEELS"
5360 PRINT " TO INPUT THE ENDING"
5370 PRINT " POINT WHEN READY PRESS"
5380 PRINT " 'RETURN' "
5390 IMAGE 4A,5T,3D.1D
5400 RCLOSE
5420 POINTER X2(U1),Y2(U1),C$
5450 MOVE 0,36.5
5460 PRINT " "
5470 ROPEN 3
5480 PRINT "COORDINATES OF THE ENDING POINT"
5490 PRINT USING 5390:"X=  ",X2(U1)
5500 PRINT USING 5390:"Y=  ",Y2(U1)
5501 GO TO 5510
5503 PRINT USING 5390:"X=  ",X1(U1)
```

```
5505 PRINT USING 5390:"Y=   ",Y1(U1)
5510 RCLOSE
5520 ROPEN 8
5530 MOVE 0,25
5540 PRINT "!!!!!!!!!!!!!!!!!!!!!!!!!!!!!!!!!!!!!!!"
5550 PRINT "!!!   ENTER ANOTHER NEW KEY#   !!!"
5560 PRINT "!!!!!!!!!!!!!!!!!!!!!!!!!!!!!!!!!!!!!!!"
5570 RCLOSE
5580 RETURN


7200 REM * THIS SUBROUTINE PLOTS A ARROW *
7250 J=18
7260 GOSUB 9000
7270 MOVE X1(U1),Y1(U1)
7280 DRAW X2(U1),Y2(U1)
7290 ROTATE T1
7295 IF L<=6 THEN 7300
7297 L=6
7300 RMOVE -L/4,L/16
7310 DRAW X2(U1),Y2(U1)
7320 RMOVE -L/4,-L/16
7330 DRAW X2(U1),Y2(U1)
7350 RETURN

9000 REM * THIS SUBROUTINE CHECKS THE CALLING ROUTINE *

9050 IF D9<>1 THEN 9100
9060 U1=U3
9070 GO TO 9110
9100 GOSUB 5070
9110 GOSUB 4720
9120 RETURN
9490 REM  * THIS SUB STORES THE DATA *
9492 RINIT
9495 KILL 17
9500 FIND 17
9505 WRITE @33:L1
9510 WRITE @33:U1
9520 FOR I5=1 TO U1
9530 WRITE @33:X1(I5),Y1(I5),X2(I5),Y2(I5),P9(I5)
9540 NEXT I5
9541 WRITE @33:S$
9542 FOR T=1 TO L1
9543 WRITE @33:H6(T)
9544 NEXT T
9545 MOVE 0,60
9550 ROPEN 7
9560 PRINT "!!!!!!!!!!!!!!!!!!!!!!!!!!!!!!!!!!!!!!!!!!!!!!!!!!!!!"
9562 PRINT "!!!     DATA HAS BEEN STORED ON FILE 17     !!!"
9564 PRINT "!!!          ENTER ANOTHOR NEW KEY#         !!!"
9565 PRINT "!!!!!!!!!!!!!!!!!!!!!!!!!!!!!!!!!!!!!!!!!!!!!!!!!!!!!"
9570 RCLOSE
9600 RETURN
9700 REM * THIS SUB RETRIEVES THE DATA FROM TAPE *
9705 RINIT
9710 FIND 17
9715 READ @33:L1
9720 READ @33:U7
9725 FOR I5=1 TO U7
9730 READ @33:X1(I5),Y1(I5),X2(I5),Y2(I5),P9(I5)
9740 NEXT I5
9742 READ @33:S$
```

```
9744 FOR T=1 TO L1
9746 READ @33;H6(T)
9748 NEXT T
9750 D9=1
9752 T=1
9760 FOR U3=1 TO U7
9770 GOS F9(U3) OF 3750,6250,6000,2620,8000,370,1280,1570,2000,2470,3880
9780 GOSUB F9(U3)-11 OF 4100,8700,7500,6600,3280,8500,7200
9782 CHARSIZE 1
9790 NEXT U3
9800 D9=0
9805 MOVE 0,60
9810 ROPEN 8
9820 PRINT "!!!!!!!!!!!!!!!!!!!!!!!!!!!!!!!!!!!!!!!!!!!!"
9823 PRINT "!!!        ENTER ANOTHER NEW KEY#        !!!"
9825 PRINT "!!!!!!!!!!!!!!!!!!!!!!!!!!!!!!!!!!!!!!!!!!!!"
9827 RCLOSE
9830 RETURN
```

6.8 P. MODULE **FOR PIPELINES DRAFTING**

Just as E. MODULE is needed for drafting of electric circuits, P. MODULE has been prepared for the common need of drafting the elements of pipelines (Fig. 6-11). Transporting fuels, water, chemical compounds, and other fluids requires that piping elements be properly assembled for delivery of these various supplies to different destinations for processing or consumption. Figure 6-12 lists some

Figure 6-11a   Pipelines. (Courtesy of McDonnell–Douglas, St. Louis, Missouri.)

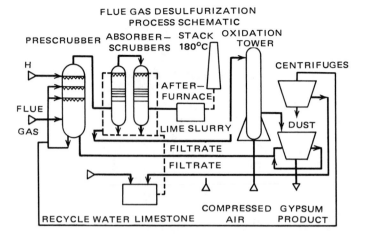

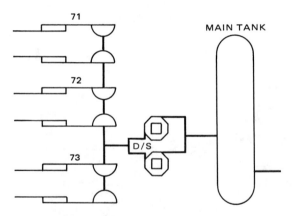

Figure 6-11b Various pipelines in industry. (Courtesy of Ramtek Corporation.)

principal elements that have been programmed in P.MODULE and also shows the plots of the geometric shapes of these elements. Not all piping elements have been incorporated in P.MODULE, those provided should, however, suffice for demonstration usage in this text. As an exercise, the readers may add other elements to P.MODULE for satisfying their particular needs in drafting of pipelines.

The user-definable keys are utilized for entering request for drawing various elements in accordance with the key table. Program P.MODULE provides refreshed messages for instructing the user to maneuver the thumbwheels and then press the "RETURN" key to enter the two sets of (x, y) coordinates for specifying

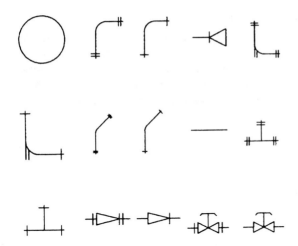

```
* PROGRAM P.MODULE *

KEY TABLE (F = FLANGED. S = SCREWED)

* 1 * ------- CIRCLE
* 2 * ------- OO-DEGREE ELBOW (S)
* 3 * ------- OO-DEGREE ELBOW (S)
* 4 * ------- PIPE PLUG
* 5 * ------- BASE (F)
* 6 * ------- BASE (S)
* 7 * ------- UNUSED
* 8 * ------- 45-DEGREE ELBOW (F)
* 9 * ------- 45-DEGREE ELBOW (S)
*10 * ------- LINE
*11 * ------- STRAIGHT TEE (F)
*12 * ------- STRAIGHT TEE (S)
*13 * ------- CONCENTRIC REDUCER (F)
*14 * ------- CONCENTRIC REDUCER (S)
*15 * ------- LOCKSHIELD VALVE (F)
*16 * ------- LOCKSHIELD VALVE (S)
*17 * ------- REVISE
*18 * ------- STORE
*19 * ------- RETRIEVE
*20 * ------- STOP
```

STEP	KEY	BEGINNING		ENDING	
		X	Y	X	Y
1	1	47.5	60.8	50.5	60.6
2	2	67.5	64.8	74.6	74.1
3	3	68.8	85.0	86.5	74.1
4	4	92.5	69.9	101.1	69.9
5	5	108.7	74.9	115.3	65.7
6	6	40.9	51.7	50.9	40.7
7	8	67.2	49.7	68.7	45.0
8	9	79.9	40.8	81.7	46.3
9	10	91.9	46.3	101.2	46.3
10	11	105.7	43.5	114.6	43.5
11	12	48.3	21.8	58.9	21.0
12	13	64.3	24.4	75.9	24.4
13	14	77.6	24.4	88.8	24.4
14	15	98.6	21.9	101.4	21.9
15	16	105.2	21.9	115.8	21.9

** DESIGN ENDS **

Figure 6-12 Sample application of Program P. MODULE lists the assignment of definable keys for drawing various piping elements.

where a selected element should be drawn. In case the last element drawn needs to be corrected, definable key 17 can be pressed to redraw a new element. The displayed pipe network composed of the drawn elements can be stored in and retrieved from a designated data file on cassette tape by pressing the user-definable keys 18 and 19, respectively. When definable key 20 is pressed, the program is ended and a list of all drawn elements will be displayed on the screen.

PROGRAM P. MODULE

```
1 REM * PROGRAM P.MODULE *
2 GO TO 100
3 CALL "WAIT",1000
4 GOSUB 3100
5 GO TO 3
8 GOSUB 980
9 GO TO 3
12 GOSUB 3050
13 GO TO 3
16 GOSUB 3320
17 GO TO 3
20 GOSUB 3450
21 GO TO 3
24 GOSUB 360
25 GO TO 3
28 GOSUB 760
29 GO TO 3
32 GOSUB 410
33 GO TO 3
36 GOSUB 3220
37 GO TO 3
40 GOSUB 890
41 GO TO 3
44 GOSUB 1790
45 GO TO 3
48 GOSUB 790
49 GO TO 3
52 GOSUB 1470
53 GO TO 3
56 GOSUB 840
57 GO TO 3
60 GOSUB 3820
61 GO TO 3
64 GOSUB 4120
65 GO TO 3
68 GOSUB 1940
69 GO TO 3
72 GOSUB 4240
73 GO TO 3
76 GOSUB 4400
77 GO TO 3
80 GOSUB 1710
89 GO TO 3
100 CHARSIZE 1
110 RINIT
120 PAGE
130 INIT
140 SET KEY
150 GOSUB 170
160 GO TO 290
170 PRINT "* PROGRAM P.MODULE *"
172 PRINT "_KEY TABLE (F=FLANGED, S=SCREWED)"
180 PRINT "_'1'------CIRCLE                  '11'------STRAIGHT TEE(F)"
```

```
190 PRINT "'2'------90-DEGREE ELBOW(F) '12'------STRAIGHT TEE(S)"
200 PRINT "'3'------90-DEGREE ELBOW(S) '13'------CONCENTRIC REDUCER(F)"
210 PRINT "'4'------PIPE PLUG          '14'------CONCENTRIC REDUCER(S)"
220 PRINT "'5'------BASE(F)            '15'------LOCKSHIELD VALVE(F)"
230 PRINT "'6'------BASE(S)            '16'------LOCKSHIELD VALVE(S)"
240 PRINT "'7'------UNUSED             '17'------REVISE    "
250 PRINT "'8'------45-DEGREE ELBOW(F) '18'------STORE     "
260 PRINT "'9'------45-DEGREE ELBOW(S) '19'------RETRIEVE  "
270 PRINT "'10'-----LINE               '20'------STOP      "
280 RETURN
290 P5=0
300 U7=0
310 U1=0
320 D9=0
330 DIM P9(100),X1(100),Y1(100),X2(100),Y2(100)
340 GOSUB 2500
350 GO TO 3
360 REM * THIS SUBROUTINE PLOTS A SCREWED BASE *
370 J=6
380 GOSUB 4170
390 GOSUB 3550
400 RETURN
410 REM * THIS SUB PLOTS THE FLANGED ELBOW (45-DEGREE) *
420 J=8
430 GOSUB 4170
440 GOSUB 460
450 RETURN
460 REM * THIS PORTION PLOTS THE SCREWED PARTS *
470 MOVE X1(U1),Y1(U1)
480 ROTATE 0
490 L=Y2(U1)-Y1(U1)
493 R5=X2(U1)-X1(U1)
500 RDRAW 0,L
510 IF T1>90 AND T1<180 OR (T1>-90 AND T1<0) THEN 520
515 T8=1
517 GO TO 620
520 T8=-1
620 FOR I=180 TO 140 STEP -5
630 RDRAW R5*(COS(I-5)-COS(I)),T8*R5*(SIN(I-5)-SIN(I))
640 NEXT I
650 IF T8=-1 THEN 670
655 RDRAW T8*L*COS(45),L*SIN(45)
657 GO TO 675
670 RDRAW T8*L*COS(45),L*SIN(45)+L*0.01
671 RMOVE 0,-L*0.01
675 ROTATE 45*T8
680 RMOVE -L/15*T8,-L/8
710 RDRAW 0,L/4
711 IF J<>8 THEN 720
713 RMOVE -L/15*T8,-L/4
715 RDRAW 0,L/4
720 MOVE X1(U1),Y1(U1)
725 ROTATE 0
730 RMOVE -L/8,L/15
740 RDRAW L/4,0
741 IF J<>8 THEN 750
743 RMOVE -L/4,L/15
745 RDRAW L/4,0
750 RETURN
760 REM * RESERVED *
770 J=7
780 RETURN
790 REM * THIS SUBROUTINE PLOTS A STRAIGHT TEE (SCREWED) *
800 J=12
810 GOSUB 4170
820 GOSUB 4600
```

```
830 RETURN
840 REM * THIS SUB PLOTS A CONCENTRIC REDUCER (SCREWED) *
850 J=14
860 GOSUB 4170
870 GOSUB 1580
880 RETURN
890 REM * THIS SUBROUTINE PLOTS A LINE *
900 J=10
910 IF D9()1 THEN 940
920 U1=U3
930 GO TO 950
940 GOSUB 2590
950 MOVE X1(U1),Y1(U1)
960 DRAW X2(U1),Y2(U1)
970 RETURN
980 REM * THIS SUBROUTINE PLOTS A 90-DEGREE ELBOW *
990 J=2
1000 GOSUB 4170

1010 GOSUB 1150
1020 MOVE X1(U1),Y1(U1)
1030 IF T9=90 THEN 1060
1040 RMOVE -L/8,L/15*2
1050 GO TO 1070
1060 RMOVE -L/8,L/15*2
1070 RDRAW L/4,0
1080 MOVE X2(U1),Y2(U1)
1090 IF T9=90 THEN 1120
1100 RMOVE -L/15*2,-L/8
1110 GO TO 1130
1120 RMOVE L/15*2,-L/8
1130 RDRAW 0,L/4
1140 RETURN
1150 REM * THIS PORTION PLOTS THE SCREWED ELBOW (90-DEGREE) *
1160 MOVE X1(U1),Y1(U1)
1170 ROTATE 0
1180 M9=TAN(T1)
1190 L=SIN(T1)*L
1200 IF M9>2 THEN 1220
1210 GO TO 1230
1220 L=L/M9
1230 T9=0
1240 RDRAW 0,0.71*L
1250 RMOVE 0,-0.01*L
1260 IF T1<0 AND T1>-90 THEN 1300
1270 IF T1<180 AND T1>90 THEN 1300
1280 RMOVE 0.3*L,0.3*L
1290 GO TO 1310
1300 T9=90
1310 FOR I=90-T9 TO 175-T9 STEP 5
1320 RDRAW 0.3*L*(COS(I+5)-COS(I)),0.3*L*(SIN(I+5)-SIN(I))
1330 NEXT I
1340 IF T9=90 THEN 1370
1350 RMOVE 0.3*L,0.3*L
1370 DRAW X2(U1),Y2(U1)
1380 IF T9=90 THEN 1410
1390 RMOVE -L/15,-L/8
1400 GO TO 1420
1410 RMOVE L/15,-L/8
1420 RDRAW 0,L/4
1430 MOVE X1(U1),Y1(U1)
1440 RMOVE -L/8,L/15
1450 RDRAW L/4,0
1460 RETURN
1470 REM * THIS SUB PLOTS A CONCENTRIC REDUCER (FLANGED) *
1480 J=13
1490 GOSUB 4170
```

```
1500 GOSUB 1580
1510 MOVE X1(U1),Y1(U1)
1520 RMOVE L/4*(2/3),-L/8
1530 RDRAW 0,L/4
1540 MOVE X2(U1),Y2(U1)
1550 RMOVE -L/4*(2/3),-L/8
1560 RDRAW 0,L/4
1570 RETURN
1580 REM * THIS PORTION PLOTS A CONCENTRIC REDUCER (SCREWED) *
1590 MOVE X1(U1),Y1(U1)
1600 ROTATE T1
1610 RDRAW L/4,0
1620 RDRAW 0,L/8
1630 RDRAW L/2,-L/8
1640 RDRAW -L/2,-L/8
1650 RDRAW 0,L/8
1660 RMOVE L/2,-L/8
1670 RDRAW 0,L/4
1680 RMOVE 0,-L/8
1690 DRAW X2(U1),Y2(U1)
1700 RETURN
1710 REM * PROGRAM ENDS *
1720 G9=8
1730 RINIT
1740 GOSUB 2230
1750 REM INIT
1760 MOVE 0,10
1770 PRINT "**  DESIGN ENDS  **"
1780 END
1790 REM * THIS SUBROUTINE PLOTS A STRAIGHT TEE (FLANGED) *
1800 J=11
1810 GOSUB 4170
1820 GOSUB 4600
1830 MOVE X1(U1),Y1(U1)
1840 RMOVE L/7,-L/8
1850 RDRAW 0,L/4
1860 MOVE X2(U1),Y2(U1)
1870 RMOVE -L/7,-L/8
1880 RDRAW 0,L/4
1890 MOVE X1(U1),Y1(U1)
1900 RMOVE L/2,0.6*L
1910 RMOVE -L/8,-L/7
1920 RDRAW L/4,0
1930 RETURN
1940 REM * THIS SUBROUTINE REVISES THE LAST DRAWN ELEMENT *
1950 RINIT
1960 MOVE 0,80
1970 MOVE 0,20
1980 U2=U1
1990 PRINT "INPUT THE NEW KEY NUMBER USING "
2000 PRINT "  NUMERIC KEYS ; NOT DEFINABLE KEY!!!"

2010 INPUT P9(U2)
2020 PRINT "ADJUST THE THUMBWHEELS TWICE!!!"
2030 PRINT "  TO INPUT THE BEGINNING AND"
2040 PRINT "  ENDING POINTS OF THE ELEMENT"
2050 POINTER X1(U2),Y1(U2),C$
2060 POINTER X2(U2),Y2(U2),C$
2070 D9=1
2080 PAGE
2090 FOR U3=1 TO U1
2100 GOSUB P9(U3) OF 3100,980,3050,3320,3450,360,760,410,3220,890,1790
2110 GOSUB P9(U3)-11 OF 790,1470,840,3820,4120,1940,4240,4400,1710
2120 NEXT U3
2130 D9=0
2140 MOVE 0,98.3
2150 GOSUB 170
```

```
2160 MOVE 0,70
2170 ROPEN 5
2180 PRINT "!!!!!!!!!!!!!!!!!!!!!!!!!!!!!!!!!!!!!!!!!!!!"
2190 PRINT "!!  ENTER ANOTHER NEW KEY#          !!"
2200 PRINT "!!!!!!!!!!!!!!!!!!!!!!!!!!!!!!!!!!!!!!!!!!!!"
2210 RCLOSE
2220 GO TO 3
2230 REM * THIS SUBROUTINE PRINTS OUT THE ELEMENT COORDIANTES *
2240 MOVE 0,70
2250 PRINT "STEP# KEY#  BEGINNING        ENDING "
2260 PRINT "              X      Y       X      Y"
2270 FOR I5=1 TO U1
2280 PRINT USING 2290:I5,P9(I5),X1(I5),Y1(I5),X2(I5),Y2(I5)
2290 IMAGE 2D,4X,2D,2X,4(3D,1D,2X)
2300 NEXT I5
2310 RETURN
2320 REM * THIS SUB FINDS THE LENGTH AND ROTATION ANGLE *
2330 SET DEGREES
2340 L=((X1(U1)-X2(U1))^2+(Y1(U1)-Y2(U1))^2)^0.5
2350 IF X1(U1)>X2(U1) AND Y1(U1)=Y2(U1) THEN 2430
2360 IF X1(U1)=X2(U1) AND Y1(U1)>Y2(U1) THEN 2450
2370 IF X1(U1)>X2(U1) AND Y1(U1)>Y2(U1) THEN 2470
2380 IF Y1(U1)=>Y2(U1) THEN 2410
2390 T1=ACS((X2(U1)-X1(U1))/L)
2400 GO TO 2490
2410 T1=ASN((Y2(U1)-Y1(U1))/L)
2420 GO TO 2490
2430 T1=180
2440 GO TO 2490
2450 T1=-90
2460 GO TO 2490
2470 T1=ATN((Y2(U1)-Y1(U1))/(X2(U1)-X1(U1)))
2480 T1=T1+180
2490 RETURN
2500 REM * THIS SUBROUTINE INSTRUCTS THE USER *
2510 ROPEN 7
2520 IF U1<>0 THEN 2580
2530 MOVE 0,75
2540 PRINT "******************************************"
2550 PRINT "**          ENTER FIRST KEY#          **"
2560 PRINT "******************************************"
2570 RCLOSE
2580 RETURN
2590 REM * INPUT THE BEGINNING AND ENDING POINTS *
2600 P5=P5+1
2610 U1=P5+U7
2620 P9(U1)=J
2630 RINIT
2640 MOVE 0,65
2650 ROPEN 1
2660 PRINT "KEY# CHOSEN: ";J
2670 PRINT "ADJUST THE THUMBWHEELS"
2680 PRINT " TO INPUT THE BEGINNING"
2690 PRINT "POINT WHEN READY PRESS"
2700 PRINT " 'RETURN' "
2710 RCLOSE
2720 POINTER X1(U1),Y1(U1),C$
2730 MOVE 0,53
2740 PRINT " "
2750 ROPEN 2
2760 PRINT "COORDINATE OF BEGINNING POINT"
2770 PRINT USING 2830:"X= ",X1(U1)
2780 PRINT USING 2830:"Y=  ",Y1(U1)
2790 PRINT "ADJUST THE THUMBWHEELS"
2795 IF J=8 OR J=9 THEN 2805
2800 PRINT "TO INPUT THE ENDING"
2801 PRINT "POINT WHEN READY PRESS"
```

```
2803 GO TO 2820
2805 PRINT "TO INPUT THE CENTER OF "
2810 PRINT "CURVATURE WHEN READY PRESS"
2820 PRINT " 'RETURN' "
2830 IMAGE 4A,5T,3D.1D
2840 RCLOSE
2850 POINTER X2(U1),Y2(U1),C$
2860 MOVE 0,36.5
2870 PRINT " "
2880 ROPEN 3
2885 IF J=8 OR J=9 THEN 2895
2890 PRINT "COORDINATE OF ENDING POINT"
2891 GO TO 2900
2895 PRINT "COORDINATE OF CENTER OF CURVATURE"
2900 PRINT USING 2830:"X=   ",X2(U1)
2910 PRINT USING 2830:"Y=   ",Y2(U1)
2920 RCLOSE
2930 ROPEN 8
2940 MOVE 0,25
2950 PRINT "!!!!!!!!!!!!!!!!!!!!!!!!!!!!!!!!!!!!"
2960 PRINT "!!!  ENTER ANOTHER NEW KEY#  !!!"
2970 PRINT "!!!!!!!!!!!!!!!!!!!!!!!!!!!!!!!!!!!!"
2980 RCLOSE
2990 RETURN
3000 REM ******************************************************************

3010 REM **           U1 IS COUNTER  ;   P9 STORE KEY #          **
3020 REM **           D9 IS INDENTIFICATION FOR WHICH CALLING    **
3030 REM **  G9 IS THE IDENTIFIER TO STORE THE DATA OR NOT       **
3040 REM **       R7 IS THE RADIUS ;                             **
3050 REM * THIS SUBROUTINE PLOTS A SCREWED ELBOW (90-DEGREE) *
3060 J=3
3070 GOSUB 4170
3080 GOSUB 1150
3090 RETURN
3100 REM * THIS SUBROUTINE PLOTS A CIRCLE ELEMENT *
3110 J=1
3120 IF D9()1 THEN 3150
3130 U1=U3
3140 GO TO 3160
3150 GOSUB 2590
3160 GOSUB 2320
3170 R7=0.5*L
3180 ROTATE T1
3190 MOVE X2(U1),Y2(U1)
3200 GOSUB 3270
3210 RETURN
3220 REM * THIS SUBROUTINE PLOTS A SCREWED ELBOW (45-DEGREE) *
3230 J=9
3240 GOSUB 4170
3250 GOSUB 460
3260 RETURN
3270 REM * THIS SUBROUTINE PLOTS A CIRCLE *
3280 FOR D5=0 TO 357 STEP 3
3290 RDRAW R7*(COS(D5+3)-COS(D5)),R7*(SIN(D5+3)-SIN(D5))
3300 NEXT D5
3310 RETURN
3320 REM * THIS SUBROUTINE PLOTS A PIPE PLUG *
3330 J=4
3340 GOSUB 4170
3350 MOVE X1(U1),Y1(U1)
3360 ROTATE T1
3370 RDRAW L/2,0
3380 RDRAW 0,L/6
3390 RDRAW 0,-L/3
3400 RMOVE 0,L/6
3410 RDRAW L/2,-L/4
```

```
3420 RDRAW 0,L/2
3430 RDRAW -L/2,-L/4
3440 RETURN
3450 REM * THIS SUBROUTINE PLOTS A FLANGED BASE *
3460 J=5
3470 GOSUB 4170
3480 GOSUB 3550
3490 RMOVE 0,-L/15
3500 RDRAW -L/4,0
3510 MOVE X2(U1),Y2(U1)
3520 RMOVE -T8*2/15*L,-L/8
3530 RDRAW 0,L/4
3540 RETURN
3550 REM * THIS PORTION PLOTS THE SCREWED BASE *
3560 MOVE X1(U1),Y1(U1)
3570 ROTATE 0
3580 L=-L*SIN(T1)
3590 M8=TAN(T1)
3600 IF M8<=2 THEN 3620
3610 L=L/M8
3620 RDRAW 0,-(L+L/8)
3630 RMOVE 0,L/8+L*0.3
3640 R7=0.3*L
3650 IF T1>0 AND T1<90 OR (T1>180 AND T1<270) THEN 3680
3660 T8=1
3670 GO TO 3690
3680 T8=-1
3690 FOR D5=180 TO 265 STEP 5
3700 RDRAW T8*R7*(COS(D5+5)-COS(D5)),R7*(SIN(D5+5)-SIN(D5))
3710 IF D5<>215 THEN 3740
3720 RDRAW 0,R7*SIN(D5+3)-(0.3-1/4)*L
3730 RMOVE 0,-(R7*SIN(D5+3)-(0.3-1/4)*L)
3740 NEXT D5
3750 DRAW X2(U1),Y2(U1)
3760 RMOVE -T8*L/15,-L/8
3770 RDRAW 0,L/4
3780 MOVE X1(U1),Y1(U1)
3790 RMOVE -L/8,-L/15
3800 RDRAW L/4,0
3810 RETURN
3820 REM * THIS SUB PLOTS A LOCKSHIELD VALVE (FLANGED) *
3830 J=15
3840 GOSUB 4170
3850 GOSUB 3920
3860 MOVE X1(U1),Y1(U1)
3870 RMOVE L/4*(2/3),-L/8
3880 RDRAW 0,L/4
3890 RMOVE L/4*(2/3)+L/2,0
3900 RDRAW 0,-L/4
3910 RETURN
3920 REM * THIS PORTION PLOTS THE LOCKSHIELD VALVE (SCREWED) *
3930 MOVE X1(U1),Y1(U1)
3940 ROTATE T1
4050 RDRAW -L/4*(3/5),0
4060 RDRAW -L/20,-L/20
4070 RMOVE (L/20+L/4*(3/5))*2,0
4080 RDRAW -L/20,L/20
4090 RDRAW -L/4*(3/4),0
4100 RETURN
4110 RETURN
4120 REM * THIS SUB PLOTS A LOCKSHIELD VALVE (SCREWED) *
4130 J=16
4140 GOSUB 4170
4150 GOSUB 3920
```

```
4160 RETURN
4170 REM * THIS SUBROUTINE CHECKS THE CALLING ROUTINE *
4180 IF D9<>1 THEN 4210
4190 U1=U3
4200 GO TO 4220
4210 GOSUB 2590
4220 GOSUB 2320
4230 RETURN
4240 REM * THIS SUB STORES THE DATA *
4250 RINIT
4260 KILL 36
4270 FIND 36
4280 WRITE @33:U1
4290 FOR I5=1 TO U1
4300 WRITE @33:X1(I5),Y1(I5),X2(I5),Y2(I5),P9(I5)
4310 NEXT I5
4320 MOVE 0,60
4330 ROPEN 7
4340 PRINT "!!!!!!!!!!!!!!!!!!!!!!!!!!!!!!!!!!!!!!!!!!!!!!!!"
4350 PRINT "!!!      DATA HAS BEEN STORED ON FILE 36     !!!"
4360 PRINT "!!!            ENTER ANOTHOR NEW KEY#         !!!"
4370 PRINT "!!!!!!!!!!!!!!!!!!!!!!!!!!!!!!!!!!!!!!!!!!!!!!!!"
4380 RCLOSE
4390 RETURN
4400 REM * THIS SUB RETRIEVES THE DATA FROM TAPE *
4410 RINIT
4420 FIND 36
4430 READ @33:U7
4440 FOR I5=1 TO U7
4450 READ @33:X1(I5),Y1(I5),X2(I5),Y2(I5),P9(I5)
4460 NEXT I5
4470 D9=1
4480 FOR U3=1 TO U7
4490 GOSUB P9(U3) OF 3100,980,3050,3320,3450,360,760,410,3220,890,1790
4500 GOSUB P9(U3)-11 OF 790,1470,840,3820,4120,1940,4240,4400,1710
4510 NEXT U3
4520 D9=0
4530 MOVE 0,60
4540 ROPEN 8
4550 PRINT "!!!!!!!!!!!!!!!!!!!!!!!!!!!!!!!!!!!!!!!!!!!!!!!!"
4560 PRINT "!!!         ENTER ANOTHER NEW KEY#        !!!"
4570 PRINT "!!!!!!!!!!!!!!!!!!!!!!!!!!!!!!!!!!!!!!!!!!!!!!!!"
4580 RCLOSE
4590 RETURN
4600 REM * THIS SUB PLOTS A SCREWED TEE *
4610 MOVE X1(U1),Y1(U1)
4620 DRAW X2(U1),Y2(U1)
4630 ROTATE T1
4640 RMOVE -L/14,-L/8
4650 RDRAW 0,L/4
4660 MOVE X1(U1),Y1(U1)
4670 RMOVE L/14,-L/8
4680 RDRAW 0,L/4
4690 MOVE X1(U1),Y1(U1)
4700 RMOVE L/2,0
4710 RDRAW 0,0.6*L
4720 RMOVE -L/8,-L/14
4730 RDRAW L/4,0
4740 RETURN
4750 IF T1<180 AND T1>90 THEN 1300
4760 RMOVE 0.3*L,0.3*L
4770 GO TO 1370
4780 DRAW X2(U1),Y2(U1)
4790 IF T9=90 THEN 1410
```

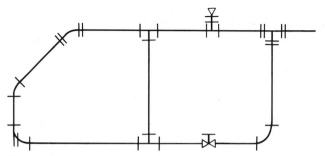

Figure 6-13   A sample pipeline for exercise of P.MODULE.

EXERCISE

Use Program P.MODULE to plot the pipe network shown in Fig. 6-13.

EXERCISE

Flowcharts are often employed for planning how a computer program should be written. The conventional symbols used for drawing a flowchart are listed in Fig. 6-14. Write a BASIC Program F.CHART to assign user-definable keys for drawing of these symbols and apply this program to draw the flow chart of solving a quadratic equation that may have real or complex roots. Refer to Program DIM.NING for incorporation of labeling as another definable-key arrangement so that instructions can be placed inside the flowchart symbols.

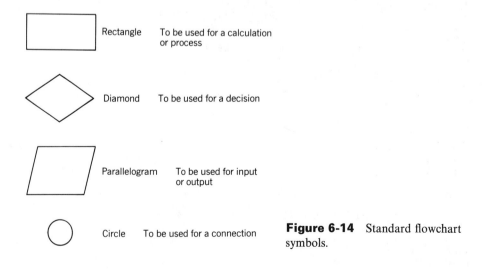

Oval   To be used for start or stop

Rectangle   To be used for a calculation or process

Diamond   To be used for a decision

Parallelogram   To be used for input or output

Circle   To be used for a connection

Figure 6-14   Standard flowchart symbols.

6.9   COPY DRAWING/CHART ON TABLET TO CRT AND DATA FILE

A basic need in CAD is to copy onto the cathode ray tube of a computer graphics display terminal a design that has been sketched on paper. If necessary, the sketch should be refined with a computer program residing in the display unit and then displayed. And it is also desirable that the display be stored in a data file of a peripheral device such as a cassette tape or a floppy disk so that the design can be redrawn, modified, stored again, retrieved, then modified again, and so on, as is often necessary in a CAD trial process.

The listed Program COPY.KEEP has been developed for such a purpose. By laying the sketch flat on the surface of the graphics tablet, the stylus can be lifted and then pressed on two points for drawing a straight line, or be kept pressing on the tablet but moved by tracing along a curve to copy it onto the screen. The tablet control device (Fig. 5-7) has to be set at "POINT" and "STREAM" modes by pushing in the respective switches in order to achieve the described line and curve copying.

Figure 6-15a is a hand-drawn sketch, whereas Figure 6-15b is a hardcopy of the display when the sketch was copied onto CRT by use of Program COPY.KEEP. Line AB was not traced but copied onto CRT as a straight line by pressing the stylus on the vertices A and B when the POINT switch of the tablet control device was pushed in. Curve DA was traced with the STREAM switch of the tablet control device being pushed in. Program COPY.KEEP has been prepared in such a way (Fig. 6-16) that unless the definable key 2 is pressed to relocate a new starting point, the program continues to draw a line from the current stylus position to the point where the stylus is moved and pressed on the tablet. Definable key 4 is assigned for labeling the characters entered from keyboard. The starting position of the label is specified by pressing the stylus. The character size can also be entered by the user from keyboard. The printout on the left edge of Fig. 6-15b, 4A, 4B, 4C, and 4D, are examples of specified character size and labels.

At present, Program COPY.KEEP can store only the copied drawing onto a data file and retrieve it to be redrawn on CRT. It is left as an exercise for the readers to adapt the REVISE key arrangement as in Program P.MODULE so

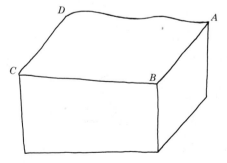

Figure 6-15a   Hand-drawn sketch of a three-dimensional solid.

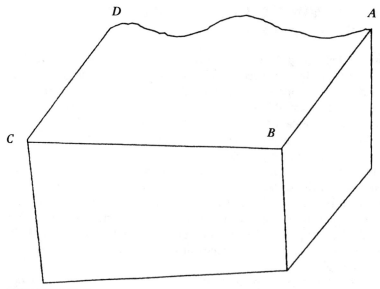

ENTER CHART OR GRAPH

4A
4B
4C
4D

USER KEY GUIDE, 2-MOVE WKD DRAWING. 3-END PROGRAM, 4-LABEL,
ENTER LOCATION W/PEN, CHARSIZE. THEN LABEL.

Figure 6-15b   Copied and modified display of the hand-drawn sketch
shown in Figure. 6-15a by use of Program COPY.KEEP.

```
*********************************
* TABLET TO DATA FILE PROGRAM *
*********************************
```

THIS PROGRAM USES THE GRAPHIC TABLET TO INPUT A CHART OR GRAPH TO A DATA FILE
FOR FUTURE DISPLAY ON GRAPHIC SYSTEM CRT.

DATA FILE SIZE SHOULD BE 20 X ESTIMATED NUMBER OF DATA POINT + 1000.
ENTER DATA FILE NUMBER & SIZE = 19,6000

OPERATING INSTRUCTIONS:

1. TO MOVE AND NOT DRAW. PRESS AND RELEASE USER DEFINABLE KEY NO. 2 BEFORE TOUCHING NEXT POINT WITH PEN.

2. TO END PROGRAM, PRESS AND RELEASE USER DEFINABLE KEY NO.3 AND TOUCH PEN TO TABLET.

3. TO PRINT A LABEL AT A SELECTED POINT, PRESS AND RELEASE KEY NO. 4. TOUCH PEN AT LOWER LEFT CORNER OF DESIRED
   LABEL LOCATION, AND ENTER CHARSIZE (1,2,3, OR 4) AND LABEL BY USE OF KEYBOARD.

4. TO ENTER A CURVE OR TO FREE HAND SKETCH, PRESS IN STREAM SWITCH ON TABLET CONTROLLER. SET STREAM FLOW TO NEAR
   LOW AND SWITCH BACK TO POINT MODE IMMEDIATELY WHEN THE CURVE IS COMPLETED TO AVOID OVERLOADING YOUR DATA FILE!

NOW ENTER THE ORIGIN ON TABLET BY PUSHING IN 'ORIGIN' BOTTON ON TABLET CONTROL BOX AND TOUCHING PEN AT THE ORIGIN
LOCATION DESIRED   RELEASE ORIGIN BUTTON. NEXT MESSAGE WILL APPEAR WHEN DATA HAS BEEN ENTERED.

TO ENTER A NEW DRAWING, ENTER N.   TO DRAW AN EXISTING DRAWING ALREADY IN DATA FILE, ENTER E. : N

ENTER WINDOW SIZE BY TOUCHING PEN AT OPPOSITE DIAGONAL CORNERS OF WINDOW WANTED.   THE LEFT LOWER AND RIGHT
UPPER CORNERS OF THE DRAWING PAPER COULD BE ENTERED.
TWO SETS OF COORDINATES SHOULD APPEAR AFTER DATA ENTRY.
  0.01
  5.45 4.1951
NOW YOU ARE READY TO TRACE THE DRAWING FROM TABLET TO CRT. WHEN READY PRESS 'RETURN'.

Figure 6-16   Instruction menu of Program COPY.KEEP.

that Program COPY.KEEP can be expanded to have the capability of adding more lines, curves, and labels to the existing file. G̲ and ↑ used in Program P.MODULE are for ringing the bell and returning the cursor to the home position, respectively.

PROGRAM COPY . KEEP

```
1 INIT
2 REM * PROGRAM COPY.KEEP *
3 GO TO 100
8 GO TO 870
12 GO TO 920
16 GO TO 950
100 SET KEY
110 SET DEGREES
120 PAGE
130 CHARSIZE 4
140 PRINT USING 170:"*****************************"
150 PRINT USING 170:"* TABLET TO DATA FILE PROGRAM *"
160 PRINT USING 170:"*****************************"
170 IMAGE 20X,31A
180 CHARSIZE 3
190 PRINT "_THIS PROGRAM USES THE GRAPHIC TABLET TO INPUT A CHART OR ";
200 PRINT "GRAPH TO A DATA FILE "
210 PRINT "FOR FUTURE DISPLAY ON GRAPHIC SYSTEM CRT."
220 REMAR "_DATA FILE SIZE SHOULD BE 20 X ESTIMATED NUMBER OF DATA ";
230 REMAR "POINT + 1000."
240 REMAR "ENTER DATA FILE NUMBER & SIZE = ";
250 F1=12
260 FIND F1
265 M=5120
270 CHARSIZE 3
280 PRINT "_OPERATING INSTRUCTIONS:"
290 CHARSIZE 2
300 PRINT "_ 1. TO MOVE AND NOT DRAW, PRESS AND RELEASE USER DEFINA"
310 PRINT "_                                                       ";
320 PRINT "BLE KEY NO. 2 BEFORE TOUCHING NEXT POINT WITH PEN._"
330 PRINT " 2. TO END PROGRAM, PRESS AND RELEASE USER DEFINABLE KEY NO."
340 PRINT "_  _____                  ";
350 PRINT " 3 AND TOUCH PEN TO TABLET."
360 PRI " 3. TO PRINT A LABEL AT A SELECTED POINT, PRESS AND RELEASE "
370 PRINT "_  _____       ";
380 PRINT " KEY NO. 4, TOUCH PEN AT LOWER LEFT CORNER OF DESIRED"
390 PRINT "    LABEL LOCATION, AND ENTER CHARSIZE (1,2,3, OR 4) AND";
400 PRINT " LABEL BY USE OF KEYBOARD."
410 PRINT "_ 4. TO ENTER A CURVE OR TO FREE HAND SKETCH, PRESS IN ";
420 PRINT "STREAM SWITCH ON TABLET CONTROLLER.  SET STREAM FLOW TO NEAR"
430 PRINT "   LOW AND SWITCH BACK TO POINT MODE IMMEDIATELY WHEN THE";
440 PRINT " CURVE IS COMPLETED TO AVOID OVERLOADING YOUR DATA FILE!"
450 PRINT "_NOW ENTER THE ORIGIN ON TABLET BY PUSHING IN `ORIGIN' ";
460 PRINT "BOTTON ON TABLET CONTROL BOX AND TOUCHING PEN AT THE ORIGIN"
470 PRINT "LOCATION DESIRED.  RELEASE ORIGIN BUTTON. NEXT MESSAGE WILL";
480 PRINT " APPEAR WHEN DATA HAS BEEN ENTERED."
490 REM *DUMMY INPUT STATEMENT FOR ORIGIN DATA*
500 INPUT @8:X,Y,Z$
510 PRINT "_TO ENTER A NEW DRAWING, ENTER N.  TO DRAW AN EXISTING ";
520 PRINT "DRAWING ALREADY IN DATA FILE, ENTER E. : ";
530 INPUT B$
540 IF B$="E" THEN 1120
550 PRINT "_ENTER WINDOW SIZE BY TOUCHING PEN AT OPPOSITE DIAGONAL ";
560 PRINT "CORNERS OF WINDOW WANTED.  THE LEFT LOWER AND RIGHT "
```

```
570 PRINT "UPPER CORNERS OF THE DRAWING PAPER COULD BE ENTERED."
572 PRINT "TWO SETS OF COORDINATES SHOULD APPEAR AFTER DATA ENTRY."
580 INPUT @8:A,B,C$
590 PRINT A;B;C$
600 INPUT @8:D,E,F$
610 PRINT D;E;F$
620 A=1.3*A
630 D=1.3*D
640 WRITE @33:A,B,D,E
660 PRINT "NOW YOU ARE READY TO TRACE THE DRAWING FROM TABLET TO CRT. ";
665 PRINT "WHEN READY PRESS 'RETURN'."
668 INPUT F$
670 PAGE
680 HOME
690 PRINT "ENTER CHART OR GRAPH"
700 MOVE 0,10
710 CHARSIZE 2
720 PRINT "                        USER KEY GUIDE:  2-MOVE W/O DRAWING, 3-END";
730 PRINT " PROGRAM, 4-LABEL: ENTER LOCATION W/PEN, CHARSIZE, "
732 PRINT "                                       THEN LABEL."
740 WINDOW A,D,B,E
750 N=0
760 INPUT @8:X,Y,Z$
770 MOVE X,Y
780 Z$="B"
790 WRITE @33:X,Y,Z$
800 P=1
810 N=N+1
820 IF N*20=>M THEN 1090
830 INPUT @8:X,Y,Z$
840 DRAW X,Y
850 WRITE @33:X,Y,Z$
860 GO TO 810
870 REM  SUBROUTINE MOVE
880 Z$="B"
890 MOVE X,Y
900 WRITE @33:X,Y,Z$
910 GO TO 810
920 Z$="END"
930 WRITE @33:X,Y,Z$
940 GO TO 1320
950 REM  THIS SUBROUTINE WRITES AND STORES LABELS.
960 P=P+1
970 PRINT "_"
980 FOR P1=1 TO P
990 PRINT "_"
1000 NEXT P1

1010 Z$="D"
1020 INPUT C1,A$
1030 MOVE X,Y
1040 CHARSIZE C1
1050 PRINT A$
1060 WRITE @33:X,Y,Z$,C1,A$
1070 N=N+2
1080 GO TO 820
1090 PRINT "_____WARNING: YOU MAY EXCEED DATA FILE SIZE!"
1100 MOVE X,Y
1110 GO TO 830
1120 REM  * SUBROUTINE DISPLAYS DRAWING CURRENTLY IN DATA FILE *
1130 READ @33:A,B,D,E
1140 WINDOW A,D,B,E
1150 PAGE
1160 P=0
```

```
1170 N=0
1180 N=N+1
1190 READ @33:X,Y,Z$
1200 IF Z$="B" THEN 1250
1210 IF Z$="D" THEN 1270
1220 IF Z$="END" THEN 1320
1230 DRAW X,Y
1240 GO TO 1180
1250 MOVE X,Y
1260 GO TO 1180
1270 MOVE X,Y
1280 READ @33:C1,A$
1290 CHARSIZE C1
1300 PRINT A$
1310 GO TO 1180
1320 WINDOW 0,130,0,130
1330 MOVE 0,4
1340 CHARSIZE 2
1350 PRINT "END OF PROGRAM"
1360 CHARSIZE 2
1370 END
```

6.10 KINEMATICS OF A CRANK-SLIDER MECHANISM

Graphical methods are frequently used in determination of the velocity and acceleration of machine and structural components. Take the simple example of the internal combustion engine of an automobile. The piston of a cylinder, the connecting rod, and the rotating crank form a crank-slider mechanism (Fig. 6-17). For simplicity in discussion, the crank is assumed to rotate at a constant angular velocity ω. It is of importance in machine design to know the velocity and acceleration of the slider in terms of ω and the lengths of the crank and the connecting rod. If the crank has an angular acceleration, the motion of the slide has to be studied at every instant as ω and the geometric configuration of the mechanism are changed.

Klein developed a graphical method for drawing the velocity and acceleration vector polygons of commonly encountered mechanism problems.[5] Program CRANK.S makes use of Klein's method to interactively draw the velocity and acceleration polygons of a crank-slider mechanism on the CRT of the computer graphics display device. Figure 6-17 was drawn by Program CRANK.S. It illustrates that the user has to maneuver the thumbwheels for interactively entering the coordinates of the two intercepts of the two circles. A line is to be drawn passing these two points to intercept the line *AC* at *D*, which needs to be labeled interactively also. Again by use of the thumbwheels to specify where the velocity and acceleration polygons are to be placed and to enter points *A* and *D*, the desired information about kinematic data of the crank-slider mechanism can completely be determined and graphically displayed on the screen.

---

[5]Referred to in F. H. Raven, "Kinematics," Chapter 12 in *Handbook of Engineering Mechanics*, edited by W. Flugge, McGraw-Hill, 1962.

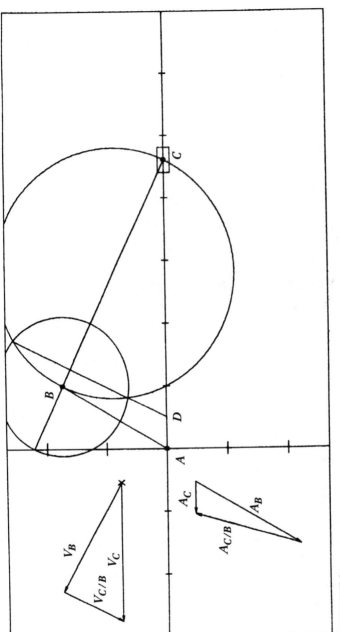

* PROGRAM CRANK-S *

GRAPHIC ANALYSIS OF CRANK-SLIDER MECHANISM
   USING THE KLEIN METHOD.
CRANK LENGTH L1. CM = 2
COUPLER LENGTH L2. CM = 4
CRANK POSITION THETA1. DEGREE = 68
CRANK ANGULAR VELOCITY V1. RPM = 5
ENTER CHARACTER SIZE FOR LABELLING = 3
VELOCITY OF B = 1.8472 CM/SEC.
ACCELERATION OF B = 8.54831392 CM/SEC12.

*** FOLLOW THE REFRESHED MESSAGE ***
*** AND TAKE ACTION AS INSTRUCTED ***

NUMERIC KEYBOARD SELECTION: 1 = MOVE, 2 = CIRCLE, 3 = LABEL, 4 = MEASURE LENGTH, 5 = MEASURE ANGLE, 6 = EXTEND LINE, 7 = VELOCITY POLYGON, 8 = ACCELERATION POLYGON,
                                         9 = REPEAT/END PROGRAM

Figure 6-17   Interactive CAD of mechanism.

Program CRANK.S has also demonstrated the use of numeric keys, instead of definable keys as in Programs E.MODULE and P.MODULE, for interactive graphics applications. This is made possible by the availability of the function VAL. As line 1630 in Program CRANK.S shows, VAL converts a character into numeric value, which can then be used for control in a GOSUB statement such as line 1640. Indeed, Program CRANK.S can be further improved. For instance, the interactive entering of the coordinates of the points *A* and *D* could certainly be replaced by internal computation of the program. The present arrangement, however, provides the user more opportunity to explore interactive graphics practices.

PROGRAM CRANK.S

```
90 PAGE
92 INIT
95 CHARSIZE 2
100 PRINT "* PROGRAM CRANK-S *"
110 PRINT "_GRAPHIC ANALYSIS OF CRANK-SLIDER MECHANISM "
120 PRINT "  USING THE KLEIN METHOD."
130 N=0
140 VIEWPORT 0,130,0,100
150 WINDOW 0,130,0,100
160 SET DEGREES
170 IF N>0 THEN 190
180 N=N+1
190 REM *CRANKSLIDER DISPLAY ROUTINE*
200 PRINT "CRANK LENGTH L1, CM = ";
210 INPUT L1
220 PRINT "COUPLER LENGTH L2, CM = ";
230 INPUT L2
240 PRINT "CRANK POSITION THETA1, DEGREE = ";
250 INPUT T1
260 IF T1=>0 THEN 280
270 T1=360-T1
280 IF T1<=360 THEN 300
290 T1=T1-360
300 PRINT "CRANK ANGULAR VELOCITY W1, RPM = ";
310 INPUT W1
320 PRINT "ENTER CHARACTER SIZE FOR LABELLING = ";
330 INPUT C
340 V8=W1*6.2832*L1/60
350 A8=V8^2/L1
360 PRINT "VELOCITY OF B = ";V8;" CM/SEC."
370 PRINT "ACCELERATION OF B = ";A8;" CM/SEC^2."
380 MOVE 70,90
390 PRINT "*** FOLLOW THE REFRESHED MESSAGE ***"
400 MOVE 70,87
410 PRINT "*** AND TAKE ACTION AS INSTRUCTED ***"
420 S1=ABS(W1)
430 S2=W1^2
440 MOVE 0,0
450 PRINT "_NUMERIC KEYBOARD SELECTION : 1=MOVE, 2=CIRCLE,";
460 PRINT " 3=LABEL, 4=MEASURE LENGTH, 5=MEASURE ANGLE, 6=EXTEND LINE,"
470 PRINT "                              7=VELOCITY POLYGON, ";
480 PRINT "8=ACCELERATION POLYGON, 9=REPEAT/END PROGRAM"
490 MOVE 0,10
500 DRAW 130,10
510 DRAW 130,80
520 DRAW 0,80
```

```
530 DRAW 0,10
540 VIEWPORT 0,130,10,80
550 REM *SCALE ROUTINE*
560 H1=-L1-1
570 H2=L1+L2+1
580 V1=(-H1+H2)*70/130/2
590 WINDOW H1,H2,-V1,V1
600 AXIS 1,1
610 REM *CRANKSLIDER DISPLAY*
620 B1=L1*COS(T1)
630 B2=L1*SIN(T1)
640 C1=B1+SQR(L2^2-B2^2)
650 MOVE 0,0
660 DRAW B1,B2
670 DRAW C1,0
680 E2=B2*C1/(C1-B1)
690 DRAW 0,E2
700 MOVE C1-L1/10,-L1/20
710 DRAW C1+L1/10,-L1/20
720 DRAW C1+L1/10,L1/20
730 DRAW C1-L1/10,L1/20
740 DRAW C1-L1/10,-L1/20
750 REM *DRAW PINS AT JOINTS*
760 CHARSIZE 1
770 H=-H1+H2
780 MOVE -0.0022*H,-0.006*H
790 PRINT "O"
800 MOVE B1-0.0022*H,B2-0.006*H
810 PRINT "O"
820 MOVE C1-0.0022*H,-0.006*H
830 PRINT "O"
840 CHARSIZE 3
850 MOVE -0.1*L1,-0.15*L1
860 PRINT "A"
870 MOVE -0.1*L1+B1,0.05*L1+B2
880 PRINT "B"
890 MOVE C1,-0.15*L1
900 PRINT "C"
910 X1=0.5*(B1+C1)
920 Y1=0.5*B2
930 X2=C1
940 Y2=0
950 GOSUB 1710
960 X1=B1
970 Y1=B2
980 X2=0
990 Y2=E2
1000 GOSUB 1710

1010 ROPEN 1
1020 MOVE H1,-V1
1030 CHARSIZE 2
1040 PRINT "ADJUST THE THUMBWHEELS TO THE INTERCEPTS OF THE ";
1050 PRINT "TWO CIRCLES FOR DRAWING A STRAIGHT LINE AND ENTENDED TO ";
1060 PRINT "THE X-AXIS."
1070 RCLOSE
1080 POINTER X1,Y1,U$
1090 POINTER X2,Y2,U$
1100 RINIT
1110 GOSUB 2010
1120 ROPEN 1
1130 MOVE H1,-V1
1140 PRINT "ADJUST THE THUMBWHEELS TO THE NEIGHBORHOOD OF THE NEWLY ";
```

```
1150 PRINT "FOUND INTERCEPT AND ";
1160 PRINT "PRESS '3' KEY TO LABEL IT WITH 'D'."
1170 RCLOSE
1180 POINTER X2,Y2,Z$
1190 GOSUB 1800
1200 GO TO 1500
1210 REM *VECTOR ARROW ROUTINE*
1220 IF L<0 THEN 1250
1230 E=-0.01*H
1240 GO TO 1260
1250 E=0.01*H
1260 ROTATE A
1270 RDRAW L,0
1280 ROTATE A+10
1290 RDRAW E,0
1300 RMOVE -E,0
1310 ROTATE A-10
1320 RDRAW E,0
1330 RMOVE -E,0
1340 IF ABS(L)<0.015*H THEN 1490
1350 ROTATE A
1360 IF A<90 THEN 1410
1370 IF A>270 THEN 1390
1380 IF A>180 THEN 1410
1390 RMOVE -0.3*L,-0.01*H
1400 GO TO 1420
1410 RMOVE -0.4*L,0.01*H
1420 CHARSIZE 3
1430 PRINT A$;
1440 CHARSIZE 1
1450 PRINT B$
1460 MOVE X2,Y2
1470 ROTATE A
1480 RMOVE L,0
1490 RETURN
1500 REM *GRAPHIC ROUTINE*
1510 X1=0
1520 Y1=0
1530 Z1=0
1540 ROPEN 1
1550 MOVE H1,-V1
1560 PRINT "ADJUST THE THUMBWHEELS TO A DESIRED LOCATION AND ";
1570 PRINT "ENTER A NUMERIC KEY FOR NEXT REQUEST ACCORDING ";
1580 PRINT "TO THE FOLLOWING "
1590 PRINT "SELECTION TABLE."
1600 POINTER X2,Y2,Z$
1610 RCLOSE
1620 RINIT
1630 Z2=VAL(Z$)
1640 GOSUB Z2 OF 1690,1710,1800,1870,2140,2010,2250,2560,2900
1650 X1=X2
1660 Y1=Y2
1670 Z1=Z2
1680 GO TO 1540
1690 MOVE X2,Y2
1700 RETURN
1710 REM   *CIRCLE ROUTINE*
1720 R=SQR((X2-X1)^2+(Y2-Y1)^2)
1730 MOVE X1+R,Y1
1740 FOR I=1 TO 360
1750 X=R*COS(I)+X1
1760 Y=R*SIN(I)+Y1
1770 DRAW X,Y
```

```
1780 NEXT I
1790 RETURN
1800 REM   *LABEL ROUTINE*
1810 CHARSIZE C
1820 MOVE X2,Y2
1830 INPUT A$
1840 CHARSIZE 2
1850 RINIT
1860 RETURN
1870 REM   *MEASURE LENGTH ROUTINE*
1880 L=SQR((X2-X1)^2+(Y2-Y1)^2)
1890 ROPEN 1
1900 MOVE H1,-V1
1910 PRINT "ADJUST THE THUMBWHEELS TO SPECIFY WHERE THE LENGTH ";
1920 PRINT " BE PRINTED."
1930 POINTER X4,Y5,U$
1940 RINIT
1950 MOVE X4,Y5
1960 CHARSIZE 2
1970 PRINT USING 1980:L," IN."
1980 IMAGE 3D.3D,4A
1990 MOVE X2,Y2
2000 RETURN
2010 REM *EXTEND LINE ROUTINE*
2020 IF Y1<Y2 THEN 2100
2030 MOVE X1,Y1
2040 X=X2-X1
2050 IF X=0 THEN 2120
2060 X=(Y2-Y1)/X
2070 X=X2-Y2/X
2080 DRAW X,0
2090 RETURN
2100 MOVE X2,Y2
2110 GO TO 2040
2120 DRAW X2,0
2130 RETURN
2140 REM   *MEASURE ANGLE ROUTINE*
2150 X=X2-X1
2160 L=SQR((X2-X1)^2+(Y2-Y1)^2)
2170 Y=Y2-Y1
2180 A=ACS(X/L)
2190 POINTER X4,Y5,U$
2200 MOVE X4,Y5
2210 CHARSIZE 1
2220 PRINT USING 2230:A," DEGREES"
2230 IMAGE 4D.3D,8A
2240 RETURN
2250 REM *VELOCITY POLYGON*
2260 CHARSIZE 3
2270 MOVE X2-0.0035*H,Y2-0.01*H
2280 PRINT "X"
2290 CHARSIZE 1
2300 MOVE X2,Y2
2310 V3=L1*W1/S1
2320 L=V3
2330 A1=T1+90
2340 A=A1
2350 A$="V"
2360 B$="B"
2370 GOSUB 1210
2380 T3=ASN(L1/L2*SIN(T1))
```

```
2390 A3=180-T3
2400 A2=A3-A1
2410 V4=V3*COS(A2)
2420 V4=V4/COS(T3)
2430 A5=90-T3
2440 V5=V3*SIN(A2+T3)/SIN(A5)
2450 A4=A3+90
2460 A=A4
2470 L=V5
2480 B$="C/B"
2490 GOSUB 1210
2500 MOVE X2,Y2
2510 L=V4
2520 A=180
2530 B$="C"
2540 GOSUB 1210
2550 RETURN
2560 REM **ACCELERATION POLYGON**
2570 MOVE 0.95*H1,-0.98*V1
2580 CHARSIZE 1
2590 ROPEN 11
2600 MOVE H1,-V1
2610 PRINT "ADJUST THUMBWHEELS TO POINTS A & D."
2620 POINTER X3,Y3,U$
2630 POINTER X4,Y4,U$
2640 RCLOSE
2650 RINIT
2660 L=ABS(X4-X3)
2670 IF T1>270 THEN 2690
2680 IF T1>90 THEN 2720
2690 A=180
2700 X5=L
2710 GO TO 2740
2720 A=0
2730 X5=-L
2740 A$="A"
2750 B$="C"
2760 MOVE X2,Y2
2770 GOSUB 1210
2780 A=T1+180
2790 B$="B"
2800 MOVE X2,Y2
2810 L=L1
2820 GOSUB 1210
2830 L=SQR((B1-X5)^2+B2^2)
2840 A=ACS((B1-X5)/L)
2850 IF T1<180 THEN 2870
2860 A=-A
2870 B$="C/B"
2880 GOSUB 1210
2890 RETURN
2900 REM *REPEAT/END PROGRAM ROUTINE*
2910 PAGE
2920 HOME
2930 CHARSIZE 2
2940 PRINT "ENTER R TO REPEAT PROGRAM, E TO END PROGRAM : ";
2950 INPUT A$
2960 IF A$="R" THEN 120
2970 PRINT "END OF PROGRAM"
2980 END
```

```
                WHEATSTONE BRIDGE PROGRAM

SPECIFY Z2 TO FIND Z1

INPUT ARE :

R3, ohm = 10000
R4, ohm = 5000
FREQUENCY, hz = 1.0E-6
                Z2  ELEMENTS :
          TYPE                    VALUE
      RESISTANCE, ohm         1.000E+001
      CAPACITANCE, forod      5.080E-009
      INDUCTANCE, henry       5.000E-002

*********************************************************
** YOU NEED A RESISTANCE OF      2.000E+001  ohm     **
** YOU NEED A CAPACITANCE OF     2.540E-009  forod   **
*********************************************************
                DESIGN END
```

Figure 6-18   A sample application of Program W. STONE.

6.11   WHEATSTONE BRIDGE— BALANCING ELECTRIC ELEMENTS

The Wheatstone bridge serves as a good example of a simple electric CAD. It is commonly used for balancing of electric elements for both d.c. and a.c. circuits. As shown in Fig. 6-18, if the circuit is balanced so that the voltage across the voltmeter is equal to zero, the relationship between the impedances Z_1 and Z_2, and the resistances R_3 and R_4 (Fig. 6-19) can be written as[6]

$$Z_1/Z_2 = R_3/R_4 \qquad \text{(a)}$$

Knowing about three elements of the circuit, Eq. (a) enables the value of the fourth element to be calculated. For example, in a d.c. circuit, a resistive Wheatstone

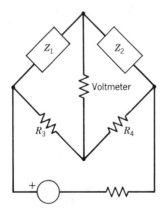

Figure 6-19   Wheatstone bridge.

---

[6] In Chapter 7, the characteristics of electric elements are introduced and the governing differential equations of the circuit are derived.

bridge involves four resistors. That is Z_1 and Z_2 in Eq. (a) become simply R_1 and R_2. But for an a.c. circuit, the impedance Wheatstone bridge may contain resistors, inductors, and capacitors. The impedance Z for a resistor R, an inductance L, and a capacitance C are equal to R, $2\pi f L j$, and $-j/2\pi f C$, respectively, where f is the frequency of the input current and $j = (-1)^{1/2}$.

Since more meaningful discussion on circuit CAD can be made only in Part Two of this book when the elements are well understood and the equations governing the circuit are solved, at this point it is more appropriate to illustrate the manipulation of complex numbers. Program W. STONE can be used with Eq. (a) by specifying elements in the Z_1 or Z_2 path to determine any unknown element in the other path that could contain a resistor, inductor, or capacitor in the circuit. In this program, the following basic algebraic operations involving complex constants are employed:

$$(a_1 + b_1 j) \pm (a_2 + b_2 j) = (a_1 \pm a_2) + (b_1 \pm b_2)j \tag{b,c}$$

$$(a_1 + b_1 j) \times (a_2 + b_2 j) = r_1 r_2 [\cos(\theta_1 + \theta_2) + \sin(\theta_1 + \theta_2)j] \tag{d}$$

$$(a_1 + b_1 j)/(a_2 + b_2 j) = (r_1/r_2)[\cos(\theta_1 - \theta_2) + \sin(\theta_1 - \theta_2)j] \tag{e}$$

where

$$r_1 = (a_1^2 + b_1^2)^{1/2} \qquad r_2 = (a_2^2 + b_2^2)^{1/2} \tag{f,g}$$

$$\theta_1 = \tan^{-1}(b_1/a_1) \qquad \theta_2 = \tan^{-1}(b_2/a_2) \tag{h,i}$$

As it is customarily said, $a_1 + b_1 j$ is the Cartesian form of a complex number whereas $r_1 \exp(\theta_1 j)$ is the polar form.

PROGRAM W. STONE

```
100 PAGE
120 PRINT "* PROGRAM W.STONE *"
130 PRINT "_SAMPLE OR YOUR OWN PROBLEM,  (0 OR 1)? : ";
140 INPUT S7
150 IF S7=1 THEN 180
160 GOSUB 1350
170 GO TO 460
180 REM  **  INPUT SECTION  **
190 PRINT "DO YOU WANT TO SPECIFY Z1 OR Z2,  (1 OR 2)? : ";
200 INPUT I7
210 PRINT "INPUT R3 & R4"
220 INPUT R3
230 INPUT R4
240 IF I7=2 THEN 270
250 PRINT "INPUT THE NO OF ELEMENTS IN Z1"
260 GO TO 280
270 PRINT "INPUT THE NO OF ELEMENTS IN Z2"
280 INPUT N
290 DIM C1(N),C2(N),C(N),D(N)
300 PRINT "USING THE FOLLOWING CODES TO"
310 IF I7=2 THEN 340
320 PRINT " INPUT THE TYPE & VALUE OF EACH ELEMENT OF Z1 SEPARATELY"
330 GO TO 350
340 PRINT " INPUT THE TYPE & VALUE OF EACH ELEMENT OF Z2 SEPARATELY"
```

```
350 PRINT "*****************************"
360 PRINT "** 1:RESISTANCE, ohm      **"
370 PRINT "** 2:CAPACITANCE, farad   ** "
380 PRINT "** 3:INDUCTANCE, henry    **"
390 PRINT "*****************************"
400 FOR I=1 TO N
410 INPUT C(I)
420 INPUT D(I)
430 NEXT I
440 PRINT "INPUT FREQUENCY, hz "
450 INPUT F5
460 REM ** CALCULATION SECTION   **
470 FOR I=1 TO N
480 GOSUB C(I) OF 910,950,990
490 NEXT I
500 REM * SUM *
510 E=0
520 F=0
530 FOR I=1 TO N
540 E=C1(I)+E
550 F=C2(I)+F
560 NEXT I
570 REM * FIND Z1 OR Z2 *
580 IF I7=2 THEN 630
590 REM * FIND Z2 *
600 A=E/(R3/R4)
610 B=F/(R3/R4)
620 GO TO 650
630 A=E*(R3/R4)
640 B=F*(R3/R4)
650 PAGE
660 GOSUB 1030
670 U$="** YOU NEED A RESISTANCE OF"
680 V$="ohm    **"
690 X$="** YOU NEED A CAPACITANCE OF"
700 Y$="farad **"
710 P$="** YOU NEED A INDUCTANCE OF"
720 Q$="henry **"
730 PRINT "_*******************************************************"
740 IF A=0 THEN 770
750 PRINT USING 880:U$,A,V$
760 GO TO 780
770 PRINT "    NO RESISTANCE NEEDED
780 IF B=0 THEN 860
790 IF B<0 THEN 830
800 L=B/(2*PI*F5)
810 PRINT USING 880:P$,L,Q$
820 GO TO 870
830 C5=1/(2*PI*F5*B)
840 PRINT USING 880:X$,-C5,Y$
850 GO TO 870
860 PRINT "    NO CAPACITANCE NEEDED            "
870 PRINT "*******************************************************"
880 IMAGE 30A,2X,3E,2X,8A
890 PRINT "_              DESIGN END"
900 END
910 REM  **  CONVERT RESISTANCE TO COMPLEX FORM  **
920 C1(I)=D(I)
930 C2(I)=0
940 RETURN
950 REM  **  CONVERT CAPACITANCE TO COMPLEX FORM  **
```

```
960 C1(I)=0
970 C2(I)=-1/(2*PI*D(I)*F5)
980 RETURN
990 REM  **  CONVERT INDUCTANCE TO COMPLEX FORM  **
1000 C1(I)=0

1010 C2(I)=2*PI*D(I)*F5
1020 RETURN
1030 REM  **  PRINTOUT ALL THE INPUT  **
1040 CHARSIZE 2
1050 PRINT "              WHEATSTONE BRIDGE PROGRAM   "
1060 PRINT " "
1070 IF I7=2 THEN 1100
1080 PRINT "_SPECIFY Z1 TO FIND Z2"
1090 GO TO 1110
1100 PRINT "SPECIFY Z2 TO FIND Z1"
1110 PRINT "_INPUT ARE :"
1120 PRINT "_R3, ohm = ";R3
1130 PRINT "R4, ohm = ";R4
1140 PRINT "FREQUENCY, hz = ";F5
1150 IF I7=2 THEN 1180
1160 PRINT "_            Z1  ELEMENTS : "
1170 GO TO 1190
1180 PRINT "_            Z2  ELEMENTS : "
1190 PRINT "        TYPE              VALUE "
1200 A$="RESISTANCE, ohm "
1210 B$="CAPACITANCE, farad "
1220 C$="INDUCTANCE, henry "
1230 FOR I=1 TO N
1240 GO TO C(I) OF 1250,1270,1290
1250 PRINT USING 1320:A$,D(I)
1260 GO TO 1310
1270 PRINT USING 1320:B$,D(I)
1280 GO TO 1310
1290 PRINT USING 1320:C$,D(I)
1300 GO TO 1310
1310 NEXT I
1320 IMAGE 4X,19A,5X,3E
1330 RETURN
1340 IF S7=1 THEN 180
1350 REM ** SAMPLE DATA **
1360 N=3
1370 DIM C1(3),C2(3),C(3),D(3)
1380 F5=0.1*10^-5
1390 I7=2
1400 R3=0.1*10^5
1410 R4=R3
1420 C(1)=1
1430 C(2)=2
1440 C(3)=3
1450 D(1)=10
1460 D(2)=0.508*10^-8
1470 D(3)=0.05
1480 RETURN
```

PART TWO

CAD/CAM OF ELEMENTS AND SYSTEMS

CHAPTER 7

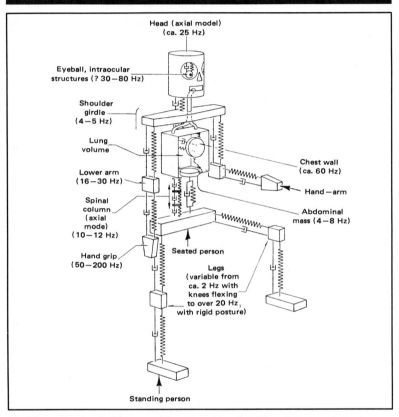

Modeling of human body system. (Courtesy of Bruel & Kjaer Instruments, Inc.)

MODELING OF ELEMENTS AND SYSTEMS

7.1 INTRODUCTION

In Part One, we discussed and dealt with interactive computer graphics devices in implementation of a number of **simple** automated designs. Almost all real physical systems, however, are rather **complex**. Unless they can be decomposed into simple elements and each element can be adequately understood (Fig. 7-1), it is difficult to design the system as a whole. In deciding whether a designed system is acceptable or not, it is necessary to check whether all design specifications have been met and also to determine the limitations of the designed system. For example, a bridge built across a stream for the convenience and safe passing of pedestrians and passenger cars may need to erect a sign of "maximum load allowed" to prevent heavy trucks from passing over it. In order to evaluate, modify, refine, and if possible optimize the design of a physical system, it is thus imperative to know how to analyze the behavior of the system under various disturbances. The disturbances could be in the form of loads, heat, fluid flow, or electromagnetic or nuclear forces.

Part Two of this book is devoted to the CAD/CAM of elements and systems. In this chapter, the modeling of engineering elements will be discussed. The physical systems are to be idealized as assemblages of simple building blocks called

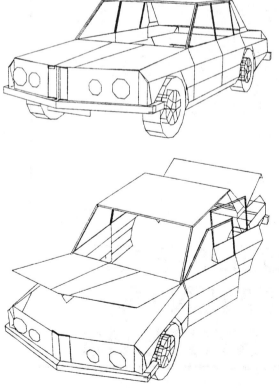

Figure 7-1 A car is modeled into elements for study of their functional and geometric relationships. The hierarchical data structure enables various geometric configurations such as opening of doors, hood, and trunk to be simulated and displayed. (Courtesy of the SAMMIE Group, the University of Nottingham, England.)

elements. The behavior of each element is to be approximately predicted by **linear** mathematical analysis. For systems that behave nonlinearly, an advanced and elaborated exposition beyond the scope of this book will have to be pursued by the reader.

Knowing how to predict the behavior of a physical system accurately and swiftly by use of a simplified model is what distinguishes an engineer from his or her peers. The system should be modeled as composed of simple elements but not oversimplified, and it should be able to yield the necessary information for making engineering decisions. A small increase in the complexity of the physical system often results in much greater increase in cost, time, and effort. Such engineering judgment on selection of a simple but adequate model for a designed system has to be built on analytical experience and can be acquired only through constant involvement with actual physical systems in the testing laboratories and production plants.

It is also important to realize the similarities among various physical systems. In this chapter the resemblance in behavior of the electrical, mechanical, fluid, and thermal systems will be discussed and the analytical solution of the mathematical equations governing the behavior of the systems will be delineated. In the ensuing chapters, methods other than the analytical solution of ordinary differential equations will be presented. These methods facilitate the incorporation of the computer-aided concept into the design and manufacturing of engineering elements and systems. Prior to discussing the engineering systems, let us first introduce the commonly encountered engineering elements.

7.2 ENERGY-DISSIPATION ELEMENTS

Engineering materials, both in the fluid and solid forms, have the tendency to resist motions relative to each other. Such resistance exists at the molecular level between atoms and at the macroscopic level between component parts of a machine or structure. **Friction** is often the term used for the resistance between moving solids, whereas **viscosity** is used for that between fluid particles. As they **rub** each other, energies are consumed or dissipated. The following describes a number of commonly employed engineering elements for which such resistance needs to be dealt with.

Electric Resistor

For control of the amount of current to pass in electric circuits, resistors are manufactured. **Ohm's law** relates the amount of current i in amperes and the voltage v in volts across a resistor with resistance R in ohms by the equation

$$i = v/R \tag{a}$$

Figure 7-2 shows a simple circuit where a physical resistor is connected to a physical battery with the connecting wires being assumed to be of zero resistance. If the resistances of the connecting wires are also considered in the case of a practical circuit, a series circuit results, as shown in Fig. 7-3.

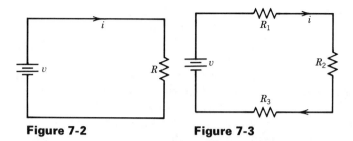

Figure 7-2   **Figure 7-3**

An equivalent resistance R_{eq}, which represents the combined effect of a series of resistors, can be derived by application of **Kirchhoff's voltage law**, which states that the total voltage drop around a circuit should be equal to the source voltage. In the case of the circuit shown in Fig. 7-3, Eq. (a) can be applied for all three resistors; since the current is the same, we arrive at

$$v = v_1 + v_2 + v_3 = iR_1 + iR_2 + iR_3$$
$$= i(R_1 + R_2 + R_3) = iR_{eq}$$

Hence

$$R_{eq} = R_1 + R_2 + R_3 \tag{b}$$

When several circuit elements, each with different resistance, are connected in parallel such as in the case of the power lines supplied into houses, by a voltage generator at the power company, **Kirchhoff's current law**, which states that the sum of the currents leaving a junction must be equal to that entering, can be applied. For the parallel circuits shown in Fig. 7-4, the current equation can be written, with the aid of Eq. (a), as

$$i = i_1 + i_2 + i_3 = \frac{v}{R_1} + \frac{v}{R_2} + \frac{v}{R_3}$$

If one is interested in finding the equivalent resistance for a parallel circuit, i.e., $R_{eq} = v/i$, it should be evident from the above equation that

$$R_{eq} = \frac{v}{i} = 1 \bigg/ \left(\frac{1}{R_1} + \frac{1}{R_2} + \frac{1}{R_3} \right) \tag{c}$$

In the general case of n resistances, $R_1, R_2, \ldots, R_n$ in series and parallel circuits, Eqs. (b) and (c) can be generalized, respectively, to the form of

$$\text{Series:} \quad R_{eq} = \sum_{i=1}^{n} R_i \tag{d}$$

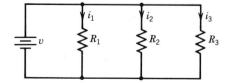

Figure 7-4

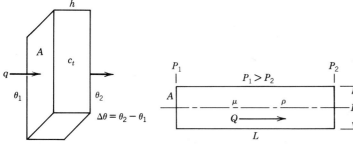

Figure 7-5 Thermal conduction.

Figure 7-6 Fluid flow.

$$\text{Parallel:} \quad R_{eq} = 1 \bigg/ \sum_{i=1}^{n} \frac{1}{R_i} \tag{e}$$

Thermal Resistance

Heat q that flows through an insulated wall (Fig. 7-5) can be ideally related to the temperature difference $\Delta\theta$ by the equation

$$\Delta\theta = R_t q \tag{f}$$

where the proportional constant R_t is called the **ideal thermal resistance** (expressed in units of kg-cal/sec-m-°C). For a wall of thickness h, surface area normal to the heat flow A, and a thermal conductivity c_t, the resistance R_t can be expressed as

$$R_t = h/c_t A \tag{g}$$

There are other types of heat flow, namely, convection and radiation. For the former, a similar linear equation can be written, whereas for the latter, the $q \sim \Delta\theta$ relationship becomes nonlinear. Here, Eqs. (f) and (g) suffice to cover the discussion required in this text.

Fluid Resistance

The most frequently encountered fluid flow problem is that of an incompressible fluid passing through a uniform tube due to the pressure difference at its two ends. Figure 7-6 shows the flow passing a tube of length L and inside diameter D. Let the pressures at the two ends be P_1 and P_2. The Hagen–Poiseuille law[1] states that the flow rate, Q, in the segment 1–2 of the tube, is linearly proportional to the pressure difference $P_1 - P_2$. That is,

$$Q = R_f(P_1 - P_2) \tag{h}$$

where R_f is the **fluid resistance** given by

$$R_f = 128\mu L/\pi D^4 \tag{i}$$

---

[1] J. F. Blackburn, G. Reethof, and J. L. Shearer, *Fluid Power Control*, John Wiley & Sons, New York, 1960.

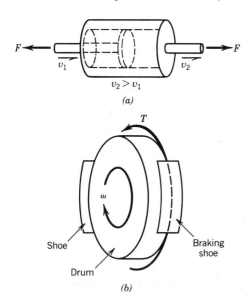

Figure 7-7 (*a*) Translational mechanical damper. (*b*) Rotational mechanical damper.

with μ being the **absolute viscosity** of the fluid (in N-sec/in^2). If the length L and inside diameter D of the tube are both in meters, the flow rate Q is in newtons per second and the pressures in newtons per square meter.

Mechanical Damper

The shock absorber installed in almost all automobiles is an example of a mechanical damper that dissipates energy when the viscous force is induced by the difference in velocities between the cylinder and the piston, as shown in Fig. 7-7a. Let the retarding force be designated as F. Ideally, the linear relationship between F and the velocities can be written as

$$F = c\,\Delta v = c(v_2 - v_1) = c\frac{d}{dt}(x_2 - x_1) \tag{j}$$

The proportional constant c is called the **translational damping coefficient**.

The automobile drum-type braking system sketched in Fig. 7-7b is another example of a mechanical damper, except in this case the equation relating the retarding torque T to the wheel angular velocity ω is

$$T = b\omega \tag{k}$$

The constant b is called the **rotational damping coefficient**.

7.3 ENERGY-STORAGE ELEMENTS

There are many types of engineering elements capable of storing energies. A water tower stores energy in the form of high pressure due to its elevated position so that the water can be delivered with adequate pressure. When a spring is com-

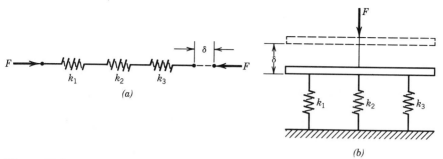

Figure 7-8

pressed by an axial force, it will return a part or all of the energy by stretching back partly or totally to its original length if the force is reduced or completely removed. The energy can also be stored in the form of heat or electric charge. The following subsections describe how these energy-storing elements should be modeled.

Coil Spring

Robert Hooke in 1678 described how metal wires could be coiled around a cylinder to produce springs and observed the fact that the displacement of a spring is proportional to the axial load applied on it. The force F and the displacement δ is related by the linear equation

$$\delta = F/k \qquad \text{(a)}$$

where k is called the **spring constant**. As we know nowadays, Eq. (a) holds only when the load does not exceed a certain limit, beyond which the behavior of the spring becomes nonlinear.

As with electrical resistors, springs in series and in parallel can be treated by consideration of an equivalent spring. Figure 7-8a shows that if the total displacement of the springs in series is δ, the displacements of each spring can be added by observing that all springs are subjected to the same axial force F. That is,

$$\delta = \delta_1 + \delta_2 + \delta_3 = F\left(\frac{1}{k_1} + \frac{1}{k_2} + \frac{1}{k_3}\right)$$

Thus, the equivalent spring constant can be easily obtained as

$$k_e = 1\left/\left(\frac{1}{k_1} + \frac{1}{k_2} + \frac{1}{k_3}\right)\right. \qquad \text{(b)}$$

In the case of a parallel system shown in Fig. 7-8b, the springs all displace δ and must share the total load F. By considering the equilibrium of forces, we obtain

$$F = F_1 + F_2 + F_3 = k_1\delta + k_2\delta + k_3\delta$$

Hence, the equivalent spring constant is

$$k_e = k_1 + k_2 + k_3 \qquad \text{(c)}$$

EXERCISE

Derive the equivalent spring constant for the mixed system shown here.

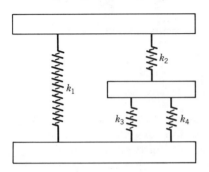

Answer: $k_e = k_1 + \dfrac{1}{1/k_2 + 1/(k_3 + k_4)}$

Electric Capacitor

Capacitors are designed for storing energy in an electric field. They are made of conducting plates sandwiched with layers of insulating sheets. Because of a potential difference in the conducting plates, flow of electric charge takes place. The larger the surface area of the plates and the narrower the gap between them, the greater the electron movement and consequently the better the charge storage capacity. An ideal capacitor has a linear relationship between the charge q and the potential difference v, that is,

$$\int i\, dt = q = Cv \qquad \text{(d)}$$

or

$$i = C\frac{dv}{dt} \qquad \text{(e)}$$

where C is the **capacitance** of the element and has units of farad (or A-sec/V).

Fluid Reservoir

Consider the case of a cylindrical tank of uniform cross-sectional area A. It is easy to observe that the fluid level h will change, depending upon the input quantity. If the flow rate is Q, then we can write

$$Q = \frac{A\, dh}{dt} \qquad \text{(f)}$$

In comparison to Eq. (e), A is evidently playing the role of the capacitance in a fluid system.

Thermal Capacitance

The capability of a material in storing heat is manifested by many daily experiences. For example, turning on a stove to boil a pot of water can be described by the equation

$$q = \frac{c_p m \, d\theta}{dt} \tag{g}$$

The change of temperature θ in the mass of water m depends on the heat flow q and on the specific heat c_p. Apparently the capacitance for a thermal system is the product of m and c_p.

Torsional Spring

When an elastic shaft is subjected to a twisting torque T at the free end as shown in Fig. 7-9a, an angle of twist θ will result. If the magnitude of T is not great enough to cause inelastic distortion, the shaft will return to its undeformed state when T is released. In that case, the shaft behaves the same as a spring and the deformation is linearly proportional to the applied torque, namely,

$$\theta = T/k_t \tag{h}$$

where k_t is the spring constant of the twisted shaft. It is apparent that because of the elastic behavior, a not severely twisted shaft is an energy-storage element.

Electric Inductor

The tendencies of all elements to continue to remain at their present status make them good energy-storage devices. Whether to retain elastic behavior or to keep the current flowing or to maintain a constant velocity, they all tend to conserve energy. Inductances and inertances are other mechanisms of storing energies. The electric inductor is to be introduced first, followed by the discussion of the inertia in fluid flows, in mechanical translation and rotation.

Coils that induce voltage when the passing current produces a magnetic field are often used as inductors. For an ideal inductor, the induced voltage is linearly proportional to the time rate of the current change. That is (Fig. 7-9e),

$$v = L \frac{di}{dt} \tag{i}$$

where L is **inductance** in units of henry (H or V-sec/A).

Fluid Inertance

As shown in Fig. 7-6, the pressure difference $\Delta P = P_1 - P_2$ may also be related to the flow rate Q in terms of the length L, cross-sectional area A, and fluid density ρ as

$$\Delta P = P_1 - P_2 = \frac{\rho L}{A} \frac{dQ}{dt} = L_f \frac{dQ}{dt} \tag{j}$$

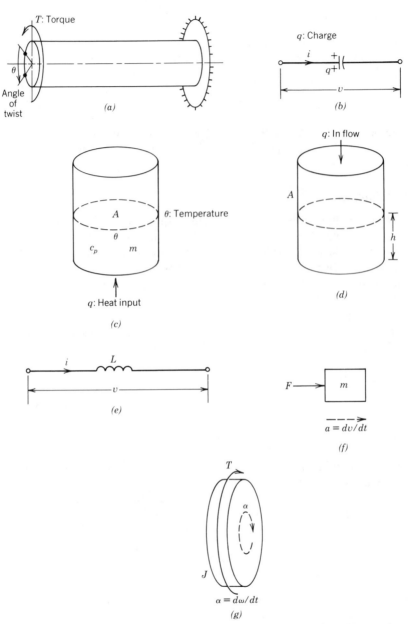

Figure 7-9   Various energy-storage elements: (a) torsional spring, (b) capacitor, (c) thermal capacitor, (d) fluid reservoir, (e) inductor, (f) translational mass, and (g) rotational inertia.

where L_f is the **fluid inductance**. Equation (j) is derived from the force–acceleration relationship for that segment of pipe. The force pushing the fluid to the right is simply $A \Delta P$, whereas the acceleration is $dv/dt = d(Q A)/dt$. Recognizing that the mass m is equal to $\rho L A$, Newton's second law consequently leads to the result in Eq. (j).

Translational and Rotational Inertia

Figure 7-9f and g depict the inertia of mechanical translation and rotation, respectively. A linear force F causes a mass m to change its velocity in the direction of the applied force, whereas a twisting torque causes the disk with a moment of inertia of mass J to change its angular velocity ω.

Newton's second law relating the acceleration a of a mass m due to application of a force F can be represented as

$$F = ma = \frac{m\, dv}{dt} \tag{k}$$

that is, to replace the acceleration by the change of velocity v in time. In so doing, we see that the mass m is the capacitance of a translating mechanical system.

In a similar manner, the applied torque T can be related to the angular acceleration α as

$$T = J\alpha = \frac{J\, d\omega}{dt} \tag{l}$$

Thus, the moment of inertia of mass, J, plays the role of capacitance in a rotational mechanical system.

7.4   ENERGY-CONVERSION ELEMENTS

There are numerous physical elements designed for converting one form of energy into another. Transformers, motors, generators, and gear trains are perhaps the best known. As sketched in Fig. 7-10a, by using different numbers of coils, N_1 and N_2, the electric voltage levels can be changed with a transformer. The voltages and numbers of coils are directly related as

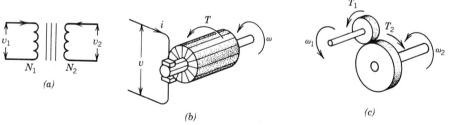

(a)   (b)   (c)

Figure 7-10   Energy-conversion elements: (a) transformer, (b) motor or generator, and (c) pairing of gears.

$$v_2/v_1 = N_2/N_1 \tag{a}$$

Since power P should remain unchanged, the currents can, as a consequence, be converted by use of the equations

$$P = i_1v_1 = i_2v_2 \tag{b}$$

These characteristics of a transformer are commercially utilized to transmit electric power from the plant to the distribution stations and finally to consumers.

The mechanical counterpart of a transformer is the pairing of gears shown in Fig. 7-10c, already discussed in Section 6.2. The gear trains transmit torque at the contact of the gear tooth. As a result, equations analogous to Eqs. (a) and (b) have been presented as Eq. (c) in Section 6.2.

For converting electrical power into mechanical power and vice versa, using the starter to turn the engine and the generator to charge up the battery in an automobile are the best examples. In order to deal with these devices (Fig. 7-10b), the following **electromechanical physical relations** are necessary:

$$v = K\omega \tag{c}$$

$$T = Ki \tag{d}$$

They relate the electrical current i and voltage v to the mechanical torque T and angular velocity ω. The proportional constant K depends on the type of electro-mechanical element involved.

In fact, there are numerous other types of energy-conversion elements. For example, the microphone converts displacements into voltage, whereas the speaker plays the opposite role. The pump propels water into an elevated tower to increase the pressure head and store hydraulic energy. The thermocouple mechanically triggers the mechanism in a thermostat to activate the furnace or turn on the air conditioner. For each arrangement, the constant K for the physical relation in conversion of energy needs to be properly obtained in mathematical analysis of the system.

7.5 SERIES LCR CIRCUIT AND IMPEDANCE

Figure 7-11 shows a circuit containing resistance, capacitance, and inductance in series, which is a very important combination of electric elements in electronic apparatus, such as radio receivers and transmitters. For analysis of such circuits, it is more convenient to introduce the term **impedance**, defined as

$$Z = V/I \tag{a}$$

where V and I are the amplitudes of the voltage and current in the general expressions $v = Ve^{j2\pi ft}$ and $i = Ie^{j2\pi ft}$. f is the frequency in hertz (cycles/sec) and $j = (-1)^{1/2}$. Since the current–voltage relationships for the inductor and capacitor are $v = L\,di/dt$ and $i = C\,dv/dt$, respectively, as has been explained in the preceding sections, we can derive an expression for the impedance Z of a LCR circuit by introducing capacitive and inductive reactances.

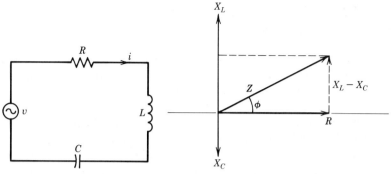

Figure 7-11   LCR circuit.   **Figure 7-12**

Let the voltage across the capacitor be $v_1 = V_1 e^{j2\pi ft}$. As in the case of a resistor, the opposition to current flow of a capacitor called **capacitive reactance** X_C can be defined as

$$X_C = |v_1/i| = |v_1/(C\, dv_1/dt)|$$
$$= |V_1 e^{j2\pi ft}/(CV_1 j2\pi f e^{j2\pi ft})|$$
$$= |1/j2\pi fC| = 1/2\pi fC \qquad \text{(b)}$$

Because of the j factor appearing in the denominator inside the absolute signs, the current i leads v_1 by 90°.

In a similar manner, **inductive reactance** can be introduced for measuring the opposition to alternating current of an inductor. It can be expressed in terms of the voltage across the inductor, v_2, as

$$X_L = |v_2/i| = \left|\left(L\frac{di}{dt}\right)\middle/ i\right| = |Lj2\pi f I e^{j2\pi ft}/I e^{j2\pi ft}|$$
$$= |Lj2\pi f| = 2\pi fL \qquad \text{(c)}$$

Because of the j factor appearing in the numerator inside the absolute signs, the current in an inductor lags the voltage by 90°.

For the reason that there are lag and lead of the current in the different elements of the LCR circuit, the impedance Z has to be evaluated by vector analysis, as illustrated in Fig. 7-12. In other words, the value of Z has to be determined by application of the Pythagorean theorem as

$$Z = [R^2 + (X_L - X_C)^2]^{1/2} \qquad \text{(d)}$$

The angle ϕ between Z and R vectors shown in Fig. 7-10 is the so-called **phase angle**, which also relates the voltage v and current i, and is to be calculated by the equation

$$\phi = \tan^{-1}[(X_L - X_C)/R] \qquad \text{(e)}$$

7.6 MODELING AND ANALOGY OF LINEAR SYSTEMS

Engineering designs can be looked at as tasks of assembling appropriate basic electric, fluid, thermal, and other elements mentioned earlier to perform certain desired functions. Once a system consisting of a set of selected elements has been decided upon, the concern will then be about its behavior in responding to various expected demands or unexpected disturbances. Realistically, all engineering systems are **nonlinear**, but within certain limits, their behavior can be adequately predicted by **linear** analyses. Consider the case of a series LCR circuit illustrated in Fig. 7-11. If it is of interest to know the change in time of the current i in the circuit when the voltage change, $v(t)$, is prescribed, we can derive a differential equation for $i(t)$ based on the Kirchhoff's voltage law

$$v_R + v_C + v_L = v(t) \tag{a}$$

where v_R, v_C, and v_L are the voltages across the resistance R, capacitance C, and inductance L, respectively. Equation (a) can be rewritten in terms of i by simply applying the relationships between the voltage and current for each element. The resulting equation is

$$iR + \frac{1}{C} \int i\,dt + L\frac{di}{dt} = v(t) \tag{b}$$

After differentiating both sides of Eq. (b) with respect to the time t, we obtain

$$L\frac{d^2i}{dt^2} + R\frac{di}{dt} + \frac{1}{C}i = \frac{dv(t)}{dt} \tag{c}$$

If the parameters L, R, and C remain constant during the period of time in which the current change of the circuit is to be investigated in response to a specified $v(t)$, Eq. (c) is a linear second-order differential equation with constant coefficients that we shall later show has an analytical solution and is readily obtainable.

Analogy of Physical Systems

Similar governing differential equations can be derived for the other systems shown in Fig. 7-13. First, consider the parallel electric circuit. Application of Kirchhoff's current law to the node 1 leads to

$$i(t) = i_R + i_L + i_C = \frac{v}{R} + \frac{1}{L} \int v\,dt + C\frac{dv}{dt}$$

where i_R, i_L, and i_C are the currents in the resistor, inductor, and capacitor, respectively. After differentiating both sides of the above equation with respect to t, we obtain

$$C\frac{d^2v}{dt^2} + \frac{1}{R}\frac{dv}{dt} + \frac{v}{L} = \frac{di}{dt} \tag{d}$$

When the mass m in the translational mechanical system is subjected to a disturbing force $F(t)$, Newton's second law can be applied to obtain

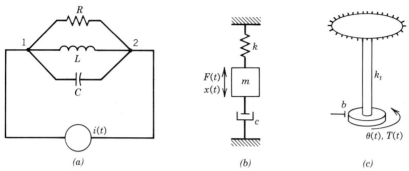

Figure 7-13   Analogy between electric and mechanical systems: (*a*) parallel circuit, (*b*) translational, and (*c*) rotational systems. Also, the series circuit has already been shown in Fig. 7-11.

$$F(t) - F_d - F_s = ma = m\frac{d^2x}{dt^2}$$

After substituting the damping force $F_d = c(dx/dt)$ and the reactive spring force $F_s = kx$ (both of which have been discussed in preceding sections) into the above equation, it becomes

$$m\frac{d^2x}{dt^2} + c\frac{dx}{dt} + kx = F(t) \tag{e}$$

It should be easy to show that the differential equation for the rotational mechanical system is

$$J\frac{d^2\theta}{dt^2} + b\frac{d\theta}{dt} + k_t\theta = T(t) \tag{f}$$

Because of the similarity of Eqs. (c) through (f), it becomes apparent that there is an analogy between electrical and mechanical systems.

Multiple-Variable Systems

Often a system involves more than one variable in the analysis of its behavior. For example, consider the dual-tank system shown in Fig. 7-14. Suppose that the inflows $Q_1(t)$ and $Q_2(t)$ are two prescribed time functions. We want to know the water level changes $h_1(t)$ and $h_2(t)$ in the two tanks for a selected valve setting that controls the flow resistances R_1, R_2, and R_3. The conservation of mass in the two tanks leads to two coupled differential equations in h_1 and h_2, namely,

$$A_1\frac{dh_1}{dt} = Q_1 - \frac{1}{R_1}(h_1 - h_3) \tag{g}$$

$$A_2\frac{dh_2}{dt} = Q_2 - \frac{1}{R_2}(h_2 - h_3) \tag{h}$$

where A_1 and A_2 are the tank areas, assumed to be constants. The pressure head at junction 3, h_3, can be related to the discharge q_3 as

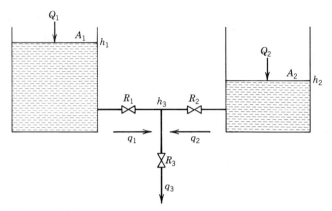

Figure 7-14   Dual-tank problem.

$$q_3 = q_1 + q_2 = \frac{1}{R_3} h_3$$

$$= \frac{1}{R_1} (h_1 - h_3) + \frac{1}{R_2} (h_2 - h_3) \tag{i}$$

Upon manipulation, h_3 can be written in terms of h_1 and h_2 as

$$h_3 = R_3(R_2 h_1 + R_1 h_2)/(R_1 R_2 + R_2 R_3 + R_3 R_1) \tag{j}$$

By eliminating h_3 from Eqs. (g) and (h), we obtain two differential equations in h_1 and h_2, that is,

$$A_1 \frac{dh_1}{dt} = Q_1 - a_1 h_1 + a_3 h_2 \tag{k}$$

$$A_2 \frac{dh_2}{dt} = Q_2 - a_2 h_2 + a_3 h_1 \tag{l}$$

where

$$a_1 = (R_2 + R_3)/\Delta \qquad a_2 = (R_1 + R_3)/\Delta \tag{m,n}$$

$$a_3 = R_3/\Delta \qquad \Delta = R_1 R_2 + R_2 R_3 + R_3 R_1 \tag{o,p}$$

Equations (k) and (l) illustrate that engineering designs often lead to a system of differential equations that govern the system's behavior in the following general form:

$$\frac{d\mathbf{X}}{dt} = F_i(t, \mathbf{X}, \mathbf{Y}, \mathbf{A}) \tag{q}$$

for $i = 1, 2, \ldots, n$ with n being the number of elements in the system response (**output**) vector $\mathbf{X} = [x_1 \quad x_2 \quad \cdots \quad x_n]^T$ and disturbance (**input**) vector $\mathbf{Y} = [y_1 \quad y_2 \quad \cdots \quad y_n]^T$. The functions F_i, for $i = 1, 2, \ldots n$ describe the forms of governing equations and $\mathbf{A} = [a_1 \quad a_2 \quad \cdots \quad a_k]^T$ is a vector containing physical parameters of the system. Referring to the dual-tank system, $n = 2$, $\mathbf{X} = [h_1\mathbf{A}_1 \quad h_2\mathbf{A}_2]^T$, $k = 3$, $\mathbf{Y} = [Q_1(t) \quad Q_2(t)]^T$, and F_1 and F_2 are the right-hand-side expressions of Eqs. (k) and (l), respectively.

In the next section we shall discuss how a system of linear first-order differential equations can be uncoupled to arrive at a high-order differential equation containing only one unknown.

7.7 ANALYTICAL SOLUTION OF SYSTEM RESPONSE (DUAL-TANK PROBLEM)

Coupled linear differential equations such as Eqs. (k) and (l) for the dual-tank problem derived in the preceding section can be decoupled and the system reduced into a higher-order differential equation of one of the variables involved. The reduction process proceeds first by expressing the equations in terms of the **differential operator** $D \equiv d/dt$ and then rearranging as

$$(A_1D + a_1)h_1 - a_3h_2 = Q_1 \tag{a}$$

$$-a_3h_1 + (A_2D + a_2)h_2 = Q_2 \tag{b}$$

Cramer's rule[2] can be conveniently employed here to decouple h_2 and h_1, yielding

$$h_1 = \begin{vmatrix} Q_1 & -a_3 \\ Q_2 & A_2D + a_2 \end{vmatrix} \Big/ \begin{vmatrix} A_1D + a_1 & -a_3 \\ -a_3 & A_2D + a_2 \end{vmatrix} \tag{c}$$

$$h_2 = \begin{vmatrix} A_1D + a_1 & Q_1 \\ -a_3 & Q_2 \end{vmatrix} \Big/ \begin{vmatrix} A_1D + a_1 & -a_3 \\ -a_3 & A_2D + a_2 \end{vmatrix} \tag{d}$$

Since D is a differential operator, the determinants in both numerators and denominators of the above equations should be expressed as product terms in the form of $(A_2D + a_2)Q_1$ instead of $Q(A_2D + a_2)$. This is because Q_1 is a time function and DQ_1 means its derivative with respect to the time t and because QD is meaningless. A similar argument applies for the time functions Q_2, h_1, and h_2. The resulting differential equations, after simplification, can be shown to be

$$(A_1A_2D^2 + a_1A_2D + a_2A_1D + a_1a_2 - a_3^2)h_1 = A_2DQ_1 + a_2Q_1 + a_3Q_2 \tag{e}$$

$$(A_1A_2D^2 + a_1A_2D + a_2A_1D + a_1a_2 - a_3^2)h_2 = A_1DQ_2 + a_3Q_1 + a_1Q_2 \tag{f}$$

In order to demonstrate the procedure for analytical solution of a second-order differential equation, we consider a specific case where

$$A_1 = A_2 = 2000 \text{ cm}^2 \qquad R_1 = R_3 = 0.01 \text{ sec/cm}^2 \qquad R_2 = 0.001 \text{ sec/cm}^2$$

From Eqs. (m), (n), and (o) in the preceding section, the coefficients a_{1-3} can be calculated to be

$$a_1 = 91.7 \qquad a_2 = 166.7 \qquad a_3 = 83.3$$

---

[2] Cramer's rule is discussed in Appendix A.

Equation (e), as a consequence, becomes

$$[(4 \times 10^6 D^2) + (517 \times 10^3 D) + (853 \times 10^1)]h_1$$
$$= 2000DQ_1 + 167Q_1 + 83.3Q_2 \tag{g}$$

We shall use this equation to delineate the procedure[3] for solving a typical linear second-order ordinary differentiation and explain the physical interpretation of the solution.

Complementary Function of Differential Equation

Suppose that initially, that is, at $t = 0$, tank 1 contains fluid at a level of $h_1 = 30$ cm and tank 2 is empty, $h_2 = 0$, and that there are no input flows to either tank, that is, $Q_1(t) = Q_2(t) = 0$. Equation (g), which is called **nonhomogeneous**, is then reduced to a **homogeneous** one with a zero term on the right-hand side, namely,

$$\frac{d^2}{dt^2} h_1 + 0.129 \frac{d}{dt} h_1 + 0.00208h_1 = 0 \tag{h}$$

Here we will first review the procedure for finding the solution of the above homogeneous equation, which is called the **complementary function** and is only a part of the **complete solution** of the nonhomogeneous equation (g). Since Eq. (h) has constant coefficients, the complementary function can be assumed to be of the form

$$h_{1c}(t) = Ce^{mt} \tag{i}$$

where the subscript c is added to indicate that it is the complementary function. $h_{1c}(t)$ and its first and second derivatives with respect to t can be substituted into Eq. (h) to yield

$$Ce^{mt}(m^2 + 0.129m + 0.00208) = 0 \tag{j}$$

In order to arrive at nontrivial solutions, Ce^{mt} should not be equal to zero and Eq. (j) leads to

$$m^2 + 0.129m + 0.00208 = 0 \tag{k}$$

The roots of this quadratic equation are

$$m_1 = -0.0189 \qquad m_2 = -0.1103 \tag{l, m}$$

Equation (k) and m_1 and m_2 are called the **characteristic equation** and **roots**, respectively. Thus, there are two independent solutions for $h_{1c}(t)$, which can now be written as

$$h_{1c}(t) = C_1 e^{-0.0189t} + C_2 e^{-0.1103t} \tag{n}$$

The constants C_1 and C_2 are to be determined by the initial conditions of the system, which in this case are $h_1 (t = 0)$ and $h_2 (t = 0)$. If $Q_1(t)$ and $Q_2(t)$ are not

[3]C. R. Wylie, Jr., *Advanced Engineering Mathematics*, 2nd ed., McGraw-Hill, New York, 1960.

identically equal to zero, these constants cannot be determined until the complete solutions of the system are worked out. Here, since Q_1 and Q_2 are assumed to be identically equal to zero for all t, we can proceed right away to find C_1 and C_2 for the reason that $h_{1c}(t)$ constitutes the complete solution of $h_1(t)$.

From Eq. (n), we can easily obtain

$$Dh_1 \equiv \frac{d}{dt} h_{1c}(t) = -0.0189C_1e^{-0.0189t} - 0.1103C_2e^{-0.1103t}$$

Recalling that $Q_1 = 0$, $A_1 = 2000$, $a_1 = 91.7$, and $a_3 = 83.3$, substitution of the above result and Eq. (n) into Eq. (a) yields

$$h_2(t) = 0.647C_1e^{-0.0189t} - 1.547C_2e^{-0.1103t} \tag{o}$$

Suppose that initially at $t = 0$, $h_1 = 30$ cm, and $h_2 = 0$. Equations (n) and (o) can be easily solved by Cramer's rule to give

$$C_1 = \begin{vmatrix} 30 & 1.000 \\ 0 & -1.547 \end{vmatrix} \bigg/ \begin{vmatrix} 1.000 & 1.000 \\ 0.647 & -1.547 \end{vmatrix} = \frac{-46.41}{-2.194} = 21.15$$

$$C_2 = \begin{vmatrix} 1.000 & 39 \\ 0.647 & 0 \end{vmatrix} \bigg/ (-2.194) = \frac{-19.41}{-2.194} = 8.85$$

Consequently, the solutions of the dual-tank system when the initial fluid levels are $h_1 = 30$ cm and $h_2 = 0$ without any inflows to the system, $Q_1 = Q_2 = 0$, are

$$h_1(t) = 21.15e^{-0.0189t} + 8.85e^{-0.1103t} \tag{p}$$

$$h_2(t) = 13.68e^{-0.0189t} - 13.68e^{-0.1103t} \tag{q}$$

It is desirable that a program be developed to analyze the general **transient** problem of the dual tanks. For $Q_1 = Q_2 = 0$, Eqs. (a) and (b) will lead to the general solutions of

$$h_1(t) = C_1e^{-m_1t} + C_2e^{-m_2t}$$

$$h_2(t) = C_3e^{-m_1t} + C_4e^{-m_2t}$$

where

$$m_{1,2} = \{-a_1A_2 \pm [a_1A_2 - 4A_1A_2(a_1a_2 - a_3^2)]^{1/2}\}/2A_1A_2$$

$$C_1 = [h_{10}(A_1m_2 + a_1) - h_{20}a_3]/A_1(m_2 - m_1)$$

$$C_2 = [h_{20}a_3 - h_{10}(A_1m_1 + a_1)]/A_1(m_2 - m_1)$$

$$C_3 = (A_1m_1 + a_1)C_1/a_3$$

$$C_4 = (A_1m_2 + a_2)C_2/a_3$$

and h_{10} and h_{20} are the **initial** fluid levels in tanks 1 and 2, respectively. The details for deriving the above results are left for the readers to work on as an exercise. A program in BASIC language called DUAL.TANK has been prepared and the

```
WHAT ARE THE INPUT
A1, A2, R1, R2, R3?
2899, 2988,.81,.891,.81
H1, H2, Del: oT, Tend?
38, 9, 1, 583

A1 = 2988
A2 = 2899
R1 = 8.91
R2 = 1.5E-3
R3 = 9.91
n1 = 39
n2 = 8
Time Increment = 1
Ending Time = 508

Characteristic Roots Are

-0.9188922662511
-0.118274488416

WANT TO RUN NEXT CASE. Y/N?
N

------END------
```

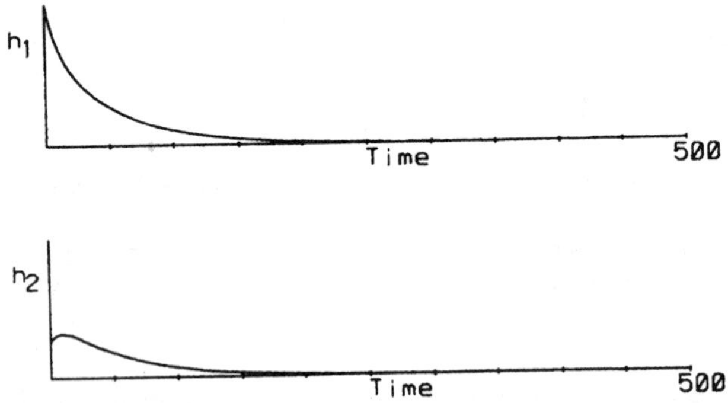

Figure 7-15

input data for deriving the results of Eqs. (p) and (q) have been used to generate
the transient-flow plot of Fig. 7-15.

```
100 REM * DUAL-TANK TRANSIENT ANALYSIS *
110 PAGE
120 PRINT "WHAT ARE THE INPUT :"
130 PRINT "A1,A2,R1,R2,R3?"
140 INPUT A1,A2,R1,R2,R3
150 PRINT "H1,H2,DeltaT,Tend?"
160 INPUT H1,H2,T,T1
170 PRINT " A1=";A1
180 PRINT "A2=";A2
190 PRINT "R1=";R1
200 PRINT "R2=";R2
210 PRINT "R3=";R3
220 PRINT "h1=";H1
230 PRINT "h2=";H2
240 PRINT "Time Increment =";T
250 PRINT "Ending Time=";T1
260 N=INT(T1/T)
270 N=N+1
```

```
280 DIM H3(N),H4(N),F1(N),F2(N)
290 M3=H1 MAX H2
300 D=R1*R2+R2*R3+R1*R3
310 Z1=(R2+R3)/D
320 Z2=(R1+R3)/D
330 Z3=R3/D
340 S1=A1*A2
350 S2=Z1*A2+Z2*A1
360 S3=Z1*Z2-Z3^2
370 M1=0.5/S1*(-S2+(S2^2-4*S1*S3)^0.5)
380 M2=0.5/S1*(-S2-(S2^2-4*S1*S3)^0.5)
390 C1=(H1*(A1*M2+Z1)-Z3*H2)/A1/(M2-M1)
400 C2=(Z3*H2-H1*(A1*M1+Z1))/A1/(M2-M1)
410 C3=(A1*M1+Z1)/Z3*C1
420 C4=(A1*M2+Z2)/Z3*C2
430 FOR I=1 TO N
440 T3=(I-1)*T
450 H3(I)=C1*EXP(M1*T3)+C2*EXP(M2*T3)
460 H4(I)=C3*EXP(M1*T3)+C4*EXP(M2*T3)
470 M=H3(I) MAX H4(I)
480 M3=M3 MAX M
490 NEXT I
500 PRINT "_Characteristic Roots Are_"
510 PRINT M1
520 PRINT M2
530 S=15/M3
540 N1=70/N
550 MOVE 50,45
560 DRAW 50,30
570 DRAW 120,30
580 FOR I=1 TO 10
590 MOVE 50+I*7,30.25
600 DRAW 50+I*7,29.75
610 NEXT I
620 CHARSIZE 4
630 MOVE 117,27
640 PRINT T1
650 MOVE 46,40
660 PRINT "h"
670 MOVE 47.5,39
680 PRINT "1"
690 MOVE 85,27
700 PRINT "Time"
710 MOVE 50,20
720 DRAW 50,5
730 DRAW 120,5
740 FOR I=1 TO 10
750 MOVE 50+I*7,5.25
760 DRAW 50+I*7,4.75
770 NEXT I
780 MOVE 117,2
790 PRINT T1;"_"
800 MOVE 46,15
810 PRINT "h "
820 MOVE 47.5,14
830 PRINT "2"
840 MOVE 85,2
850 PRINT "Time__"
860 MOVE 50,S*H3(1)+30
870 DRAW 50+N1,S*H3(2)+30
880 MOVE 50,S*H4(1)+5
890 DRAW 50+N1,S*H4(2)+5
```

```
900 FOR I=3 TO N
910 MOVE 50+(I-2)*N1,S*H3(I-1)+30
920 DRAW 50+(I-1)*N1,S*H3(I)+30
930 MOVE 50+(I-2)*N1,S*H4(I-1)+5
940 DRAW 50+(I-1)*N1,S*H4(I)+5
950 NEXT I
960 CHARSIZE 2
970 MOVE 0,10
980 PRINT "WANT TO RUN NEXT CASE, Y/N?"
990 INPUT A$
1000 IF A$="Y" THEN 10

1010 PRINT "___ ---------- END ---------"
1020 END
```

Particular Integral Of Differential Equation

In the cases where the **forcing functions** $Q_1(t)$ and $Q_2(t)$ in Eqs. (a) and (b) are not equal to zero, the complete solutions have to be written in the form of

$$h_1(t) = h_{1c}(t) + h_{1p}(t) \tag{r}$$

$$h_2(t) = h_{2c}(t) + h_{2p}(t) \tag{s}$$

where $h_{1c}(t)$ and $h_{2c}(t)$ denote the complementary function, which we have just completed discussing and the newly introduced $h_{1p}(t)$ and $h_{2p}(t)$ are the **particular integrals** of the system in response to the nonzero $Q_1(t)$ and $Q_2(t)$. As we are dealing mainly with linear ordinary differential equations with constant coefficients, the following table can be used as a guide in selecting trial forms for the particular integrals, depending on the forms of the nonhomogeneous terms, which in our case are Q_1 and Q_2:

Nonhomogeneous Term	**Particular Integral**
A specified constant C	An unknown constant A
Ct^n (n a positive integer)	$\sum_{i=0}^{n} A_i x^i$, A_i to be determined
Ce^{mt} (m either real or complex)	Ae^{mt}
$C \cos kx$ or $C \sin kx$	$A_1 \cos kx + A_2 \sin kx$

Consider the case of $Q_1 = 20 \text{ cm}^3/\text{sec}$ and $Q_2 = 500 \text{ cm}^3/\text{sec}$. Eq. (g) becomes

$$\frac{d^2}{dt^2} h_1 + 0.129 \frac{d}{dt} h_1 + 0.00208 h_1 = 0.0112 \tag{t}$$

The nonhomogeneous term is a constant equal to 0.0112; accordingly we select

$$h_{1p}(t) = A$$

Apparently, the first and second time derivatives of $h_{1p}(t)$ are equal to zero, and Eq. (t) reveals that $0.00208A = 0.0112$ or $A = 5.38$. As a result, the complete solution for $h_1(t)$ is changed from Eq. (n), consisting of the complementary function alone to

$$h_1(t) = h_{1c}(t) + h_{1p}(t) = C_1 e^{-0.0189t} + C_2 e^{-0.1103t} + 5.38 \qquad \text{(u)}$$

And instead of Eq. (o), the complete solution for $h_2(t)$ derived from Eqs. (a) and (u) is

$$h_2(t) = 0.647 C_1 e^{-0.0189t} - 1.547 C_2 e^{-0.1103t} + 5.92 \qquad \text{(v)}$$

If the same initial conditions of $h_1 = 30$ cm and $h_2 = 0$ are adopted, Eqs. (u) and (v) can be solved by Cramer's rule to obtain $C_1 = 235.50$ and $C_2 = -220.88$. The resulting complete solutions of the dual-tank system are

$$h_1(t) = 235.50 e^{-0.0189t} - 220.88 e^{-0.1103t} + 5.38 \qquad \text{(w)}$$

$$h_2(t) = 152.37 e^{-0.0189t} + 341.71 e^{-0.1103t} + 5.92 \qquad \text{(x)}$$

A program similar to the DUAL.TANK transient analysis can be developed for the forced steady-state flow. The problem can, however, be more generally treated for all types of forcing functions $Q_1(t)$ and $Q_2(t)$ by numerical approximations. This problem will be taken up again when finite-difference methods are discussed in Chapter 10. For the same reason the procedure for complete solution of linear ordinary differential equations with constant coefficients is only briefly reviewed and not exhaustively explored.

EXERCISE

For a general first-order system, the governing differential equation can be written as

$$a \frac{d}{dt} x(t) + b x(t) = y(t)$$

When the system is subjected to a sinusoidal disturbance $y(t) = c \sin (\omega t + \theta)$, there is frequently a need in analysis and design of physical systems to determine the forced response $x(t)$ of the system that is the particular integral of the above equation. Since we are involved in this text only with linear systems, we shall consider a class of problems where a, b, and c are constants. As an exercise, show that the particular integral for $x(t)$ is

$$x(t) = M \sin (\omega t + \theta + \phi)$$

where

$$M = c/[a^2 \omega^2 + b^2]^{1/2}$$

$$\phi = \tan^{-1} (-\omega a/b)$$

The meanings of ω, M, and ϕ will be explained in detail in the next chapter when the frequency response of a system is discussed. The constant θ is added to facilitate the possible consideration of $y(t)$, which happens to be a cosine function.

Program SYSTEM.1D has been prepared for the general need of solving the first-degree systems. Sample results and listing are presented in Fig. 7-16.

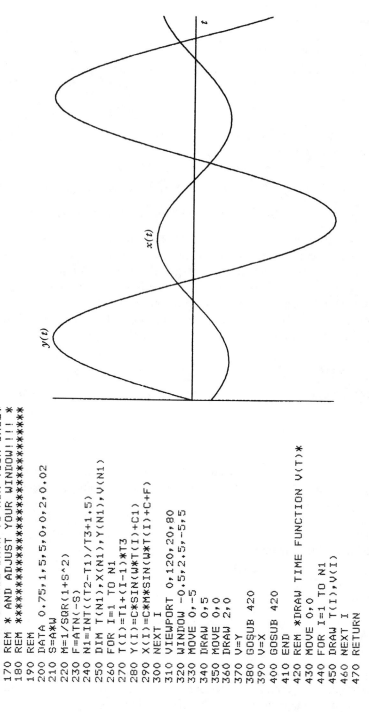

```
100 REM * PROGRAM SYSTEM.1D *
110 INIT
120 PAGE
130 READ A,B,C,W,C1,T1,T2,T3
140 REM
150 REM *****************************
160 REM *CHANGE DATA TO RUN YOUR CASE*
170 REM * AND ADJUST YOUR WINDOW!!!! *
180 REM *****************************
190 REM
200 DATA 0.75,1.5,5,0,0,2,0.02
210 S=A*W
220 M=1/SQR(1+S^2)
230 F=ATN(-S)
240 N1=INT((T2-T1)/T3+1.5)
250 DIM T(N1),X(N1),Y(N1),V(N1)
260 FOR I=1 TO N1
270 T(I)=T1+(I-1)*T3
280 Y(I)=C*SIN(W*T(I)+C1)
290 X(I)=C*M*SIN(W*T(I)+C+F)
300 NEXT I
310 VIEWPORT 0,120,20,80
320 WINDOW -0.5,2.5,-5,5
330 MOVE 0,-5
340 DRAW 0,5
350 MOVE 0,0
360 DRAW 2,0
370 V=Y
380 GOSUB 420
390 V=X
400 GOSUB 420
410 END
420 REM *DRAW TIME FUNCTION V(T)*
430 MOVE 0,0
440 FOR I=1 TO N1
450 DRAW T(I),V(I)
460 NEXT I
470 RETURN
```

Figure 7-16   Steady-state response $y(t)$ of a first-order system to sinusoidal disturbance $x(t)$.

7.8 DESIGN SPECIFICATIONS IN TIME DOMAIN

Dynamic systems are often tested by use of a unit step function, $u(t)$,[4] to study
the response in the time domain. The response curve exhibits a number of important performance characteristics of the system, and based on these the design specifications can be set and checked. Here we shall use the translational mechanical
system shown in Fig. 7-13b to describe the time-domain specifications, even
though it can be any dynamic system. The system response in this case is the displacement function $x(t)$, when the disturbing forcing function, $F(t)$, is $u(t)$. The
differential equation governing $x(t)$ is

$$m\frac{d^2x}{dt^2} + c\frac{dx}{dt} + kx = u(t) \tag{a}$$

By following the procedure outlined in the preceding section, we can show that
the analytical steady-state solution is

$$x(t) = 1 - \frac{\omega_n}{\omega_d} e^{-at} \sin(\omega_d t + \phi) \tag{b}$$

where

$$\omega_n = (k/m)^{1/2} \qquad \omega_d = (4km - c^2)^{1/2}/2m \tag{c,d}$$

$$a = c/2m \qquad \phi = \tan^{-1}(\omega_d/a) \tag{e,f}$$

The parameters ω_n, ω_d, and a are called **undamped natural frequency, damped
natural frequency**, and **damping coefficient**, respectively. A typical plot of $x(t)$,
which in this case is called the **unit-step response**, for $m = k = 1$, $c = 0.4$ with
proper units, of course, is presented in Fig. 7-17. A number of specification terms
in time domain can be defined by use of Fig. 7-17. They are described as follows:

1. Overshoot, x_h, is the difference of maximum transient response and steady-
state response.
2. Rise time, t_r, is the time required for the system to respond from 10 to 90%
of the steady-state result.
3. Settlement time, t_s, is the time required for the system to settle its response to
within a specified percentage of the steady-state result, usually 2 or 5%.
4. Delay time, t_d, is the time required for reaching 50% of the steady-state response.

Since the input is a unit-step function and hence bounded, the overshoot is a
measure of **relative stability** to assess quantitatively whether or not the system
response is reasonably bounded. The rise time, settlement time, and delay time
measure the **speed of response** of the system.

Figure 7-17 is plotted by Program 2D.USR (two-dimensional unit-step re-

[4] See Appendix C for the definition of a unit step function.

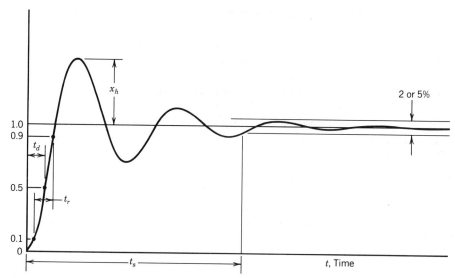

Figure 7-17 Design specifications in time domain: (a) x_h = overshoot, (b) t_d = delay time, (c) t_r = rise time, and (d) t_s = settlement time.

sponse), which can be employed for obtaining the steady-state response of any physical systems with a governing equation similar to Eq. (a).

EXERCISE

Denote the voltage across the capacitor in Fig. 7-11 as $v^*(t)$. Show that the governing differential equation for $v^*(t)$ is

$$LC \frac{d^2v^*}{dt^2} + RC \frac{dv^*}{dt} + v^* = v(t)$$

For finding the unit-step response of $v^*(t)$, Program 2D. USR can readily be utilized by replacing M, C, K, and $x(t)$ with LC, RC, unity, and v^*, respectively, and letting $v(t)$ be $u(t)$. Choose appropriate values for LC and RC such that $(RC)^2 > 4LC$ and run Program 2D. USR to display $v^*(t)$.

PROGRAM 2D. USR

```
100 REM * PROGRAM 2D. USR *
110 INIT
120 PAGE
130 READ K, M, C, T1, T2, T3
140 DATA 1, 1, 0. 4, 0, 30, 0. 1
150 W0=SOR(K/M)
160 W=1/2/M*SOR(4*K*M−C↑2)
170 A=C/2/M
180 0=ATN(W/A)
```

```
190 NI = INT((T2 - T1)/T3 + 1.5)
200 DIM T(NI), V(NI)
210 FOR I = I TO NI
220 T(I) = T1 + (I-1)*T3
230 V(I) = 1 - WO/W*EXP(-A*T(I))*SIN(W*T(I)+O)
240 NEXT I
250 VIEWPORT 0, 120, 20, 80
260 WINDOW -0.5, 30.5, 0, 2
270 MOVE 0, 1
280 DRAW 30, 1
290 MOVE 0, -5
300 DRAW 0, 5
310 MOVE 0, 0
320 DRAW 30, 0
330 GOSUB 350
340 END
350 REM*DRAW TIME FUNCTION V (T)*
360 MOVE 0, 0
370 FOR I = I TO NI
380 DRAW T (I), V(I)
390 NEXT I
400 RETURN
```

7.9 WHEATSTONE BRIDGE PROBLEM— FORMULATION OF MATRIX GOVERNING EQUATION

To further demonstrate how the governing mathematical equations for an engineering system are to be derived, we recall the Wheatstone bridge electric system (Fig. 7-18), discussed early in Chapter 6. Here, the impedances Z_1 and Z_2 are specifically defined as combinations of capacitor C, resistors R_1 and R_5, and inductors L_1 and L_2. Suppose that the source current is $i_{7,1} = Ie^{j2\pi ft}$, where subscript 7,1 indicates that the direction of the current is from the node 7 to the node 1, $j = (-1)^{1/2}$, and f is the frequency of the alternating current. To find the particular solution of the currents and voltages throughout the circuit, we assume the voltages at the nodes to be

$$v_i = V_i e^{j2\pi ft} \qquad i = 1, 2, \ldots, 7 \tag{a}$$

Generally, the amplitudes V_i contain j and are complex. The governing equations for V_i can be formulated by application of Kirchhoff's current law at the seven nodes. For example, for node 1 we may write

$$i_{7,1} = i_{1,2} + i_{1,4}$$

Or by use of the current-voltage equations for the elements involved in the circuit, the above equation can be written as

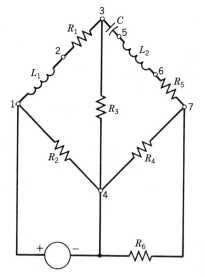

Figure 7-18   Three-loop, seven-node circuit of a Wheatstone bridge.

$$\frac{1}{L_1}\int (v_1 - v_2)\,dt + \frac{1}{R_2}(v_1 - v_4) = I e^{j2\pi f t}$$

Upon substitution, Eq. (a) can be simplified to

$$\left(\frac{1}{R_2} - \frac{1}{2\pi f L_1}j\right)V_1 + \left(\frac{1}{2\pi f L_1}j\right)V_2 + \left(-\frac{1}{R_2}\right)V_4 = I \qquad \text{(b)}$$

Notice that the factor $e^{j2\pi f t}$, common to all terms, has been dropped. Following a similar derivation, consideration of node 2 will yield

$$\left(\frac{1}{2\pi f L_1}j\right)V_1 + \left(\frac{1}{R_1} - \frac{1}{2\pi f L_1}j\right)V_2 + \left(-\frac{1}{R_1}\right)V_3 = 0 \qquad \text{(c)}$$

For node 3, the equation involving a capacitor is

$$\frac{1}{R_1}(v_2 - v_3) = \frac{1}{R_3}(v_3 - v_4) + C\frac{d}{dt}(v_3 - v_5)$$

The equation in terms of the amplitudes, after simplification, is

$$\left(-\frac{1}{R_1}\right)V_2 + \left(\frac{1}{R_1} + \frac{1}{R_3} + 2\pi f C j\right)V_3 + \left(-\frac{1}{R_3}\right)V_4 + (-2\pi f C j)V_5 = 0 \quad \text{(d)}$$

Having demonstrated the derivation of governing equations based on Kirchhoff's current law for the nodes 1, 2, and 3, the following results for the remaining nodes can be easily obtained by a similar procedure:

$$\left(-\frac{1}{R_2}\right)V_1 + \left(-\frac{1}{R_3}\right)V_3 + \left(\frac{1}{R_2} + \frac{1}{R_3} + \frac{1}{R_4}\right)V_4 + \left(-\frac{1}{R_4}\right)V_7 = 0 \qquad \text{(e)}$$

$$(-2\pi f C j)V_3 + \left(2\pi f C - \frac{1}{2\pi f L_2}\right)j V_5 + \left(\frac{1}{2\pi f L_2}j\right)V_6 = 0 \qquad \text{(f)}$$

$$\left(\frac{1}{2\pi f L_2}j\right)V_5 + \left(\frac{1}{R_3} - \frac{1}{2\pi f L_2}j\right)V_6 + \left(-\frac{1}{R_3}\right)V_7 = 0 \qquad \text{(g)}$$

and

$$\left(-\frac{1}{R_6}\right)V_1 + \left(-\frac{1}{R_4}\right)V_4 + \left(-\frac{1}{R_5}\right)V_6 + \left(\frac{1}{R_4} + \frac{1}{R_5} + \frac{1}{R_6}\right)V_7 = 0 \qquad \text{(h)}$$

Hence, there are seven linear algebraic equations, (b) through (h), which suffice for the solution of the seven amplitudes V_1 through V_7.

EXERCISE

Derive the governing matrix equation for the circuit shown below.

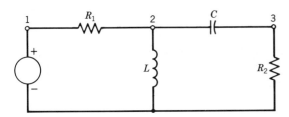

7.10 CONCLUDING REMARKS

In this chapter, modeling of commonly used physical elements, when their constitutive relations are linear, has been delineated. These elements were then assembled to form various physical systems, either completely electrical, mechanical, fluid, thermal, or coupled, and it was shown that the systems are generally governed by linear differential equations with constant coefficients. Because of the similarity in these governing equations, analogous solution and interpretation can be made among these systems.

The complementary solution of the governing differential equation of a system corresponds to the reaction of the system in response to the initial disturbances. Due to the viscous nature of most physical systems, the system will eventually settle down to its equilibrium configuration. Therefore, such responses are "in passing" from the disturbed state to the final equilibrated state and, hence, are called the **transient responses**. Lowering the mass of the translational mechanical system shown in Fig. 7-13b by a small amount and letting it oscillate up and down is a typical example of transient motion. The mass will eventually return to its undisturbed position.

The particular integral of the governing differential equation of a system is the system's response to a continuous disturbance. Because of the continuous nature of the problem, the question to be answered is then about the eventual status of the system. Or, more precisely, the question is "What will be the configuration of the system when enough time has passed?" Such a response of a system hence is called a **steady-state response**. Again referring to the translational mechanical system shown in Fig. 7-13b, the steady-state response of the system is concerned with the displacement $x(t)$ when $F(t)$, which may simply be a sinusoidal function, is continually acting on the mass.

In this chapter, it has also been shown that differential equations and linear algebraic equations are both involved in the analysis of physical systems. We shall demonstrate in the next chapter how the Laplace transform technique can best be utilized to relate these two types of equations, which govern the behavior of linear physical systems. This leads to the study of the system's transfer function and its characteristic roots. A need thus arises for the solution of higher-order polynomials, and to that end, Bairstow's method is presented in Appendix B. The Laplace transform itself is covered in Appendix C. In Chapter 9 the finite-element nodal analysis will be presented to further elaborate in a systematic fashion the analogy of complex physical systems.

EXERCISE

An insulation material has a thermal conductivity c_t and is used to wrap around a cylindrical tank of outside diameter D. If the thickness of this insulation material covering the tank is h, what is the difference in the thermal resistance R_t if heat flow changes from inward to outward?

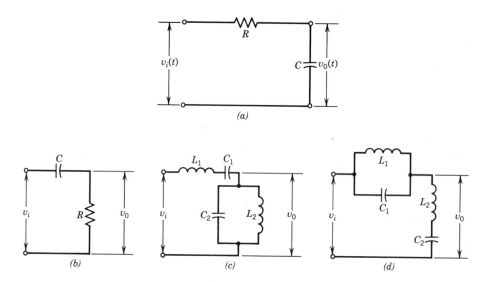

EXERCISE

Electric filters are commonly used for selecting a particular range of frequencies for transmission of audio–video signals. The circuits shown above are (a) low-pass, (b) high-pass, (c) band-pass, and (d) band-reject filters. The last one rejects a certain range of frequencies. The differential equation governing $v_0(t)$ for the low-pass filter will be derived and discussed in Chapter 8. Derive the governing differential equations for the other three filters.

CHAPTER 8

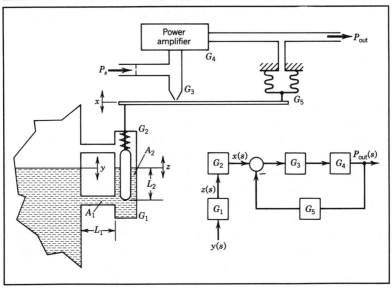

Schematic and block diagram of a pneumatic level transmitter system. (Adapted from Llyod and Anderson, *Industrial Process Control*, Courtesy of Fisher Controls International, Inc.)

MANIPULATION
OF SYSTEM'S
TRANSFER FUNCTION

8.1 INTRODUCTION

In Chapter 7 it was demonstrated that the behavior of linear physical systems is governed by either differential or matrix equations. Laplace transform enables the linear differential equations to be converted into linear algebraic equations, thereby making it possible to discuss the governing equations of physical systems all in the form of linear algebraic equations.

The formal presentation of Laplace transform is presented in Appendix C. In this chapter the application of Laplace transform for physical elements and systems is emphasized. The concept of transfer function can be introduced as a consequence of transforming the time domain of the system's dynamics into the frequency domain. With the aid of block diagrams, the application of the principle of superposition for linear elements greatly facilitates the derivation of the system's transfer function.

The computer-aided aspect of manipulating the element and system transfer functions is the central theme of this chapter. How to simplify the transfer function of a physical system by the operator of a computer graphics system interactively will be demonstrated. Bode analysis of the frequency response of physical systems is to be employed in the trial design process. Because of the developed interactive capability, altering the element characteristics and examining the effects on the system response become spontaneous, giving the designer-operator a real "feel" of what he or she is trying to accomplish.

The derivation and manipulation of matrix equations that govern the dynamic response of physical systems will be further elaborated in Chapter 9, where finite-element nodal analyses are to be discussed. Appendixes A and B should also be referred to often for the solution of matrix equations and polynomials, respectively. Particularly, Bairstow's method for finding real and/or complex roots is important in dealing with the characteristic roots of the systems involved in the present chapter.

8.2 TRANSFER FUNCTION

Let us introduce the concept of transfer function by consideration of a simple electric system, shown in Fig. 8-1. It is a simple RC circuit. As will become clear later, this system can be used to maintain approximately the same incoming voltage level at lower frequencies but attenuate the voltage level at higher frequencies. A so-called **low-pass filter** may be built this way for transmission of audio or

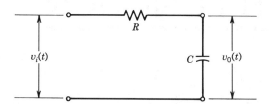

Figure 8-1 A RC circuit.

video signals. For such a system, Kirchhoff's current law provides the differential equation

$$\frac{1}{R}(v_i - v_0) = C \frac{d}{dt} v_0$$

The change in time t of the output voltage $v_0(t)$ can hence be related to the varying input voltage $v_i(t)$ in the form of

$$RC \frac{d}{dt} v_0 + v_0 = v_i \tag{a}$$

The Laplace transform of Eq. (a) can be easily obtained by use of the formulas listed in Table C-1 of Appendix C. That is,

$$RC[sV_0(s) - v_0(t = 0)] + V_0(s) = V_i(s)$$

For steady-state analysis, the initial-condition term $v_0(t = 0)$ can be dropped since it will not affect the long-term response of the system, and the above equation can be reduced to

$$(RCs + 1)V_0(s) = V_i(s) \tag{b}$$

Hence, as far as the steady-state solution is concerned, it is clear that the output voltage is

$$v_0(t) = L^{-1}[V_0(s)] = L^{-1}[V_i(s)/(RCs + 1)] \tag{c}$$

But here we are also interested in a new term called the **transfer function** of the system, which is defined in this case as

$$G(s) = V_0(s)/V_i(s) = \frac{1}{RCs + 1} \tag{d}$$

In the general case, the transfer function of a system with an input function $i(t)$ and an output function $o(t)$ is defined to be

$$G(s) = L[o(t)]/L[i(t)] = O(s)/I(s) \tag{e}$$

In order to appreciate the role that transfer functions play in helping analyze the steady-state response of a system it represents, let us consider a simple sinusoidal input voltage $v_i(t) = K \sin \omega t$ for the RC circuit in Fig. 8-1, where K is the amplitude in volts and ω is the frequency in radians per second. From Table C-1 in Appendix C, we find $V_i(s) = K\omega/(s^2 + \omega^2)$, and Eq. (c) gives

$$v_0(t) = L^{-1}[K\omega/(s^2 + \omega^2)(RCs + 1)] \tag{f}$$

By partial fraction expansions, we may write $v_0(t)$ in terms of the undetermined constants A, B, and D as

$$v_0(t) = L^{-1}\left[\frac{As + B}{s^2 + \omega^2} + \frac{D}{RCs + 1}\right] \tag{g}$$

Matching of the numerator terms in Eqs. (f) and (g) yields

$$RCA + D = 0, \qquad A + RCB = 0, \qquad B + \omega^2 D = K\omega$$

The solutions are[1]

$$A = -KRC\omega/\Delta, \qquad B = K\omega/\Delta$$

$$D = K(RC)^2\omega/\Delta, \qquad \Delta = 1 + (RC\omega)^2$$

Upon substituting the above results in Eq. (g), we obtain

$$v_0(t) = L^{-1}\left[(-KRC\omega/\Delta)\frac{s}{s^2 + \omega^2} + (K/\Delta)\frac{\omega}{s^2 + \omega^2} + (KRC\omega/\Delta)\frac{1}{s - (-1/RC)}\right]$$

$$= -(KRC\omega/\Delta)\cos\omega t + (K/\Delta)\sin\omega t + (KRC\omega/\Delta)\exp(-t/RC)$$

The terms in the last right-side expression are derived by simply looking up the required inversions from the $f(t)$ column in Table C-1. For steady-state analysis, the last term $(KRC\omega/\Delta)\exp(-t/RC)$ can be dropped since we are interested only in $v_0(t)$ when t is sufficiently large. As a result, the solution can be written as

$$v_0(t) = MK\sin(\omega t + \phi) \tag{h}$$

where

$$M(\omega) = [1 + (RC\omega)^2]^{-1/2} \tag{i}$$

$$\phi(\omega) = \tan^{-1}(-RC\omega) \tag{j}$$

M and ϕ are called the **magnification** or **magnitude ratio** and **phase angle**, respectively, of the system.

8.3 FREQUENCY RESPONSE

It is of practical interest to investigate $|M(\omega)|$ over a selected range of the frequency spectrum. For example, a low-pass RC filter shown in Fig. 8-1 would require that $|M(\omega)|$ be large in lower range and small in higher range of ω. Study of the magnification ratio $M(\omega)$ and phase angle $\phi(\omega)$ relative to the change in frequency ω probes into the frequency response of the system, for which the transfer function plays an important role. To show that, we compare Eqs. (i) and (j) with Eq. (d). Doing so reveals that if s is assigned to be $j\omega$, then

$$|G(s = j\omega)| = |M(\omega)| \tag{k}$$

This result can be easily proven, first from Eq. (d) that

$$G(j\omega) = 1/(1 + RC\omega j)$$

$$= (1 - RC\omega j)/[1 + (RC\omega)^2] \tag{l}$$

---

[1] Cramer's rule explained in Appendix A can be readily employed for the solution.

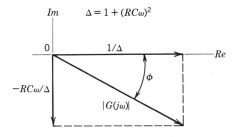

Figure 8-2 Graphical representation of the transfer function $G(j\omega)$ on a complex plane, for the RC circuit shown in Fig. 8-1.

Equation (l) is the rationalized form of $G(j\omega)$. The polar form of $G(j\omega)$, as illustrated in Fig. 8-2, has a magnitude of $M(\omega)$ and an angle equal to $\phi(\omega)$.

The argument that $M(\omega)$ and $\phi(\omega)$ can be derived directly from the transfer function is in fact supported by Euler's formula relating sinusoidal and complex exponential functions:

$$e^{j\omega t} = \cos \omega t + j \sin \omega t \qquad (m)$$

The input function $v_i(t) = K\sin \omega t$ can be treated as $v_i(t) = \text{Im}[Ke^{j\omega t}]$, where Im denotes the "imaginary part of." Actually, s is defined in the Laplace transform pair as a complex number. That means in general that s can be taken as equal to $a + j\omega$ with a being a real constant. In other words, our discussion thus far has covered only the special case of $a = 0$.

For $v_i(t) = 5 \sin \omega t$, $RC = 0.212 \times 10^{-5}$ sec, $|M(\omega)|$ and $\phi(\omega)$ are plotted in Fig. 8-3 for $f(= 2\pi\omega)$ in the range of 10^4 to 10^7 Hz. It shows that $M(\omega)$ drops by

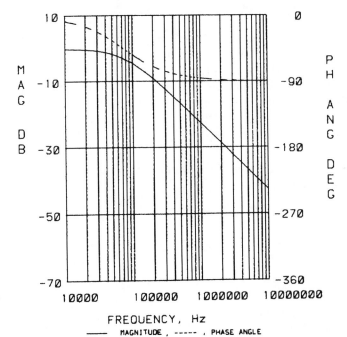

FREQUENCY, Hz

—— MAGNITUDE , ----- , PHASE ANGLE

Figure 8-3

3 dB (decibel) at $f = 75,000$ Hz, which is known as the **cutoff frequency**. The unit of decibel is defined as

$$M(\omega) \text{ in dB} = 20 \log_{10} M(\omega) \tag{n}$$

A drop of 3 dB means $M(\omega) = 10^{(-3/20)} = 0.708$, or the output voltage amplitude is about 71% of the input voltage amplitude. Program BODE is made available for the general need of plotting $M(\omega)$ and $\phi(\omega)$; it will be discussed in a later section.

EXERCISE

Show that the magnification ratio for a translational–mechanical system, shown in Fig. 7-13b, is

$$M(\omega) = \{[1 - (\omega/\omega_n)^2]^2 + (2c\omega/C_c\omega_n)^2\}^{-1/2}$$

where ω_n and C_c are called **natural frequency** and **critical damping coefficient** of the system, respectively, related to the system parameters m, c, and k by the equations

$$\omega_n = (k/m)^{1/2} \qquad C_c = 2(km)^{1/2}$$

Plot $M(\omega)$ versus ω/ω_n for $C/C_c = 1$, 0.5, and 0.25 over the range of $\omega/\omega_n = 0$ to 2.5.

8.4 BLOCK-DIAGRAM MANIPULATION OF TRANSFER FUNCTIONS

For ease in manipulation of compounded systems, it has been found that **block diagrams** of transfer functions are helpful graphical representations. In the case of an RC circuit shown in Fig. 8-1, the block diagram is given in Fig. 8-4. As another example, the hydraulic integrator shown in Fig. 8-5a converts the input flow rate $q(t)$ into mechanical displacement $x(t)$. The conservation of mass requires

$$q(t) = A \frac{d}{dt} x(t) \tag{a}$$

where A is the *uniform*, cross-sectional area of the cylinder. Application of Laplace transform to Eq. (a) yields the transfer function

$$G(s) = X(s)/Q(s) = 1/As \tag{b}$$

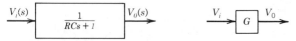

Figure 8-4   Block-diagram representation of the RC circuit shown in Fig. 8-1.

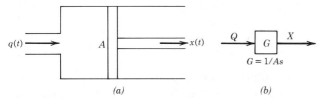

Figure 8-5   Hydraulic integrator and its block diagram representation.

Thus, for representing the hydraulic integrator, the block diagram is simply the one shown in Fig. 8-5b.

In practical cases, a system involves many physical elements. Each element may have a simple transfer function, but when they are interconnected, a systematic way of manipulating the constituent transfer functions to arrive at an overall transfer function for the entire system must be developed. To illustrate this need, let us consider an electric motor. Figure 8-6a shows that a motor can be treated as two separate subsystems, electric and mechanical. The input voltage $v(t)$ generates a torque T proportional to the current i passing the inductor, say, $T = Ki$. This torque is resisted by the demand torque T_d and the torsional spring and damper, which reduce the torque by amounts equal to $b\omega$ and $J\alpha$, respectively. ω and α are the angular velocity and acceleration of the shaft. Returning to the electric subsystem, the rotation in turn causes a voltage drop equal to $K\omega$ in the circuit. The differential equations for the electric motor can be shown to be

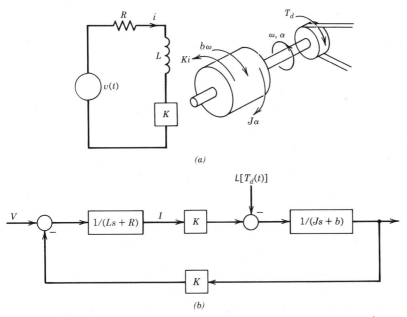

Figure 8-6   (a) An electric motor and (b) its block-diagram representation.

$$L\frac{di}{dt} + Ri + K\omega = v \qquad \text{(b)}$$

$$Ki - J\alpha - b\omega = T_d \qquad \text{(c)}$$

The block-diagram representation for the above coupled system is presented in Fig. 8-6b. The problem that needs to be solved is the reduction of this compound diagram into a simple $G(s)$ relating $\Omega(s)$ to $V(s)$, which are the Laplace transforms of $\omega(t)$ and $v(t)$, respectively. Notice that the circles used in Fig. 8-6b are called **summing points**, the arrows represent unidirectional signal flow, and the solid dots are called **takeoff points**. Along with the blocks, they are the four types of elements for construction of block diagrams.

Since the system is linear, the demand torque T_d can be treated separately from $v(t)$. In other words, the block diagram can be simplified by excluding T_d to find $G^*(s)$ relating $\Omega(s)$ and $V(s)$ and then excluding $v(t)$ to find $G^{**}(s)$ relating $\Omega(s)$ and $L[T_d(t)]$. The overall transfer function is $G(s) = G^*(s) - G^{**}(s)$.

Major transformation formulas for simplifying the block diagrams are presented in Fig. 8-7. The proof of these results is quite straightforward. Consider the case of cascade blocks shown in Fig. 8-7a. By definition of transfer functions, we can have $M = G_1U$ and $X = G_2M$. When M is eliminated, $X = G_2G_1U$. And because both G_1 and G_2 are functions of s and thus commutative, $X = G_2G_1U = G_1G_2U$. For the feedback loop shown in Fig. 8-7c, we have

$$M = U + G_2X \qquad X = G_1M$$

Substitution of M into the second equation yields

$$X = G_1U + G_1G_2X$$

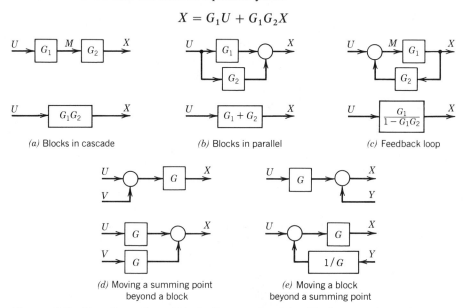

(a) Blocks in cascade   (b) Blocks in parallel   (c) Feedback loop

(d) Moving a summing point beyond a block   (e) Moving a block beyond a summing point

Figure 8-7 Transformation of block diagrams involving blocks, summing points, takeoff points, and arrows.

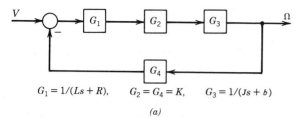

$$G_1 = 1/(Ls + R), \qquad G_2 = G_4 = K, \qquad G_3 = 1/(Js + b)$$

(a)

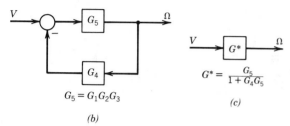

$$G_5 = G_1 G_2 G_3$$

(b)

$$G^* = \frac{G_5}{1 + G_4 G_5}$$

(c)

Figure 8-8 Simplification of the block diagrams for the electric motor from (a) to (b) to (c) according to the transformation formulas in Fig. 8-7.

Solving for X, one finally arrives at the result

$$X = \frac{G_1}{1 - G_1 G_2} U$$

The other formulas in Fig. 8-7 are just as easy to prove.

Figure 8-8 shows the step-by-step simplification of the block diagram of Fig. 8-6b and the final derivation of the transfer function $G^*(s)$ relating $\Omega(s)$ and $V(s)$ by setting $T_d = 0$, which is

$$G^*(s) = \Omega(s)/V(s) = G_5/(1 + G_4 G_5)$$

$$= G_1 G_2 G_3/(1 + G_1 G_2 G_3 G_4) = K/[Ls + R)(Js + b) + K^2] \qquad (d)$$

EXERCISE

By setting $v(t) = 0$, simplify the block diagram of Fig. 8-6b and show that

$$G^{**}(s) = \Omega(s)/L[T_d(t)] = -(Ls + R)/[(Ls + R)(Js + b) + K^2]$$

EXERCISE

Simplify the block diagram for the pneumatic level transmitter system shown in Fig. 8-9 so that the system's transfer function can be described as $G(s) = P_{out}(s)/Y(s)$.

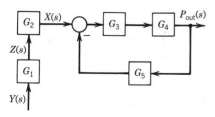

Figure 8-9 A block diagram representing a pneumatic control of the outflow pressure P_{out} by the fluid level y.

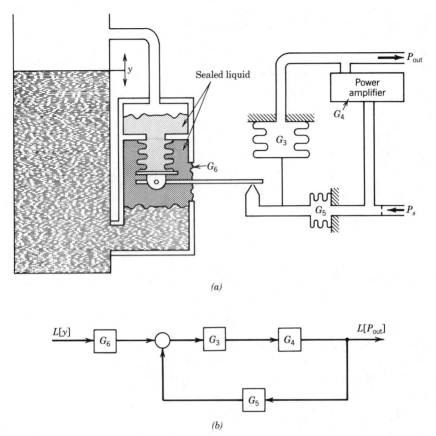

(a)

(b)

Figure 8-10   (*a*) Schematic of a pneumatic level transmitter with pressure difference sensor, and (*b*) its block-diagram representation. (Courtesy of Fisher Controls International, Inc.)

Derive an expression for $G(s)$ in terms of G_1 through G_5. When a pressure difference sensor is employed in the system as shown in Fig. 8-10*a*, the block diagram is reduced to Fig. 8-10*b*. How should the previously derived $G(s)$ be changed to reflect this alteration?

8.5   DESIGN SPECIFICATIONS IN FREQUENCY DOMAIN

Plotting the magnitude ratio $M(\omega)$ and phase angle $\phi(\omega)$ of a system's frequency response over a specified range of frequency (ω) spectrum enables the other characteristics of the system to be examined in the frequency domain, in addition to those in the time domain already discussed in Chapter 7. In order to determine whether or not a system is performing satisfactorily and is relatively stable—that is, to show how close it is to becoming unstable—a number of parameters have to be quantified, including bandwidth, gain margin, phase margin, cutoff rate, resonant frequency, and peak.

Bandwidth is a range of frequencies over which a system will respond satisfactorily. For example, the $M(\omega)$ curve for an RC circuit shown in Fig. 8-3 has a drop of 3 dB at $f_c = 2\pi\omega_c = 75,000$ Hz. The output voltage will thus be maintained at a relatively high level for $0 \le f \le f_c$ and lower for $f > f_c$. The system hence satisfies the low-pass requirement.

Cutoff frequency is ω_c in the above paragraph. It is the frequency at which the magnitude ratio is 0.707 or drops approximately 3 dB below its maximum value. ω_c is in units of radians per second, whereas f_c is in cycles per second or hertz. f is called simply **frequency**, whereas ω is called **angular frequency**. As a design specification, the rate at which the magnitude ratio decreases, called **cutoff rate**, is often given in units of decibels per octave or decibels per decade. An octave is the change in frequency by a factor of 2, whereas a decade is by a factor of 10. For example, the cutoff rate of 2 dB/decade requires $M(\omega)$ to decrease by 2 dB from $f = 10$ to $f = 10^2$ or $f = 10^2$ to $f = 10^3$ Hz and so on.

The frequency $\omega = \omega_p$ at which the magnitude of the system's transfer function $G(j\omega)$, or $M(\omega)$, reaches a maximum is called the **resonance frequency**, whereas $M(\omega_p)$ is called the **resonance peak**. Apparently, $M(\omega)$ will become infinity if its denominator expression is equal to zero. For example, the system transfer function shown in Fig. 8-8c will lead to this situation for some values of ω that satisfy the equation

$$1 + G_4(j\omega)G_5(j\omega) = 0$$

The above equation is called the **characteristic equation** of the system and its roots are called **characteristic roots**.[2] Definitely, one does not want to have the characteristic roots (frequencies) falling within the required bandwidth.

Referring to Fig. 8-11, the frequency $\omega = \omega_\pi$ at which the phase angle of the system's transfer function $\phi(\omega)$ is equal to $\pm 180°$ is called the **phase crossover frequency**, and the reciprocal of $M(\omega_\pi)$ or $|G(j\omega_\pi)|$ is called the **gain margin** G_M. If $|G(j\omega)|$ is in decibels, then $G_M = -|G(\omega_\pi)|$ or the negative value of the $|G(\omega_\pi)|$ reading from Fig. 8-11. The reason for the minus sign is that when G_M is measured in decibels, it follows that

$$\begin{aligned} G_M \text{ in dB} &= 20 \log_{10}(1/|G(j\omega)|) \\ &= 20[\log_{10} 1 - \log_{10}|G(j\omega)|)] \\ &= -20 \log_{10}|G(j\omega)| = -|G(j\omega)| \text{ in dB} \end{aligned}$$

The frequency $\omega = \omega_1$ at which $|G(j\omega_1)| = 1$ is called the **gain crossover frequency**. The **phase margin** is defined as

$$\phi_M = \phi(\omega_1) - \phi(\omega_\pi)$$

---

[2] They are also called poles, as will be discussed in the section on Bode analysis. Since they are polynomials of ω, Bairstow's method explained in Appendix B can be employed for iterative solution of the roots.

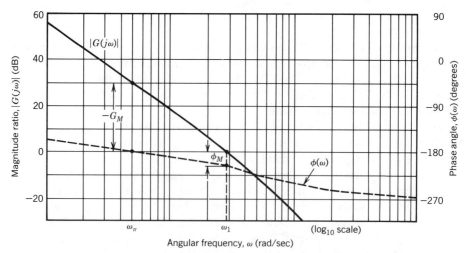

Figure 8-11   Gain crossover frequency ω_1, gain margin G_M, phase crossover frequency ω_ϕ, and phase margin ϕ_M.

It should be clear that the bandwidth and cutoff rate are measures of the speed of the system's response; the resonance peak quantifies whether the system response is bounded or stable. Gain and phase margins determine how close the system is to being unstable. Both of them (in dB and degrees, respectively) must be positive to assure relative stability of the system. Simple examples of adding compensation elements to a system for achieving better stability will be presented after the Bode analysis has been discussed.

EXERCISE

Use Fig. 8-12 and 8-13 in Section 8.6 to find the values of gain and phase margins accurate to two significant digits by direct measurement from the BODE plots. Also give the estimated values of the gain and phase crossover frequencies, in units of both radians per second and cycles per second or hertz.

EXERCISE

Given a system's transfer function $G(s)$, the resonance peak M_p and resonance frequency ω_p can be determined by direct measurement from the magnitude ratio $M(\omega)$ versus ω curve. However, better accuracy of these data has to be achieved by an analytical solution. First, $M(\omega) = |G(j\omega)|$ is to be expressed in polar form; the derivative of $M(\omega)$ is then set equal to zero to solve for ω_p and consequently the value of M_p can be computed. As an example, consider the case of $G(s) = 1/[(s + 1)^2 + 2^2]$. It yields

$$M(\omega) = 1/|[(j\omega + 1)^2 + 4]| = 1/|[(5 - \omega^2) + (2\omega)j]|$$
$$= 1/[(5 - \omega^2)^2 + (2\omega)^2]^{1/2} = 1/[\omega^4 - 6\omega^2 + 5]^{1/2}$$

$$\frac{dM(\omega)}{d\omega} = -(\omega^4 - 6\omega^2 + 5)^{-3/2}(4\omega^3 - 12\omega)/2$$

The only acceptable solution for the resonance frequency is $\omega_p = 3^{1/2}$ and the corresponding resonance peak is $M_p = \frac{1}{4}$.

Write a BASIC program to implement the above computation and also display the $M(\omega) \sim \omega$ graph for visual verification of the computed results.

EXERCISE

Describe how the crossover frequencies and gain and phase margins can be analytically computed by writing $M(\omega)$ and $\phi(\omega)$ in polar forms. Also write a BASIC program for these calculations by use of $G(s)$ given in the above exercise.

8.6 BODE ANALYSIS

For a linear engineering system subjected to a time-dependent disturbance, $x_f(t)$, the differential equation governing the system's response, $x_r(t)$, may in general be written as[3]

$$\sum_{i=0}^{N} A_i D^i x_r(t) = \sum_{k=0}^{M} B_k D^k x_f(t) \tag{a}$$

where $D^i \equiv d^i/dt^i$ for $i \neq 0$ and $D^0 \equiv 1$ are the differentiation operators introduced for simplicity of expressions. In Eq. (a), the coefficients A_i and B_k are related to the physical characteristics of the system. When all initial conditions are assumed equal to zero and the Laplace transform technique is applied, the transfer function of Eq. (a) is

$$G(s) = X_r(s)/X_f(s) = \sum_{k=0}^{M} B_k s^k \bigg/ \sum_{i=0}^{N} A_i s^i \tag{b}$$

The right-hand side of Eq. (b) can be factorized so that the transfer function becomes a fraction consisting of mostly linear terms and quadratic terms. That is,

$$G(s) = \frac{K \prod_{i=1}^{n_z} (s + z_i) \prod_{k=1}^{n_{zc}} (s^2 + 2\zeta_{zk}\omega_{nzk}s + \omega_{nzk}^2)}{s^l \prod_{i=1}^{n_p} (s + p_i) \prod_{k=1}^{n_{pc}} (s^2 + 2\zeta_{pk}\omega_{npk}s + \omega_{npk}^2)} \tag{c}$$

where $K = B_M/A_N$ is called the **gain** constant, $-z_i$ and $-p_i$ are called **zeros** and **poles**, and l is called the **order** of the pole at the origin of the system. The coefficients ζ and ω_n appearing in the quadratic factors associated with the zeros, subscripted z, and with the poles, subscripted p, are the **damping ratios** and **natural frequencies** of the system. In Eq. (c), n_z, n_{zc}, n_p, and n_{pc} are the total numbers of zeros, complex zeros, poles, and complex poles of the system, respectively.

---

[3] S. J. Zitek and Y. C. Pao, "Computer Graphics Aids Bode Analysis," *Instruments & Control Systems*, September 1975, pp. 31–36.

The symbols $\sum$ (summation) and $\prod$ (product) used in Eqs. (a), (b), and (c) perhaps should be explained for clarity. For readers who are not familiar with these symbols, they are defined for the following abbreviations:

$$\sum_{i=1}^{n} f_i = f_1 + f_2 + \cdots + f_n$$

and

$$\prod_{i=1}^{n} f_i = f_1 f_2 \cdots f_n$$

The **Bode analysis** of a system is to study the system's transfer function for a selected range of the frequency domain. That is, if the system is excited by a sinusoidal forcing function $x_f(t) = A_f \sin \omega t$ for $\omega_{min} \le \omega \le \omega_{max}$, how the system will respond is of concern. To find the system's response, Eq. (c) can be modified by letting $s = j\omega$, where $j = \sqrt{-1}$ for consideration of sinusoidal excitations of the system. As a result, Eq. (c) becomes

$$G(j\omega) = \frac{\left[K_B \prod_{i=1}^{n_z} \left(1 + \frac{j\omega}{z_i}\right) \prod_{k=1}^{n_{zc}} \left(1 + 2j\zeta_{zk} \frac{\omega}{\omega_{nzk}} - \frac{\omega^2}{\omega_{nzk}^2}\right) \right]}{\left[(j\omega)^l \prod_{i=1}^{n_p} \left(1 + \frac{j\omega}{p_i}\right) \prod_{k=1}^{n_{pc}} \left(1 + 2j\zeta_{pk} \frac{\omega}{\omega_{npk}} - \frac{\omega^2}{\omega_{npk}^2}\right) \right]} \tag{d}$$

where K_B is called the **Bode gain** to be calculated with the formula

$$K_B = K \prod_{i=1}^{n_z} z_i \prod_{k=1}^{n_{zc}} \omega_{nzk}^2 \Big/ \prod_{i=1}^{n_p} p_i \prod_{k=1}^{n_{zp}} \omega_{npk}^2 \tag{e}$$

If the system response is expressed as $x_r(t) = A_r \sin(\omega t + \phi)$, the magnitude ratio A_r/A_f and the phase shift ϕ are both dependent on the excitation frequency ω. It is usually desirable to plot these two parameters versus ω over the selected frequency range $(\omega_{min}, \omega_{max})$. The magnitude ratio and phase shift expressions can be derived as

$$|G(j\omega)| = \frac{K_B \prod_{i=1}^{n_z} \left|1 + \frac{j\omega}{z_i}\right| \prod_{k=1}^{n_{zc}} \left|1 + 2j\zeta_{zk} \frac{\omega}{\omega_{nzk}} - \frac{\omega^2}{\omega_{nzk}^2}\right|}{|(j\omega)^l| \prod_{i=1}^{n_p} \left|1 + \frac{j\omega}{p_i}\right| \prod_{k=1}^{n_{pc}} \left|1 + 2j\zeta_{pk} \frac{\omega}{\omega_{npk}} - \frac{\omega^2}{\omega_{npk}^2}\right|} \tag{f}$$

$$\phi = \sum_{i=1}^{n_z} \tan^{-1}\left(\frac{\omega}{z_i}\right) + \sum_{k=1}^{n_{zc}} \tan^{-1}\left(\frac{2\zeta_{zk}\omega\omega_{nzk}}{\omega_{nzk}^2 - \omega^2}\right) - l \times 90°$$

$$- \sum_{i=1}^{n_p} \tan^{-1}\left(\frac{\omega}{p_i}\right) - \sum_{k=1}^{n_{pc}} \tan^{-1}\left(\frac{2\zeta_{pk}\omega\omega_{npk}}{\omega_{npk}^2 - \omega^2}\right) \tag{g}$$

A program called BODE has been developed for plotting of $|G(j\omega)|$ and $\phi(\omega)$ over a specified range of three-decade frequency spectrum. Figures 8-12 and 8-13 are presented as sample plots, respectively, for the transfer functions

```
NO. OF POLES=4
NO. OF ZEROS=2
NO. OF COMPLEX POLES=0
NO. OF COMPLEX ZEROS=0
ORDER L=1
BODE GAIN=20000
LOWER FREQUENCY=10
POLES:
P(1)=4.6
P(2)=500
P(3)=0.7
P(4)=700
ZEROS:
Z(1)=7
Z(2)=70
```

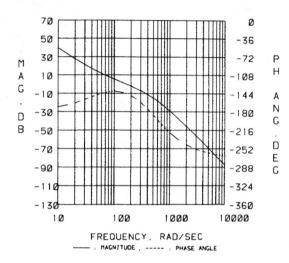

```
**    DESIGN ENDS    **
```

Figure 8-12 Sample Bode plot involving real zeros and poles.

$$G(j\omega) = \frac{20{,}000(j\omega/7 + 1)(j\omega/70 + 1)}{j\omega(j\omega/4.6 + 1)(j\omega/500 + 1)(j\omega/0.7 + 1)(j\omega/700 + 1)}$$

and

$$G_c(j\omega) = \frac{10(1 + j\omega)}{(j\omega)^2[1 + j\omega/4 - (\omega/4)^2]}$$

G_c is a transfer function with a complex pole. Program menu and user's response are illustrated in Fig. 8-14 for consideration of the case of G described above. Program BODE has been found to be an effective tool for interactive compensation of systems, a subject to be discussed in the next section.

PROGRAM BODE

```
100 REM * PROGRAM BODE - Bode analysis *
110 INIT
120 PAGE
130 REM * D7=1 ----- PLOT THE MAGNITUDE *
140 REM * D7=2 ----- PLOT THE PHASE     *
```

```
NO. OF POLES=0
NO. OF ZEROS=1
NO. OF COMPLEX POLES=1
NO. OF COMPLEX ZEROS=0
ORDER L=2
BODE GAIN=10
LOWER FREQUENCY=0.1
ZEROS:
Z(1)=1
COMPLEX POLES
ZETA (1)=0.5
OM(1)=4
```

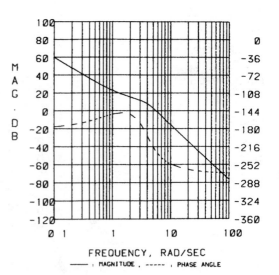

FREQUENCY, RAD/SEC

———— , MAGNITUDE , ----- , PHASE ANGLE

```
**      DESIGN ENDS      **
```

Figure 8-13   Sample Bode plot involving complex pole.

```
150 REM * D7=3 ---- TERMINATE THE PROGRAM *
160 D7=1
170 PRINT "DO YOU WANT TO INPUT YOUR OWN DATA OR USE A SAMPLE RUN?"
180 PRINT "ANSWER Y OR N : ";
190 INPUT Q$
200 IF Q$="Y" THEN 320
210 PRINT "DATA SET `BODE' OR `LOW-PASS FILTER', ENTER 1 OR 2? ";
220 INPUT I8
230 IF I8=1 THEN 260
240 GOSUB 3450
250 GO TO 580
260 READ P1,Z1,C1,C2,O1,B7,F1,I9
270 DATA 4,2,0,0,1,20000,10,2
280 DIM P(P1),Z(Z1)
290 READ P,Z
300 DATA 4.6,500,500,0.5,50,5
310 GO TO 580
320 PRINT "NUMBERS OF POLES, ZEROS, COMPLEX POLES = ";
330 INPUT P1,Z1,C1
340 PRINT "NUMBER OF COMPLEX ZERO, ORDER L, GAIN, LOWEST FREQUENCY = "
350 INPUT C2,O1,B7,F1
360 PRINT "INPUT FREQUENCY UNIT `Hz' OR `Rad/s', ENTER 1 or 2? ";
370 INPUT I9
```

```
3570 REM   * THIS SUB PLOTS THE MAGNITUDE CURVE *
3580 C7=0
3590 FOR W=T1 TO 10*T1 STEP 0.5*T1
3600 GOSUB 2120
3610 M=LGT(1+0.5*C7)+L9
3620 IF W=F1 THEN 3660
3630 IF D7=2 AND INT((L9*18+C7+1)/2)-(L9*18+C7+1)/2<>0 THEN 3660
3640 DRAW D1,M
3650 GO TO 3670
3660 MOVE D1,M
3670 C7=C7+1
3680 NEXT W
3690 T1=T1*10
3700 L9=L9+1
3710 IF L9=3 THEN 3730
3720 GO TO 3580
3730 RETURN
```

```
DO YOU WANT TO INPUT YOUR OWN DATA OR USE A SAMPLE RUN?
ANSWER Y OR N.
Y
NUMBERS OF POLES, ZEROS, COMPLEX POLES =4,2,0
NUMBER OF COMPLEX ZERO, ORDER L, GAIN, LOWEST FREQUENCY =
0.1,20000,10
INPUT FREQUENCY UNIT 'Hz' OR 'Rad/s' (1 or 2)?
2
INPUT THE POLE VALUES 4.6,500,0.7,700
INPUT THE REAL ZERO VALUES 7,70
```

Figure 8-14   User's menu and response for Program BODE to generate the magnitude ratio and phase angle plots shown in Fig. 8-12.

```
380 IF P1=0 THEN 400
390 DIM P(P1)
400 IF Z1=0 THEN 420
410 DIM Z(Z1)
420 IF C1=0 THEN 440
430 DIM X1(C1),X2(C1)
440 IF C2=0 THEN 460
450 DIM Y1(C2),Y2(C2)
460 IF P1=0 THEN 490
470 PRINT "INPUT THE POLE VALUES ";
480 INPUT P
490 IF Z1=0 THEN 520
500 PRINT "INPUT THE REAL ZERO VALUES ";
510 INPUT Z
520 IF C1=0 THEN 550
530 PRINT "INPUT THE COMPLEX POLE VALUES ";
540 INPUT X1,X2
550 IF C2=0 THEN 580
560 PRINT "INPUT THE COMPLEX ZERO VALUES ";
570 INPUT Y1,Y2
580 CHARSIZE 3
590 W=F1
600 GOSUB 2120
610 M7=M
620 M1=M
630 W=F1*1000
640 GOSUB 2120
```

```
650 M8=M
660 M2=M
670 M1=M1+ABS((M7-M8)*1.1)
680 M2=M2-ABS((M7-M8)*1.3)
690 I1=0.015767*(M1-M2)
700 WINDOW -4.5,4.5,M2,M1
710 PAGE
720 J=1
730 M3=INT(M7+ABS((M7-M8)*0.2))
740 M4=INT(M8-ABS((M7-M8)*0.2))
750 M3=(INT(M3/10)+1)*10
760 M4=(INT(M4/10)-1)*10
770 REM! M3 IS POSITIVE Y; M4 IS NEGATIVE Y
780 IF (M3-M4)/20-INT((M3-M4)/20)=0 THEN 800
790 M3=M3+10
800 FOR I=1 TO 10
810 IF J<>1 THEN 840
820 D1=LGT(I)
830 GO TO 880
840 IF J<>2 THEN 870
850 D1=LGT(I)+1
860 GO TO 880
870 D1=LGT(I)+2
880 MOVE D1,M4
890 DRAW D1,M3
900 NEXT I
910 J=J+1
920 IF J<>4 THEN 800
930 I1=0.015767*(M1-M2)
940 FOR I=M4 TO M3 STEP 20
950 MOVE 0,I
960 DRAW 3,I
970 IF I=0 THEN 1020
980 IF I>0 THEN 1050
990 MOVE -0.52,I-I1
1000 PRINT I

1010 GO TO 1070
1020 MOVE -0.3,I-I1
1030 PRINT 0
1040 GO TO 1070
1050 MOVE -0.41,I-I1
1060 PRINT I
1070 NEXT I
1080 HOME
1090 FOR X=0 TO 3 STEP 1
1100 IF X=1 THEN 1130
1110 MOVE X-0.2,M4-3*I1
1120 GO TO 1140
1130 MOVE X-0.13,M4-3*I1
1140 PRINT F1*10^X
1150 NEXT X
1160 GOSUB 3270
1170 HOME
1180 PRINT "NO. OF POLES=";P1
1190 PRINT "NO. OF ZEROS=";Z1
1200 PRINT "NO. OF COMPLEX POLES=";C1
1210 PRINT "NO. OF COMPLEX ZEROS=";C2
1220 PRINT "ORDER L=";O1
1230 PRINT "BODE GAIN=";B7
1240 PRINT "LOWER FREQUENCY=";F1
1250 IF P1=0 THEN 1300
```

```
1260 PRINT "POLES:"
1270 FOR I=1 TO P1
1280 PRINT "P(";I;")=";P(I)
1290 NEXT I
1300 IF Z1=0 THEN 1350
1310 PRINT "ZEROS:"
1320 FOR I=1 TO Z1
1330 PRINT "Z(";I;")=";Z(I)
1340 NEXT I
1350 IF C1=0 THEN 1470
1360 IF Z1>P1 THEN 1380
1370 GO TO 1380
1380 PRINT "COMPLEX POLES"
1390 FOR I=1 TO C1
1400 IF P1>Z1 THEN 1440
1410 PRINT "ZETA (";I;")=";X1(I)
1420 PRINT "OM(";I;")=";X2(I)
1430 GO TO 1460
1440 PRINT "ZETA (";I;")=";X1(I)
1450 PRINT "OM(";I;")=";X2(I)
1460 NEXT I
1470 IF C2=0 THEN 1570
1480 PRINT "COMPLEX ZEROS"
1490 FOR I=1 TO C2
1500 IF P1>Z1 THEN 1540
1510 PRINT "ZETA (";I;")=";Y1(I)
1520 PRINT "OM(";I;")=";Y2(I)
1530 GO TO 1560
1540 PRINT "ZETA (";I;")=";Y1(I)
1550 PRINT "OM(";I;")=";Y2(I)
1560 NEXT I
1570 L9=0
1580 T1=F1
1590 GOSUB 3570
1600 GO TO 1610
1610 D7=D7+1
1620 IF D7=2 THEN 1570
1630 MOVE LGT(4),M4-6*I1
1640 IF I9=2 THEN 1670
1650 PRINT "FREQUENCY, Hz"
1660 GO TO 1680
1670 PRINT "FREQUENCY, RAD/SEC"
1680 MOVE LGT(1.5),M4-7*I1
1690 CHARSIZE 1
1700 RMOVE 0.1,0
1710 RDRAW 0.3,0
1720 RMOVE 0,-0.45*I1
1730 PRINT " : MAGNITUDE"
1740 RMOVE 0.8,0-0.15*I1
1750 PRINT " , ----- : PHASE ANGLE"
1760 CHARSIZE 4
1770 MOVE -0.7,M3-8*I1
1780 PRINT "M"
1790 RMOVE 4.6,I1
1800 PRINT "P"
1810 RMOVE 0,-2*I1
1820 PRINT "H"
1830 RMOVE 0,-2*I1
1840 PRINT " "
1850 RMOVE 0,-2*I1
1860 PRINT "A"
```

```
1870 RMOVE 0,-2*I1
1880 PRINT "N"
1890 RMOVE 0,-2*I1
1900 PRINT "G"
1910 RMOVE 0,-2*I1
1920 PRINT ","
1930 RMOVE 0,-2*I1
1940 PRINT "D"
1950 RMOVE 0,-2*I1
1960 PRINT "E"
1970 RMOVE 0,-2*I1
1980 PRINT "G"
1990 MOVE -0.7,M3-10*I1
2000 PRINT "A"


2010 MOVE -0.7,M3-12*I1
2020 PRINT "G"
2030 MOVE -0.7,M3-14*I1
2040 PRINT ","
2050 MOVE -0.7,M3-16.5*I1
2060 PRINT "D"
2070 MOVE -0.7,M3-18.5*I1
2080 PRINT "B"
2090 MOVE -4.1,M2+20
2100 PRINT "**    DESIGN ENDS    **"
2110 END
2120 REM! THIS SUB FINDS THE MAGNITUDE IN DB.
2130 REM! FINAL TERM E+FJ
2140 GOSUB 2230
2150 E=(A*C+B*D)/(C^2+D^2)
2160 F=(B*C-A*D)/(C^2+D^2)
2170 IF D7=1 THEN 2200
2180 GOSUB 3070
2190 IF D7=2 THEN 2220
2200 G=(E^2+F^2)^0.5
2210 M=20*LGT(G)
2220 RETURN
2230 REM! THIS SUB GENERATES THE FUNCTION AS FOLLOWS:
2240 REM! CALCULATE (A+BJ)*(C+DJ) THEN STORES A+BJ
2250 REM!     THEN MULTIPLY (A+BJ)*(E+FJ) AND STORE AS A+BJ
2260 REM!        DOING THIS UNTIL ALL ARE MULTIPLIED
2270 REM!    A1+B1J    ARE     POLES IN DENOMENATER
2280 REM!    A3+B3J    ARE COMPLEX POLES IN DENOMENATER
2290 REM!    A2+B2J    ARE     ZEROS IN NUMERATOR
2300 REM!    A4+B4J    ARE COMPLEX ZEROS IN NUMERATOR
2310 REM!    A+BJ      IS THE FINAL TERM IN NUMERATOR
2320 REM!    C+DJ      IS THE FINAL TERM IN DENOMENATOR.
2330 REM!    B7 IS THE BODE GAIN.
2340 REM!    O1 IS THE ORDER.
2350 IF I9=2 THEN 2370
2360 W=W*2*PI
2370 IF Z1=0 THEN 2470
2380 A2=1
2390 B2=W/Z(1)
2400 K=2
2410 IF K>Z1 THEN 2490
2420 S2=A2
2430 A2=S2*1-B2*W/Z(K)
2440 B2=B2*1+S2*W/Z(K)
```

```
2450 K=K+1
2460 GO TO 2410
2470 A2=1
2480 B2=0
2490 IF P1=0 THEN 2590
2500 A1=1
2510 B1=W/P(1)
2520 K=2
2530 IF K>P1 THEN 2610
2540 S1=A1
2550 A1=S1*1-B1*W/P(K)
2560 B1=B1*1+S1*W/P(K)
2570 K=K+1
2580 GO TO 2530
2590 A1=1
2600 B1=0
2610 IF C2=0 THEN 2710
2620 A4=1-(W/Y2(1))^2
2630 B4=2*Y1(1)*W/Y2(1)
2640 K=2
2650 IF K>C2 THEN 2730
2660 S4=A4
2670 A4=S4*(1-(W/Y2(K))^2)-B4*(2*Y1(K)*W/Y2(K))
2680 B4=B4*(1-(W/Y2(K))^2)-S4*(2*Y1(K)*W/Y2(K))
2690 K=K+1
2700 GO TO 2650
2710 A4=1
2720 B4=0
2730 IF C1=0 THEN 2820
2740 A3=1-(W/X2(1))^2
2750 B3=2*X1(1)*W/X2(1)
2760 IF K>C1 THEN 2840
2770 S3=A3
2780 A3=S3*(1-(W/X2(K))^2)-B3*(2*X1(K)*W/X2(K))
2790 B3=B3*(1-(W/X2(K))^2)-S3*(2*X1(K)*W/X2(K))
2800 K=K+1
2810 GO TO 2760
2820 A3=1
2830 B3=0
2840 A=B7*(A2*A4-B1*B4)
2850 B=B7*(A4*B2+B4*A2)
2860 A5=A1*A3-B1*B3
2870 A6=A1*B3+B1*A3
2880 REM! DEAL WITH ORDER
2890 O2=O1-INT(O1/4)*4
2900 IF O1-INT(O1/2)*2<>0 THEN 2990
2910 C=A5*W^O1
2920 D=A6*W^O1
2930 IF O2=0 THEN 2960
2940 C=-C
2950 D=-D
2960 IF I9=2 THEN 2980
2970 W=W/2/PI
2980 RETURN
2990 C=-A6*W^O1
3000 D=A5*W^O1

3010 IF O2=1 THEN 3040
3020 C=-C
3030 D=-D
```

```
3040 IF I9=2 THEN 3060
3050 W=W/2/PI
3060 RETURN
3070 REM! THIS SUB FINDS THE PHASE ANGLE
3080 REM! THE PHASE ANGLE OF E+FJ
3090 REM! PHASE ANGLE H7
3100 SET DEGREES
3110 IF E=0 THEN 3160
3120 H7=ATN(F/E)
3130 IF E=)0 THEN 3150
3140 H7=H7+180
3150 GO TO 3200
3160 IF F=)0 THEN 3190
3170 H7=-90
3180 GO TO 3200
3190 H7=90
3200 IF H7=)0 THEN 3220
3210 H7=H7+360
3220 GOSUB 3240
3230 RETURN
3240 REM *    THIS SUB SCALES THE PHASE ANGLE.   *
3250 M=H7*(K7*20/360)+M4
3260 RETURN
3270 REM *    THIS SUB LABELS THE PHASE-ANAGLE AXIS.    *
3280 K7=ABS(M4-M3)/20
3290 IF K7)12 THEN 3370
3300 IF K7=7 THEN 3350
3310 IF K7=11 THEN 3330
3320 GO TO 3380
3330 K7=10
3340 GO TO 3380
3350 K7=6
3360 GO TO 3380
3370 K7=12
3380 FOR K8=0 TO K7 STEP 1
3390 MOVE 3.1,M4+20*K8-I1
3400 K9=360/K7*K8-360
3410 PRINT USING 3430:K9
3420 NEXT K8
3430 IMAGE 4D
3440 RETURN
3450 REM  *  DATA OF LOW-PASS-FILTER  *
3460 P1=1
3470 Z1=0
3480 C1=0
3490 C2=0
3500 O1=0
3510 B7=1
3520 F1=10000
3530 I9=1
3540 DIM P(P1)
3550 P(1)=1/2.12E-6
3560 RETURN
```

EXERCISE

Apply Program BODE for generating the magnitude ratio and phase angle graphs for the filter circuits shown in Fig. 8-15 by selecting appropriate values for the resistance, capacitances, and inductances.

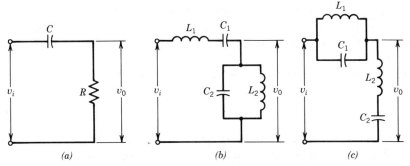

Figure 8-15 Filter circuits: (a) high-pass, (b) band-pass, and (c) band-reject.

8.7 COMPENSATION OF SYSTEM

A number of simple elements called **compensators**, which may be added to the system to adjust its performance so that the design specifications can be met, are shown in Fig. 8-16. The lead compensator is usually used for increasing the gain (or phase margin), or bandwidth of the system. Its transfer function can be easily shown to be

$$G_{\text{Lead}}(j\omega) = \frac{R_2}{R_1 + R_2} \frac{1 + j\omega/(1/R_1 C)}{1 + j\omega/(1/R_1 C + 1/R_2 C)} \tag{a}$$

The **lead compensator** thus has a Bode gain of $R_2/(R_1 + R_2)$, which is less than

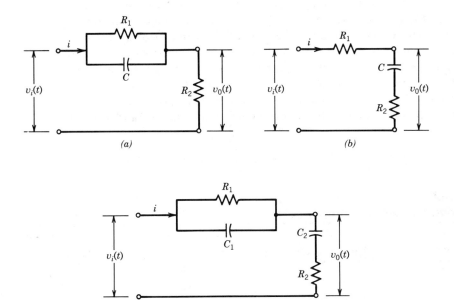

Figure 8-16 (a) Lead, (b) lag, and (c) lead–lag compensators.

one. Consequently, additional modification will be necessary to adjust the system's gain if a lead compensator is adopted. Equation (a) also shows that the lead compensator has a zero equal to minus $1/R_1C$ and a pole equal to minus $1/R_1C + 1/R_2C$.

The **lag compensator** shown in Fig. 8-13b has a transfer function

$$G_{Lag}(j\omega) = \frac{1 + j\omega/(1/R_2C)}{1 + j\omega/[1/(R_1 + R_2)C]} \tag{b}$$

Therefore, it has a zero equal to minus $1/R_2C$ and a pole equal to minus $1/(R_1 + R_2)C$ and will not alter the system's gain. The **lag–lead compensator** shown in Fig. 8-16c is usually arranged so that the product of the two zeros is equal to the product of the two poles in the transfer function

$$G_{Lead-Lag}(j\omega) = \frac{[1 + j\omega/(1/R_1C_1)][1 + j\omega/(1/R_2C_2)]}{(1 + j\omega/p_1)(1 + j\omega/p_2)} \tag{c}$$

where

$$p_1p_2 = 1/R_1C_1R_2C_2 \qquad p_1 + p_2 = \frac{1}{R_1C_1} + \frac{1}{R_2C_2} + \frac{1}{R_2C_1} \tag{d,e}$$

When the resistances R_1 and R_2 and the capacitances C_1 and C_2 are so chosen such that Eqs. (d) and (e) are satisfied, the addition of a lead–lag compensator to the system will also, like the lag compensator, not alter the gain of the system's frequency response.

The addition of a lag compensator to a system provides an attenuation effect, making the system more sluggish. The bandwidth is usually decreased. The lead–lag compensator gives the advantages of both lead and lag compensators and very little of their undesirable characteristics. Figures 8-17 through 8-20 produced by Program BODE are presented to demonstrate the consequence of using various compensators for improvement of the gain and phase margins.

Table 8-1 summarizes the results of various attempts to change the gain and phase margins of the uncompensated system, which has a transfer function

$$G(j\omega) = \frac{20,000}{j\omega(j\omega/4.6 + 1)(j\omega/500 + 1)}$$

TABLE 8-1
Various Trials of Compensation

Figure Number	Type of Compensator	Gain Margin (dB)	Phase Margin (degree)
8-17	None	−31.95	−29.12
8-18	Lead	−6.80	−20.33
8-19	Lead	1.35	2.94
8-12	Lead–lag	−38.48	35.57
8-20	Lead–lag	12.83	35.03

NO. OF POLES=2
NO. OF ZEROS=0
NO. OF COMPLEX POLES=0
NO. OF COMPLEX ZEROS=0
ORDER L=1
BODE GAIN=20000
LOWER FREQUENCY=10
POLES:
P(1)=4.6
P(2)=500

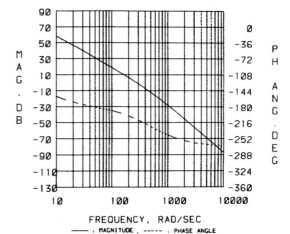

**   DESIGN ENDS   **

Figure 8-17 An uncompensated system with a transfer function
$G(j\omega) = 20{,}000/j\omega(j\omega/4.6 + 1)(j\omega/500 + 1)$.

NO. OF POLES=3
NO. OF ZEROS=1
NO. OF COMPLEX POLES=0
NO. OF COMPLEX ZEROS=0
ORDER L=1
BODE GAIN=20000
LOWER FREQUENCY=10
POLES:
P(1)=4.6
P(2)=500
P(3)=500
ZEROS:
Z(1)=50

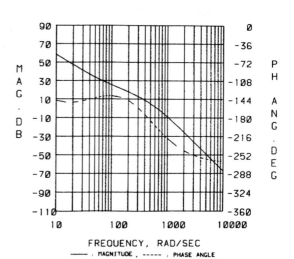

**   DESIGN ENDS   **

Figure 8-18 Compensated system with a transfer function $G_{new} = GG_L$ obtained by
adding a new compensator with $G_L(j\omega) = (j\omega/50 + 1)/(j\omega/500 + 1)$.

NO. OF POLES=3
NO. OF ZEROS=1
NO. OF COMPLEX POLES=0
NO. OF COMPLEX ZEROS=0
ORDER L=1
BODE GAIN=20000
LOWER FREQUENCY=10
POLES:
P(1)=4.6
P(2)=500
P(3)=1000
ZEROS:
Z(1)=100

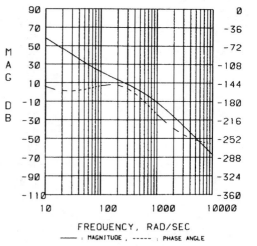

** DESIGN ENDS **

Figure 8-19 Compensated system with a transfer function $G_{new} = GG_{L'}$ obtained by adding a new lead compensator with $G_{L'} = (j\omega/100 + 1)/(j\omega/1000 + 1)$.

NO. OF POLES=4
NO. OF ZEROS=2
NO. OF COMPLEX POLES=0
NO. OF COMPLEX ZEROS=0
ORDER L=1
BODE GAIN=20000
LOWER FREQUENCY=10
POLES:
P(1)=4.6
P(2)=500
P(3)=500
P(4)=0.5
ZEROS:
Z(1)=50
Z(2)=5

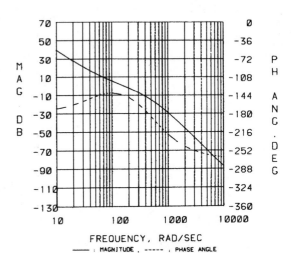

** DESIGN ENDS **

Figure 8-20 Compensated system with a transfer function $G_{new} = GG_{L-L}$ by adding a lead–lag compensator with $G_{L-L} = (j\omega/5 + 1)(j\omega/50 + 1)/(j\omega/0.5 + 1)(j\omega/500 + 1)$.

The lead compensator G_L is not as good as the lead compensator $G_{L'}$ in increasing the margins from negative values to positive values in order to assure the relative stability of the system. The use of lead–lag compensator

$$G_{L-L} = (j\omega/7 + 1)(j\omega/70 + 1)/(j\omega/0.7 + 1)(j\omega/700 + 1),$$

with the addition of the Bode plot of the compensated system shown in Fig. 8-12, improves only the phase margin, not the gain margin. The lead–lag compensator

$$G_{L-L} = (j\omega/5 + 1)(j\omega/50 + 1)/(j\omega/0.5 + 1)(j\omega/500 + 1)$$

appears to have given the best result in improving the system's relative stability.

The values of gain and phase margins listed in Table 8-1 are computed by use of an iterative scheme called **successive linear interpolation**, which is to be discussed in Chapter 10. Less accurate values of these margins can be obtained by direct measurement from the Bode plots presented in Figs. 8-12 and 8-17 through 8-20.

EXERCISE

Modify the Program BODE so that the range of frequency can be expanded from the presently arranged three decades to any specified number of decades. Also, a provision should be added in the program to label the frequency axis with an appropriate scaling factor to be calculated depending on the range of frequency being investigated. For example, in Fig. 8-3 the new horizontal label could be arranged as 1 to 1000 Frequency, Hz ($\times 10^4$) instead of the present label, 10,000 to 10,000,000. Since Program BODE can provide plots for the frequency range both in radians per second and hertz, the new provision should also accommodate both options.

EXERCISE

Figure 8-21 shows a sketch of the instrument that can be employed for measuring displacement $x(t)$, velocity $v(t)$, and acceleration $a(t)$. Depending on the measured input function, $x(t)$, $v(t)$, or $a(t)$, it is called a **seismograph, velocity meter**, or **accelerometer**, respectively. Derive the differential equations for the relative displacement of the measuring mass, $x_r(t)$ and show that the transfer functions for the three

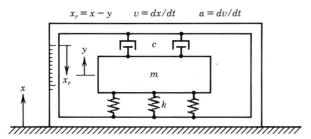

Figure 8-21 A sketch for seismograph or velocity meter or accelerometer for measuring $x(t)$, $v(t)$, or $a(t)$.

instruments are

$$G_1(s) = X_r(s)/X(s) = ms^2/(ms^2 + cs + k)$$
$$G_2(s) = X_r(s)/V(s) = ms/(ms^2 + cs + k)$$
$$G_3(s) = X_r(s)/A(s) = m/(ms^2 + cs + k)$$

where $X_r(s)$, $X(s)$, $V(s)$, and $A(s)$ are the Laplace transforms of $x_r(t)$, $x(t)$, $v(t)$, and $a(t)$, respectively, m is the instrument mass inside the case that is mounted on the foundation, and c and k are the overall damping and spring constants of the system.

As a seismograph, it is desirable that the instrument mass stay in space with very little motion, whereas x_r readily represents the motion of foundation x. This requires a heavy mass with light springs. Choose appropriate values for m, c, and k and generate BODE plots to confirm this expectation. $|G_1(j\omega)|$ should approach unity for $\omega > \omega_n$ where ω_n is the natural frequency equal to $(k/m)^{1/2}$.

For a velocity meter, very heavy damping is usually required in order to assure a wide range of frequencies that can be accurately measured. Let $m = 1$, $k = 1$, and $c = 0.1, 0.4$, and 1 and use Program BODE to plot $|G_2(j\omega)|$ and to examine the change in the measurable frequency range.

By use of Program BODE, show that the spring element used in an accelerometer should be sufficiently stiff to result in a natural frequency $\omega_n = (k/m)^{1/2}$ greater than four times the highest frequency to be measured.

CHAPTER 9

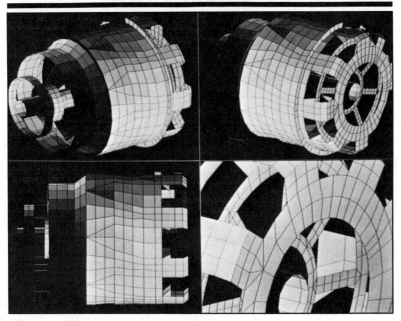

(Courtesy of PDA Engineering, Santa Ana, California.)

INTRODUCTORY
FINITE-ELEMENT
MATRIX ANALYSES

9.1 INTRODUCTION

There is a class of engineering problems that lead to the solution of a system of linear algebraic equations. A most notable example is the finite-element analysis of the two-dimensional and three-dimensional structures. In this chapter the formulation of the finite element analysis is explained in detail by use of a two-dimensional truss problem. It is also discussed that in an analogous manner, analyses of fluid flow and electric networks result in similar matrix equations as well.

Matrix algebra, which facilitates the manipulation of the systems of linear algebraic equations, as well as Cramer's rule and Gaussian elimination, which are fundamental tools for solving matrix equations, are all delineated in Appendix A. There are other methods of solving matrix equations, but the readers are referred to other textbooks on numerical methods so as not to overextend this text beyond the intended introductory level.

In this chapter the main emphasis is placed on the mathematical modeling and numerical solution of engineering systems based on the so-called **nodal analysis**.

Only **introductory** materials on the finite element method are presented. Besides stressing the analogy among different engineering systems in finite-element formulation and solution of the problems and the "computer-aided" involvements, no attempt is made to cover this fast growing subject in such a short chapter. However, appropriate references are given in the footnotes of ensuing pages when finite-element method and computer programs are discussed.

9.2 FINITE-ELEMENT ANALYSIS

Practical designs inevitably involve complicated geometric shapes and different materials. Consider the case of building a house. Rooms of varied sizes made of wood, concrete, metal, and many other materials are involved. In order to quantitatively assess whether or not a selected design is structurally safe, it is necessary to decompose a complicated structure into basic elements, each being of simple geometric shape and made of a single material. Detailed analysis of each element will then be possible. By preserving the physical interrelationships of the elements based on the ways that they are assembled together, an approximate but relatively accurate evaluation of the behavior of the structure as a whole can thereby be realized. It was through this concept that finite-element methods were developed first for aerospace structural analyses in the 1950s and more recently for many other fields.[1]

Here, the finite-element method is to be introduced with the aid of a simple **truss** problem. Figure 9-1 shows a structure representing a crane designed for lifting weights, which is constructed with straight uniform bars riveted or pinned together. These bars, called **truss members**, are assumed to be capable of carrying tensile or compressive **axial** forces only. Consider a typical truss member, say the

---

[1] O. C. Zienkiewicz, *The Finite Element Method*, McGraw-Hill, New York, 1977.

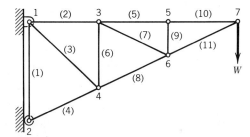

Figure 9-1

member 3-6 in Fig. 9-1. In order to formulate the finite-element analysis in a systematic manner, the two-dimensional forces, F_x and F_y, and displacements, δ_x and δ_y, at the joints (called **nodes**) are all directed in the positive direction, as shown in Fig. 9-2. As shown in Fig. 9-1, the truss member 3-6 is the seventh **element** of the structure; superscripts (7) are therefore added to F_{3x}, F_{3y}, F_{6x}, and F_{6y} to indicate that these nodal forces are associated with the member (7). Each element has its nodal forces, and when the members are connected, all the nodal forces combined together at the joint should be equal to the actual loads applied at that joint. For example, if the weight W is the only load applied at the node 7 as shown in Fig. 9-1, and if the structure is to remain stationary, then the horizontal and vertical forces at the node 7 must be in equilibrium. That is,

$$F_{7x}^{(10)} + F_{7x}^{(11)} = 0$$
$$F_{7y}^{(10)} + F_{7y}^{(11)} = -W$$

In general, the element forces and the applied forces, denoted as F_{ix} and F_{iy} without superscripts, at a typical ith node are related by the equations

$$\sum_{j=1}^{m} F_{ix}^{(j)} = F_{ix} \quad \text{and} \quad \sum_{j=1}^{m} F_{iy}^{(j)} = F_{iy} \tag{a}$$

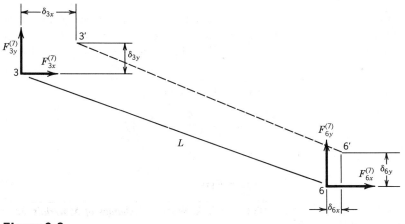

Figure 9-2

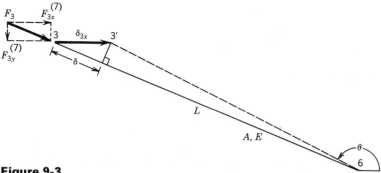

Figure 9-3

where m is the number of elements sharing the ith node. It should be noted that unlike F_{ix} and F_{iy}, the element forces carry superscripts (j), indicating their association with the jth element; δ_{ix} and δ_{iy} do not need superscripts since they are common to all elements joined at the ith node.

To derive equations relating the nodal forces $F_{ix}^{(j)}$ and $F_{iy}^{(j)}$ and the nodal displacements δ_{ix} and δ_{iy}, we consider a specific displacement, say δ_{3x}, and find out what should be the force F_{3x} required for producing this displacement. As shown in Fig. 9-3, if the joint 3 is to move to 3' for a resulting displacement of δ_{3x}, the element 3-6 is shortened by δ. From elementary mechanics of materials,[2] it is well known that if a uniform bar of length L, cross-sectional area A, with a modulus of elasticity E is subjected to a compressive force P, the bar will be shortened by an amount that can be calculated by the formula

$$\delta = PL/AE \tag{b}$$

In most practical cases involving trusses, the deformation δ is small (less than 0.2% of length L). Geometrically, δ can be obtained by projecting δ_{3x} onto the line 3-6 in Fig. 9-3. That is,

$$\delta = \delta_{3x} \cos(\pi - \theta) = -\delta_{3x} \cos\theta \tag{c}$$

where θ is the orientation of the element 3-6. Since a truss member is to carry axial load only, the force F_3 shown in Fig. 9-3 has a horizontal component $F_{3x}^{(7)}$, which can then be related to δ_{3x} with the aid of Eqs. (b) and (c), and noticing that F_3 is P, as

$$F_{3x}^{(7)} = F_3 \cos(\pi - \theta) = -F_3 \cos\theta$$

$$= -\frac{\delta AE}{L}\cos\theta = \left(\frac{AE}{L}\cos^2\theta\right)\delta_{3x} \tag{d}$$

Equation (d) is usually written as

$$F_{3x}^{(7)} = k_{11}^{(7)}\delta_{3x} \tag{e}$$

---

[2] A. Higdon, E. H. Ohlsen, W. B. Stiles, and J. A. Weese, *Mechanics of Materials*, John Wiley & Sons, New York, 1967.

where the **stiffness coefficient** relating the nodal force $F_{3x}^{(7)}$ to the nodal displacement δ_{3x} is introduced as

$$k_{11}^{(7)} = \frac{AE}{L} \cos^2 \theta \tag{f}$$

In a similar manner, the forces in the x direction and applied at node 3 required to produce displacements δ_{3y}, δ_{6x}, and δ_{6y} can be derived. The total force $F_{3x}^{(7)}$ can then be written as

$$k_{11}^{(7)}\delta_{3x} + k_{12}^{(7)}\delta_{3y} + k_{13}^{(7)}\delta_{6x} + k_{14}^{(7)}\delta_{6y} = F_{3x}^{(7)} \tag{g}$$

Three similar equations will also result when the forces $F_{3y}^{(7)}$, $F_{6x}^{(7)}$, and $F_{6y}^{(7)}$ are considered. In matrix notation, the resulting equations can be written as

$$[k^{(7)}]\{\delta\}_{3,6} = \{F^{(7)}\} \tag{h}$$

where the stiffness matrix $[k^{(7)}]$ can be shown to be

$$[k^{(7)}] = \frac{AE}{L} \begin{bmatrix} m^2 & mn & -m^2 & -mn \\ & n^2 & -mn & -n^2 \\ & & m^2 & mn \\ \text{Symmetric} & & & n^2 \end{bmatrix} \tag{i}$$

and

$$m = \cos \theta^{(7)} \qquad n = \sin \theta^{(7)} \tag{j}$$

The member orientation, θ, can be directly specified or calculated from the nodal coordinates, such as

$$\theta^{(7)} = \tan^{-1}\left[(y_6 - y_3)/(x_6 - x_3)\right] \tag{k}$$

The nodal displacement and force vectors in Eq. (h) are defined as

$$\{\delta\}_{3,6} = [\delta_{3x} \quad \delta_{3y} \quad \delta_{6x} \quad \delta_{6y}]^T \tag{l}$$

$$\{F^{(7)}\} = [F_{3x}^{(7)} \quad F_{3y}^{(7)} \quad F_{6x}^{(7)} \quad F_{6y}^{(7)}]^T \tag{m}$$

The superscript T denotes the transposition of a matrix, which requires interchange of the rows and columns.

When all elements are assembled to form the truss structure, the problem becomes one of how each element's stiffness matrix will contribute to the **structural stiffness matrix** equation

$$[K]\{\delta\} = \{F\} \tag{n}$$

where for convenience the displacement and force vectors have been renamed as

$$\{\delta\} = [\delta_{1x} \quad \delta_{1y} \quad \delta_{2x} \quad \delta_{2y} \quad \cdots \quad \delta_{Nx} \quad \delta_{Ny}]^T$$
$$= [\delta_1 \quad \delta_2 \quad \delta_3 \quad \delta_4 \quad \cdots \quad \delta_{2N-1} \quad \delta_{2N}]^T \tag{o}$$

$$\{F\} = [F_{1x} \quad F_{1y} \quad F_{2x} \quad F_{2y} \quad \cdots \quad F_{Nx} \quad F_{Ny}]^T$$
$$= [F_1 \quad F_2 \quad F_3 \quad F_4 \quad \cdots \quad F_{2N-1} \quad F_{2N}]^T \tag{p}$$

The structural stiffness matrix, $[K]$ of order $2N \times 2N$ with N being the number of nodes of the structure, is to be formed from properly combining the element structural matrices $[k^{(j)}]$ for $j = 1, 2, \ldots, M$ with M being the number of elements of the structure. This formation is to be carried out based on how the nodes are shared by the elements and also by recognizing the fact that the element forces are related to the applied forces according to Eq. (a).

9.3 A NUMERICAL EXAMPLE

Perhaps a simple example will help present all details involved in the solution of the deformation and internal forces in a truss structure. Consider the simple problem of three truss members ($M = 3$) connected at three nodes ($N = 3$) and subjected to the loads as shown in Fig. 9-4. First, the element stiffness matrices can be calculated using Eq. (i) in the preceding section and noting that $\theta^{(1)} = 60°$, $\theta^{(2)} = 150°$, and $\theta^{(3)} = 0°$. They are

$$[k^{(1)}] = \frac{AE}{2} \begin{bmatrix} 0.250 & 0.433 & -0.250 & -0.433 \\ & \boxed{0.750} & -0.433 & -0.750 \\ & & 0.250 & \boxed{0.433} \\ \text{Symmetric} & & & 0.750 \end{bmatrix} \quad (q)$$

$$[k^{(2)}] = \frac{AE}{3.464} \begin{bmatrix} 0.750 & \boxed{-0.433} & -0.750 & 0.433 \\ & 0.250 & 0.433 & -0.250 \\ & & 0.750 & -0.433 \\ \text{Symmetric} & & & 0.250 \end{bmatrix} \quad (r)$$

$$[k^{(3)}] = \frac{AE}{4} \begin{bmatrix} 1.000 & 0 & -1.000 & 0 \\ & 0 & 0 & 0 \\ & & \boxed{1.000} & 0 \\ \text{Symmetric} & & & 0 \end{bmatrix} \quad (s)$$

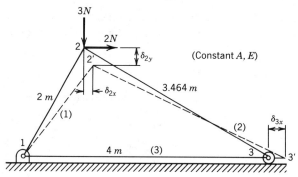

Figure 9-4

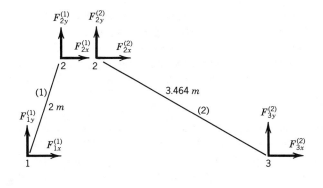

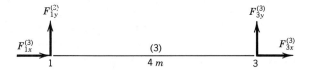

Figure 9-5

The relationships between the element forces and the actual applied forces can be explicitly written out as, referring to Figs. 9-4 and 9-5,

$$\begin{Bmatrix} F_{1x}^{(1)} + F_{1x}^{(3)} \\ F_{1y}^{(1)} + F_{1y}^{(3)} \\ F_{2x}^{(1)} + F_{2x}^{(2)} \\ F_{2y}^{(1)} + F_{2y}^{(2)} \\ F_{3x}^{(2)} + F_{3x}^{(3)} \\ F_{3y}^{(2)} + F_{3y}^{(3)} \end{Bmatrix} = \begin{Bmatrix} F_{1x} \\ F_{1y} \\ F_{2x} \\ F_{2y} \\ F_{3x} \\ F_{3y} \end{Bmatrix} = \begin{Bmatrix} ? \\ ? \\ 2 \\ -3 \\ 0 \\ ? \end{Bmatrix} \qquad (t)$$

As can be observed from the above expression, the reactions F_{1x}, F_{1y}, and F_{3y} are to be determined by the boundary conditions, which are $\delta_{1x} = \delta_{1y} = \delta_{3y} = 0$. The treatment of boundary conditions will be further elaborated upon later.

The structural stiffness matrix equation for this three-member truss system is to be obtained by properly combining the stiffnesses (q), (r), and (s), and by use of Eq. (t). The final form is

$$AE \begin{bmatrix} 0.375 & 0.217 & -0.125 & -0.217 & -0.250 & 0 \\ & 0.375 & -0.217 & -0.375 & 0 & 0 \\ & & 0.342 & \boxed{0.092} & -0.217 & 0.125 \\ & & & 0.447 & 0.125 & -0.072 \\ & & & & \boxed{0.467} & -0.125 \\ \text{Symmetric} & & & & & 0.072 \end{bmatrix} \begin{Bmatrix} \delta_{1x} \\ \delta_{1y} \\ \delta_{2x} \\ \delta_{2y} \\ \delta_{3x} \\ \delta_{3y} \end{Bmatrix} = \begin{Bmatrix} F_{1x} \\ F_{1y} \\ 2 \\ -3 \\ 0 \\ F_{3y} \end{Bmatrix} \qquad (u)$$

Since $F_{2x} = F_{2x}^{(1)} + F_{2x}^{(2)} = 2N$, the circled elements in Eqs. (q) and (r) should be added to yield the circled element in Eq. (u). Likewise, the blocked elements in

Eqs. (r) and (s) are added to yield the blocked element in Eq. (u). Notice that the circled element $(AE/2) \times 0.433$ in Eq. (q) is the stiffness coefficient relating δ_{2y} to $F_{2x}^{(1)}$ and that the circled element $(AE/3.464) \times (-0.433)$ is the stiffness coefficient relating δ_{2y} to $F_{2x}^{(2)}$, which is located at the first row and second column in Eq. (r) but has to be relocated to the third row and fourth column as far as Eq. (u) is concerned.

Figure 9-4 shows that joint 1 is hinged while node 3 is free to move horizontally. So the **boundary conditions** are $\delta_{1x} = \delta_{1y} = \delta_{3y} = 0$. Equation (u) thus needs to be converted in order to have the unknown forces F_{1x}, F_{1y}, and F_{3y} appear on the left-hand sides of the equations and to move δ_{1x}, δ_{1y}, and δ_{3y} to the right-hand sides. An alternative procedure without involving the matrix partitioning and interchanging a part of $\{\delta\}$ and $\{F\}$ is to replace the equations connected with the prescribed boundary conditions, in this case the first, second, and sixth equations of (u), by letting the corresponding F elements be equal to zero and setting the affected diagonal elements in the stiffness matrix $[K]$ equal to 10^8. The modified structural stiffness matrix equation, after incorporation of the boundary conditions, should be

$$
AE \begin{bmatrix}
1 \times 10^8 & 0.217 & -0.125 & -0.217 & -0.250 & 0 \\
 & 1 \times 10^8 & -0.217 & -0.375 & 0 & 0 \\
 & & 0.342 & 0.092 & -0.217 & 0.125 \\
 & & & 0.447 & 0.125 & -0.072 \\
 & & & & 0.467 & -0.125 \\
\text{Symmetric} & & & & & 1 \times 10^8
\end{bmatrix}
\begin{Bmatrix}
\delta_1 \\ \delta_2 \\ \delta_3 \\ \delta_4 \\ \delta_5 \\ \delta_6
\end{Bmatrix}
=
\begin{Bmatrix}
0 \\ 0 \\ 2 \\ -3 \\ 0 \\ 0
\end{Bmatrix}
$$

$$(v)$$

Notice that the elements in $\{\delta\}$ have been renamed according to Eq. (o) and that letting the diagonal elements of $[K]$ equal 1×10^8 has the effect of suppressing the other elements in the same row so that the corresponding displacements as a consequence must be approximately equal to zero to satisfy the boundary conditions.

Once the displacements $\{\delta\}$ are solved from Eq. (v), individual nodal forces of the truss members can be obtained by use of the respective element stiffness matrix equations, such as Eq. (h). For the truss problem of Fig. 9-4, it can be shown that the displacement vector is

$$\{\delta\} = [0 \quad 0 \quad 16.5 \quad -13.3 \quad 11.2 \quad 0]^T/AE$$

By use of Eqs. (q), (r), and (s), and also the element stiffness equations $[k^{(1)}]\{\delta\}_{1,2} = \{F^{(1)}\}$, $[k^{(2)}]\{\delta\}_{2,3} = \{F^{(2)}\}$, and $[k^{(3)}]\{\delta\}_{1,3} = \{F^{(3)}\}$, we obtain the nodal forces of all three elements in Fig. 7-5 as

$$
\begin{aligned}
\{F^{(1)}\} &= [F_{1x}^{(1)} \quad F_{1y}^{(1)} \quad F_{2x}^{(1)} \quad F_{2y}^{(1)}]^T \\
&= [0.799 \quad 1.384 \quad -0.799 \quad -1.384]^T
\end{aligned}
$$

$$
\begin{aligned}
\{F^{(2)}\} &= [F_{2x}^{(2)} \quad F_{2y}^{(2)} \quad F_{3x}^{(2)} \quad F_{3y}^{(2)}]^T \\
&= [2.799 \quad -1.616 \quad -2.799 \quad 1.616]^T
\end{aligned}
$$

$$\{F^{(3)}\} = \begin{bmatrix} F^{(3)}_{1x} & F^{(3)}_{1y} & F^{(3)}_{3x} & F^{(3)}_{3y} \end{bmatrix}^T$$
$$= \begin{bmatrix} -2.799 & 0 & 2.799 & 0 \end{bmatrix}^T$$

And from the above results, we can easily determine the support reactions to be

$$F_{1x} = F^{(1)}_{1x} + F^{(3)}_{1x} = 0.799 - 2.799 = -2N$$

$$F_{1y} = F^{(1)}_{1y} + F^{(3)}_{1y} = 1.384 + 0 = 1.384N$$

$$F_{3y} = F^{(2)}_{3y} + F^{(3)}_{3y} = 1.616 + 0 = 1.616N$$

These results are in agreement with those obtained from solving the three equations of equilibrium based on the elementary approach dealing with the two-dimensional static problems.[3]

9.4 COMPUTER PROGRAM TRUSS. FE

Knowing the nodal coordinates of a truss member, the elemental stiffness matrix $[k]$ defined by Eq. (i) in Section 9.2 can be easily coded. Subroutine EKTRUSS has thus been developed. For assemblying all of the elemental stiffness matrices to form the structural matrix $[K]$, a subroutine ASK is made available for the general case of a truss structure comprised of $N2$ elements. It should, however, be noted that the structural stiffness matrix is generated and stored in a rectangular form[4] rather than the conventional square form. Only the elements on and above the diagonal of the matrix are kept by taking advantage of its symmetry to reduce the computer storage requirement. The solution of the structural stiffness matrix equation $[K]\{\delta\} = \{F\}$, Eq. (n), is handled by the subroutine BANDEQ, which is based on the Gaussian elimination method and explained in detail in Appendix A.

Program TRUSS. FE makes use of these subroutines to solve the three-member truss problem depicted in Fig. 9-4. The computer results printed below show perfect agreement in the nodal displacements, $\{\delta\}$, with the manually calculated counterparts presented earlier.

```
* PROGRAM TRUSS.FE - FINITE ELEMENT ANALYSIS OF TRUSS *

NODAL COORDINATES :
    1           0.0000          0.0000
    2           1.0000          1.7320
    3           4.0000          0.0000

ELEMENT NODES :
    1       1       2
    2       3       2
    3       1       3

STIFFNESS MATRIX :

 0.375008250454   0.216514289786   -0.125008250454   -0.216514289786
-0.25             0
 0.375002749000   -0.216514289786   -0.375002749000    0
 0                 0
```

---

[3] J. L. Meriam, *Engineering Mechanics*, Vol. 1, *Statics*, John Wiley & Sons, New York, 1978.

[4] Y. C. Pao, "On Computations Involving Stiffness Matrices Stored in Rectangular Form," *International Journal for Numerical Methods in Engineering*, Vol. 9, 1975, pp. 250–252.

```
0.341519364627    0.0915152065366    -0.216511114173    0.124999083249
0
0.447168887305    0.124999083249     -0.0721661373059   0
0
0.466511114173    -0.124999083249    0                  0
0
0.0721661373059   0                  0                  0
0
```

LOADS :

```
0                 0                  2                  -3
0                 0
```

BOUNDARIES AT WHICH NO DISPLACEMENT PERMITTED *

```
1                 2                  6
```

DISPLACEMENTS :
```
    0.00000    0.00000
   16.49525  -13.21446
   11.19630    0.00000
```

PROGRAM TRUSS.FE

```
100 PAGE
110 PRINT "* PROGRAM TRUSS.FE - FINITE ELEMENT ANALYSIS OF TRUSS *"
120 INIT
130 REM INPUT
140 GOSUB 470
150 PRINT "_NODAL COORDINATES :"
160 FOR I1=1 TO N1
170 PRINT USING 180:I1,X(I1),Y(I1)
180 IMAGE 4D,4X,10D.4D,10D.4D
190 NEXT I1
200 PRINT "_ELEMENT NODES :"
210 FOR I2=1 TO N2
220 PRINT USING 230:I2,N(1,I2),N(2,I2)
230 IMAGE 4D,4X,2(6D)
240 NEXT I2
250 DIM K(N3,N4),K1(4,4),D(N3)
260 REM GENERATE STIFFNESS MATRIX
270 GOSUB 720
280 PRINT "_STIFFNESS MATRIX : "
290 PRINT K.
300 PRINT "_LOADS :"
310 PRINT R1
320 PRINT "_BOUNDARIES AT WHICH NO DISPLACEMENT PERMITTED *"
330 PRINT U
340 FOR L1=1 TO U1
350 K(U(L1),1)=K(U(L1),1)*10^8
360 R1(U(L1))=0
370 NEXT L1
380 FOR L1=1 TO N3
390 D(L1)=R1(L1)
400 NEXT L1
410 REM SOLVE FOR DISPLACEMENTS
420 GOSUB 900
430 PRINT "_DISPLACEMENTS :"
440 PRINT USING 450:D
```

```
450 IMAGE  3(5D.5D,5D.5D/)
460 END
470 REMARK   SUBROUTINE INPUT
480 N1=3
490 N2=3
500 N3=N1*2
510 N4=6
520 DIM X(N1),Y(N1),N(2,N2),R1(N3),U(3)
530 R1=0
540 FOR L1=1 TO N2
550 FOR L2=1 TO 2
560 READ N(L2,L1)
570 NEXT L2
580 NEXT L1
590 DATA 1,2,3,2,1,3
600 READ Y
610 DATA 0,1.732,0
620 READ X
630 DATA 0,1,4
640 R1(3)=2
650 R1(4)=-3
660 U1=3
670 FOR L1=1 TO U1
680 READ U(L1)
690 NEXT L1
700 DATA 1,2,6
710 RETURN
720 REMARK      SUBROUTINE ASK - TRUSS VERSION
730 K=0
740 FOR I2=1 TO N2
750 REM GENERATE ELEMENT STIFFNESS MATRICES
760 GOSUB 1120
770 FOR L1=1 TO 4
780 I=2*N(L1/2+0.4,I2)+(-1+-1^L1)/2
790 FOR L2=1 TO 4
800 IF L2<L1 THEN 860
810 J=2*N(L2/2+0.4,I2)+(-1+-1^L2)/2
820 IF J=>I THEN 850
830 K(J,I-J+1)=K1(L1,L2)+K(J,I-J+1)
840 GO TO 860
850 K(I,J-I+1)=K1(L1,L2)+K(I,J-I+1)
860 NEXT L2
870 NEXT L1
880 NEXT I2
890 RETURN
900 REMARK      SUBROUTINE BANDEQ
910 FOR I3=1 TO N3
920 FOR I4=2 TO N4
930 IF K(I3,I4)=0 THEN 1010
940 T=K(I3,I4)/K(I3,1)
950 FOR J4=I4 TO N4
960 IF K(I3,J4)=0 THEN 980
970 K(I3+I4-1,J4-I4+1)=K(I3+I4-1,J4-I4+1)-T*K(I3,J4)
980 NEXT J4
990 K(I3,I4)=T
1000 D(I3+I4-1)=D(I3+I4-1)-T*D(I3)

1010 NEXT I4
1020 D(I3)=D(I3)/K(I3,1)
1030 NEXT I3
1040 FOR I3=2 TO N3
```

```
1050 J=N3+1-I3
1060 FOR I4=2 TO N4
1070 IF K(J,I4)=0 THEN 1090
1080 D(J)=D(J)-K(J,I4)*D(J+I4-1)
1090 NEXT I4
1100 NEXT I3
1110 RETURN
1120 REM    SUBROUTINE  EKTRUSS
1130 A=1
1140 M=1
1150 SET DEGREES
1160 T1=ATN((Y(N(2,I2))-Y(N(1,I2)))/(X(N(2,I2))-X(N(1,I2))))
1170 M5=COS(T1)
1180 N5=SIN(T1)
1190 L=SQR((X(N(1,I2))-X(N(2,I2)))^2+(Y(N(1,I2))-Y(N(2,I2)))^2)
1200 K1(1,1)=M5^2
1210 K1(1,2)=M5*N5
1220 K1(1,3)=-(M5^2)
1230 K1(1,4)=-M5*N5
1240 K1(2,2)=N5^2
1250 K1(2,3)=-M5*N5
1260 K1(2,4)=-(N5^2)
1270 K1(3,3)=M5^2
1280 K1(3,4)=M5*N5
1290 K1(4,4)=N5^2
1300 K1(2,1)=K1(1,2)
1310 K1(3,1)=K1(1,3)
1320 K1(4,1)=K1(1,4)
1330 K1(3,2)=K1(2,3)
1340 K1(4,2)=K1(2,4)
1350 K1(4,3)=K1(3,4)
1360 T9=A*M/L
1370 K1=K1*T9
1380 RETURN
```

EXERCISE

The 24-node 45-element two-dimensional truss shown here is constructed with bars of lengths 3, 4, 5, 6, and 8 m of the same material and are joined by smooth pins. The structure is supported by a hinge and a roller and is subjected to a concentrated load of 100 KN as shown. Modify the program TRUSS. FE so that it can be applied for determination of the axial forces in this structure.

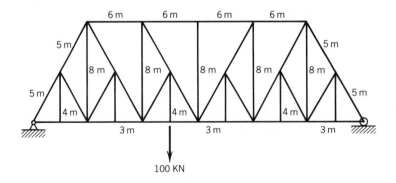

100 KN

9.5 PREPROCESSING AND POSTPROCESSING

Prior to the analysis of a structure by application of the finite element method, a decision has to be made on how the structure should be partitioned into a finite number of elements and on the types of elements to be used. Since the finite-element method will lead to a stiffness matrix equation for solving the displacements at the partitioning nodes, the nodes and subsequently the elements must be sequentially numbered. In order to examine the partitioning pattern to decide whether it is adequate or should be modified, a plot or display of the partitioning with labeling of the nodes and elements is necessary. Programs prepared for implementing such a need are called **preprocessors**.

To view the partitioning pattern of a three-dimensional structure often requires a proper rotation so that the constituent elements can be displayed for easy inspection. Consider the case of a half tube of 60 units long and inner and outer radii equal to 20 and 25 units, respectively. If it is displayed with the axial cutting plane coincident with the screen, very little about the finite-element partitioning pattern will be revealed, as evidenced by Fig. 9-6. If the half tube is rotated about the z, y,

```
100 REM * PROGRAM HALF.TUBE
110 SET DEGREES
120 DIM X(114),Y(114),Z(114)
130 N=0
140 FOR I=30 TO -30 STEP -30
150 FOR R=20 TO 25 STEP 5
160 FOR C=0 TO 180 STEP 10
170 N=N+1
180 X(N)=R*COS(C)
190 Y(N)=I
200 Z(N)=-R*SIN(C)
210 NEXT C
220 NEXT R
230 NEXT I
240 VIEWPORT 50,130,20,100
250 WINDOW -50,50,-50,50
260 FOR I=1 TO 38
270 MOVE X(I),Y(I)
280 DRAW X(76+I),Y(76+I)
290 NEXT I
300 FOR J=1 TO 96 STEP 19
310 I1=J+1·
320 I2=J+18
330 MOVE X(J),Y(J)
340 FOR I=I1 TO I2
350 DRAW X(I),Y(I)
360 NEXT I
370 NEXT J
380 FOR J=1 TO 77 STEP 38
390 I2=J+18
400 FOR I=J TO I2
410 MOVE X(I),Y(I)
420 DRAW X(I+19),Y(I+19)
430 NEXT I
440 NEXT J
450 END
```

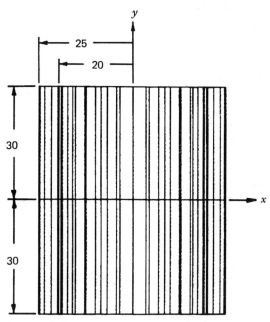

Figure 9-6

and x axes by $-30°$, $-20°$, and $20°$, respectively, and in that order, the partitioning pattern of dividing the half tube into 36 longitudinal strips can then be easily comprehended, as demonstrated by Fig. 9-7. The respective programs producing

```
100 REM * PROGRAM HALF.TUBE
110 SET DEGREES
120 DIM XO(114),YO(114),ZO(114),X(114),Y(114),Z(114)
130 N=0
140 FOR I=30 TO -30 STEP -30
150 FOR R=20 TO 25 STEP 5
160 FOR C=0 TO 180 STEP 10
170 N=N+1
180 XO(N)=R*COS(C)
190 YO(N)=I
200 ZO(N)=-R*SIN(C)
210 NEXT C
220 NEXT R
230 NEXT I
240 DATA 20,-20,-30
250 READ C1,C2,C3
260 GOSUB 490
270 VIEWPORT 50,130,20,100
280 WINDOW -50,50,-50,50
285 PAGE
290 FOR I=1 TO 38
300 MOVE X(I),Y(I)
310 DRAW X(76+I),Y(76+I)
320 NEXT I
330 FOR J=1 TO 96 STEP 19
340 I1=J+1
350 I2=J+18
360 MOVE X(J),Y(J)
370 FOR I=I1 TO I2
380 DRAW X(I),Y(I)
390 NEXT I
400 NEXT J
410 FOR J=1 TO 77 STEP 38
420 I2=J+18
430 FOR I=J TO I2
440 MOVE X(I),Y(I)
450 DRAW X(I+19),Y(I+19)
460 NEXT I
470 NEXT J
480 END
490 REM * CALL SUB TRANSF BUT WITHOUT TRANSLATION & SCALING
500 K1=COS(C1)
510 K2=COS(C2)
520 K3=COS(C3)
530 S1=SIN(C1)
540 S2=SIN(C2)
550 S3=SIN(C3)
560 FOR I=1 TO N
570 X(I)=K3*K2*XO(I)+(-S3*K1+K3*S2*S1)*YO(I)+(S3*S1+K3*S2*K1)*ZO(I)
580 Y(I)=S3*K2*XO(I)+(K3*K1+S2*S3*S1)*YO(I)+(-K3*S1+S3*S2*K1)*ZO(I)
590 Z(I)=-S2*XO(I)+K2*S1*YO(I)+K2*K1*ZO(I)
600 NEXT I
610 RETURN
```

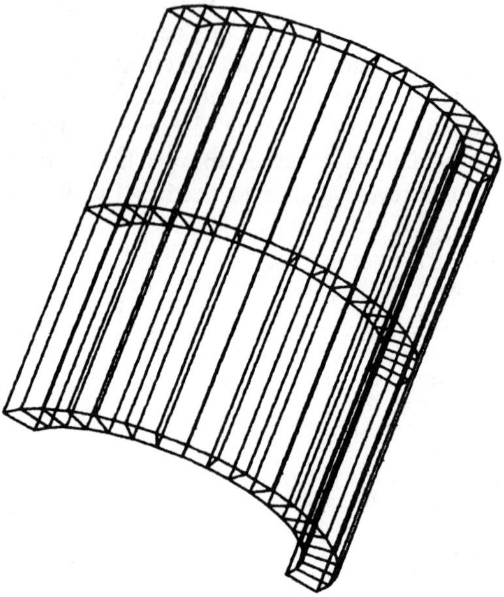

Figure 9-7

these two plots are listed alongside. A program called PLOTFE for plotting
general finite-element partitioning has been prepared and listed in Fig. 9-8 with a
sample plot of a connected left and right ventricular cross section of a dog heart.[5]

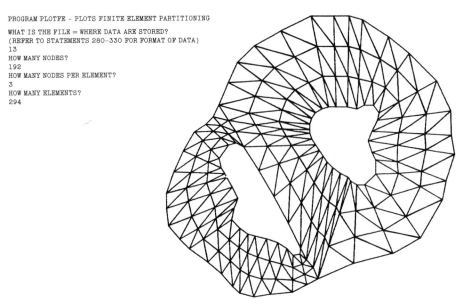

```
PROGRAM PLOTFE - PLOTS FINITE ELEMENT PARTITIONING

WHAT IS THE FILE = WHERE DATA ARE STORED?
(REFER TO STATEMENTS 280-330 FOR FORMAT OF DATA)
13
HOW MANY NODES?
192
HOW MANY NODES PER ELEMENT?
3
HOW MANY ELEMENTS?
294
```

Figure 9-8   Finite-element partitioning of a connected left and right ventricular cross
section of a dog heart reconstructed by the X-ray tomographic technique.

```
100 REM * PROGRAM PLOTFE - PLOTS FINITE ELEMENT PARTITIONING *
110 REM    READ FROM DATA FILE #K, N1 (# OF NODES),
120 REM    THEIR COORDINATES (X,Y), N2 (# OF ELEMENTS),
130 REM    N5 (# OF NODES PER ELEMENT) AND N MATRIX
140 REM    (ELEMENT NODES OF SIZE N5XN2).
150 INIT
160 PAGE
170 PRINT "PROGRAM PLOTFE - PLOTS FINITE ELEMENT PARTITIONING"
180 PRINT " WHAT IS THE FILE # WHERE DATA ARE STORED ?"
190 PRINT "(REFER TO STATEMENTS 280-330 FOR FORMAT OF DATA)"
200 INPUT K
210 PRINT "HOW MANY NODES ?"
220 INPUT N1
230 PRINT "HOW MANY NODES PER ELEMENT ?"
240 INPUT N5
250 PRINT "HOW MANY ELEMENTS ?"
260 INPUT N2
270 DIM X(N1),Y(N1),N(N5,N2)
280 FIND K
290 FOR I=1 TO N1
300 READ @33:J,X(I),Y(I)
```

[5] Y. C. Pao and E. L. Ritman, "Stress Analysis of Connected Right and Left Ventricules by
Idealized Geometry and Finite Element Models," *The 1980 Advances in Bioengineering*,
ASME Publication G00176, 1980, pp. 49–52.

```
310 NEXT I
320 FOR I=1 TO N2
330 READ @33:J,N(1,I),N(2,I),N(3,I)
340 NEXT I
345 PAGE
350 REM SCALING THEN PLOT
360 GOSUB 390
370 GOSUB 550
380 END
390 REM * CALL MAX,MIN TO FIND MAX & MIN OF X & Y
400 X1=X(1)
410 X2=X(1)
420 Y1=Y(1)
430 Y2=Y(1)
440 FOR I1=2 TO N1
450 IF X(I1)>X1 THEN 470
460 X1=X(I1)
470 IF X(I1)<X2 THEN 490
480 X2=X(I1)
490 IF Y(I1)>Y1 THEN 510
500 Y1=Y(I1)
510 IF Y(I1)<Y2 THEN 530
520 Y2=Y(I1)
530 NEXT I1
540 RETURN
550 REM * SUBROUTINE DRAW ELEMENTS
560 VIEWPORT 50,130,20,100
570 W=X2-X1 MAX Y2-Y1
580 WINDOW X1,X1+W,Y1,Y1+W
590 FOR I2=1 TO N2
600 MOVE X(N(1,I2)),Y(N(1,I2))
610 FOR I1=2 TO N5
620 DRAW X(N(I1,I2)),Y(N(I1,I2))
630 NEXT I1
640 DRAW X(N(1,I2)),Y(N(1,I2))
650 NEXT I2
660 RETURN


100 REM * PROGRAM LVRV STRESS PLOT *
110 REM      20-LEVEL CONTOURS
120 CHARSIZE 2
130 PAGE
140 INIT
150 V$=",-1)(+=IJ2LCTOXN#$*@"
160 VIEWPORT 40,130,5,95
170 PRINT "CONTOUR PLOT OF LVRV STRESSES"
180 PRINT "_INPUT THE CHARACTER SIZE 1,2,3, OR, 4 FOR PLOTTING :
190 INPUT C
200 IF C<3 THEN 230
210 H=2,5+(C-3)*0,3
220 GO TO 240
230 H=1,5+(C-1)*0,1
240 H1=H*0,625
250 REM INPUT, SCALING THEN PLOT
260 GOSUB 300
270 GOSUB 670
280 GOSUB 1830
290 END
300 REMARK  SUBROUTINE INPUT
```

```
310 T$="LVRV 20-LEVEL (,-1)(+=I]2LCTOXN#$*@) STRESS CONTOURS"
320 N1=192
330 N2=294
340 DIM X(N1),Y(N1),N(3,N2),S(N1)
350 N3=3
360 PRINT "_INPUT 1 OR 0 FOR PRINT OR NO PRINT : ";
370 INPUT W1
380 PRINT "_INPUT 1/0 FOR PLOT/NO PLOT OF PARTITIONING : ";
390 INPUT W2
400 PRINT "_MAKE NODAL COORDINATES AND ELEMENT NODES READY"
410 PRINT "   FROM FILE #48 AND STRESS VALUES AS WELL?"
420 PRINT "   TYPE IN Y/N : ";
430 INPUT A$
440 IF A$="Y" THEN 460
450 GO TO 400
460 PRINT "_PROCESSING, DO NOT INTERRUPT !!!"
470 GOSUB 2080
480 GO TO 490
490 IF W1=0 THEN 270
500 PRINT "  NODE            X             Y             Z"
510 FOR I1=1 TO N1
520 PRINT USING 530:I1,X(I1),Y(I1),S(I1)
530 IMAGE 4D,3(10D.4D)
540 NEXT I1
550 PRINT "_UNIT ON HOLD FOR CHECKING THE DATA"
560 PRINT "WHEN READY TO CONTINUE, PRESS ANY KEY."
570 INPUT K$
580 PRINT "ELEMENT      NODE NUMBERS"
590 FOR I2=1 TO N2
600 PRINT USING 610:I2,N(1,I2),N(2,I2),N(3,I2)
610 IMAGE 4D,3(6D)
620 NEXT I2
630 PRINT "_UNIT ON HOLD FOR CHECKING THE DATA"
640 PRINT "WHEN READY TO CONTINUE, PRESS ANY KEY."
650 INPUT K$
660 RETURN
670 REMARK  SUBROUTINE SCALE
680 X1=X(1)
690 X2=X(1)
700 Y1=Y(1)
710 Y2=Y(1)
720 S1=S(1)
730 S2=S(1)
740 FOR I1=1 TO N1
750 IF X(I1)>X1 THEN 770
760 X1=X(I1)
770 IF X(I1)<X2 THEN 790
780 X2=X(I1)
790 IF Y(I1)>Y1 THEN 810
800 Y1=Y(I1)
810 IF Y(I1)<Y2 THEN 830
820 Y2=Y(I1)
830 IF S(I1)>S1 THEN 850
840 S1=S(I1)
850 IF S(I1)<S2 THEN 870.
860 S2=S(I1)
870 NEXT I1
880 W=X2-X1 MAX Y2-Y1
890 K0=S1
900 K9=S2
910 S4=S2-S1
```

```
920 FOR I1=1 TO N1
930 S(I1)=(S(I1)-S1)/(S2-S1)*19
940 NEXT I1
950 RETURN
960 REM * SUBROUTINE CONTOUR *
970 REM     PLOT 20-LEVEL S VALUES OF THE I2ND ELEMENT
980 REM     WITH CHARACTERS SPECIFIED BY V$ AND SIZE C.
990 J1=N(1,I2)
1000 J2=N(2,I2)

1010 J3=N(3,I2)
1020 X1=X(J3) MIN X(J1) MIN X(J2)
1030 X2=X(J3) MAX X(J1) MAX X(J2)
1040 Y1=Y(J3) MIN Y(J1) MIN Y(J2)
1050 Y2=Y(J3) MAX Y(J1) MAX Y(J2)
1060 S1=S(J3) MIN S(J1) MIN S(J2)
1070 S2=S(J3) MAX S(J1) MAX S(J2)
1080 M1=INT(X1/H1+0.5)
1090 M2=INT(X2/H1+0.5)
1100 FOR I=M1 TO M2
1110 X0=I*H1
1120 Z1=Y2
1130 Z2=Y1
1140 T1=S2
1150 T2=S1
1160 IF ABS(X0-X(J1))+ABS(X0-X(J2))>ABS(X(J2)-X(J1)) THEN 1310
1170 IF X(J2)=X(J1) THEN 1270
1180 Z=Y(J1)+(Y(J2)-Y(J1))*(X0-X(J1))/(X(J2)-X(J1))
1190 T=S(J1)+(S(J2)-S(J1))*(X0-X(J1))/(X(J2)-X(J1))
1200 IF Z<Y1 THEN 1230
1210 Z1=Z1 MIN Z
1220 T1=T1 MIN T
1230 IF Z>Y2 THEN 1310
1240 Z2=Z2 MAX Z
1250 T2=T2 MAX T
1260 GO TO 1310
1270 Z1=Z1 MIN Y(J2) MIN Y(J1)
1280 T1=T1 MIN S(J2) MIN S(J1)
1290 Z2=Z2 MAX Y(J2) MAX Y(J1)
1300 T2=T2 MAX S(J2) MAX S(J1)
1310 IF ABS(X0-X(J2))+ABS(X0-X(J3))>ABS(X(J2)-X(J3)) THEN 1460
1320 IF X(J3)=X(J2) THEN 1420
1330 Z=Y(J2)+(Y(J3)-Y(J2))*(X0-X(J2))/(X(J3)-X(J2))
1340 T=S(J2)+(S(J3)-S(J2))*(X0-X(J2))/(X(J3)-X(J2))
1350 IF Z<Y1 THEN 1380
1360 Z1=Z1 MIN Z
1370 T1=T1 MIN T
1380 IF Z>Y2 THEN 1460
1390 Z2=Z2 MAX Z
1400 T2=T2 MAX T
1410 GO TO 1460
1420 Z1=Z1 MIN Y(J3) MIN Y(J2)
1430 T1=T1 MIN S(J3) MIN S(J2)
1440 Z2=Z2 MAX Y(J3) MAX Y(J2)
1450 T2=T2 MAX S(J3) MAX S(J2)
1460 IF ABS(X0-X(J1))+ABS(X0-X(J3))>ABS(X(J1)-X(J3)) THEN 1610
1470 IF X(J1)=X(J3) THEN 1570
1480 Z=Y(J3)+(Y(J1)-Y(J3))*(X0-X(J3))/(X(J1)-X(J3))
1490 T=S(J3)+(S(J1)-S(J3))*(X0-X(J3))/(X(J1)-X(J3))
1500 IF Z<Y1 THEN 1530
1510 Z1=Z1 MIN Z
1520 T1=T1 MIN T
```

```
1530 IF Z>Y2 THEN 1610
1540 Z2=Z2 MAX Z
1550 T2=T2 MAX T
1560 GO TO 1610
1570 Z1=Z1 MIN Y(J3) MIN Y(J1)
1580 T1=T1 MIN S(J3) MIN S(J1)
1590 Z2=Z2 MAX Y(J3) MAX Y(J1)
1600 T2=T2 MAX S(J3) MAX S(J1)
1610 L1=INT(Z1/H+0.5)
1620 L2=INT(Z2/H+0.5)
1630 FOR J=L1 TO L2
1640 Y0=J*H
1650 IF Y0>Z2 OR Y0<Z1 THEN 1800
1660 IF Z2>Z1 THEN 1690
1670 S0=T2
1680 GO TO 1700
1690 S0=T1+(Y0-Z1)/(Z2-Z1)*(T2-T1)
1700 IF S0>19 THEN 1750
1710 IF S0<0 THEN 1730
1720 GO TO 1760
1730 S0=0
1740 GO TO 1760
1750 S0=19
1760 MOVE X0-0.5*H1,Y0-H*0.65
1770 P$=SEG(V$,S0,1)
1780 PRINT USING 1790;P$
1790 IMAGE 1A
1800 NEXT J
1810 NEXT I
1820 RETURN
1830 REMARK    SUBROUTINE WINDOWING
1840 CHARSIZE C
1850 PAGE
1860 H=H/90*W
1870 H1=H1/90*W
1880 WINDOW X1,X1+1.05*W,Y1,Y1+W*1.05
1890 FOR I2=1 TO N2
1900 IF W2=0 THEN 1950
1910 MOVE X(N(3,I2)),Y(N(3,I2))
1920 FOR I5=1 TO 3
1930 DRAW X(N(I5,I2)),Y(N(I5,I2))
1940 NEXT I5
1950 GOSUB 960
1960 NEXT I2
1970 HOME
1980 CHARSIZE 2
1990 PRINT "NUMBER OF NODES= ";N1
2000 PRINT "NUMBER OF ELEMENTS= ";N2

2010 PRINT "_MIN CONTOUR VALUE = ";K0
2020 PRINT "   (CHARACTER .)"
2030 PRINT "_MAX CONTOUR VALUE = ";K9
2040 PRINT "   (CHARACTER @)"
2050 PRINT "_";T$
2060 HOME
2070 RETURN
2080 REM * LVRV ELEMENT NODES, NODAL COORDINATES AND
2090 REM          CIRCUMFERENTIAL STRESSES.
2100 FIND 48
2110 FOR I=1 TO 192
2120 READ @33:I2,X(I),Y(I)
2130 NEXT I
2140 FOR J=1 TO 294
```

```
2150 READ @33:I2,N(1,J),N(2,J),N(3,J)
2160 NEXT J
2170 FOR I=1 TO 192
2180 READ S(I)
2190 NEXT I
2200 RETURN
2210 DATA 3.36,2.67,2.07,1.4,1.64,1.54,1.36,1.21,0.991,0.703
2220 DATA 0.596,0.762,0.879,1.12,1.17,1.37,1.72,2.25,1.48,1.44
2230 DATA 2.46,2.32,1.98,2.29,1.19,1.32,1.35,3.03,3.33,3.55
2240 DATA 0.957,0.912,0.858,0.723,0.718,0.696,0.63,0.622,0.528,0.49
2250 DATA 0.418,0.457,0.536,0.621,0.63,0.671,0.65,0.508,0.194,-0.109
2260 DAT 0.125,0.255,0.137,-0.096,-0.058,-0.287,-0.067,0.153,0.408,0.485
2270 DATA 0.211,0.302,0.245,0.254,0.431,0.278,0.35,0.477,0.472,0.455
2280 DATA 0.524,0.5,0.452,0.415,0.279,0.277,0.223,0.006,-0.29,-0.17
2290 DATA -0.148,-0.586,-0.259,-0.056,-0.087
2300 DATA 0.159,-0.137,0.413,0.097,-0.037
2310 DATA 0.056,-0.037,0.037,-0.013,0.104,0.132,0.196,0.473,0.523,0.607
2320 DATA 0.613,0.586,0.454,0.35,0.202,0.167,0.148,-0.068,0.002,-0.233
2330 DATA 0.739,-0.379,-0.261,0.184,0.418,0.256,0.285,0.24,0.275,-0.038
2340 DATA -0.247,-0.237,0.305,0.529,1.28,0.909,0.767,0.003,0.008,0.012
2350 DATA -0.194,0.608,0.583,0.341,0.283,0.778,0.863,-0.083,0.013,0.193
2360 DATA 0.485,0.628,0.657,1.57,1.07,-0.083,-0.201,-0.132,-0.212,1.49
2370 DATA 0.68,1.61,0.818,-0.015,-0.202,0.075,-0.771,-0.238,-0.185,1.09
2380 DAT 0.666,-0.112,-0.128,0.004,-0.417,-0.128,-0.127,0.979,0.6,-0.052
2390 DATA -0.143,0.036,-0.406,-0.233,-0.052,0.781,0.615,-0.041,-0.112,0
2400 DATA 0.365,-0.223,-0.138,0.488,0.536
2410 DATA -0.038,-0.119,0.084,0.065,-0.215
2420 DATA -0.116,0.041
```

After solving the stiffness matrix equation of a structure to obtain the nodal displacements, the deformed shape of the structure can be plotted or displayed. Also, the strains and subsequently the stresses can be computed throughout the structure by utilizing the nodal displacements along with the stiffness matrices of all the constituent elements. The stress and strain distributions need to be plotted or displayed for study in most cases. Programs prepared for producing these needs are called **postprocessors**.

Figure 9-9 shows a typical stress distribution in the connected left and right canine ventricular cross section plotted earlier in Fig. 9-8. This stress plot is obtained by application of Program CONTOUR, discussed in Chapter 4. There are numerous preprocessors and postprocessors available in the market; Fig. 9-10 exemplifies applications of these processors in the industry.

9.6 PROGRAM FINITE. E

There are numerous computer programs available in the market on finite-element analyses. Fong's article[6] provides an extensive, up-to-date evaluation of the most commonly used programs. For elementary two-dimensional finite-element analyses, Program FINITE. E is provided herein. It is an extended version

---

[6] H. H. Fong, "An Evaluation of Eight U.S. General Purpose Finite-Element Computer Programs," Paper No. 82-0699-CP, the 23rd AIAA/ASME/ASCE/AHS Structures, Structural Dynamics and Materials Conference, May 10–12, 1982, New Orleans, Louisiana.

```
 CONTOUR PLOT OF LVRV STRESSES
 INPUT THE CHARACTER SIZE 1,2,3, OR, 4 FOR PLOTTING : 4
 INPUT 1 OR 9 FOR PRINT OR NO PRINT : 9
 INPUT 1/9 FOR PLOT/NO PLOT OF PARTITIONING : 9
 MAKE NODAL COORDINATES AND ELEMENT NODES READY
   FROM FILE=13 AND STRESS VALUES FROM FILE
K15 ? TYPE IN Y/N : Y

 PROCESSING. DO NOT INTERRUPT !!!

 NUMBER OF NODES=182
 NUMBER OF ELEMENTS=284

 MIN CONTOUR VALUE=-8.77
   (CHARACTER .)

 MAX CONTOUR VALUE = 3.55
   (CHARACTER O)
 LVRV 20-LEVEL 1, -1, ><=112LCTOXIN=8*81 STRESS CONTOURS

                        >><><<<<<<<++
                     >>>>>><<<<<<<<++++
                  >>>>><>>><<<<<<<<+++++++
                >>>>>><>><<<<<+<<+<++++++++++
              1>>>>>>>><<<<<+<<+<++<++++++++++
             111>>>>>><<<<+<<+<++<+<<<++++<++++
           11111>>><<+<++++<+<++<<<<<<++<+<+++++
           111111>><<<+++==I=I+=+++++<++++++++++
          1111--1>><<+=+=I+=I=I+=++<++=<<<<<<<++++
         1--11-111>><++II=]]I+=++=++<=++++++++<++++
         <<<----11>><+IIL2]                 +==++++++++<<<
       11<++<>111111><=IL                   II==++++<<<<<
      >11>>11----111><=I]                  ]II==+++<<<<<<
    +<1111>1--..--1>>+I2C                  ]II=++++<<<<<>>>
   =<11111     --.>-111><=]L               2]I==++<<<<<>>>>
  =+>111>        -11111111>=]2             2]I==+++<<<<<>>>
 =+>11111         ->1111111<<+I             LI==+++<<<<<>>
 +<>11111>          >><>>11111<=IC           ]I==+++<<<<<>>
+<>11111111           >>>111>1-1<=LON*   OCL]I=++<<<<>>>
<>>111111111           >1<>111-1>+]LOXXOCL2]==++<>>>>>
<>>111111111            <>>>11-1><=]LLLLL]I==++<>>>>>>1
+<>11111111>             >>>>11>>><==II====+++<<>>>>1
<<>>1111111                1>>111>><<++=+++++<<>>>>>1
+<<>1111111><               >>1><>>><<<<<<<<<<<<>>>>>>
  <>111111-1>---            >>>><>>>>>>>>>>>>>>>>>>>
    >>>1>1111--1-.I>1+    <><>>11>>>>>>>>>>>>1
       <1++11111--<I<+<=I><>1111>>>>>11
          >>==111>+]I=+<+=<<>1111>
           +<<=>11<==++<<++<11
              1<I>=]I=I+++<>
               1<=2=]I==
```

Figure 9-9 Twenty-level stress contours plot of the connected left and right ventricular cross section shown in Fig. 9-8.

of FEAMAX and FEAMPS developed by the author for studies of cardiopulmonary dynamics.[7]

In brief, subroutine ASK assembles the stiffness matrix $[K]$ by taking the elemental stiffness matrices $[k]$ provided by subroutine EKORDB. Subroutine

[7] Y. C. Pao, "Finite Elements in Stress Analysis and Estimation of Mechanical Properties of Working Heart," in *Finite Elements in Biomechanics*, edited by R. H. Gallagher, B. Simon, P. Johnson, and J. Gross, John Wiley & Sons, 1982, Chapter 8, pp. 127–142.

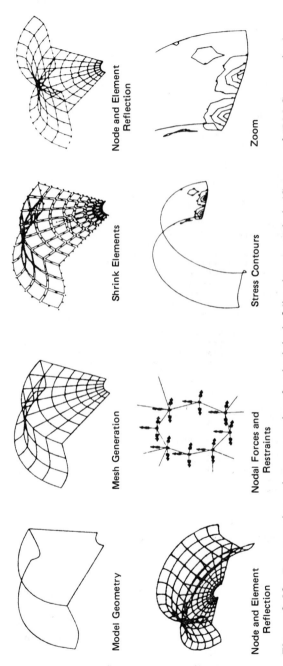

Node and Element Reflection

Zoom

Shrink Elements

Stress Contours

Mesh Generation

Nodal Forces and Restraints

Model Geometry

Node and Element Reflection

Figure 9-10   Preprocessing and postprocessing of a wheel rim in finite-element analysis. (Courtesy of IBM Corporation.)

ELASTY defines the stiffness of elements and subroutine LOAD specifies the nodal load vector $\{F\}$. Solution of the matrix equation $[K]\{\delta\} = \{F\}$ for the nodal displacements $\{\delta\}$ is handled by subroutine BANDEQ, which is explained in detail in Appendix A. Statements for preprocessing and postprocessing are included in the main program.

For demonstration purposes, a sample case of a simply supported beam with a concentrated load applied at midlength is presented with program messages, user's responses, and computed results and plots.

PROGRAM FINITE.E

```
* PROGRAM FINITE.E *

  HAS THE OPTIONS OF AXISYMMMETRIC, PLANE-STRESS,
  AND PLANE-STRAIN FINITE-ELEMENT ANALYSES.

  ENTER 1 OR 2 FOR A SAMPLE OR USER'S OWN CASE : 1

  A SAMPLE PLANE-STRESS PROBLEM:

  A SIMPLY SUPPORTED BEAM SUBJECTED TO A UNIT LOAD
  AT MID-SPAN.

  YOUNG'S MODULUS = 50

  POISSON'S RATIO = 0.3

  WANT TO PRINT NODAL COORDINATES? Y/N : Y

  NODE                    X                    Y
   1                  0.0000               2.0000
   2                  0.0000               0.0000
   3                 10.0000               2.0000
   4                 10.0000               0.0000
   5                 20.0000               2.0000
   6                 20.0000               0.0000
   7                 30.0000               2.0000
   8                 30.0000               0.0000
   9                 40.0000               2.0000
  10                 40.0000               0.0000

  WANT TO PRINT ELEMENTAL NODES? Y/N : Y

  ELEMENT              NODES
   1            2      3      1
   2            2      4      3
   3            4      5      3
   4            4      6      5
   5            8      5      6
   6            8      7      5
   7           10      7      8
   8           10      9      7

  WANT TO PRINT BOUNDARY CONDITIONS? Y/N : Y
```

```
BOUNDARY CONDITIONS ARE DS(I)=0, FOR I=
3
4
20

WANT TO PRINT LOADS? Y/N : Y

LOADS ARE:
NODE         FX           FY
  1      0.00000      0.00000
  2      0.00000      0.50000
  3      0.00000      0.00000
  4      0.00000      0.00000
  5      0.00000     -1.00000
  6      0.00000      0.00000
  7      0.00000      0.00000
  8      0.00000      0.00000
  9      0.00000      0.00000
 10      0.00000      0.50000

PLOT UNDEFORMED PARTITIONING? Y/N : N

ENTER 1/2/3 FOR AXISYMMETRIC/PLANE-STRESS/PLANE-STRAIN : 2

DO NOT INTERRUPT, COMPUTING !!!!

WANT TO PRINT NODAL DISPLACEMENTS? Y/N : Y

NODAL DISPLACEMENTS
        U              V
   0.19097        0.00139
   0.00000        0.00000
   0.16457       -0.96734
   0.01972       -0.97040
   0.09848       -1.46678
   0.09848       -1.45526
   0.03239       -0.96734
   0.17725       -0.97040
   0.00600        0.00139
   0.19697        0.00000

WANT TO PLOT DEFORMED SHAPE? Y/N : N

WANT TO PRINT ELEMENTAL STRESSES AND STRAINS? Y/N : Y

ELEMENT STRESSES AND STRAINS
#     SX        SY        SXY       SNX       SNY       SNXY
1  -0.13357  -0.00534  -0.02671  -0.00264   0.00069  -0.00139
2   0.13357   0.11663  -0.47329   0.00197   0.00153  -0.02461
3  -0.33790  -0.02481   0.43242  -0.00661   0.00153   0.02249
4   0.33790  -0.18648  -0.93242   0.00788  -0.00576  -0.04849
5   0.33790  -0.18648   0.93242   0.00788  -0.00576   0.04849
6  -0.33790  -0.02481  -0.43242  -0.00661   0.00153  -0.02249
7   0.13357   0.11663   0.47329   0.00197   0.00153   0.02461
8  -0.13357  -0.00534   0.02671  -0.00264   0.00069   0.00139

WANT TO PRINT NODAL STRESSES AND STRAINS? Y/N : Y

NODAL STRESSES AND STRAINS
NODE  STX       STY       STXY      SNX       SNY       SNXY
 1   -0.1336   -0.0053   -0.0267   -0.0026    0.0007   -0.0014
```

2	0.0000	0.0556	-0.2500	-0.0003	0.0011	-0.0130
3	-0.1126	0.0288	-0.0225	-0.0024	0.0013	-0.0012
4	0.0445	-0.0316	-0.3244	0.0011	-0.0009	-0.0169
5	0.0000	-0.1056	0.0000	0.0006	-0.0021	0.0000
6	0.3379	-0.1865	0.0000	0.0079	-0.0058	0.0000
7	-0.1126	0.0288	0.0225	-0.0024	0.0013	0.0012
8	0.0445	-0.0316	0.3244	0.0011	-0.0009	0.0169
9	-0.1336	-0.0053	0.0267	-0.0026	0.0007	0.0014
10	0.0000	0.0556	0.2500	-0.0003	0.0011	0.0130

WANT TO PRINT NODAL PRINCIPAL STRESSES? Y/N : Y

NODAL PRINCIPAL STRESSES

NODE	S1	S2	SZ	SXY	ANGLE
1	-0.13891	0.00000	-0.04167	0.09724	11.30994
2	0.27937	-0.22372	0.01669	0.26267	131.82485
3	-0.11613	0.03233	-0.02514	0.09099	8.83293
4	0.33314	-0.32017	0.00389	0.32925	-41.65646
5	-0.10565	0.00000	-0.03169	0.07395	90.00000
6	0.33790	-0.18648	0.04542	0.29247	0.00000
7	-0.11613	0.03233	-0.02514	0.09099	-8.83293
8	0.33314	-0.32017	0.00389	0.32925	41.65646
9	-0.13891	0.00000	-0.04167	0.09724	-11.30994
10	0.27937	-0.22372	0.01669	0.26267	48.17515

COMPUTATION HAS ENDED!!!

* PROGRAM FINITE.E *

   HAS THE OPTIONS OF AXISYMMMETRIC, PLANE-STRESS,
   AND PLANE-STRAIN FINITE-ELEMENT ANALYSES.

ENTER 1 OR 2 FOR A SAMPLE OR USER'S OWN CASE : 1

A SAMPLE PLANE-STRESS PROBLEM:

   A SIMPLY SUPPORTED BEAM SUBJECTED TO A UNIT LOAD
   AT MID-SPAN.

YOUNG'S MODULUS = 50

POISSON'S RATIO = 0.3

WANT TO PRINT NODAL COORDINATES? Y/N : N

WANT TO PRINT LOADS? Y/N : N

PLOT UNDEFORMED PARTITIONING? Y/N : Y

UNIT ON HOLD, PRESS ANY KEY TO CONTINUE !!

ENTER 1/2/3 FOR AXISYMMETRIC/PLANE-STRESS/PLANE-STRAIN : 2

DO NOT INTERRUPT, COMPUTING !!!!

WANT TO PRINT NODAL DISPLACEMENTS? Y/N : N

WANT TO PLOT DEFORMED SHAPE? Y/N : Y

```
UNIT ON HOLD, PRESS ANY KEY TO CONTINUE !!
100 PAGE
110 PRINT "* PROGRAM FINITE.E *"
120 PRINT "_  HAS THE OPTIONS OF AXISYMMMETRIC, PLANE-STRESS,"
130 PRINT "  AND PLANE-STRAIN FINITE-ELEMENT ANALYSES."
140 REM   N1=# OF NODES, N2=# OF ELEMENTS, N3=# OF EQUATIONS
150 REM   N4=BANDWIDTH, N5=# OF NODE PER ELEMENT.
160 INIT
170 PRINT "_ENTER 1 OR 2 FOR A SAMPLE OR USER'S OWN CASE : ";
180 INPUT Z9
190 REM X,Y,N,R1,U ARE DIMENSIONED IN SUB. INPUT
200 GOSUB 770
210 DIM K(N3,N4),K1(6,6),D(N3),D1(3,6),A1(3),B1(3)
220 DIM C1(3),B(3,6),B2(6,3),Q(3,3)
230 DIM N0(N1),E(3),S(3),E1(3,N1),S1(3,N1),E2(3,N2)
240 DIM D0(6),X3(N1),Y3(N1)
250 N0=0
260 E1=0
270 S1=0
280 PRINT "_PLOT UNDEFORMED PARTITIONING? Y/N : ";
290 INPUT Z$
300 IF Z$="N" THEN 340
310 REM * FIND MAX & MIN OF X & Y THEN PLOT ELEMENTS
320 GOSUB 3210
330 GOSUB 3370
340 PRINT "_ENTER 1/2/3 FOR AXISYMMETRIC/PLANE-STRESS/PLANE-STRAIN : ";
350 INPUT E3
360 PRINT "_DO NOT INTERRUPT, COMPUTING !!!!"
370 REM * ASSEMBLE K,THEN APPLY B.C.TO ARRAYS K,R1
380 GOSUB 1890
390 IF U1=0 THEN 440
400 FOR L1=1 TO U1
410 K(U(L1),1)=K(U(L1),1)*10^8
420 R1(U(L1))=0
430 NEXT L1
440 FOR L1=1 TO N3
450 D(L1)=R1(L1)
460 NEXT L1
470 REM * CALL BANDEQ TO SOLVE KX=D
480 GOSUB 2090
490 PRINT "_WANT TO PRINT NODAL DISPLACEMENTS? Y/N : ";
500 INPUT Z$
510 IF Z$="N" THEN 560
520 PRINT "_NODAL DISPLACEMENTS"
530 PRINT "        U              V"
540 PRINT USING 550:D
550 IMAGE 10(5D.5D,5D.5D/)
560 PRINT "_WANT TO PLOT DEFORMED SHAPE? Y/N : ";
570 INPUT Z$
580 IF Z$="N" THEN 650
590 FOR I1=1 TO N1
600 X(I1)=X(I1)+D(2*I1-1)
```

```
610 Y(I1)=Y(I1)+D(2*I1)
620 NEXT I1
630 GOSUB 3210
640 GOSUB 3370
650 PRINT "_WANT TO PRINT ELEMENTAL STRESSES AND STRAINS? Y/N : ";
660 INPUT Z$
670 IF Z$="N" THEN 750
680 REMARK    CALCULATION OF STRAINS AND STRESSES (ELEMENT & NODAL.)
690 GOSUB 3540
700 PRINT "_WANT TO PRINT NODAL PRINCIPAL STRESSES? Y/N : ";
710 INPUT Z$
720 IF Z$="N" THEN 750
730 REMARK    CALCULATION OF PRINCIPAL STRESSES
740 GOSUB 3990
750 PRINT "_COMPUTATION HAS ENDED!!!"
760 END
770 REMARK   SUBROUTINE INPUT
780 GO TO Z9 OF 790,1070
790 PRINT "_A SAMPLE PLANE-STRESS PROBLEM:"
800 PRINT "_ A SIMPLY SUPPORTED BEAM SUBJECTED TO A UNIT LOAD
810 PRINT "   AT MID-SPAN."
820 DATA 50,0.3,10,8,20,8,3
830 READ M,P,N1,N2,N3,N4,N5
840 PRINT "_YOUNG'S MODULUS = ";M
850 PRINT "_POISSON'S RATIO = ";P
860 DIM X(N1),Y(N1),N(3,N2),R1(N3),U(10)
870 R1=0
880 FOR L1=1 TO N2
890 FOR L2=1 TO 3
900 READ N(L2,L1)
910 NEXT L2
920 NEXT L1
930 DATA 2,3,1,2,4,3,4,5,3,4,6,5,8,5,6,8,7,5,10,7,8,10,9,7
940 READ Y
950 DATA 2,0,2,0,2,0,2,0,2,0
960 READ X
970 DATA 0,0,10,10,20,20,30,30,40,40
980 R1(4)=0.5
990 R1(10)=-1
1000 R1(20)=0.5

1010 U1=3
1020 FOR L1=1 TO U1
1030 READ U(L1)
1040 NEXT L1
1050 DATA 3,4,20
1060 GO TO 1550
1070 PRINT "_NUMBER OF NODES = ";
1080 INPUT N1
1090 PRINT "NUMBER OF ELEMENTS = ";
1100 INPUT N2
1110 N3=2*N1
1120 PRINT "NUMBER OF BANDS = ";
1130 INPUT N4
1140 DIM X(N1),Y(N1),N(3,N2),R1(N3),U(10)
1150 R1=0
1160 PRINT "_ENTER THE X,Y FOR EACH NODE AS CALLED FOR"
1170 FOR I1=1 TO N1
1180 PRINT "AT NODE ";I1;"   ";
1190 INPUT X(I1),Y(I1)
1200 NEXT I1
```

```
1210 PRINT "_ENTER THE N5 NODES OF EACH ELEMENT AS CALLED FOR"
1220 FOR I2=1 TO N2
1230 PRINT "THE NODES OF ELEMENT ";I2;" ARE    ";
1240 INPUT N(1,I2),N(2,I2),N(3,I2)
1250 NEXT I2
1260 PRINT "_HOW MANY LOADS?  ";
1270 INPUT T5
1280 PRINT "FOR EACH LOAD, ENTER:  1) THE NODE NUMBER WHERE APPLIED"
1290 PRINT "                       2) WHETHER IN X OR Y (1 OR 2)"
1300 PRINT "                       3) THE SIGNED MAGNITUDE"
1310 FOR L1=1 TO T5
1320 PRINT "FOR ";L1;" LOAD  ";
1330 INPUT T1,T2,T3
1340 T4=2*T1
1350 IF T2<>1 THEN 1370
1360 T4=T4-1
1370 R1(T4)=T3
1380 NEXT L1
1390 PRINT "_HOW MANY BOUNDARY CONDITIONS? :  ";
1400 INPUT U1
1410 PRINT "FOR EACH B.C., ENTER:  1) THE NODE NUMBER WHERE APPLIED"
1420 PRINT "                       2) WHETHER IN X OR Y (1 OR 2)"
1430 FOR L1=1 TO U1
1440 PRINT "FOR ";L1;" B.C.  ";
1450 INPUT T1,T2
1460 T4=2*T1
1470 IF T2<>1 THEN 1490
1480 T4=T4-1
1490 U(L1)=T4
1500 NEXT L1
1510 PRINT "_YOUNG'S MODULUS = ";
1520 INPUT M
1530 PRINT "_POISSON'S RATIO = ";
1540 INPUT P
1550 PRINT "_WANT TO PRINT NODAL COORDINATES? Y/N :  ";
1560 INPUT Z$
1570 IF Z$="N" THEN 1780
1580 PRINT "_NODE                    X              Y"
1590 FOR I1=1 TO N1
1600 PRINT USING 1610:I1,X(I1),Y(I1)
1610 IMAGE 4D,4X,10D.4D,10D.4D
1620 NEXT I1
1630 PRINT "_WANT TO PRINT ELEMENTAL NODES? Y/N :  ";
1640 INPUT Z$
1650 IF Z$="N" THEN 1830
1660 PRINT "_ELEMENT            NODES"
1670 FOR I2=1 TO N2
1680 PRINT USING 1690:I2,N(1,I2),N(2,I2),N(3,I2)
1690 IMAGE 4D,4X,3(6D)
1700 NEXT I2
1710 PRINT "_WANT TO PRINT BOUNDARY CONDITIONS? Y/N :  ";
1720 INPUT Z$
1730 IF Z$="N" THEN 1920
1740 PRINT "_BOUNDARY CONDITIONS ARE DS(I)=0, FOR I="
1750 FOR L1=1 TO U1
1760 PRINT U(L1)
1770 NEXT L1
1780 PRINT "_WANT TO PRINT LOADS? Y/N :  ";
1790 INPUT Z$
1800 IF Z$="N" THEN 1880
1810 PRINT "_LOADS ARE:"
```

```
1820 PRINT "NODE          FX          FY"
1830 FOR L1=1 TO N3 STEP 2
1840 L2=(L1+1)/2
1850 PRINT USING 1860:L2,R1(L1),R1(L1+1)
1860 IMAGE 4D,2(5D.5D)
1870 NEXT L1
1880 RETURN
1890 REM * SUBROUTINE ASK - ASSEMBLE K MATRIX
1900 K=0
1910 FOR I2=1 TO N2
1920 Z=2
1930 GOSUB 2310
1940 FOR L1=1 TO 6
1950 T=INT(L1/2.4+1)
1960 I=2*N(T,I2)+(-1+-1^L1)/2
1970 FOR L2=1 TO 6
1980 IF L2<L1 THEN 2050
1990 T=INT(L2/2.4+1)
2000 J=2*N(T,I2)+(-1+-1^L2)/2

2010 IF J=>I THEN 2040
2020 K(J,I-J+1)=K1(L1,L2)+K(J,I-J+1)
2030 GO TO 2050
2040 K(I,J-I+1)=K1(L1,L2)+K(I,J-I+1)
2050 NEXT L2
2060 NEXT L1
2070 NEXT I2
2080 RETURN
2090 REM * CALL SUBROUTINE BANDEQ TO SOLVE K(N3,N4)X(N3)=D(N3)
2100 FOR I3=1 TO N3
2110 FOR I4=2 TO N4
2120 IF K(I3,I4)=0 THEN 2200
2130 T=K(I3,I4)/K(I3,1)
2140 FOR J4=I4 TO N4
2150 IF K(I3,J4)=0 THEN 2170
2160 K(I3+I4-1,J4-I4+1)=K(I3+I4-1,J4-I4+1)-T*K(I3,J4)
2170 NEXT J4
2180 K(I3,I4)=T
2190 D(I3+I4-1)=D(I3+I4-1)-T*D(I3)
2200 NEXT I4
2210 D(I3)=D(I3)/K(I3,1)
2220 NEXT I3
2230 FOR I3=2 TO N3
2240 J=N3+1-I3
2250 FOR I4=2 TO N4
2260 IF K(J,I4)=0 THEN 2280
2270 D(J)=D(J)-K(J,I4)*D(J+I4-1)
2280 NEXT I4
2290 NEXT I3
2300 RETURN
2310 REMARK    SUBROUTINE EKORDB (K1 OR B)
2320 A1(1)=X(N(2,I2))*Y(N(3,I2))-X(N(3,I2))*Y(N(2,I2))
2330 A1(2)=X(N(3,I2))*Y(N(1,I2))-X(N(1,I2))*Y(N(3,I2))
2340 A1(3)=X(N(1,I2))*Y(N(2,I2))-X(N(2,I2))*Y(N(1,I2))
2350 B1(1)=Y(N(2,I2))-Y(N(3,I2))
2360 B1(2)=Y(N(3,I2))-Y(N(1,I2))
2370 B1(3)=Y(N(1,I2))-Y(N(2,I2))
2380 C1(1)=X(N(3,I2))-X(N(2,I2))
2390 C1(2)=X(N(1,I2))-X(N(3,I2))
2400 C1(3)=X(N(2,I2))-X(N(1,I2))
2410 A=(A1(1)+A1(2)+A1(3))/2
```

```
2420 T=0.5/A
2430 FOR L=1 TO 3
2440 L2=2*L
2450 L1=L2-1
2460 B(1,L1)=B1(L)*T
2470 B(1,L2)=0
2480 B(2,L1)=0
2490 B(2,L2)=C1(L)*T
2500 B(3,L1)=C1(L)*T
2510 B(3,L2)=B1(L)*T
2520 NEXT L
2530 GOSUB 2610
2540 IF Z=0 THEN 2600
2550 D1=Q MPY B
2560 IF Z=1 THEN 2600
2570 B2=TRN(B)
2580 K1=B2 MPY D1
2590 K1=K1*A
2600 RETURN
2610 REM * SUBROUTINE ELASTY *
2620 REM     GENERATES ELASTICITY MATRIX Q FOR FINITE
2630 REM     ELEMENT METHOD USING YOUNG'S MODULUS M AND
2640 REM      POISSON'S RATIO P.  THERE ARE THREE
2650 REM    OPTIONS : E3=1 - AXISYMMETRIC
2660 REM                E3=2 - PLANE STRESS
2670 REM                E3=3 - PLANE STRAIN
2680 Q=0
2690 GO TO E3 OF 2850,2770,2700.
2700 T1=M/(1-P-2*P^2)
2710 Q(1,1)=(1-P)*T1
2720 Q(1,2)=P*T1
2730 Q(2,1)=Q(1,2)
2740 Q(2,2)=Q(1,1)
2750 Q(3,3)=M/(2+2*P)
2760 GO TO 2970
2770 Q(1,1)=1
2780 Q(1,2)=P
2790 Q(2,1)=Q(1,2)
2800 Q(2,2)=1
2810 Q(3,3)=(1-P)/2
2820 T1=M/(1-P^2)
2830 Q=Q*T1
2840 GO TO 2970
2850 Q(1,1)=1
2860 Q(1,2)=P/(1-P)
2870 Q(1,3)=P/(1-P)
2880 Q(2,2)=1
2890 Q(2,3)=P/(1-P)
2900 Q(3,3)=1
2910 Q(4,4)=(1-2*P)/(2*(1-P))
2920 Q(2,1)=Q(1,2)
2930 Q(3,1)=Q(1,3)
2940 Q(3,2)=Q(2,3)
2950 T1=M*(1-P)/(1+P)/(1-2*P)
2960 Q=Q*T1
2970 RETURN
2980 REMARK    SUBROUTINE PSTRESS
2990 REM   ISOTROPIC ONLY
3000 P0=P*(S(1)+S(2))

3010 C0=(S(1)+S(2))/2
3020 C2=SQR((S(1)-S(2))^2+4*S(3)^2)
```

```
3030 P1=C0+0.5*C2
3040 P2=C0-0.5*C2
3050 P3=ATN(2*S(3)/(S(1)-S(2)))*28.6479
3060 P4=0.5*(S(2)-S(1))*SIN(2*P3)+S(3)*COS(2*P3)
3070 IF ABS(P1)=>ABS(P2) THEN 3110
3080 T=P1
3090 P1=P2
3100 P2=T
3110 P4=ABS(P4)
3120 A3=ABS(P1-P0)
3130 IF P4=>A3 THEN 3150
3140 P4=A3
3150 A3=ABS(P2-P0)
3160 IF P4=>A3 THEN 3180
3170 P4=A3
3180 IF ABS(S(1))>ABS(S(2)) THEN 3200
3190 P3=P3+90
3200 RETURN
3210 REM * CALL MAX,MIN TO FIND MAX & MIN OF X & Y
3220 X1=X(1)
3230 X2=X(1)
3240 Y1=Y(1)
3250 Y2=Y(1)
3260 FOR I1=2 TO N1
3270 IF X(I1)>X1 THEN 3290
3280 X1=X(I1)
3290 IF X(I1)<X2 THEN 3310
3300 X2=X(I1)
3310 IF Y(I1)>Y1 THEN 3330
3320 Y1=Y(I1)
3330 IF Y(I1)<Y2 THEN 3350
3340 Y2=Y(I1)
3350 NEXT I1
3360 RETURN
3370 REM * SUBROUTINE DRAW ELEMENTS
3380 PAGE
3390 VIEWPORT 50,130,20,100
3400 W=X2-X1 MAX Y2-Y1
3410 WINDOW X1,X1+W,Y1,Y1+W
3420 FOR I2=1 TO N2
3430 MOVE X(N(1,I2)),Y(N(1,I2))
3440 FOR I1=2 TO N5
3450 DRAW X(N(I1,I2)),Y(N(I1,I2))
3460 NEXT I1
3470 DRAW X(N(1,I2)),Y(N(1,I2))
3480 NEXT I2
3490 HOME
3500 PRINT "UNIT ON HOLD, PUSH ANY KEY TO CONTINUE !!";
3510 INPUT A$
3520 PAGE
3530 RETURN
3540 REM SUBROUTINE FOR CALCULATION OF ELEMENTAL AND NODAL STRAINS AND"
3550 REM    STRESS.  NO(I) KEEPS # OF ELEMENTS SHARING THE ITH NODE.
3560 PRINT "_ELEMENT STRESSES AND STRAINS"
3570 PRI " #     SX        SY        SXY       SNX       SNY       SNXY".
3580 FOR I2=1 TO N2
3590 FOR L5=1 TO 3
3600 NO(N(L5,I2))=NO(N(L5,I2))+1
3610 NEXT L5
3620 FOR L5=1 TO 6
3630 DO(L5)=D(2*N(L5/2+0.4,I2)-(1--1^L5)/2)
3640 NEXT L5
```

```
3650 REMARK   CALL SUB K1 OR B FOR MATRIX B
3660 Z=0
3670 GOSUB 2310
3680 E=B MPY DO
3690 S=Q MPY E
3700 FOR L5=1 TO 3
3710 E2(L5,I2)=E(L5)
3720 FOR L6=1 TO 3
3730 E1(L5,N(L6,I2))=E1(L5,N(L6,I2))+E(L5)
3740 S1(L5,N(L6,I2))=S1(L5,N(L6,I2))+S(L5)
3750 NEXT L6
3760 NEXT L5
3770 PRINT USING 3780:I2,S(1),S(2),S(3),E(1),E(2),E(3)
3780 IMAGE 2D,3(4D.5D),3(4D.5D)
3790 NEXT I2
3800 PRINT "_WANT TO PRINT NODAL STRESSES AND STRAINS? Y/N : ";
3810 INPUT Z$
3820 IF Z$="N" THEN 3980
3830 PRINT "_NODAL STRESSES AND STRAINS"
3840 PRI "NODE  STX          STY          STXY          SNX          SNY          SN>
3850 FOR I1=1 TO N1
3860 IF NO(I1)<>0 THEN 3880
3870 STOP
3880 FOR L5=1 TO 3
3890 E1(L5,I1)=E1(L5,I1)/NO(I1)
3900 S1(L5,I1)=S1(L5,I1)/NO(I1)
3910 NEXT L5
3920 E4=E1(1,I1)
3930 E5=E1(2,I1)
3940 E6=E1(3,I1)
3950 PRINT USING 3960:I1,S1(1,I1),S1(2,I1),S1(3,I1),E4,E5,E6
3960 IMAGE 2D,3(5D.4D),3(5D.4D)
3970 NEXT I1
3980 RETURN
3990 REMARK SUB PRINCIPAL STRESS
4000 PRINT "_NODAL PRINCIPAL STRESSES"
4010 PRINT "NODE          S1          S2          SZ          SXY          ANGL>
4020 FOR I1=1 TO N1
4030 S(2)=S1(2,I1)
4040 S(3)=S1(3,I1)
4050 S(1)=S1(1,I1)
4060 GOSUB 2980
4070 PRINT USING 4080:I1,P1,P2,P0,P4,P3
4080 IMAGE 4D,5(5D.5D)
4090 NEXT I1
4100 RETURN
```

9.7 NETWORK OF FLUID FLOW

The water distribution system in a city represents a fluid flow network with supplies from one or several reservoirs and perhaps pumps. Individual households and high-rise office buildings use the water supply through the flow paths, but some may have their own pumps to supplement the total need and to increase the water pressure. For analysis of such a complex system, the concept of finite element can also be employed. Consider a fluid flow network as sketched in Fig. 9-11. Imagine each junction to be a pump or faucet. A positive or negative value of the flow rate, Q, at that junction will indicate whether it is a **source** or a **sink**.

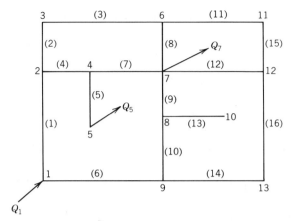

Figure 9-11

The flow path between two junctions is to be considered as an element. There are 13 junctions (nodes), 16 flow paths (elements with parenthesized numbers), one source Q_1, and two sinks Q_5 and Q_7 in the fluid flow network in Fig. 9-11. Knowing the supply Q_1 and demands Q_5 and Q_7, the problem is to determine the pressure and flow distributions in the network. Also of concern is how these pressure and flow distributions will be affected if the supplies and demands are varied.

In order to formulate the problem, take a typical path 2-4, namely, element 4 in Fig. 9-11. Let the flow rate be designated as $Q^{(4)}$ and the pressures at the junctions 2 and 4 be P_2 and P_4. According to the Hagen–Poiseuille law,[8] the elemental fluidity equation can be written as

$$\frac{128\mu L_4}{\pi D_4^4}(P_4 - P_2) = Q^{(4)} \tag{a}$$

where μ is the absolute viscosity of the fluid and L_4 and D_4 are the length and inside diameter of the fourth element, respectively. When the principle of conservation of mass is applied for the fluid at junction 2, an equation can be derived by combining Eq. (a) for element 4 and for the counterparts for elements 1 and 2, as they all share this junction. First, the entering flow rate is to be calculated as

$$Q^{(1)} = \frac{128\mu L_1}{\pi D_1^4}(P_1 - P_2)$$

whereas the outgoing flow rates are $Q^{(4)}$ and $Q^{(2)}$ calculated with the equation

$$Q^{(2)} = \frac{128\mu L_2}{\pi D_2^4}(P_2 - P_3)$$

Unlike junctions 1, 5, and 7 shown in Fig. 9-11, where there are net flow rates as either a supply source (Q_1) or as drains (Q_5 and Q_7), at junction 2 the net flow

[8] J. L. Shearer, A. T. Murphy, and H. H. Richardson, *Introduction to System Dynamics*, Addison-Wesley, Reading, Massachusetts, 1971.

rate should be equal to zero. Hence, the equation relating pressure at junction 2 to that at its neighboring junctions 1, 3, and 4 is

$$Q^{(1)} = Q^{(2)} + Q^{(4)}$$

or

$$\frac{128\mu L_1}{\pi D_1^4}(P_1 - P_2) = \frac{128\mu L_2}{\pi D_2^4}(P_2 - P_3) + \frac{128\mu L_4}{\pi D_4^4}(P_2 - P_4)$$

After simplification, the system's fluidity equation for junction 2 is obtained as

$$-\frac{L_1}{D_1^4}P_1 + \left(\frac{L_1}{D_1^4} + \frac{L_2}{D_2^4} + \frac{L_4}{D_4^4}\right)P_2 - \frac{L_2}{D_2^4}P_3 - \frac{L_4}{D_4^4}P_4 = 0 \qquad (b)$$

Similar equations can be derived by consideration of all the junctions. For example, the equation for junction 1 is

$$\left(\frac{L_1}{D_1^4} + \frac{L_6}{D_6^4}\right)P_1 - \frac{L_1}{D_1^4}P_2 - \frac{L_6}{D_6^4}P_6 = \frac{\pi}{128\mu}Q_1 \qquad (c)$$

Here, the right-hand side is not equal to zero because there is a source at junction 1.

The fluid flow network will have N junctions in general; the unknowns P_i for $i = 1, 2, \ldots, N$ can be solved from the so-called system's **fluidity matrix** equation

$$[F]\{P\} = \{Q\} \qquad (d)$$

Equation (d) is derived from all equations of the form (c) described above properly arranged. Once the end pressures of a flow path are known, the flow rate of the path can then be readily solved by the elementary fluidity equation, such as Eq. (a) for the path 2-4 in Fig. 9-11.

9.8 PROGRAM PIPEFLOW

A numerical example can help further explain the specific details involved in analysis of a pipe network. In Fig. 9-12 the supply of 4000 gal/min of a certain fluid at node 11 provides the various needs at different nodes in the network and two main outlets at nodes 37 and 46. For simplicity of discussion, only two different kinds of demands, q and Q, equal to 10 and 20 gal/min, respectively, are considered. All pipes are assumed to be of equal length (500 ft), and their diameters are as specified. There is a total of 46 nodes and 52 two-node elements. To help solve this problem, a general BASIC program PIPEFLOW has been developed. On the attached sample printout, the ordering of the nodes for all 52 of these elements is tabulated.

The convention by which the first and second nodes of an element are defined can be explained with the aid of the assumed flow directions in all loops of the network. The direction of flow in each loop can be assumed as either clockwise or counterclockwise, but the connected loops should have correct designations so

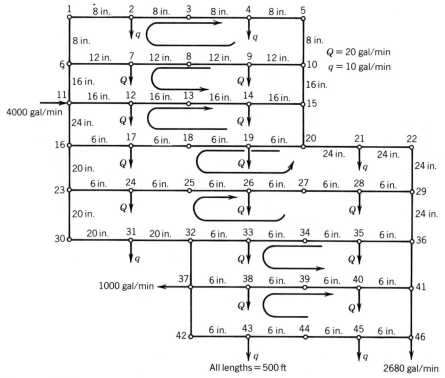

Figure 9-12

as not to cause conflict in the flow direction in the elements common to the loops. For example, if the top two loops in Fig. 9-12 are assumed to be *both clockwise,* then the flow will be from node 8 to node 7 according to the top loop but from node 7 to node 8 according to the lower loop. Hence, conflicting flow directions occur. If the lower loop is assumed to be counterclockwise as shown in Fig. 9-12, then there is no conflict in all of the elements 6, 7, 8, and 9, which are common to the two loops. The element numbers are not shown in Fig. 9-12 so as not to over-crowd the figure; they are listed in the sample printout.

Once the flow direction in each element is assumed, the *first* and *second* nodes can be precisely defined because the flow will always be directed from the former to the latter. For example, nodes 1 and 2 are the first and second nodes of element 1, whereas nodes 46 and 45 are the first and second of element 51, respectively, based on the flow directions assumed in the top and bottom loops. Whether the assumed flow direction in an element is correct or not could easily be verified when the flow rate in that element is determined. Since the sample output indicates that the flow rate in element 51 is −128.12 gal/min, it points to the fact that in reality the flow should be from node 45 to node 46. All the minus flow rate results listed in the printout require such corrections. Two obvious examples are elements 11 and 17 near the inlet flow at node 11; the flows must definitely be from that node toward nodes 6 and 16, respectively.

For each element, the elemental fluidity matrix $[f]$ can be easily evaluated to be

$$128\mu L\begin{bmatrix} 1 & -1 \\ -1 & 1 \end{bmatrix}\Big/\pi D^4$$

The network fluidity matrix $[F]$; in this case, is of the order of 46 by 46. Every $[f]$ will contribute to different elements in $[F]$. For example, the four elements of $[f]$ for pipe element 5, which connects nodes 1 and 6, should contribute as a part of the elements F_{11}, F_{16}, F_{61}, and F_{66}. This assemblying task is carried out in sub-routine SK (lines 1260–1420, Program PIPEFLOW) where a rectangular matrix rather than a square is formed to take advantage of the symmetry of $[F]$.

It should be noted that in Program PIPEFLOW, the nodal pressures have been computed with $128\mu/\pi$ being set equal to 10^4 solely for convenience. This value has no effect on the determined values of the flow rates in the network.

PROGRAM PIPEFLOW

```
* PROGRAM PIPEFLOW *

WANT TO PRINT INPUT DATA? ENTER 1 (YES) OR 0 (NO) : 1

WANT TO PRINT NODAL PRESSURES? ENTER 1/0 FOR YES/NO : 1

INPUT OPTION: SAMPLE OR YOUR OWN PROBLEM? (1 OR 2) : 1
```

ELEM	NODE1	NODE2	LENGTH	DIAMETER
1	1	2	500.00	8.00
2	2	3	500.00	8.00
3	3	4	500.00	8.00
4	4	5	500.00	8.00
5	6	1	500.00	8.00
6	7	6	500.00	12.00
7	8	7	500.00	12.00
8	9	8	500.00	12.00
9	10	9	500.00	12.00
10	5	10	500.00	8.00
11	6	11	500.00	16.00
12	11	12	500.00	16.00
13	12	13	500.00	16.00
14	13	14	500.00	16.00
15	14	15	500.00	16.00
16	15	10	500.00	16.00
17	16	11	500.00	24.00
18	17	16	500.00	6.00
19	18	17	500.00	6.00
20	19	18	500.00	6.00
21	20	19	500.00	24.00
22	15	20	500.00	24.00
23	21	20	500.00	24.00
24	22	21	500.00	24.00
25	16	23	500.00	20.00
26	23	24	500.00	6.00
27	24	25	500.00	6.00
28	25	26	500.00	6.00
29	26	27	500.00	6.00
30	27	28	500.00	6.00
31	28	29	500.00	6.00
32	29	22	500.00	24.00
33	30	23	500.00	20.00
34	31	30	500.00	20.00
35	32	31	500.00	20.00

36	33	32	500.00	6.00
37	34	33	500.00	6.00
38	35	34	500.00	6.00
39	36	35	500.00	6.00
40	29	36	500.00	24.00
41	32	37	500.00	24.00
42	37	38	500.00	6.00
43	38	39	500.00	6.00
44	39	40	500.00	6.00
45	40	41	500.00	6.00
46	41	36	500.00	24.00
47	42	37	500.00	6.00
48	43	42	500.00	6.00
49	44	43	500.00	6.00
50	45	44	500.00	6.00
51	46	45	500.00	6.00
52	41	46	500.00	24.00

PRESS ANY KEY TO CONTINUE : C

THE NODAL NET FLOW RATES :

0	-10	0	-10
0	0	-20	0
-20	0	4000	-20
0	-20	0	0
-20	0	-20	0
-10	0	0	-20
0	-20	0	-20
0	0	-10	0
-20	0	-20	0
-1000	-20	0	-20
0	0	-10	
-10	-2680		

PRESS ANY KEY TO CONTINUE : C

NODE	PRESSURE
1	3803.44
2	3797.51
3	3791.66
4	3785.81
5	3780.04
6	3809.38
7	3799.77
8	3790.99
9	3782.22
10	3774.27
11	3934.65
12	3862.58
13	3793.13
14	3723.69
15	3656.86
16	2279.50
17	2281.12
18	2282.78
19	2284.45
20	2724.16
21	2231.16
22	1744.79
23	1282.06
24	1278.04
25	1274.08
26	1270.11
27	1266.20
28	1262.29
29	1258.43
30	780.39
31	278.72

32	-219.74	20	642.65
33	-218.90	21	662.65
34	-218.01	22	1405.63
35	-217.12	23	-742.97
36	-216.18	24	-732.97
37	-1467.93	25	3117.03
38	-1463.42	26	1549.32
39	-1458.86	27	1529.32
40	-1454.29	28	1529.32
41	-1449.68	29	1509.32
42	-1474.72	30	1509.32
43	-1481.52	31	1489.32
44	-1488.29	32	-732.97
45	-1495.06	33	-1567.71
46	-1501.80	34	-1567.71
		35	-1557.71

PRESS ANY KEY TO CONTINUE : C

ELEM	FLOW		
1	724.17	36	323.37
2	714.17	37	343.37
3	714.17	38	343.37
4	704.17	39	363.37
5	724.17	40	2222.29
6	-231.61	41	1881.07
7	-211.61	42	-1740.38
8	-211.61	43	-1760.38
9	-191.61	44	-1760.38
10	704.17	45	-1780.38
11	-955.78	46	-1858.93
12	549.84	47	-2621.45
13	529.84	48	-2621.45
14	529.84	49	-2611.45
15	509.84	50	-2611.45
16	-895.78	51	-2601.45
17	-2494.37	52	78.55
18	622.65		
19	642.65		

LISTING OF PROGRAM PIPEFLOW

```
100 INIT
110 PAGE
120 PRINT "* PROGRAM PIPEFLOW *"
130 REMARK  SUB MENU
140 GOSUB 1650
150 REMARK  DIMENSION FOR X,Y,N,R1,U    IN SUB INPUT
160 GOSUB 460
170 DIM K(N3,N4),K1(2,2),D(N3)
180 REMARK  CALL SUB ASSEMBLE K
190 GOSUB 1260
200 FOR L1=1 TO N3
210 D(L1)=R1(L1)
220 NEXT L1
230 IF W1=0 THEN 280
240 PRINT "_THE NODAL NET FLOW RATES :_"
250 PRINT R1
260 PRINT " PRESS ANY KEY TO CONTINUE : ";
270 INPUT A$
280 REMARK  CALL SUB BANDED EQN SOLVER
290 GOSUB 1430
300 IF W2=0 THEN 380
310 PRINT "_NODE        PRESSURE"
320 FOR I1=1 TO 46
330 PRINT USING 340:I1,D(I1)
340 IMAGE 4D,4X,8D.2D
350 NEXT I1
360 PRINT "_PRESS ANY KEY TO CONTINUE : ";
370 INPUT A$
380 REMARK
```

```
390 REM   SUB FLOW CALC.  DIM FOR Q0 ALSO
400 GOSUB 1830
410 PRINT "_ELEM              FLOW   "
420 FOR I2=1 TO N2
430 PRINT USING 340;I2,Q0(I2)
440 NEXT I2
450 END
460 REMARK  SUBROUTINE INPUT
470 GO TO Z9 OF 480,1140
480 N1=46
490 N2=52
500 N3=N1
510 N4=8
520 DIM N(2,N2),R1(N3),U(10)
530 DIM L5(N2),D5(N2)
540 R1=0
550 FOR L1=1 TO N2
560 FOR L2=1 TO 2
570 READ N(L2,L1)
580 NEXT L2
590 NEXT L1
600 DATA 1,2,2,3,3,4,4,5,6,1,7,6,8,7,9,8,10,9,5,10
610 DATA 6,11,11,12,12,13,13,14,14,15,15,10,16,11,17,16,18,17,19,18
620 DATA 20,19,15,20,21,20,22,21,16,23,23,24,24,25,25,26,26,27,27,28
630 DATA 28,29,29,22,30,23,31,30,32,31,33,32,34,33,35,34,36,35,29,36
640 DATA 32,37,37,38,38,39,39,40,40,41,41,36,42,37,43,42,44,43,45,44
650 DATA 46,45,41,46
660 FOR I2=1 TO N2
670 L5(I2)=500
680 NEXT I2
690 FOR L2=1 TO 22
700 READ U2
710 D5(U2)=6
720 NEXT L2
730 DATA 18,19,20,26,27,28,29,30,31,36,37,38,39,42,43,44,45,47,48,49,50
740 DATA 51
750 FOR L2=1 TO 6
760 READ U2
770 D5(U2)=8
780 NEXT L2
790 DATA 1,2,3,4,5,10
800 FOR L2=1 TO 4
810 READ U2
820 D5(U2)=12
830 NEXT L2
840 DATA 6,7,8,9
850 FOR L2=1 TO 6
860 READ U2
870 D5(U2)=16
880 NEXT L2
890 DATA 11,12,13,14,15,16
900 FOR L2=1 TO 4
910 READ U2
920 D5(U2)=20
930 NEXT L2
940 DATA 25,33,34,35
950 FOR L2=1 TO 10
960 READ U2
970 D5(U2)=24
980 NEXT L2
990 DATA 17,21,22,23,24,32,40,41,46,52
```

```
1000 R1(11)=4000
1010 R1(37)=-1000
1020 R1(46)=-2680
1030 FOR L2=1 TO 6
1040 READ U2
1050 R1(U2)=-10
1060 NEXT L2
1070 DATA 2,4,21,31,43,45
1080 FOR L2=1 TO 13
1090 READ U2
1100 R1(U2)=-20
1110 NEXT L2
1120 DATA 7,9,12,14,17,19,24,26,28,33,35,38,40
1130 IF W1=0 THEN 1250
1140 PRINT
1150 A$="ELEM      NODE1      NODE2"
1160 B$="       LENGTH      DIAMETER"
1170 PRINT USING 1180;A$,B$
1180 IMAGE 24A,26A
1190 FOR I2=1 TO N2
1200 PRINT USING 1210;I2,N(1,I2),N(2,I2),L5(I2),D5(I2)
1210 IMAGE 4D,2(10D),2(10D,2D)
1220 NEXT I2
1230 PRINT "_PRESS ANY KEY TO CONTINUE : ";
1240 INPUT A$
1250 RETURN
1260 REMARK     SUBROUTINE ASK
1270 K=0
1280 FOR I2=1 TO N2
1290 GOSUB 1730
1300 FOR L1=1 TO 2
1310 I=N(L1,I2)
1320 FOR L2=1 TO 2
1330 IF L2<L1 THEN 1390
1340 J=N(L2,I2)
1350 IF J=>I THEN 1380
1360 K(J,I-J+1)=K1(L1,L2)+K(J,I-J+1)
1370 GO TO 1390
1380 K(I,J-I+1)=K1(L1,L2)+K(I,J-I+1)
1390 NEXT L2
1400 NEXT L1
1410 NEXT I2
1420 RETURN
1430 REMARK     SUBROUTINE BANDEQ
1440 FOR I3=1 TO N3
1450 FOR I4=2 TO N4
1460 IF K(I3,I4)=0 THEN 1540
1470 T=K(I3,I4)/K(I3,1)
1480 FOR J4=I4 TO N4
1490 IF K(I3,J4)=0 THEN 1510
1500 K(I3+I4-1,J4-I4+1)=K(I3+I4-1,J4-I4+1)-T*K(I3,J4)
1510 NEXT J4
1520 K(I3,I4)=T
1530 D(I3+I4-1)=D(I3+I4-1)-T*D(I3)
1540 NEXT I4
1550 D(I3)=D(I3)/K(I3,1)
1560 NEXT I3
1570 FOR I3=2 TO N3
1580 J=N3+1-I3
1590 FOR I4=2 TO N4
1600 IF K(J,I4)=0 THEN 1620
```

```
1610 D(J)=D(J)-K(J,I4)*D(J+I4-1)
1620 NEXT I4
1630 NEXT I3
1640 RETURN
1650 REMARK  SUBROUTINE  MENU
1660 PRINT "_WANT TO PRINT INPUT DATA? ENTER 1 (YES) OR 0 (NO) : ";
1670 INPUT W1
1680 PRINT "_WANT TO PRINT NODAL PRESSURES? ENTER 1/0 FOR YES/NO : ";
1690 INPUT W2
1700 PRINT "_INPUT OPTION: SAMPLE OR YOUR OWN PROBLEM? (1 OR 2) : ";
1710 INPUT Z9
1720 RETURN
1730 REM  SUBROUTINE   EK PIPE FLOW
1740 REM   A=128*MU/PI
1750 A=1000
1760 K1(1,1)=1
1770 K1(1,2)=-1
1780 K1(2,1)=-1
1790 K1(2,2)=1
1800 T9=A*L5(I2)/D5(I2)^4
1810 K1=K1*T9
1820 RETURN
1830 REM  SUBROUTINE  FLOW CALCULATION
1840 DIM Q0(N2)
1850 FOR I2=1 TO N2
1860 Q0(I2)=L5(I2)/D5(I2)^4*(D(N(1,I2))-D(N(2,I2)))*A
1870 NEXT I2
1880 RETURN
```

9.9   ADMITTANCE MATRIX OF ELECTRIC NETWORK

Formulation of the nodal equations similar to finite-element structural problems can be found in network analysis. Suppose that an electric passive network has n nodes. Figure 9-13 displays a sketch of the connections of node j to its neighboring nodes, isolated from the remainder of the network. I_j is denoted as the input current, where i_{jk} represents the outgoing current in the circuit connecting the jth and kth nodes, whereas k may range from 1 to n (not necessarily including all of them) and k cannot be equal to j. In Chapter 6 we defined the impedance Z of a circuit. Here, it is more convenient to introduce its reciprocal, Y, called **admittance**. For a

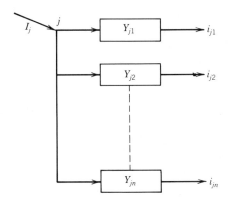

Figure 9-13

typical circuit connecting the jth and kth nodes, its current can be related to the voltages at nodes, V_j and V_k, as

$$i_{jk} = Y_{jk}(V_j - V_k) \tag{a}$$

When Kirchhoff's current law is applied to node j, the input and output currents satisfy the equation

$$\sum_{k=1}^{n} i_{jk} = \sum_{k=1}^{n} Y_{jk}(V_j - V_k) = I_j \tag{b}$$

Since there may not be a connecting circuit between nodes j and m, the situation can be circumvented in using the complete form of Eq. (b) by realizing that Y_{jm} could be simply set equal to zero. For instance, Y_{jj} is always equal to zero.

Equation (b) can be simplified to the form

$$-\sum_{k=1}^{j-1} Y_{jk} V_k + \left(\sum_{k=1}^{n} Y_{jk} \right) V_j - \sum_{k=j+1}^{n} Y_{jk} V_k = I_j \tag{c}$$

The above expression can be further simplified to

$$\sum_{k=1}^{n} y_{jk} V_k = I_j \tag{d}$$

where the new variables y_{jk} are defined by

$$y_{jk} = -Y_{jk} \qquad \text{for } j \neq k \tag{e}$$

$$y_{jj} = \sum_{k=1}^{n} Y_{jk} \tag{f}$$

y_{jj} is known as the **self-admittance** of node j. If Eq. (d) is employed for all nodes, we arrive at a matrix equation

$$[Y^*]\{V\} = \{I\} \tag{g}$$

where $[Y^*]$, $\{V\}$, and $\{I\}$ are of order $n \times n$, $n \times 1$, and $n \times 1$, respectively. Y^* is called the **indefinite nodal admittance matrix** of the network. So when the current inputs at all of the n nodes and the admittances Y_{jk} are specified, the voltages at these nodes can be readily determined from the network's admittance matrix equation (g).

As an example, consider the network shown in Fig. 9-14. It should be easy to observe that there are four nodes in the circuit. By designating nodes 1 through 4 as indicated, the indefinite nodal admittance matrix is therefore of the order of 4 by 4. Its elements can be explicitly defined as follows based on the elements' characteristics modeled in Chapter 7.

$$y_{1,1}^* = Y_{1,1} + Y_{1,2} + Y_{1,3} + Y_{1,4}$$
$$= 0 + 2\pi f C_1 + (1/R_2) + (1/R_1)$$
$$y_{1,2}^* = -Y_{1,2} = -2\pi f C_1$$

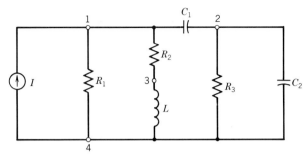

Figure 9-14

$$y_{1,3}^* = -Y_{1,3} = -1/R_2$$

$$y_{1,4}^* = -Y_{1,4} = -1/R_1$$

$$y_{2,2}^* = Y_{2,1} + Y_{2,2} + Y_{2,3} + Y_{2,4}$$
$$= 2\pi f C_1 + 0 + 0 + (1/R_3)$$

$$y_{2,3}^* = -Y_{2,3} = 0$$

$$y_{2,4}^* = -Y_{2,4} = -1/R_3$$

$$y_{3,3}^* = Y_{3,1} + Y_{3,2} + Y_{3,3} + Y_{3,4}$$
$$= (1/R_2) + 0 + 0 + (\tfrac{1}{2}\pi f L)$$

$$y_{3,4}^* = -Y_{3,4} = -\tfrac{1}{2}\pi f L$$

$$y_{4,4}^* = Y_{4,1} + Y_{4,2} + Y_{4,3} + Y_{4,4}$$
$$= (1/R_1) + 2\pi f C_2 + (\tfrac{1}{2}\pi f L) + 0$$

Notice that the indefinite nodal admittance matrix is symmetric and f in the above equations is the frequency of the input current. Having studied the possible involvement of resistors, capacitors, and inductors in a typical network, the readers should be able to deal with specific cases when the numeric data are provided and to actually carry out the solution of Eq. (g).

In fact, involvement of the frequency f in the elements of $[Y^*]$ leads to various phase shifts in the voltages and currents in the pathway elements, as has been discussed in Chapter 7. Consequently, $[Y^*]$ in general cases is a matrix with complex elements and the solution of Eq. (g) requires complex programming. Below is a BASIC program called C. TRID, which solves a complex matrix equation by Cholesky's decomposition method translated from a FORTRAN version published earlier by the author.[9]

---

[9] Y. C. Pao. "On Triangular Decomposition of Nonpositive Definite Symmetric Matrices Using Complex FORTRAN Programming," *International Journal for Numerical Methods in Engineering*, Vol. 15, 1980, pp. 611–616.

PROGRAM C. TRID

COEFFICIENT MATRIX

ROW NO. 1			ROW NO. 7	
REAL	IMAGINARY		REAL	IMAGINARY
5.00	0.00		9.00	0.00
5.00	0.00		7.00	0.00
3.00	0.00		6.00	0.00
0.00	0.00		0.00	0.00
1.00	0.00		3.00	0.00
ROW NO. 2			ROW NO. 8	
REAL	IMAGINARY		REAL	IMAGINARY
6.00	0.00		11.00	0.00
0.00	0.00		0.00	0.00
4.00	0.00		4.00	0.00
3.00	0.00		2.00	0.00
2.00	0.00		1.00	0.00
ROW NO. 3			ROW NO. 9	
REAL	IMAGINARY		REAL	IMAGINARY
7.00	0.00		7.00	0.00
6.00	0.00		5.00	0.00
0.00	0.00		4.00	0.00
5.00	0.00		3.00	0.00
4.00	0.00		0.00	0.00
ROW NO. 4			ROW NO. 10	
REAL	IMAGINARY		REAL	IMAGINARY
8.00	0.00		0.00	0.00
4.00	0.00		7.00	0.00
0.00	0.00		6.00	0.00
3.00	0.00		0.00	0.00
2.00	0.00		0.00	0.00
ROW NO. 5			ROW NO. 11	
REAL	IMAGINARY		REAL	IMAGINARY
9.00	0.00		11.00	0.00
7.00	0.00		8.00	0.00
0.00	0.00		0.00	0.00
5.00	0.00		0.00	0.00
3.00	0.00		0.00	0.00
ROW NO. 6			ROW NO. 12	
REAL	IMAGINARY		REAL	IMAGINARY
11.00	0.00		13.00	0.00
8.00	0.00		0.00	0.00
0.00	0.00		0.00	0.00
6.00	0.00		0.00	0.00
5.00	0.00		0.00	0.00

CONSTANT VECTOR: SOLUTION:

REAL	IMAGINARY		REAL	IMAGINARY
10.00	0.00		1.00	0.00
−7.00	0.00		−1.00	0.00
6.00	0.00		2.00	0.00
8.00	0.00		−2.00	0.00
−9.00	0.00		3.00	0.00
33.00	0.00		−3.00	0.00
34.00	0.00		4.00	0.00
−10.00	0.00		−4.00	0.00
31.00	0.00		5.00	0.00
−45.00	0.00		−5.00	0.00
7.00	0.00		6.00	0.00
−40.00	0.00		−6.00	0.00

```
100 INIT
110 CHARSIZE 2
120 PAGE
130 PRINT "* PROGRAM C.TRID *"
140 REM * INPUT
150 GOSUB 1060
160 PRINT "_COEFFICIENT MATRIX :_"
170 FOR I=1 TO N3
180 PRINT " ROW NO. ";I
190 PRINT "     REAL  IMAGINARY"
200 FOR J=1 TO N4
210 PRINT USING 220:A1(I,J);A2(I,J)
220 IMAGE    2(5D.2D)
230 NEXT J
240 NEXT I
250 PRINT "_CONSTANT VECTOR :_"
260 PRINT "     REAL  IMAGINARY"
270 FOR I=1 TO N3
280 PRINT USING 220:C1(I);C2(I)
290 NEXT I
300 GOSUB 380
310 PRINT "_SOLUTION :_"
320 PRINT "     REAL  IMAGINARY"
330 FOR N=1 TO 12
340 PRINT USING 350:X1(N),X2(N)
350 IMAGE    5D.2D,8D.2D
360 NEXT N
370 END
380 REM *SUBROUTINE C.TRID *
390 DIM X1(N3),X2(N3)
400 REM *COMPLEX A,C,SUM,X AND A=A1+A2*'I', ETC*
410 T0=SQR(SQR(A1(1,1)^2+A2(1,1)^2))
420 T1=ATN(A2(1,1)/A1(1,1))
430 IF A1(1,1)>0 THEN 450
440 T1=T1+PI
450 A1(1,1)=T0*COS(T1/2)
460 A2(1,1)=T0*SIN(T1/2)
470 FOR J=2 TO N4
480 T0=A1(1,J)
490 T1=A1(1,1)^2+A2(1,1)^2
500 A1(1,J)=(A1(1,J)*A1(1,1)+A2(1,J)*A2(1,1))/T1
```

```
510 A2(1,J)=(A2(1,J)*A1(1,1)-T0*A2(1,1))/T1
520 NEXT J
530 T1=A1(1,1)^2+A2(1,1)^2
540 X1(1)=(C1(1)*A1(1,1)+C2(1)*A2(1,1))/T1
550 X2(1)=(C2(1)*A1(1,1)-C1(1)*A2(1,1))/T1
560 FOR I=2 TO N3
570 M=1 MAX I+1-N4
580 FOR J=1 TO N4
590 S1=A1(I,J)
600 S2=A2(I,J)
610 FOR K=M TO I-1
620 IF I+J-K>N4 THEN 650
630 S1=S1-(A1(K,I+1-K)*A1(K,I+J-K)-A2(K,I+1-K)*A2(K,I+J-K))
640 S2=S2-(A2(K,I+1-K)*A1(K,I+J-K)+A1(K,I+1-K)*A2(K,I+J-K))
650 NEXT K
660 IF J=1 THEN 710
670 T1=A1(I,1)^2+A2(I,1)^2
680 A1(I,J)=(S1*A1(I,1)+S2*A2(I,1))/T1
690 A2(I,J)=(S2*A1(I,1)-S1*A2(I,1))/T1
700 GO TO 770
710 T0=SQR(SQR(S1^2+S2^2))
720 T1=ATN(S2/S1)
730 IF S1>0 THEN 750
740 T1=T1+PI
750 A1(I,1)=T0*COS(T1/2)
760 A2(I,1)=T0*SIN(T1/2)
770 NEXT J
780 S1=C1(I)
790 S2=C2(I)
800 FOR K=M TO I-1
810 S1=S1-(A1(K,I+1-K)*X1(K)-A2(K,I+1-K)*X2(K))
820 S2=S2-(A2(K,I+1-K)*X1(K)+A1(K,I+1-K)*X2(K))
830 NEXT K
840 T1=A1(I,1)^2+A2(I,1)^2
850 X1(I)=(S1*A1(I,1)+S2*A2(I,1))/T1
860 X2(I)=(S2*A1(I,1)-S1*A2(I,1))/T1
870 NEXT I
880 T0=X1(N3)
890 T1=A1(N3,1)^2+A2(N3,1)^2
900 X1(N3)=(X1(N3)*A1(N3,1)+X2(N3)*A2(N3,1))/T1
910 X2(N3)=(X2(N3)*A1(N3,1)-T0*A2(N3,1))/T1
920 FOR L=1 TO N3-1
930 I=N3-L
940 M=N3 MIN I+N4-1
950 S1=X1(I)
960 S2=X2(I)
970 FOR K=I+1 TO M
980 S1=S1-(A1(I,K-I+1)*X1(K)-A2(I,K-I+1)*X2(K))
990 S2=S2-(A2(I,K-I+1)*X1(K)+A1(I,K-I+1)*X2(K))
1000 NEXT K

1010 T1=A1(I,1)^2+A2(I,1)^2
1020 X1(I)=(S1*A1(I,1)+S2*A2(I,1))/T1
1030 X2(I)=(S2*A1(I,1)-S1*A2(I,1))/T1
1040 NEXT L
1050 RETURN
1060 REM *SUBROUTINE INPUT - DATA TAKEN FROM PAO'S PAPER*
1070 N3=12
1080 N4=5
1090 DIM A1(N3,N4),A2(N3,N4),C1(N3),C2(N3)
```

```
1100 A2=0
1110 C2=0
1120 READ A1
1130 DATA 5,4,3,0,1,6,0,4,3,2,7,6,0,5,4,8,4,0,3,2,9,7,0,5,3,11,8,0,6,5
1140 DATA 9,7,6,0,3,11,0,4,2,1,7,5,4,3,0,9,7,6,0,0,11,8,0,0,0,13,0,0,0,0
1150 READ C1
1160 DATA 10,-7,6,8,-9,33,34,-19,31,-45,7,-49
1170 RETURN
```

EXERCISE

The Wheatstone bridge shown in Fig. 9-15 has been discussed in the preceding chapters in derivation of governing differential equations of the electric system and in transformation into algebraic equations by application of the Laplace transform. Here, the bridge can be analyzed by the admittance-matrix method. Modify Program C.TRID to generate the y_{ij}^* elements automatically based on the elemental specifications of whether it is a resistor, an inductor, or a capacitor to arrive at $1/R$, $2\pi f C j$, or $-j/2\pi f L$, respectively, where f is the frequency of the input current and $j = (-1)^{1/2}$. Solve for V_{1-7} if $I_1 = 1$ and all $I_i = 0$ for $i = 2, 3, \ldots, 7$ by running the modified program with the values of $f = 0.1 \times 10^{-5}$ Hz, $C =$

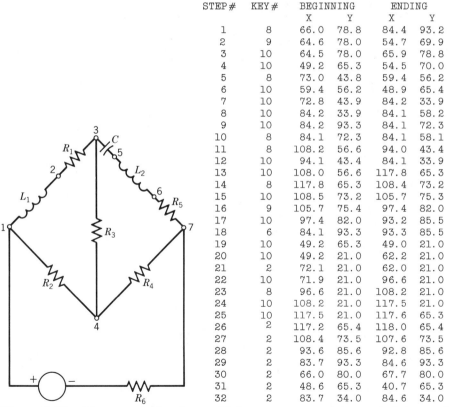

STEP #	KEY #	BEGINNING		ENDING	
		X	Y	X	Y
1	8	66.0	78.8	84.4	93.2
2	9	64.6	78.0	54.7	69.9
3	10	64.5	78.0	65.9	78.8
4	10	49.2	65.3	54.5	70.0
5	8	73.0	43.8	59.4	56.2
6	10	59.4	56.2	48.9	65.4
7	10	72.8	43.9	84.2	33.9
8	10	84.2	33.9	84.1	58.2
9	10	84.2	93.3	84.1	72.3
10	8	84.1	72.3	84.1	58.1
11	8	108.2	56.6	94.0	43.4
12	10	94.1	43.4	84.1	33.9
13	10	108.0	56.6	117.8	65.3
14	8	117.8	65.3	108.4	73.2
15	10	108.5	73.2	105.7	75.3
16	9	105.7	75.4	97.4	82.0
17	10	97.4	82.0	93.2	85.5
18	6	84.1	93.3	93.3	85.5
19	10	49.2	65.3	49.0	21.0
20	10	49.2	21.0	62.2	21.0
21	2	72.1	21.0	62.0	21.0
22	10	71.9	21.0	96.6	21.0
23	8	96.6	21.0	108.2	21.0
24	10	108.2	21.0	117.5	21.0
25	10	117.5	21.0	117.6	65.3
26	2	117.2	65.4	118.0	65.4
27	2	108.4	73.5	107.6	73.5
28	2	93.6	85.6	92.8	85.6
29	2	83.7	93.3	84.6	93.3
30	2	66.0	80.0	67.7	80.0
31	2	48.6	65.3	40.7	65.3
32	2	83.7	34.0	84.6	34.0

Figure 9-15 Wheatstone bridge, drawn with Program E.MODULE interactively by use of thumbwheels and definable keys.

0.508×10^{-8} F, $L_1 = 0.0001$ H, $L_2 = 0.05$ H, $R_1 = R_5 = R_6 = 10$ ohms, $R_2 = R_4 = 0.1 \times 10^5$ ohm, and $R_3 = 0.1 \times 10^7$ ohm.

EXERCISE

Apply finite-element nodal analysis for the circuit shown below to determine the voltages at the nodes numbered 1 through 7.

Answer: $v_1 = 0.24994$, $v_2 = 0.15524$, $v_3 = 0.29156$, $v_4 = 0.38150$, $v_5 = 0.14569$, $v_6 = 5.9972$, and $v_7 = 0.15879$, all in volts.

$R_{01} = 0.01$
$R_{03} = 500$
$R_{05} = 25$
$R_{06} = 0.1$
$R_{07} = 10$
$R_{12} = 70$
$R_{23} = 1000$
$R_{24} = 40$
$R_{25} = 150$
$R_{27} = 60$
$R_{34} = 125$
$R_{45} = 75$
$R_{46} = 200$
$R_{47} = 12$
$R_{57} = 5$ all in ohms

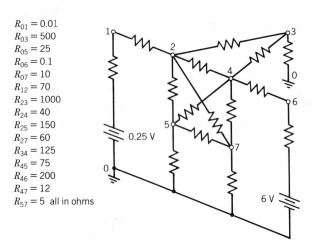

CHAPTER 10

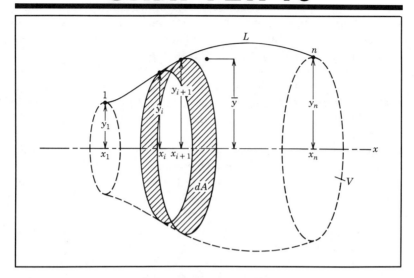

ELEMENTARY NUMERICAL METHODS OF SOLUTION

10.1 INTRODUCTION

In this chapter approximate methods will be introduced for solution of differential equations that govern the response of various physical systems as described in the previous chapters. For dynamic analysis of physical elements and systems, we are primarily interested in their behavior as time passes. In other words, it is the **temporal** changes of elemental and system behavior that need to be approximately determined when they are subjected to disturbances. Hence, the problem is how to predict the condition at the next instant if the present condition of the element or system is known. That is, we need to develop a **forward looking** procedure for depicting the behavior of an element or a system.

The forward-difference numerical method is to be presented as an **introductory tool** for solving the governing differential equations. Many other numerical methods are available for such a purpose but are beyond the scope of this book. For the reader who wishes to pursue these methods, there are numerous books on numerical methods for engineering analyses.[1] A number of practical applications are presented in this chapter to demonstrate the use of the forward-difference approximation for CAD.

In Chapter 8 the characteristic equations for linear engineering systems were derived in the form of polynomials. When the polynomials are of higher order, numerical solution of its roots needed in Bode analysis of the system has to be devised. This problem is treated separately in Appendix B. The Bairstow method based on the Newton–Raphson iterative process is discussed in detail therein.

Transcendental equations such as $a_1 \cos x + a_2 \cos (x + a_3) = a_4$ and

$$\tan^{-1} a_1 x + \tan^{-1} (a_2 x + a_3)/(a_4 x^2 + a_5 x + a_6) = a_7$$

with the a's being constants, which arise in the synthesis of four-bar linkage, in Bode analysis of systems, and in other studies, are to be treated in this chapter. Iterative methods for finding the x values that satisfy these transcendental equations within prescribed accuracy will be introduced. Successive substitution and linear interpolation are the main topics delineated with the aid of a number of practical examples.

Numerical integration of areas and volumes, which is a fundamental need in engineering design, is also discussed in this chapter and is illustrated with CAD examples.

10.2 FORWARD-DIFFERENCE APPROXIMATIONS

In Chapter 7 we solved the dual-tank problem analytically for constant inflows Q_1 and Q_2. For a general case of Q_1 and Q_2 both being arbitrary functions of time t, the solution of the governing differential equations for tank levels h_1 and h_2 will not be as easy as demonstrated for the special case. As it is our primary objective in this text to employ the computer as a tool for design and analysis,

---

[1] M. L. James, G. M. Smith, and J. C. Wolford, *Applied Numerical Methods for Digital Computation with FORTRAN and CSMP.* 2nd ed., Harper & Row, New York, 1977.

the approximated solution of the problem is proposed to be used for the general cases. Finite-difference approximation for the derivative terms in the governing differential equations is to be utilized for obtaining the numerical solution in a step-by-step manner.

Let us recall the governing differential equations of the dual-tank problem. The equations for tank levels h_1 and h_2 derived in Chapter 8 are, Fig. 10-1,

$$A_1 \frac{dh_1}{dt} + a_1 h_1 - a_3 h_2 = Q_1(t)$$

$$-a_3 h_1 + A_2 \frac{dh_2}{dt} + a_2 h_2 = Q_2(t)$$

The constants A_1 and A_2 are the uniform cross-sectional areas of tanks 1 and 2, respectively, and a_1, a_2, and a_3 are related to the fluid resistances R_1, R_2, and R_3 according to Eqs. (k)–(n) in Section 7.6. For numerical solution of $h_1(t)$ and $h_2(t)$, the above two equations will be written as

$$\frac{dh_1}{dt} = [Q_1(t) - a_1 h_1 + a_3 h_2]/A_1 = f_1(t) \tag{a}$$

$$\frac{dh^2}{dt} = [Q_2(t) + a_3 h_1 - a_2 h_2]/A_2 = f_2(t) \tag{b}$$

$f_1(t)$ and $f_2(t)$ are introduced to facilitate the numerical derivation. The finite-difference method to be discussed here is the forward-difference formula that approximates the derivatives in Eqs. (a) and (b) at $t = t_i$ by

$$\left.\frac{dh_1}{dt}\right|_{at\ t_i} \simeq [h_1(t_{i+1}) - h_1(t_i)]/\Delta t \tag{c}$$

$$\left.\frac{dh_2}{dt}\right|_{at\ t_i} \simeq [h_2(t_{i+1}) - h_2(t_i)]/\Delta t \tag{d}$$

t_{i+1} in the above expressions is to be understood as the instant when time t_i is incremented by an amount Δt. For practical purposes we are interested in knowing the level changes $h_1(t)$ and $h_2(t)$ within a selected period of concern, say from $t = 0$ to $t = t_e$. Also, perhaps the limiting case of $t \to \infty$ should be explored.

Suppose that an ending time t_e and increment Δt have been chosen. Knowing the initial levels h_{10} and h_{20} together with prescribed inflow rates $Q_1(t)$ and $Q_2(t)$, we want to determine $h_1(t_i)$ and $h_2(t_i)$ for $0 \le t_i \le t_e$. The results can be tabulated by carrying out n steps where

$$n = t_e/\Delta t \tag{e}$$

Upon substituting Eqs. (c) and (d) into Eqs. (a) and (b), we arrive at

$$h_1(t_{i+1}) = h_1(t_i) + (\Delta t)f_1(t_i) \tag{f}$$

$$h_2(t_{i+1}) = h_2(t_i) + (\Delta t)f_2(t_i) \tag{g}$$

where

$$t_i = (i - 1)\Delta t \tag{h}$$

and $i = 1, 2, \ldots, n$. For example, initially $h_1(t_1) = h_1(0) = h_{10}$ and $h_2(t_1) = h_2(0) = h_{20}$, and $Q_1(t_1)$ and $Q_2(t_1)$ can be computed so that $f_1(t_1)$ and $f_2(t_1)$ are completely defined. Thus, $h_1(t_2)$ and $h_2(t_2)$ can be calculated in a straightforward manner in accordance with Eqs. (f) and (g). Once $h_1(t_2)$ and $h_2(t_2)$ are determined, Eqs. (f) and (g) allow $h_1(t_3)$ and $h_2(t_3)$ to be found. Consequently, the process can be repeated until $h_1(t_e)$ and $h_2(t_e)$ are obtained.

Program DUAL.TANK has been modified to incorporate the successive generation of new $h_1(t_{i+1})$ and $h_2(t_{i+1})$ values from $h_1(t_i)$ and $h_2(t_i)$, and $Q_1(t_i)$ and $Q_2(t_i)$. The loop including statements 1570–1640 in the program implements the calculation based on Eqs. (f) and (g). Input data are shown in the user's menu. The results also show the steady-state solutions of $h_1(t \to \infty) = 5.4$ and $h_2(t \to \infty) = 5.7$. Notice that these results are more accurate than those presented in Chapter 7, where three significant figures were maintained throughout the analytical derivation. Here, 12 significant figures are kept throughout the computation as evidenced by the printout of the characteristic roots equal to -0.0188922662511 and -0.110274400416.

PROGRAM DUAL.TANK

```
10 REM * PROGRAM DUAL.TANK *
20 PAGE
30 CHARSIZE 4
40 MOVE 70,70
50 DRAW 75,70
60 MOVE 75,71.25
70 DRAW 75,68.75
80 DRAW 80,71.25
90 DRAW 80,68.75
100 DRAW 75,71.25
110 MOVE 80,70
120 DRAW 90,70
130 MOVE 90,71.25
140 DRAW 90,68.75
150 DRAW 95,71.25
160 DRAW 95,68.75
170 DRAW 90,71.25
180 MOVE 95,70
190 DRAW 100,70
200 MOVE 85,70
210 DRAW 85,60
220 MOVE 86.25,60
230 DRAW 83.75,60
240 DRAW 86.25,55
250 DRAW 83.75,55
260 DRAW 86.25,60
270 MOVE 85,55
280 DRAW 85,48
290 DRAW 84.5,49.5
300 DRAW 85.5,49.5
310 DRAW 85,48
320 MOVE 70,65
330 DRAW 70,90
```

```
340 MOVE 70,65
350 DRAW 50,65
360 DRAW 50,90
370 MOVE 70,86
380 DRAW 50,86
390 MOVE 100,65
400 DRAW 100,90
410 MOVE 100,65
420 DRAW 120,65
430 DRAW 120,90
440 MOVE 120,77
450 DRAW 100,77
460 MOVE 73,65
470 DRAW 82,65
480 DRAW 78,64.5
490 DRAW 78,65.5
500 DRAW 82,65
510 MOVE 88,65
520 DRAW 92,65.5
530 DRAW 92,64.5
540 DRAW 88,65
550 DRAW 97,65
560 MOVE 59,93
570 PRINT "Q"
580 MOVE 58,93
590 DRAW 56.5,89
600 DRAW 56.75,91
610 DRAW 57.75,90.5
620 DRAW 56.5,89
630 MOVE 66,87
640 PRINT "A"
650 CHARSIZE 3
660 MOVE 67.5,86.5
670 PRINT "1"
680 MOVE 60.5,92.5
690 PRINT "1"
700 MOVE 71,86
710 PRINT "h"
720 MOVE 72.5,85
730 PRINT "1"
740 MOVE 77,71.25
750 PRINT "R"
760 MOVE 78.5,70.25
770 PRINT "1"
780 MOVE 77,62.5
790 PRINT "q"
800 MOVE 78.5,61.5
810 PRINT "1"
820 MOVE 84,72
830 PRINT "h"
840 MOVE 85.5,71
850 PRINT "3"
860 MOVE 87,56
870 PRINT "R"
880 MOVE 88.5,55.5
890 PRINT "3"
900 MOVE 87,49
910 PRINT "q"
920 MOVE 88.5,48
930 PRINT "3"
940 MOVE 92,71.25
```

```
950 PRINT "R"
960 MOVE 93.5,70.75
970 PRINT "2"
980 MOVE 93,62.5
990 PRINT "q"
1000 MOVE 94.5,61.5

1010 PRINT "2"
1020 CHARSIZE 4
1030 MOVE 103.5,84.5
1040 PRINT "2"
1050 MOVE 102.5,84
1060 DRAW 104.5,80
1070 DRAW 103,81.5
1080 DRAW 104.125,82.25
1090 DRAW 104.5,80
1100 MOVE 114.5,77.5
1110 PRINT "2"
1120 MOVE 121,77
1130 PRINT "h"
1140 MOVE 122.5,76
1150 PRINT "2"
1160 CHARSIZE 4
1170 MOVE 102,85
1180 PRINT "Q"
1190 MOVE 113,78
1200 PRINT "A"
1210 HOME
1220 CHARSIZE 3
1230 PRINT "DO WANT TO SEE A SAMPLE CASE, Y/N?"
1240 INPUT Q$
1250 IF Q$="N" THEN 1290
1260 READ A1,A2,R1,R2,R3,Q1,Q2,H1,H2,T,T1
1270 DATA 2000,2000,0.01,1.0E-3,0.01,20,500,30,0,5,500
1280 GO TO 1340
1290 PRINT "WHAT ARE THE INPUT :"
1300 PRINT "A1,A2,R1,R2,R3?"
1310 INPUT A1,A2,R1,R2,R3
1320 PRINT "Q1,Q2,H1,H2,DeltaT,Tend?"
1330 INPUT Q1,Q2,H1,H2,T,T1
1340 PRINT " _A1=";A1
1350 PRINT "A2=";A2
1360 PRINT "R1=";R1
1370 PRINT "R2=";R2
1380 PRINT "R3=";R3
1390 PRINT "Q1=";Q1
1400 PRINT "Q2=";Q2
1410 PRINT "h1=";H1
1420 PRINT "h2=";H2
1430 PRINT "Time Increment =";T
1440 PRINT "Ending Time=";T1
1450 N=INT(T1/T)
1460 IF N=T1/T THEN 1480
1470 N=N+1
1480 N=1+N
1490 DIM H3(N),H4(N),F1(N),F2(N)
1500 H3(1)=H1
1510 H4(1)=H2
1520 M1=H1 MAX H2
1530 D=R1*R2+R2*R3+R1*R3
1540 Z1=(R2+R3)/D
```

```
1550 Z2=(R1+R3)/D
1560 Z3=R3/D
1570 FOR I=1 TO N-1
1580 F1(I)=Q1-Z1*H3(I)+Z3*H4(I)
1590 F2(I)=Q2-Z2*H4(I)+Z3*H3(I)
1600 H3(I+1)=1/A1*F1(I)+H3(I)
1610 H4(I+1)=1/A2*F2(I)+H4(I)
1620 M=H3(I+1) MAX H4(I+1)
1630 M1=M1 MAX M
1640 NEXT I
1650 S1=A1*A2
1660 S2=Z1*A2+Z2*A1
1670 S3=Z1*Z2-Z3^2
1680 C1=0.5/S1*(-S2+(S2^2-4*S1*S3)^0.5)
1690 C2=0.5/S1*(-S2-(S2^2-4*S1*S3)^0.5)
1700 P1=(Z2*Q1+Z3*Q2)/S3
1710 P2=(Z1*Q2+Z3*Q1)/S3
1720 PRINT "_Characteristic Roots Are_"
1730 PRINT C1
1740 PRINT C2
1750 PRINT "_Steady State Solutions Are_"
1760 PRINT "   h1 = ";P1
1770 PRINT "   h2 = ";P2
1780 S=15/M1
1790 N1=70/N
1800 MOVE 50,45
1810 DRAW 50,30
1820 DRAW 120,30
1830 FOR I=1 TO 10
1840 MOVE 50+I*7,30.25
1850 DRAW 50+I*7,29.75
1860 NEXT I
1870 CHARSIZE 4
1880 MOVE 117,27
1890 PRINT T1
1900 MOVE 46,40
1910 PRINT "h"
1920 MOVE 47.5,39
1930 PRINT "1"
1940 MOVE 85,27
1950 PRINT "Time"
1960 MOVE 50,20
1970 DRAW 50,5
1980 DRAW 120,5
1990 FOR I=1 TO 10
2000 MOVE 50+I*7,5.25

2010 DRAW 50+I*7,4.75
2020 NEXT I
2030 MOVE 117,2
2040 PRINT T1;"_"
2050 MOVE 46,15
2060 PRINT "h "
2070 MOVE 47.5,14
2080 PRINT "2"
2090 MOVE 85,2
2100 PRINT "Time__"
2110 MOVE 50,S*H3(1)+30
2120 DRAW 50+N1,S*H3(2)+30
2130 MOVE 50,S*H4(1)+5
2140 DRAW 50+N1,S*H4(2)+5
```

```
2150 FOR I=3 TO N
2160 MOVE 50+(I-2)*N1,S*H3(I-1)+30
2170 DRAW 50+(I-1)*N1,S*H3(I)+30
2180 MOVE 50+(I-2)*N1,S*H4(I-1)+5
2190 DRAW 50+(I-1)*N1,S*H4(I)+5
2200 NEXT I
2210 MOVE 0,10
2220 PRINT "WANT TO RUN NEXT CASE, Y/N?"
2230 INPUT A$
2240 IF A$="Y" THEN 10
2250 PRINT "_ ------- END ---------"
2260 END
```

EXERCISE

Experiment with Program DUAL.TANK by adjusting the values of Q_1, Q_2, R_1, R_2, and R_3 to investigate (1) the time required for h_1 and h_2 to attain approximately the same level relative to the R_1/R_2 ratio if Q_1 and Q_2 are equal to zero, R_3 is very large, causing q_3 to be almost equal to zero, and initially $h_1 = 30$ and $h_2 = 0$; (2) how the difference in Q_1 and Q_2 should be changed so that the steady-state values of h_1 and h_2 shown in Fig. 10-1 can be nearly equal to each other within 2%; and (3) the feedback problem of dividing the outflow q_3 into Q_1 and Q_2. For the third problem, draw the block diagram and modify Program DUAL.TANK to accommodate the general analysis where Q_1 and Q_2 are both sinusoidal functions.

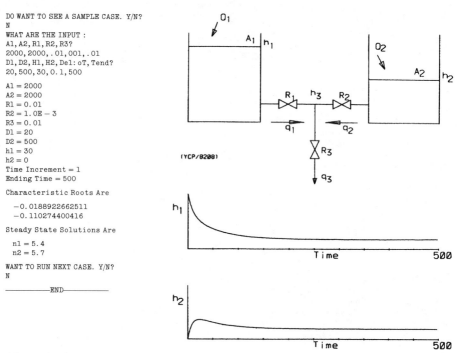

```
DO WANT TO SEE A SAMPLE CASE. Y/N?
N
WHAT ARE THE INPUT :
A1,A2,R1,R2,R3?
2000,2000,.01,001,.01
D1,D2,H1,H2,Del:oT,Tend?
20,500,30,0.1,500

A1 = 2000
A2 = 2000
R1 = 0.01
R2 = 1.0E-3
R3 = 0.01
D1 = 20
D2 = 500
h1 = 30
h2 = 0
Time Increment = 1
Ending Time = 500

Characteristic Roots Are

  -0.0188922662511
  -0.110274400416

Steady State Solutions Are

  n1 = 5.4
  n2 = 5.7

WANT TO RUN NEXT CASE. Y/N?
N

----------END----------
```

Figure 10-1

10.3 MIXING-TANK PROBLEM

Consider the problem of mixing two substances in a tank as sketched in Fig. 10-2. To be more specific, let the substances be water and salt. It is of interest to study the changes over time of the mixture in the tank and the concentration of salt in the outflow. Let the input flows of water and salt and the output of the mixture be designated Q_w, Q_s, and Q_m, respectively. And let W_m and f be the weight and concentration of salt of the mixture in the tank, respectively. In general, Q_w, Q_s, Q_m, W_m, and f are functions of time t.

The principle of the conservation of total mass can be applied for derivation of the governing equations for W_m and f. First consider the time rate of change of W_m in the tank. It must be the difference of the inflow and outflow. That is, we may write

$$\frac{dW_m}{dt} = Q_w + Q_s - Q_m \tag{a}$$

Assuming that the water and salt are well mixed, the time rate of change of the salt in the tank can be derived in a similar manner as

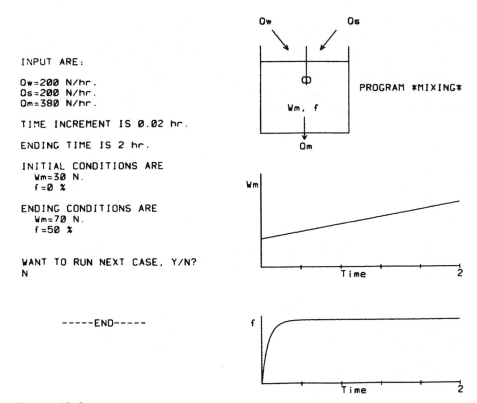

```
INPUT ARE:

Qw=200 N/hr.
Qs=200 N/hr.
Qm=380 N/hr.

TIME INCREMENT IS 0.02 hr.

ENDING TIME IS 2 hr.

INITIAL CONDITIONS ARE
  Wm=30 N.
  f=0 %

ENDING CONDITIONS ARE
  Wm=70 N.
  f=50 %

WANT TO RUN NEXT CASE, Y/N?
N


    -----END-----
```

Figure 10-2

$$\frac{d(f W_m)}{dt} = Q_s - f Q_m \tag{b}$$

We observe that Eq. (b) is nonlinear because of the product terms $f W_m$ and $f Q_m$. Again, we here seek a numerical solution by application of finite-difference approximation.

Suppose that we are interested in knowing W_m and f at various times t_i for $0 \le t_i \le t_e$ with the ending time t_e prescribed. A decision has to be made regarding what time increment Δt these results should be tabulated or plotted for. Also, the inflows $Q_w(t)$ and $Q_s(t)$ should be prescribed. The step-by-step procedure for calculation of W_m and f at the instant t_i is to use the following two equations successively for $i = 1, 2, \ldots, n$:

$$(W_m)_i = (W_m)_{i-1} + \Delta t[(Q_w)_{i-1} + (Q_s)_{i-1} - (Q_m)_{i-1}] \tag{c}$$

$$f_i = \{ f_{i-1}(W_m)_{i-1} + \Delta t[(Q_s)_{i-1} - f_{i-1}(Q_m)_{i-1}] \}/(W_m)_i \tag{d}$$

where $(W_m)_i \equiv W_m(t_i), f_i \equiv f(t_i)$, and so on. Equations (c) and (d) are derived from Eqs. (a) and (b), respectively, by application of the forward-difference approximation. Equations (c) and (d) are to be utilized n times until t_n is equal to or exceeds the required ending time t_e.

Since $t_i = (i - 1)\Delta t$, for $i = 1$, all of the terms with a subscript 0 on the right-hand side of Eqs. (c) and (d) must be known in order to carry out the first step of this successive process. These so-called **initial conditions**, namely, the initial weight and salt concentration of the mixture in the tank and the initial inflows of water and salt, and the initial outflow, have to be specified. A sample case of constant inflows and outflow has been worked out and delineated in Fig. 10-2 by use of the Program MIXING.

PROGRAM MIXING

```
100 PAGE
105 REM * PROGRAM MIXING *
110 CHARSIZE 3
120 INIT
130 GOSUB 260
140 GOSUB 500
150 GOSUB 560
160 GOSUB 1110
170 GOSUB 1310
180 GOSUB 1510
190 MOVE 0,35
200 PRINT "WANT TO RUN NEXT CASE, Y/N?"
210 INPUT A$
220 IF A$="Y" THEN 100
230 MOVE 10,20
240 PRINT "------END------"
250 END
260 REM *INPUT DATA*
270 PRINT "INPUT TIME DATA IN HOURS AND"
280 PRINT "   FLOW DATA IN N/hr., AND"
290 PRINT "   WEIGHT DATA IN NEWTONS."
300 PRINT "TIME INCREMENT = ";
```

```
310 INPUT T1
320 PRINT "ENDING TIME = ";
330 INPUT T2
340 J7=T2/T1+1
350 DIM W1(J7),F1(J7)
360 PRINT "Qw = ";
370 INPUT Q1
380 PRINT "Qs = ";
390 INPUT Q2
400 PRINT "Qm = ";
410 INPUT Q3
420 PRINT "INITIAL Wm = ";
430 INPUT W7
440 W1(1)=W7
450 PRINT "INITIAL f = ";
460 INPUT F7
470 F1(1)=F7
480 PAGE
490 RETURN
500 REM * calculate the Wm and f *
510 FOR J=2 TO J7
520 W1(J)=W1(J-1)+T1*(Q1+Q2-Q3)
530 F1(J)=(F1(J-1)*W1(J-1)+T1*(Q2-F1(J-1)*Q3))/W1(J)
540 NEXT J
550 RETURN
560 REM *PLOTTING*
570 GOSUB 970
580 MOVE 60,90
590 RDRAW 0,-22
600 RDRAW 22,0
610 RDRAW 0,22
620 RMOVE 0,-4
630 RDRAW -22,0
640 MOVE 70.5,80
650 PRINT "0"
660 MOVE 71.5,80
670 PRINT "0"
680 MOVE 71.5,80
690 RDRAW 0,10
700 MOVE 60,95
710 PRINT "Qw"
720 MOVE 85,78
730 PRINT "PROGRAM *MIXING*"
740 MOVE 63,94
750 RDRAW 4,-5
760 RDRAW -1.25,1.25/3^0.5
770 RMOVE 1.25,-1.25/3^0.5
780 RDRAW -0.35,1.35
790 MOVE 82,95
800 PRINT "Qs"
810 MOVE 79,94
820 RDRAW -4,-5
830 RDRAW 0.35,1.35
840 RMOVE -0.35,-1.35
850 RDRAW 1.25,1.25/3^0.5
860 MOVE 67,73
870 PRINT "Wm, f"
880 MOVE 71,72
890 RDRAW 0,-5.5
900 RDRAW -1/3^0.5,1
910 RMOVE 1/3^0.5,-1
```

```
920 RDRAW 1/3^0.5,1
930 RMOVE -1,-3
940 RMOVE -1/3^0.5,-1
950 PRINT "Qm"
960 RETURN
970 REM*FIND THE MXIMUM*
980 M1=W1(1)
990 FOR J1=2 TO J7
1000 IF M1<W1(J1) THEN 1020

1010 GO TO 1030
1020 M1=W1(J1)
1030 NEXT J1
1040 M3=F1(1)
1050 FOR J1=2 TO J7
1060 IF M3<F1(J1) THEN 1080
1070 GO TO 1090
1080 M3=F1(J1)
1090 NEXT J1
1100 RETURN
1110 REM * PLOT Wm VS t  *
1120 MOVE 60,35
1130 DRAW 60,58
1140 MOVE 60,35
1150 DRAW 110,35
1160 MOVE 56.5,54
1170 PRINT "Wm"
1180 MOVE 80,32
1190 PRINT "Time"
1200 S1=50/T2*T1
1210 S2=16/M1
1220 FOR I3=1 TO J7
1230 IF I3=1 THEN 1260
1240 DRAW S1*(I3-1)+60,W1(I3)*S2+35
1250 GO TO 1270
1260 MOVE 60,W1(1)*S2+35
1270 NEXT I3
1280 K5=1
1290 GOSUB 1670
1300 RETURN
1310 REM * PLOT f VS t   *
1320 MOVE 60,6
1330 DRAW 60,23
1340 MOVE 60,6
1350 DRAW 110,6
1360 MOVE 57.5,20
1370 PRINT "f"
1380 MOVE 80,3
1390 PRINT "Time"
1400 S2=16/M3
1410 FOR I3=1 TO J7
1420 IF I3=1 THEN 1450
1430 DRAW S1*(I3-1)+60,F1(I3)*S2+6
1440 GO TO 1470
1450 MOVE 60,F1(1)*S2+6
1460 GO TO 1470
1470 NEXT I3
1480 K5=2
1490 GOSUB 1670
1500 RETURN
1510 REM * MENU *
```

```
1520 MOVE 0,85
1530 PRINT "INPUT ARE:"
1540 PRINT " "
1550 PRINT "Qw=";Q1;" N/hr."
1560 PRINT "Qs=";Q2;" N/hr." ,
1570 PRINT "Qm=";Q3;" N/hr."
1580 PRINT "_TIME INCREMENT IS ";T1;" hr."
1590 PRINT "_ENDING TIME IS ";T2;" hr."
1600 PRINT "_INITIAL CONDITIONS ARE "
1610 PRINT "   Wm=";W7;" N, "
1620 PRINT "    f=";F7;" %"
1630 PRINT "_ENDING CONDITIONS ARE"
1640 PRINT "   Wm=";W1(J7);" N, "
1650 PRINT "    f=";INT(F1(J7)*100+0.5);" %"
1660 RETURN
1670 REM * THIS SUB PLOTS THE TICS *
1680 FOR I=1 TO 5
1690 IF K5=2 THEN 1730
1700 MOVE 60+I*10,34.5
1710 DRAW 60+I*10,35.5
1720 GO TO 1760
1730 MOVE 60+I*10,5.5
1740 DRAW 60+I*10,6.5
1750 GO TO 1760
1760 NEXT I
1770 IF K5=2 THEN 1800
1780 MOVE 108,32
1790 GO TO 1810
1800 MOVE 108,3
1810 PRINT T2
1820 RETURN
```

10.4 HEAT EXCHANGER

Another important element in the process control is the heat exchanger. A simplified schematic of the element is presented in Fig. 10-3. A fluid is flowing into the tank at a rate of Q_i and out of the tank at a rate of Q_o. Let the weight of fluid in the tank be W. Meanwhile the fluid is to be heated by another fluid passing through the pipe inside the tank. For simplicity of discussion, imagine the fluids to be water and steam, and let the inflow and outflow temperature of the water be T_i and T, respectively, and let T_s be the temperature of the steam, which is assumed to be a constant. Again, the principle of conservation of mass can be applied for the water to yield a differential equation for the weight of water in the tank. The equation is

$$\frac{dW}{dt} = Q_i - Q_o \tag{a}$$

In order to derive the governing equation for the water temperature T, we apply the **principle of conservation of energy**. Here, the energy gained by the water from the steam is expected to cause the water temperature to rise. The differential equation derived from this energy consideration is

$$C_p \frac{d}{dt} WT = C_p(Q_i T_i - Q_o T) + UA(T_s - T) \tag{b}$$

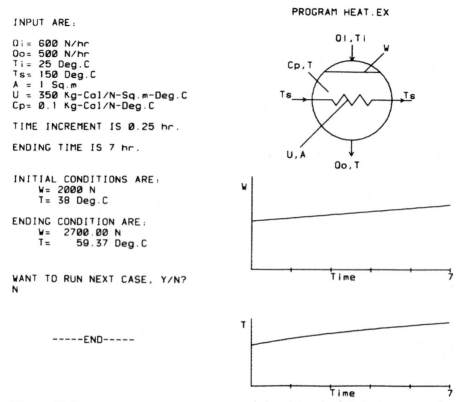

```
                                               PROGRAM HEAT.EX
INPUT ARE:

Qi = 600 N/hr
Qo= 500 N/hr
Ti= 25 Deg.C
Ts= 150 Deg.C
A = 1 Sq.m
U = 350 Kg-Col/N-Sq.m-Deg.C
Cp= 0.1 Kg-Col/N-Deg.C

TIME INCREMENT IS 0.25 hr.

ENDING TIME IS 7 hr.


INITIAL CONDITIONS ARE:
   W= 2000 N
   T= 38 Deg.C

ENDING CONDITION ARE:
   W=   2700.00 N
   T=     59.37 Deg.C


WANT TO RUN NEXT CASE, Y/N?
N


-----END-----
```

Figure 10-3

where C_p and U are the heat capacity and overall heat transfer coefficient of water, respectively, and A is the total heating area of the steam pipe. It is worth noting that the left-hand side of Eq. (b) is the time rate of energy change, $C_p Q_i T_i$ is the inflow energy, $C_p Q_o T$ is the outflow energy, and $U A(T_s - T)$ is the energy gain from the steam.

Since Eq. (b) is nonlinear, a finite-difference approximation has to be employed for numerical solution of the problem. The step-by-step procedure for calculation of $W(t)$ and $T(t)$ is to successively apply the following two equations:

$$W_j = W_{j-1} + \Delta t (Q_i - Q_o) \tag{c}$$

$$T_j = T_{j-1} + \Delta t (Q_i C_p T_j - Q_o C_p T_{j-1} + Q_{s,j-1})/C_p T_{j-1} W_j \tag{d}$$

where

$$Q_{s, j-1} = U A(T_s - T_{j-1}) \tag{e}$$

The details for implementation of this numerical procedure are similar to those for the mixing-tank problem discussed earlier. A sample case is presented in Fig. 10-3, showing the user's menu and input and output data obtained by application of Program HEAT.EX.

PROGRAM HEAT.EX

```
100 PAGE
110 CHARSIZE 3
120 INIT
130 GOSUB 270
140 GOSUB 610
150 GOSUB 700
160 GOSUB 1380
170 GOSUB 1520
180 GOSUB 1710
190 GOSUB 1910
200 MOVE 0,31
210 PRINT "WANT TO RUN NEXT CASE, Y/N?"
220 INPUT A$
230 IF A$="Y" THEN 100
240 MOVE 10,16
250 PRINT "-------END-------"
260 END
270 REM *INPUT DATA*
280 PRINT "PROGRAM HEAT.EX"
290 PRINT "(HEAT EXCHANGER ANALYSIS)"
300 PRINT " "
310 PRINT "INPUT DATA:"
320 PRINT " "
330 PRINT "Time Increment, hr = ";
340 INPUT T1
350 PRINT "Ending Time, hr = ";
360 INPUT T2
370 J7=T2/T1+1
380 DIM W1(J7),F1(J7)
390 PRINT "Qi, N/hr = ";
400 INPUT Q1
410 PRINT "Qo, N/hr = ";
420 INPUT Q2
430 PRINT "Ti, Deg.C = ";
440 INPUT T3
450 PRINT "Ts, Deg.C = ";
460 INPUT T4
470 PRINT "A, Sq.m = ";
480 INPUT A1
490 PRINT "Initial W, N = ";
500 INPUT W7
510 W1(1)=W7
520 PRINT "Initial T, Deg.C = ";
530 INPUT F7
540 F1(1)=F7
550 PRINT "U, Kg-Cal/hr-Sq.m-Deg.C = ";
560 INPUT U
570 PRINT "Cp, Kg-Cal/N-Deg.C = ";
580 INPUT C1
590 PAGE
600 RETURN
610 REM * calculat the Wm and Ti(F1) *
620 FOR J=2 TO J7
630 Q5=U*A1*(T4-F1(J-1))
640 W1(J)=W1(J-1)+T1*(Q1-Q2)
650 F1(J)=F1(J-1)+T1*((Q1*C1*T3-Q2*C1*F1(J-1)+Q5)/W1(J-1))/F1(J-1)/C1
660 NEXT J
670 F9=F1(J7)
680 W9=W1(J7)
690 RETURN
```

```
700 REM *PLOTTING*
710 MOVE 85,77
720 SET DEGREES
730 FOR S1=0 TO 360 STEP 3
740 Y=SIN(S1)*10+77
750 X=COS(S1)*10+85
760 IF S1=0 THEN 790
770 DRAW X,Y
780 GO TO 800
790 MOVE X,Y
800 NEXT S1
810 X1=COS(45)*10+85
820 Y1=SIN(45)*10+77
830 X2=COS(135)*10+85
840 Y2=SIN(135)*10+77
850 MOVE X1,Y1
860 DRAW X2,Y2
870 MOVE 70,77
880 RDRAW 3.5,0
890 RDRAW -3.5/4,3.5/4/3^0.5
900 RMOVE 3.5/4,-3.5/4/3^0.5
910 RDRAW -3.5/4,-3.5/4/3^0.5
920 RMOVE 3.5/4,3.5/4/3^0.5
930 RDRAW 6.5,0
940 RDRAW 1,-(2^0.5)
950 RDRAW 2,2^0.5*2
960 RDRAW 2,-(2^0.5)*2
970 RDRAW 2,2^0.5*2
980 RDRAW 2,-(2^0.5)*2
990 RDRAW 1,2^0.5
1000 RDRAW 8,0

1010 RDRAW -3.5/4,3.5/4/3^0.5
1020 RMOVE 3.5/4,-3.5/4/3^0.5
1030 RDRAW -3.5/4,-3.5/4/3^0.5
1040 RMOVE 3.5/4,3.5/4/3^0.5
1050 RDRAW 2,0
1060 MOVE 85,91
1070 DRAW 85,87
1080 RDRAW -1/3^0.5,1
1090 RMOVE 1/3^0.5,-1
1100 RDRAW 1/3^0.5,1
1110 MOVE 85,67
1120 DRAW 85,63
1130 RDRAW -1/3^0.5,1
1140 RMOVE 1/3^0.5,-1
1150 RDRAW 1/3^0.5,1
1160 MOVE 81,91.5
1170 PRINT "Qi,Ti"
1180 MOVE 81,59.5
1190 PRINT "Qo,T"
1200 MOVE 67,77
1210 PRINT "Ts"
1220 MOVE 98,77
1230 PRINT "Ts"
1240 MOVE 84,77
1250 RDRAW -12,-12
1260 RMOVE -3,-3
1270 PRINT "U,A"
1280 MOVE 79,79
1290 RDRAW -5,5
```

```
1300 RMOVE -5,0.5
1310 PRINT "Cp,T"
1320 MOVE 88,77+10*SIN(45)
1330 RDRAW 5,5
1340 PRINT "W"
1350 MOVE 70,98
1360 PRINT "PROGRAM HEAT.EX"
1370 RETURN
1380 REM*FIND THE MAXIMUM*
1390 M1=W1(1)
1400 FOR J1=2 TO J7
1410 IF M1<W1(J1) THEN 1430
1420 GO TO 1440
1430 M1=W1(J1)
1440 NEXT J1
1450 M3=F1(1)
1460 FOR J1=2 TO J7
1470 IF M3<F1(J1) THEN 1490
1480 GO TO 1500
1490 M3=F1(J1)
1500 NEXT J1
1510 RETURN
1520 REM * PLOT Wm VS t  *
1530 MOVE 60,35
1540 DRAW 60,58
1550 MOVE 60,35
1560 DRAW 110,35
1570 MOVE 57.5,54
1580 PRINT "W"
1590 MOVE 80,32
1600 PRINT "Time "
1610 FOR I3=1 TO J7
1620 W1(I3)=W1(I3)*16/M1+35
1630 IF I3=1 THEN 1660
1640 DRAW 50/T2*T1*(I3-1)+60,W1(I3)
1650 GO TO 1670
1660 MOVE 60,W1(1)
1670 NEXT I3
1680 K5=1
1690 GOSUB 2180
1700 RETURN
1710 REM * PLOT f VS t    *
1720 MOVE 60,6
1730 DRAW 60,23
1740 MOVE 60,6
1750 DRAW 110,6
1760 MOVE 57.5,20
1770 PRINT "T"
1780 MOVE 80,3
1790 PRINT "Time"
1800 FOR I3=1 TO J7
1810 F1(I3)=F1(I3)*16/M3+6
1820 IF I3=1 THEN 1850
1830 DRAW 50/T2*T1*(I3-1)+60,F1(I3)
1840 GO TO 1870
1850 MOVE 60,F1(1)
1860 GO TO 1870
1870 NEXT I3
1880 K5=2
1890 GOSUB 2180
1900 RETURN
```

```
1910 REM * MENU *
1920 MOVE 0,95
1930 PRINT "INPUT ARE:"
1940 PRINT " "
1950 PRINT "Qi= ";Q1;" N/hr"
1960 PRINT "Qo= ";Q2;" N/hr"
1970 PRINT "Ti= ";T3;" Deg.C"
1980 PRINT "Ts= ";T4;" Deg.C"
1990 PRINT "A = ";A1;" Sq.m"
2000 PRINT "U = ";U;" Kg-Cal/N-Sq.m-Deg.C"

2010 PRINT "Cp= ";C1;" Kg-Cal/N-Deg.C"
2020 PRINT " "
2030 PRINT "TIME INCREMENT IS ";T1;" hr."
2040 PRINT " "
2050 PRINT "ENDING TIME IS ";T2;" hr."
2060 PRINT " "
2070 PRINT " "
2080 PRINT "INITIAL CONDITIONS ARE:"
2090 PRINT "    W= ";W7;" N"
2100 PRINT "    T= ";F7;" Deg.C"
2110 PRINT " "
2120 PRINT "ENDING CONDITION ARE:"
2130 PRINT USING 2140:W9
2140 IMAGE "    W= ",5D.2D," N"
2150 PRINT USING 2160:F9
2160 IMAGE "    T= ",5D.2D," Deg.C"
2170 RETURN
2180 REM * THIS SUB PLOTS THE TICS *
2190 FOR I=1 TO 5
2200 IF K5=2 THEN 2240
2210 MOVE 60+I*10,34.5
2220 DRAW 60+I*10,35.5
2230 GO TO 2270
2240 MOVE 60+I*10,5.5
2250 DRAW 60+I*10,6.5
2260 GO TO 2270
2270 NEXT I
2280 IF K5=2 THEN 2310
2290 MOVE 108,32
2300 GO TO 2320
2310 MOVE 108,3
2320 PRINT T2
2330 RETURN
```

10.5 BEAM ANALYSIS AND DESIGN

In structural analysis, the following flexural formula relating the deflection $y(x)$ of a loaded beam with stiffness EI to the internal moment distribution $M(x)$ is well known:[2]

$$\frac{d^2y}{dx^2} = \frac{M}{EI} \tag{a}$$

where E is Young's modulus and I is the moment of inertia, both discussed in earlier chapters. In general cases, both E and I may change along the longitu-

---

[2] Hidgen et al., *op. cit.*, p. 268.

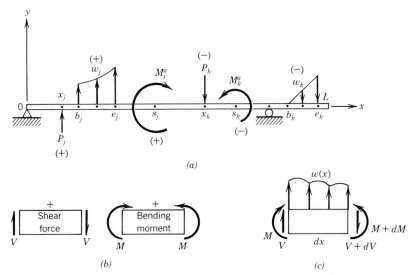

Figure 10-4 (a) Sign conventions for the applied distributed loads ws, concentrated loads Ps, and bending moments M^as; (b) sign conventions for the internal shear force V and bending moment M; and (c) equilibrium of forces and moments on a small segment of the beam.

dinal axis of the beam, which is designated as the x axis. As the beam may be subjected to three different types of loading, namely, (1) distributed load $w(x)$, (2) concentrated loads P_1 at location $x = x_1$, P_2 at location $x = x_2, \ldots, P_n$ at location $x = x_n$, and (3) bending moments M_1^a at location $x = s_1$, M_2^a at location $x = s_2, \ldots, M_m^a$ at location $x = s_m$, where the superscript a has been added to indicate the externally applied moments. In Fig. 10-4, the sign conventions for the externally applied loads and the internal bending moment M and shear force V are shown along with those for the slope $\theta = dy/dx$ and deflection y. Upward y, w, and P are considered as positive and counterclockwise M^a as positive. As far as the internal shear forces are concerned, the upward V on the left of a cut-through is considered as positive and downward V on the right of a cut-through as positive. The internal moment M causing the beam to be bent convex downward as shown in Fig. 10-4b is considered as positive.

Consideration of the equilibra of the forces and moments on a very small segment dx of the beam as sketched in Fig. 10-4c leads to, respectively,

$$\frac{dV}{dx} = w \quad \text{and} \quad \frac{dM}{dx} = V \tag{b,c}$$

If the slope θ is to replace dy/dx in Eq. (a), we can also have the equations

$$\frac{d\theta}{dx} = M/EI \quad \text{and} \quad \frac{dy}{dx} = \theta \tag{d,e}$$

For a given beam of length L, the distributions of w, V, M, θ, and y can now be

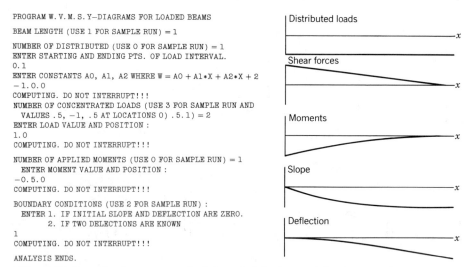

```
PROGRAM W. V. M. S. Y—DIAGRAMS FOR LOADED BEAMS

BEAM LENGTH (USE 1 FOR SAMPLE RUN) = 1

NUMBER OF DISTRIBUTED (USE 0 FOR SAMPLE RUN) = 1
ENTER STARTING AND ENDING PTS. OF LOAD INTERVAL.
0.1
ENTER CONSTANTS A0, A1, A2 WHERE W = A0 + A1*X + A2*X + 2
-1.0.0
COMPUTING. DO NOT INTERRUPT!!!
NUMBER OF CONCENTRATED LOADS (USE 3 FOR SAMPLE RUN AND
   VALUES .5, -1, .5 AT LOCATIONS 0) .5.1) = 2
ENTER LOAD VALUE AND POSITION :
1.0
COMPUTING. DO NOT INTERRUPT!!!

NUMBER OF APPLIED MOMENTS (USE 0 FOR SAMPLE RUN) = 1
   ENTER MOMENT VALUE AND POSITION :
-0.5.0
COMPUTING. DO NOT INTERRUPT!!!

BOUNDARY CONDITIONS (USE 2 FOR SAMPLE RUN) :
   ENTER 1. IF INITIAL SLOPE AND DEFLECTION ARE ZERO.
         2. IF TWO DELECTIONS ARE KNOWN
1
COMPUTING. DO NOT INTERRUPT!!!

ANALYSIS ENDS.
```

Figure 10-5a Fixed support at the left end of the beam; deflection and slope equal to zero at $x = 0$.

approximately determined by the finite-difference approach. That is, to approximate Eqs. (b)–(e) with

$$V_{i+1} = V_i + w_i \Delta x \tag{f}$$

$$M_{i+1} = M_i + V_i \Delta x \tag{g}$$

$$\theta_{i+1} = \theta_i + M_i \Delta x / E_i I_i \tag{h}$$

$$y_{i+1} = y_i + \theta_i \Delta x \tag{i}$$

where Δx is an increment properly chosen so that the variations of the applied w, P, and M^a can be fully taken into consideration as well as the specified boundary conditions. Suppose that the beam is to be investigated at l stations. Then the subscript i in Eqs. (f)–(i) will take on 1 through l and $\Delta x = L/(l-1)$. Notice that in Eq. (h) Young's modulus E and moment of inertia I also carry subscript i to allow them also to vary along the x axis.

Program W. V. M. S. Y has been developed for beam analysis. Figure 10-5a and b show sample applications of the program for a uniform beam of length equal to unity and carrying a uniformly distributed load equal to unity across the entire beam. At present, the program has no provision for variation of E and I. Creating one is left as an exercise. Also, only two types of distributed loads can be considered by the program—the linear and parabolic types described by the equation

$$w(x) = \sum_{k=0}^{i} a_k x^k, \qquad \text{for} \qquad i = 1 \text{ or } 2$$

They are to be specified by giving the beginning and ending locations b_j and e_j shown in Fig. 10-4a and the coefficients a_k in the above equation of the distributed

```
PROGRAM W. V. M. S. Y—DIAGRAMS FOR LOADED BEAMS

BEAM LENGTH (USE 1 FOR SAMPLE RUN) = 1

NUMBER OF DISTRIBUTED (USE 0 FOR SAMPLE RUN) = 1
ENTER STARTING AND ENDING PTS. OF LOAD INTERVAL.
0.1
ENTER CONSTANTS A0, A1, A2 WHERE W = A0 + A1 + X + A2 + X + 2
—1.0.0
COMPUTING. DO NOT INTERRUPT!!!
NUMBER OF CONCENTRATED LOADS (USE 3 FOR SAMPLE RUN AND
   VALUES .5, —1, .5 AT LOCATIONS 0, .5, 1) = 2
ENTER LOAD VALUE AND POSITION:
0.5.0
ENTER LOAD VALUE AND POSITION.
0.5.1
COMPUTING. DO NOT INTERRUPT!!!

NUMBER OF APPLIED MOMENTS (USE 0 FOR SAMPLE RUN) = 0
COMPUTING. DO NOT INTERRUPT!!!

BOUNDARY CONDITIONS (USE 2 FOR SAMPLE RUN) :

   ENTER 1. IF INITIAL SLOPE AND DEFLECTION ARE ZERO.
        2. IF TWO DEFLECTIONS ARE KNOWN
2
   ENTER FIRST DEFLECTION VALUE AND LOCATION
      (USE 0.0 FOR SAMPLE RUN)
0.0
   ENTER SECOND DEFLECTION VALUE AND LOCATION
      (USE 0.1 FOR SAMPLE RUN)
0.1
COMPUTING. DO NOT INTERRUPT!!!

ANALYSIS ENDS.
```

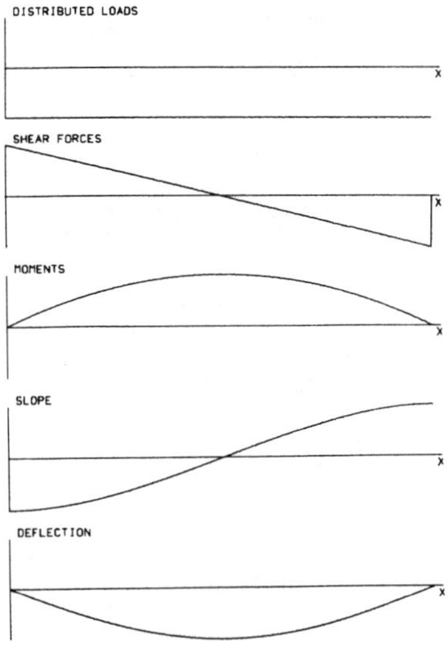

Figure 10-5b   Simply supported beam. Deflections are equal to zero at both ends of the beam, $x = 0$ and $x = 1$.

loads. The magnitude of the concentrated loads P_i for $i = 1, 2, \ldots, n$ and their locations x_i, and the applied moments M_j^a for $j = 1, 2, \ldots, m$ and their locations s_j should also be prescribed.

PROGRAM W. V. M. S. Y

```
100 PAGE
110 CHARSIZE 4
120 PRINT "* PROGRAM W.V.M.S.Y - DIAGRAMS FOR LOADED BEAMS "
130 INIT
140 CHARSIZE 2
150 N7=501
160 DIM X(N7),W(N7),V(N7),M(N7),O(N7),Y(N7)
170 DIM B(5),Z(N7)
180 W=0
190 V=0
200 M=0
210 O=0
220 Y=0
230 GOSUB 1610
240 GOSUB 1420
250 PRINT "COMPUTING, DO NOT INTERRUPT!!!"
260 E1=1
270 FOR K=2 TO N7
280 O(K)=O(K-1)+0.5*(M(K-1)+M(K))*X0/E1
290 Y(K)=Y(K-1)+0.5*(O(K-1)+O(K))*X0
300 NEXT K
310 IF B(5)=2 THEN 530
320 GO TO 680
330 FOR K=2 TO N7
```

```
340 IF B(4)>X(K) THEN 370
350 RO=B(3)-((B(4)-X(K-1))/(X(K)-X(K-1))*(O(K)-O(K-1))+O(K-1))
360 GO TO 380
370 NEXT K
380 FOR K=1 TO N7
390 O(K)=O(K)+RO
400 NEXT K
410 FOR K=2 TO N7
420 Y(K)=Y(K-1)+O(K-1)*XO
430 NEXT K
440 FOR K=2 TO N7
450 IF B(2)>X(K) THEN 480
460 CO=B(1)-((B(2)-X(K-1))/(X(K)-X(K-1))*(Y(K)-Y(K-1))+Y(K-1))
470 GO TO 490
480 NEXT K
490 FOR K=1 TO N7
500 Y(K)=Y(K)+CO
510 NEXT K
520 GO TO 680
530 FOR K=1 TO N7-1
540 IF B(4)<X(K) OR B(4)>X(K+1) THEN 570
550 C1=B(3)-Y(K)
560 T1=X(K)
570 IF B(2)<X(K) OR B(2)>X(K+1) THEN 600
580 CO=B(1)-Y(K)
590 TO=X(K)
600 NEXT K
610 RO=(C1-CO)/(T1-TO)
620 FOR K=1 TO N7
630 O(K)=O(K)+RO
640 NEXT K
650 FOR K=2 TO N7
660 Y(K)=Y(K-1)+O(K-1)*XO+CO
670 NEXT K
680 GOSUB 1100
690 HOME
700 PRINT "            "
710 PRINT "ANALYSIS ENDS."
720 END
730 REM *SUBROUTINE PLOT
740 VIEWPORT 60,125,81,96
750 Z=W
760 GOSUB 1310
770 WINDOW 0,1.02*L,-T,T
780 MOVE 0.02*L,T
790 PRINT "DISTRIBUTED LOADS"
800 MOVE 1.01*L,-0.25*T
810 PRINT "X"
820 AXIS
830 DRAW X,W
840 MOVE 0,-T
850 RETURN
860 VIEWPORT 60,125,62,77
870 Z=V
880 GOSUB 1310
890 WINDOW 0,1.02*L,-T,T
900 MOVE 0.02*L,T
910 PRINT "SHEAR FORCES"
920 MOVE L*1.01,-0.25*T
930 PRINT "X"
940 AXIS
```

```
950 DRAW X,V
960 MOVE 0,-T
970 RETURN
980 VIEWPORT 60,125,43,58
990 Z=M
1000 GOSUB 1310

1010 WINDOW 0,1.02*L,-T,T
1020 MOVE 0.02*L,T
1030 PRINT "MOMENTS"
1040 MOVE 1.01*L,-0.25*T
1050 PRINT "X"
1060 AXIS
1070 DRAW X,M
1080 MOVE 0,-T
1090 RETURN
1100 VIEWPORT 60,125,24,39
1110 Z=0
1120 GOSUB 1310
1130 WINDOW 0,1.02*L,-T,T
1140 MOVE 0.02*L,T
1150 PRINT "SLOPE"
1160 MOVE L*1.01,-0.25*T
1170 PRINT "X"
1180 AXIS
1190 DRAW X,0
1200 VIEWPORT 60,125,5,20
1210 Z=Y
1220 GOSUB 1310
1230 WINDOW 0,1.02*L,-T,T
1240 MOVE 0.02*L,T
1250 PRINT "DEFLECTION"
1260 MOVE L*1.01,-0.25*T
1270 PRINT "X"
1280 AXIS
1290 DRAW X,Y
1300 RETURN
1310 REM *SUBROUTINE MIN,MAX
1320 Z1=Z(1)
1330 Z2=Z(1)
1340 FOR K=2 TO N7
1350 Z1=Z1 MIN Z(K)
1360 Z2=Z2 MAX Z(K)
1370 NEXT K
1380 T=ABS(Z1) MAX ABS(Z2)
1390 IF T>1.0E-5 THEN 1410
1400 T=1
1410 RETURN
1420 REM *SUBROUTINE BOUNDARY CONDITIONS
1430 PRINT
1440 PRINT "BOUNDARY CONDITIONS (USE 2 FOR SAMPLE RUN) :"
1450 PRINT
1460 PRINT " ENTER 1, IF INITIAL SLOPE AND DEFLECTION ARE ZERO."
1470 PRINT "      2, IF TWO DELECTIONS ARE KNOWN"
1480 INPUT B(5)
1490 IF B(5)=1 THEN 1580
1500 PRINT
1510 PRINT " ENTER FIRST DEFLECTION VALUE AND LOCATION"
1520 PRINT "   (USE 0,0 FOR SAMPLE RUN)"
1530 INPUT B(1),B(2)
1540 PRINT " ENTER SECOND DEFLECTION VALUE AND LOCATION"
```

```
1550 PRINT "  (USE 0,1 FOR SAMPLE RUN)"
1560 INPUT B(3),B(4)
1570 RETURN
1580 O(1)=0
1590 Y(1)=0
1600 RETURN
1610 REM * SUBROUTINE INPUT
1620 PRINT "_BEAM LENGTH (USE 1 FOR SAMPLE RUN) = ";
1630 INPUT L
1640 X0=L/(N7-1)
1650 FOR K=1 TO N7
1660 X(K)=(K-1)*X0
1670 NEXT K
1680 PRINT "_NUMBER OF DISTRIBUTED (USE 0 FOR SAMPLE RUN) = ";
1690 INPUT N1
1700 IF N1=0 THEN 1910
1710 DIM X1(N1),X2(N1),A0(N1),A1(N1),A2(N1)
1720 FOR K2=1 TO N1
1730 PRINT "ENTER STARTING AND ENDING PTS. OF LOAD INTERVAL. "
1740 INPUT X1(K2),X2(K2)
1750 PRINT "ENTER CONSTANTS A0,A1,A2 WHERE W=A0+A1*X+A2*X^2  "
1760 INPUT A0(K2),A1(K2),A2(K2)
1770 NEXT K2
1780 DEF FNW(X)=A0(K2)+A1(K2)*X+A2(K2)*X^2
1790 K2=1
1800 PRINT "COMPUTING, DO NOT INTERRUPT!!!"
1810 FOR K=1 TO N7
1820 IF X(K)<X1(K2) THEN 1870
1830 IF X(K)>X2(K2) THEN 1860
1840 W(K)=FNW(X(K))
1850 GO TO 1880
1860 K2=K2+1
1870 W(K)=0
1880 NEXT K
1890 GOSUB 740
1900 PRINT
1910 PRINT "NUMBER OF CONCENTRATED LOADS (USE 3 FOR SAMPLE RUN AND "
1920 PRINT "  VALUES .5,-1,.5 AT LOCATIONS 0,.5,1) = ";
1930 INPUT N2
1940 IF N2=0 THEN 2070
1950 DIM X3(N2),V1(N2)
1960 FOR K2=1 TO N2
1970 PRINT " ENTER LOAD VALUE AND POSITION :"
1980 INPUT V1(K2),X3(K2)
1990 FOR K=2 TO N7
2000 IF X3(K2)>X(K) THEN 2050

2010 T=(X3(K2)-X(K-1))/(X(K)-X(K-1))
2020 V(K-1)=(1-T)*V1(K2)+V(K-1)
2030 V(K)=T*V1(K2)+V(K)
2040 GO TO 2060
2050 NEXT K
2060 NEXT K2
2070 PRINT "COMPUTING, DO NOT INTERRUPT!!!"
2080 FOR K=2 TO N7
2090 V(K)=V(K-1)+0.5*(W(K-1)+W(K))*X0+V(K)
2100 NEXT K
2110 GOSUB 860
2120 PRINT
2130 PRINT "NUMBER OF APPLIED MOMENTS (USE 0 FOR SAMPLE RUN) = ";
2140 INPUT N3
```

```
2150 IF N3=0 THEN 2280
2160 DIM X4(N3),M1(N3)
2170 FOR K2=1 TO N3
2180 PRINT " ENTER MOMENT VALUE AND POSITION :"
2190 INPUT M1(K2),X4(K2)
2200 FOR K=2 TO N7
2210 IF X4(K2)>X(K) THEN 2260
2220 T=(X4(K2)-X(K-1))/(X(K)-X(K-1))
2230 M(K-1)=(1-T)*M1(K2)+M(K-1)
2240 M(K)=T*M1(K2)+M(K)
2250 GO TO 2270
2260 NEXT K
2270 NEXT K2
2280 PRINT "COMPUTING, DO NOT INTERRUPT!!!"
2290 FOR K=2 TO N7
2300 M(K)=M(K-1)+0.5*(V(K-1)+V(K))*X0+M(K)
2310 NEXT K
2320 GOSUB 980
2330 RETURN
```

EXERCISE

Modify Program W. V. M. S. Y to accommodate for variable I analysis. As the moment of inertia for a rectangular cross section of width b and height h is known to be $I = bh^3/12$, apply the expanded program to study the change in y at the midlength $x = L/2$ of the beam if the weight of beam is to be redistributed. Let the total weight W be held constant and also the height h of the beam. One way to investigate the problem is to partition the length equally into 3, 5, 7, ..., n (odd) segments and increase width b linearly in steps from the support at the ends to the midlength of the beam as sketched in Fig. 10-6. If this variation tends to increase y at the midlength, reverse the width increase by making the beam wider at the ends.

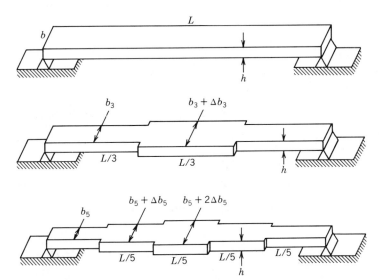

Figure 10-6 Trial designs of redistributing the beam weight W by maintaining same height h and length L but varying the width in steps.

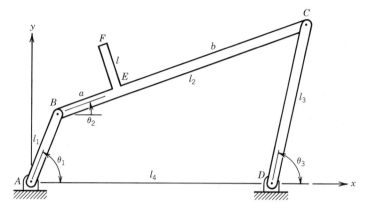

Figure 10-7

10.6 ITERATIVE PROCESS FOR FOUR-BAR LINKAGE SYNTHESIS

A commonly encountered problem in machine design is to determine the proper dimensions of a four-bar mechanism leading to a desired motion. Figure 10-7 is a sketch of a four-bar mechanism. The cranks AB and CD are to rotate about their respective axes at A and D. The distance AD is fixed but usually treated as a fixed link. The fourth bar BC is called the coupler link. If crank AB of length l_1 is serving as an input crank and rotating at a constant angular velocity ω_1, it is desirable to determine the oscillatory motions of the crank CD and link BC, such as their respective angular velocities ω_2 and ω_3, and their respective angles θ_2 and θ_3 measured from the x axis, which is often directed along the fixed link AD.

Based on the projected lengths of the linkage on the coordinate axes shown in Fig. 10-7, the following two equations can be written:

$$l_1 \cos \theta_1 + l_2 \cos \theta_2 - l_3 \cos \theta_3 = l_4 \tag{a}$$

$$l_1 \sin \theta_1 + l_2 \sin \theta_2 = l_3 \sin \theta_3 \tag{b}$$

Upon eliminating θ_2 terms in the above equations by use of the relationship $\sin^2 \theta_2 + \cos^2 \theta_2 = 1$, we obtain

$$l_2^2 = l_1^2 + l_3^2 + l_4^2 - 2l_1 l_3 \cos (\theta_1 - \theta_3) + 2l_4(l_3 \cos \theta_3 - l_1 \cos \theta_1) \tag{c}$$

We then introduce the dimensionless parameters

$$L_1 = l_4/l_3 \qquad L_2 = l_4/l_1 \tag{d,e}$$

$$L_3 = (l_1^2 - l_2^2 + l_3^2 + l_4^2)/2l_1 l_3 \tag{f}$$

Equation (c) can be simplified into the form of the well-known Freudenstein's equation[3]

---

[3] See, for example, J. E. Shigley and J. J. Uicker, Jr., *Theory of Mechanines and Mechanisms*, McGraw-Hill, New York, 1980.

AB=L1=1   BC=L2=2   CD=L3=2   AD=L4=2

BE=a=1   EF=L=0.5

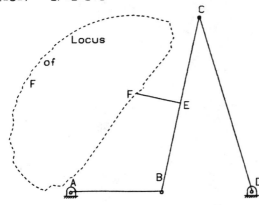

Figure 10-8   Four-bar linkage CAD program.

$$L_3 + L_2 \cos \theta_3 - L_1 \cos \theta_1 - \cos (\theta_1 - \theta_3) = 0 \qquad \text{(g)}$$

Given an angle θ_1 of the input link AB, the output angle θ_3 of the link CD can be numerically determined by an iterative method, such as **successive substitution**. By selecting an initial guess, $\theta_3^{(0)}$, the solution can be obtained with the **iterative** formula derived from Eq. (g)

$$\theta_3^{(k+1)} = \cos^{-1} \left\{ \frac{1}{L_2} [L_1 \cos \theta_1 - L_3 + \cos (\theta_1 - \theta_3^{(k)})] \right\} \qquad \text{(h)}$$

where $k = 0, 1, 2, \dots$ is the iteration counter. The iteration is to be terminated when

$$\left| \theta_3^{(k+1)} - \theta_3^{(k)} \right| < \varepsilon \qquad \text{(i)}$$

with ε being a prescribed accuracy tolerance.

Often an additional link, EF shown in Fig. 10-7, is attached to the coupler link BC. The location E, which is a and b distances measured from the pivots B and C, respectively, and the length l are to be determined so that a desired locus of the end point F can be attained.

A four-bar linkage CAD program has been developed for various experiments with the locus of F. Figure 10-8 shows a trial run of the program. When all geometrical configurations of the five-bar linkage are superimposed on a single plot, the result is as shown in Fig. 10-9. Notice that the increment for θ_1, as printed in Fig. 10-10, is taken as 5°, which results in a rather unsmooth locus for the endpoint F. In actual computation, the successive substitution scheme failed to converge when a smaller increment for θ_1 was adopted. On many occasions, a different numerical method may have to be attempted for computer solution of engineering problems.

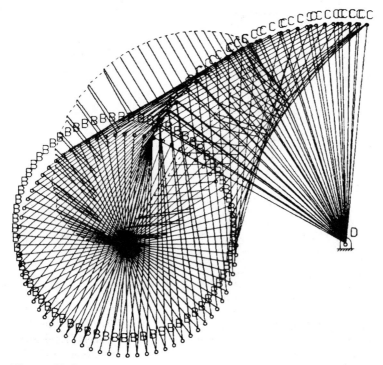

Figure 10-9

```
****************************************************
*                                                  *
*     CAD EXERCISE OF A FOUR-BAR MECHANISM          *
*                                                  *
****************************************************
INPUT THE LENGTHS L1, L2, L3, L4, L, A
FOR DEMONSTRATION, USE 1,2,2,2,0.5,1
1,2,2,2,.5,1
INPUT THE INITIAL ANGLE OF THETA1
FOR DEMONSTRATION, USE 0.
0
TYPE IN 'YES' IF YOU WANT TO SPECIFY THE INCREMENT OF THETA1
OTHERWISE TYPE IN 'NO', IT WILL BE SET EQUAL TO 5 DEGREES.
NO
INPUT THE INITIAL GUESS OF THETA3
FOR DEMONSTRATION, USE 100.
100
DO YOU WANT TO DISPLAY EVERY GEOMETRY WHEN IT IS GENERATED?
TYPE IN YES/NO
NO
```

Figure 10-10   Program menu and user's response.

PROGRAM FOUR.BAR

```
100 PAGE
110 INIT
120 CHARSIZE 4
130 PRINT "*********************************************"
140 PRINT "*                                           *"
150 PRINT "*      CAD EXERCISE OF A FOUR-BAR MECHANISM  *"
160 PRINT "*                                           *"
170 PRINT "*********************************************"
180 DIM F3(361),F4(361)
190 PRINT "INPUT THE LENGTHS L1, L2, L3, L4, L, A"
200 PRINT "FOR DEMONSTRATION, USE 1,2,2,2,0.5,1"
210 INPUT L1,L2,L3,L4,L,A
220 SET DEGREES
230 PRINT "INPUT THE INITIAL ANGLE OF THETA1"
240 PRINT "FOR DEMONSTRATION, USE 0."
250 INPUT I1
260 PRINT "TYPE IN 'YES' IF YOU WANT TO SPECIFY THE INCREMENT OF THETA1"
270 PRINT "OTHERWISE TYPE IN 'NO', IT WILL BE SET EQUAL TO 5 DEGREES."
280 INPUT A$
290 IF A$="NO" THEN 330
300 PRINT "INPUT INCREMENT OF THETA1"
310 INPUT I2
320 GO TO 340
330 I2=5
340 PRINT "INPUT THE INITIAL GUESS OF THETA3"
350 PRINT "FOR DEMONSTRATION, USE 100."
360 INPUT T3
370 PRINT "DO YOU WANT TO DISPLAY EVERY GEOMETRY WHEN IT IS GENERATED?"
380 PRINT "TYPE IN YES/NO"
390 INPUT D$
400 PAGE
410 GOSUB 1570
420 GOSUB 1300
430 RINIT
440 M=0
450 FOR S1=I1 TO I1+360 STEP I2
460 M=M+1
470 IF D$="YES" THEN 490
480 ROPEN 1
490 GOSUB 680
500 GOSUB 1180
510 GOSUB 790
520 IF D$="YES" THEN 540
530 RCLOSE
540 NEXT S1
550 VISIBILITY 1,1
560 N=0
570 DASH 85
580 MOVE F3(1),F4(1)
590 FOR I=1 TO M
600 N=N+1
610 DRAW F3(N),F4(N)
620 NEXT I
630 DRAW F3(1),F4(1)
640 FIX 1
650 VISIBILITY 1,0
660 HOME
670 END
680 REM *THIS SUBROUTINE DOES THE ITERATION FOR FINDING THETA3*
690 T1=0.5*(COS(S1)-1.25+COS(S1-T3))
```

```
700 IF T1>1 OR T1<-1 THEN 1140
710 N3=ACS(T1)
720 IF N3>10 AND T3<>0 THEN 750
730 IF ABS(N3-T3)<0.005 THEN 780
740 GO TO 760
750 IF ABS((N3-T3)/T3)<0.02 THEN 780
760 T3=N3
770 GO TO 690
780 RETURN
790 REM *THIS SUBROUTINE DOES THE PLOTTING*
800 GOSUB 1180
810 V2=L4-L1*COS(S1)+L3*COS(T3)
820 V2=V2/L2
830 IF ABS(V2)>1 THEN 1120
840 T2=ACS(V2)
850 IF L1*COS(S1)+L2*COS(T2)-L3*COS(T3)<>L4 THEN 1120
860 MOVE 0,0
870 B1=L1*COS(S1)
880 B2=L1*SIN(S1)
890 DRAW B1,B2
900 C1=L3*COS(T3)+L4
910 C2=L3*SIN(T3)
920 E1=B1+A/L2*(C1-B1)
930 E2=B2+A/L2*(C2-B2)
940 F1=E1+L*COS(T2+90)
950 F2=E2+L*SIN(T2+90)
960 MOVE E1,E2
970 DRAW F1,F2
980 MOVE 0,0.035*L1
990 PRINT "A"
1000 MOVE L4,0

1010 DRAW C1,C2
1020 DRAW B1,B2
1030 MOVE L1*(COS(S1)-0.05),L1*(SIN(S1)+0.09)
1040 PRINT "B"
1050 MOVE L4*1.025,0.035*L1
1060 PRINT "D"
1070 MOVE L3*COS(T3)+L4,L3*(SIN(T3)+0.01)
1080 PRINT "C"
1090 F3(M)=F1
1100 F4(M)=F2
1110 RETURN
1120 PAGE
1130 MOVE 0,B7
1140 PRINT "ITERATION FOR THETA3 FAILED"
1150 MOVE 0,0.8*B7
1160 PRINT "WHEN THETA1 =";S1
1170 END
1180 REM *DRAW THE PINS*
1190 MOVE L4+L3*COS(T3)-0.008*B7,L3*SIN(T3)-0.02*B7
1200 CHARSIZE 1
1210 PRINT "o"
1220 MOVE L1*COS(S1)-0.006*B7,L1*SIN(S1)-B7*0.02
1230 PRINT "o"
1240 MOVE -0.008*B7,-B7*0.02
1250 PRINT "o"
1260 MOVE L4-0.008*B7,-B7*0.02
1270 PRINT "o"
1280 CHARSIZE 4
1290 RETURN
```

```
1300 REM *DRAW THE SUPPORTS AT A AND D*
1310 MOVE L4*0.025,-L4*0.025
1320 DRAW L4*0.025,0
1330 FOR C1=1 TO 180 STEP 10
1340 DRAW L4*0.025*COS(C1),L4*0.025*SIN(C1)
1350 NEXT C1
1360 DRAW -L4*0.025,-L4*0.025
1370 MOVE L4*1.025,-0.025*L4
1380 DRAW L4*1.025,0
1390 FOR C4=1 TO 180 STEP 10
1400 DRAW L4*0.025*COS(C4)+L4,L4*SIN(C4)*0.025
1410 NEXT C4
1420 DRAW L4*.975,-L4*0.025
1430 MOVE -L4*0.04,-0.025*L4
1440 DRAW L4*0.04,-0.025*L4
1450 MOVE L4-L4*0.04,-0.025*L4
1460 DRAW L4*1.04,-0.025*L4
1470 FOR C7=-1 TO 29 STEP 6
1480 MOVE -L4*0.035+L4*C7/500,-0.035*L4
1490 DRAW -L4*0.025+C7/500*L4,-0.025*L4
1500 NEXT C7
1510 FOR C8=-1 TO 29 STEP 6
1520 MOVE L4*0.965+L4*C8/500,-0.035*L4
1530 DRAW L4*0.975+L4*C8/500,-0.025*L4
1540 NEXT C8
1550 CHARSIZE 4
1560 RETURN
1570 REM *THIES SUBROUTINE DOES THE SCALING*
1580 IF L1>L2 THEN 1610
1590 B7=L2
1600 GO TO 1620
1610 B7=L1
1620 IF B7>L3 THEN 1640
1630 L3=B7
1640 IF B7>L4 THEN 1670
1650 B7=L4
1660 WINDOW 0.8*-B7,2*B7,-B7/1.3,1.8*B7/1.3
1670 RETURN
1680 PAGE
1690 HOME
1700 END
```

10.7 SUCCESSIVE LINEAR INTERPOLATION

Let us recall the definitions of gain and phase margins depicted in Fig. 8-11. In order to calculate G_m and ϕ_m, the crossover frequencies ω_π and ω_1 must first be determined. When the Bode plots of magnitude ratio and phase angle versus the frequency ω are constructed, the points on the curves are usually generated by use of a constant increment in ω. Consequently, ω_π and ω_1 and their corresponding magnitude ratios and phase angles are often not among those selected ω values. For example, consider the case of Fig. 8-13, for which the transfer function of the system is

$$G(j\omega) = 10(1 + j\omega)/(j\omega)^2[1 + j\omega/4 - (\omega/4)^2] \tag{a}$$

And, the expression for the phase angle, according to Eq. (g) in Section 8.6, is

$$\phi(\omega) = \tan^{-1}\omega - 180° - \tan^{-1}[4\omega/(4^2 - \omega^2)] \tag{b}$$

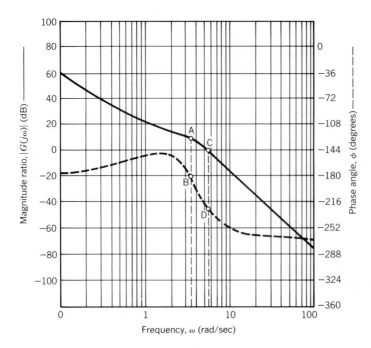

Figure 10-11

Figure 10-11 is a modified version of Fig. 8-13 with additional markings. It shows that ω_π lies between $\omega = 3$ and $\omega = 4$. We shall call these two values ω_l and ω_u to indicate that they are the lower and upper bounds of the solution ω_s, which yields $\phi_s \equiv \phi(\omega_s) = -180°$. And we shall designate ϕ_l and ϕ_u to be the corresponding ϕ values at ω_l and ω_u, recpectively.

The method of successive linear interpolation is graphically explained in Fig. 10-12. It calls for continuous guessing of the location of ω_s by finding the inter-

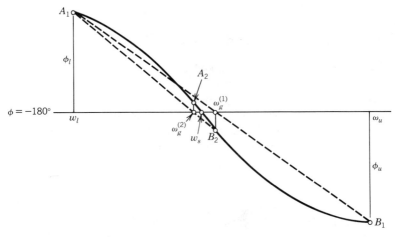

Figure 10-12

$$-180° + \epsilon > \phi(\omega_g^{(i)}) > -180° - \epsilon$$

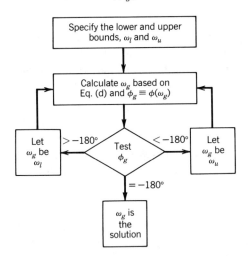

Figure 10-13

cept of the $\phi = -180°$ line and the line connecting the points A_1 and B_1. These guessed locations of ω_s are denoted as $\omega_g^{(i)}$ with the superscript i indicating the number of guessings performed. Since $A_1 B_1$ is a straight line, the relationship between two similar triangles can thus be utilized to arrive at a formula for ω_g. That is,

$$\omega_g = [\omega_l \phi_u - \omega_u \phi_l - 180(\omega_u - \omega_l)]/(\phi_u - \phi_l) \qquad (c)$$

For the transfer function given in Eq. (a), we have $\omega_l = 3$ and $\omega_u = 4$. It is easy to obtain from Eq. (b) that $\phi_l = -168°$ and $\phi_u = -194°$. The first guessed ω_s value can then be calculated by use of Eq. (c). It is $\omega_g^{(1)} = 3.85$. Let us designate the corresponding value of the phase angle as $\phi_g^{(1)} \equiv \phi(\omega_g^{(1)})$, which can be shown equal to $-190°$. Graphically, this is represented by the point B_2 on the $\phi(\omega)$ curve.

Since B_2 and B_1 are both below the $\phi = -180°$ line, $\omega_g^{(1)} = 3.85$ can be used to replace ω_u as the new upper bound for searching of the solution ω_s. The linear interpolation procedure is to be continued by connecting the points A_1 and B_2 and finding the new intercept $\omega_g^{(2)}$. Figure 10-13 is a flow chart summarizing the successive linear interpolation process. The termination occurs when $\phi(\omega_g^{(i)})$ is approximately equal to $-180°$. If a tolerance ε is prescribed, we require

$$-180° + \varepsilon > \phi(\omega_g^{(i)}) > -180° - \varepsilon \qquad (d)$$

As far as the solution of ω_π for Fig. 10-11 is concerned, it can be shown that $\omega_g^{(2)} = 3.46$ and $\phi_g^{(2)} = -179.89°$. Hence, the solution is obtained after two linear interpolations that $\omega_s = \omega_g^{(2)} = 3.46$ radians/sec and $\phi(\omega_g^{(2)})$ is off from the required $-180°$ value only by 0.01°. For this simple case, analytical solution is also possible. From Eq. (b), we see that $\phi(\omega_\pi) = -180°$ requires

$$\tan^{-1} \omega = \tan^{-1} 4\omega/(4^2 - \omega^2)$$

$$\omega = 4\omega/(16 - \omega^2)$$

$$\omega^2 = 12$$

$$\omega = 2(3)^{1/2} = 3.464$$

However, the analytical solution for a general case of $\phi(\omega)$ expressed as Eq. (g) in Section 8.6 is not as easy to obtain. The successive linear interpolation method is usually preferred for approximate solution of the problem, particularly when the bounds of the solution are readily available from a graph such as the Bode plots.

EXERCISE

For the general problem of finding the solution x_s that makes a given function $f(x)$ approximately equal to a specified value f_s, show that Eq. (c) should be modified so that the guessed x_g value can be calculated by

$$x_g = [f_s(x_u - x_l) + x_l f_u - x_u f_l]/(f_u - f_l) \tag{e}$$

where x_l and x_u are the lower and upper bounds of the solution x_s, and f_l and f_u are the corresponding values of $f(x)$ at the bounds. Modify the flowchart in Fig. 10-13 to accommodate for the general problem of solving $f(x_s) = f_s$ by use of Eq. (e). Also write a general BASIC Program LINTPOL containing a statement DEF, which defines the given function $f(x)$, and apply the program to find the gain crossover frequency for the Bode plot shown in Fig. 8-20.

10.8 CALCULATION OF AREAS AND VOLUMES

In Chapter 6 the numerical procedure for calculation of area, centroid, and moments of inertia was presented and a Program A. C. I. developed for the reader to use interactively. Here, the procedure is to be extended further to cover the general need of calculating the lengths, areas, and volumes involved in engineering design and manufacturing. As shown in Fig. 10-14, many industrial parts are rotationally

Figure 10-14 Assortment of rotationally molded plastic vessels. (Courtesy of Snyder Industries, Inc., Lincoln, Nebraska.)

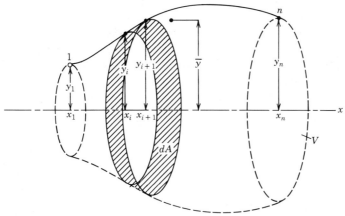

Figure 10-15   Bodies of rotational symmetry.

symmetric. For calculation of surface areas and volumes of the bodies of revolution, the theorems of Pappus–Buldinus can be employed. The theorems can be described by the following two formulas:

$$A = 2\pi \bar{y} L \qquad V = 2\pi \bar{y} A \qquad\qquad\qquad \text{(a,b)}$$

Referring to Fig. 10-15, the surface area A and volume V are generated by revolving the curve $1-n$ about the x axis. The curve $1-n$, called the **generator** of the body of revolution, should be coplanar, which means that all points should lie on a same plane; $\bar{y}$ in Eqs. (a) and (b) is the centroid and L is the length of the curve $1-n$. The first task of finding $\bar{y}$ is to integrate dL over the interval $x_1 \le x \le x_n$, since

$$\bar{y} = \left(\int_{x_1}^{x_n} y \, dL \right) \bigg/ L \qquad\qquad\qquad \text{(c)}$$

$$L = \int_{x_1}^{x_n} dL \qquad\qquad\qquad \text{(d)}$$

Again, the concept of finite-difference can be applied—that is, to convert Eqs. (c) and (d) into approximated formulas

$$\bar{y} = \left\{ \sum_{i=1}^{n-1} \frac{1}{2} (y_i + y_{i+1})[(x_{i+1} - x_i)^2 + (y_{i+1} - y_i)^2]^{1/2} \right\} \bigg/ L \qquad\qquad \text{(e)}$$

where

$$L = \sum_{i=1}^{n-1} [(x_{i+1} - x_i)^2 + (y_{i+1} - y_i)^2]^{1/2} \qquad\qquad\qquad \text{(f)}$$

In deriving Eqs. (e) and (f), the interval $x_1 \le x \le x_n$ has been partitioned into n stations by defining the in-between stations x_i for $i = 2, 3, \ldots, n - 1$ and evaluating their respective curve values, y_i. The arc length dL is then approximated by the cord length $\Delta L = [(x_{i+1} - x_i)^2 + (y_{i+1} - y_i)^2]^{1/2}$ calculated on the basis of the Pythagorean theorem.

It is common practice to partition the x interval in equal increments to expedite the computation. That is, the in-between stations are so chosen that $\Delta x = x_{i+1} - x_i = x_i - x_{i-1}$ for $i = 2, 3, \ldots, n$. And it is wise to organize the **numerical integration** in a more systematic way. For handling Eq. (d), we first define the **integrand function**, $f(x)$, by the following process:

$$L = \int_{x_1}^{x_2} dL = \int_{x_1}^{x_2} [(dx)^2 + (dy)^2]^{1/2}$$

$$= \int_{x_1}^{x_2} [1 + (dy/dx)^2]^{1/2}\, dx = \int_{x_1}^{x_2} f(x)\, dx \qquad \text{(g)}$$

That is

$$f(x) = [1 + (dy/dx)^2]^{1/2} \qquad \text{(h)}$$

Then the approximated evaluation of L is to be carried out with the formula, based on Eq. (g) and use of equal increment Δx,

$$L = \sum_{i=1}^{n-1} \frac{1}{2}(f_i + f_{i+1})\Delta x = \frac{\Delta x}{2}\left(f_1 + 2\sum_{i=2}^{n-1} f_i + f_n \right) \qquad \text{(i)}$$

The graphical interpretation of Eq. (i) is shown in Fig. 10-16. Because L corresponds to the area under the $f(x)$ curve and is to be approximately determined as the sum of the trapezoidal strips, this method is called the **trapezoidal rule**. As it is evident that the accuracy of the result L depends on the number, $n - 1$ in Eq. (i), of **linear** segments adopted to approximate the generator $1-n$, automatically we think of the question, "What if parabolic segments are used; will it improve the result?" That leads to **Simpson's rule**, which deals with instances when every three points, say (x_1, f_1), (x_2, f_2), and (x_3, f_3), are fitted by a second-degree, parabolic equation. The formula becomes

$$L = \frac{\Delta x}{3}\left[(f_1 + 4f_2 + f_3) + (f_3 + 4f_4 + f_5) + \cdots + (f_{n-2} + 4f_{n-1} + f_n) \right]$$

$$= \frac{\Delta x}{3}\left[f_1 + 4\sum_{\substack{i=2 \\ (\text{even})}}^{n-1} f_i + 2\sum_{\substack{i=3 \\ (\text{odd})}}^{n-2} f_i + f_n \right] \qquad \text{(j)}$$

Notice that for application of the Simpson's rule, n must be odd.

Evaluation of the volume is an extension of the above formulation of one-dimensional integration over an x interval to that of two-dimensional integration over a specified area. The problem can be described as that of knowing the integrand function $z(x, y)$, which represents a three-dimensional surface over a projected area bounded by $x_1 \le x \le x_m$ and $y_1 \le y \le y_n$, and requiring the volume be calculated by use of the integral

$$V = \int_{y_1}^{y_n} \int_{x_1}^{x_m} z(x, y)\, dx\, dy \qquad \text{(k)}$$

Again, the trapezoidal rule can be extended to this problem by imagining the volume to be made up of columns with a base area $\Delta x\, \Delta y$ and variable heights

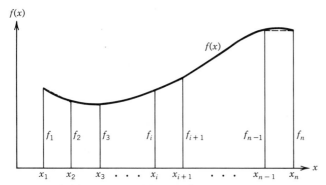

Figure 10-16 Trapezoidal-rule evaluation of area under the curve $f(x)$ for x in the range of $x_1 \leq x \leq x_n$.

at its four edges. It suggests that equal increments Δx and Δy should be used to partition the intervals $x_1 \leq x \leq x_m$ and $y_1 \leq y \leq y_n$, respectively. For a typical base area with four vertices (x_i, y_j), (x_{i+1}, y_j), (x_{i+1}, y_{j+1}), and (x_i, y_{j+1}), the contribution of the column volume to the entire volume is to be approximated as

$$\Delta V = \tfrac{1}{4}(z_{i,j} + z_{i+1,j} + z_{i+1,j+1} + z_{i,j+1})\Delta x \, \Delta y \tag{l}$$

where $z_{i,j} \equiv z(x_i, y_j)$, $z_{i+1,j} \equiv z(x_{i+1}, y_j)$, and so on. By taking into account all of the column volumes, the trapezoidal rule for two-dimensional integration can be shown to be

$$V = \tfrac{1}{4}\Delta x \, \Delta y [M_x][M_z][M_y] \tag{m}$$

where

$$[M_x]_{1 \times m} = \begin{bmatrix} 1 & 2 & 2 & \cdots & 2 & 2 & 1 \end{bmatrix} \tag{n}$$

$$[M_z]_{m \times n} = \begin{bmatrix} z_{1,1} & z_{1,2} & \cdots & z_{1,n} \\ z_{2,1} & z_{2,2} & \cdots & z_{2,n} \\ \vdots & \vdots & \vdots & \vdots \\ z_{m,1} & z_{m,2} & \cdots & z_{m,n} \end{bmatrix} \tag{o}$$

$$[M_y]_{n \times 1} = \begin{bmatrix} 1 & 2 & 2 & \cdots & 2 & 2 & 1 \end{bmatrix}^T \tag{p}$$

The matrices $[M_x]$ and $[M_y]$ contain, as elements, the coefficients 1 and 2 required for the trapezoidal rule, whereas the matrix $[M_z]$ contains the values of z computed at all partitioning points (x_i, y_j) over the entire projected area bounded by $x_1 \leq x \leq x_m$ and $y_1 \leq y \leq y_n$.

10.9 CONCLUDING REMARKS

In this chapter we have presented a number of elementary numerical methods for digital computer solution of engineering problems. The example problems discussed and the methods of solution used all point out the fact that in order to

```
100 REM * PROGRAM GENERATOR *
110 INIT
120 PAGE
130 MOVE 65,50
140 SCALE 0.2,0.2
150 N=126
160 X0=12.5/(N-1)
170 DIM X(N),Y(N),F(N)
180 FOR I=1 TO N
190 X(I)=X0*(I-1)
200 IF X(I)>1 THEN 230
210 Y(I)=SQR(1-(1-X(I))^2)
220 GO TO 270
230 IF X(I)>11 THEN 260
240 Y(I)=1+0.5/10*(X(I)-1)
250 GO TO 270
260 Y(I)=SQR(1.5^2-(X(I)-11)^2)
270 NEXT I
280 FOR I=1 TO N-1
290 F(I)=SQR(1+((Y(I+1)-Y(I))/X0)^2)
300 NEXT I
310 F(N)=0
320 MOVE 0,2
330 DRAW 0,0
340 DRAW 15,0
350 DRAW X,Y
360 S=(F(1)+F(N))/2
370 FOR I=1 TO N-2
380 S=S+F(I+1)
390 NEXT I
400 S=S*X0
410 MOVE 0,0
420 PRINT "    LENGTH OF THE GENERATOR = ";S
430 END
```

LENGTH OF THE GENERATOR = 13.7074081052

Figure 10-17   Program GENERATOR.

improve the accuracy of the result, more advanced methods will have to be learned. The reader, if interested in furthering this field of study, ought to have a more in-depth course on numerical methods in engineering analysis.

Perhaps more CAD programs will be helpful for this book. Programs GENER-ATOR and VOLUME for calculation of the length of a specified generator and for calculation of a volume defined by a three-dimensional surface $z = f(x, y)$ and bounded by specified boundaries are presented in Figs. 10-17 and 10-18, respectively. The shape of the generator used in GENERATOR consists of two circular arcs of radii equal to 1 and 1.5, and a tangent line to these two circles. In VOLUME, an eighth of a sphere of radius equal to 5 is computed.

```
100 REM * PROGRAM VOLUME *
110 INIT
120 DEF FNZ(T)=SQR(5^2-X^2-Y^2)
130 X0=0.1
140 Y0=0.1
150 M=51
160 N=51
170 DIM M1(1,M),M3(M,N),M2(N,1),V(1,1),DO(1,N)
180 FOR I=1 TO M
190 X=(I-1)*X0
200 FOR J=1 TO N
210 Y=(J-1)*Y0
220 IF 5^2-X^2-Y^2<0 THEN 250
230 M3(I,J)=FNZ(0)
240 GO TO 260
250 M3(I,J)=0
260 NEXT J
270 NEXT I
280 M1=2
290 M2=2
300 M1(1,1)=1
310 M1(1,M)=1
320 M2(1,1)=1
330 M2(N,1)=1
340 DO=M1 MPY M3
350 V=DO MPY M2
360 V=V*(X0*Y0/4)
370 PRINT "COMPUTED VOLUME IS ";V
380 END
RUN

COMPUTED VOLUME IS
 65.4327187241
```

Figure 10-18   Program VOLUME.

In Program VOLUME, the command DEF is used. In fact, earlier in Program W. V. M. S. Y, DEF has been adopted in line 1780. DEF defines a function and has the general form

DEF FN?(X) = an expression involves X, other constants and variables

where ? should be one of the 26 alphabetic characters A through Z. Hence, a program can have a maximum of 26 functions. Line 1840 in Program W. V. M. S. Y shows the application of function FNW defined by line 1780. Since only one variable is allowed in DEF, lines 120 and 230 in Program VOLUME show how multivariables should be handled.

EXERCISE

Modify Programs GENERATOR and VOLUME so that the total wall volumes of two thin-walled shells, one with a generator shown in Fig. 10-17 and the other spherical, can be calculated. Let the thickness be 0.1 and all units be in centimeters; print out the two wall volumes.

CHAPTER 11

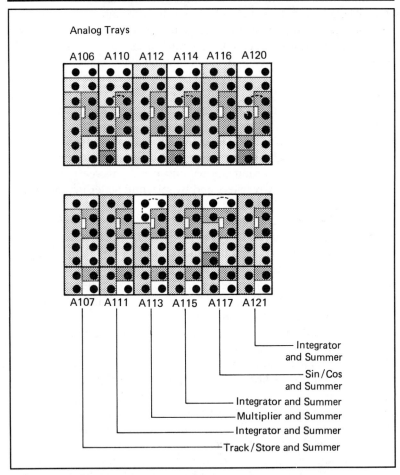

(Courtesy of Electronic Associates, Inc.)

ANALOG AND HYBRID COMPUTER APPLICATION

11.1   INTRODUCTION

Hybrid computation combines the computing techniques of digital and analog computers. The advantage of such hybridization is to join forces of the best features of both digital and analog computers. In digital computers, all variables are handled in discrete form throughout the computation. The accuracy of the data depends on the number of digits that can be manipulated in the computer memory. Digital computers used to carry out serial operations can now perform parallel operations because of the incorporation of array processors. Digital computers also have the attributes of memorizing numerical and nonnumerical data and utilizing them for logical control and decision making. Digital computers, however, require approximate lengthy arithmetic operations for commonly encountered mathematic operations such as integration and differentiation. Even though the accuracy of the computed solution can be improved by increasing the number of bits in the computer memory registers and by selecting better numerical methods, the solution times are relatively long.

Analog computer systems handle all variables in continuous form. The accuracy of numerical data is thus limited only by the quality of the circuit components adopted in the system. It allows parallel operations by running the computational elements simultaneously and in **real time**. As was pointed out in Chapter 7, there exist analogies among physical systems; analogous circuit elements can thus be substituted for corresponding elements of the physical system under study so long as the transfer functions are the same. The major disadvantages of analog computers include the limited ability to make logical decisions, handling of nonnumerical data, and storage of numerical data.

It is best to blend analog and digital computers together and to take advantage of their attributes. Of course, how they should be combined depends on the specific application. In this chapter the basic elements of analog computer systems are discussed and the nonlinear operations of multiplication and division of two variables as well as generation of nonlinear functions by utilization of analog computers are particularly emphasized.

Digital-to-analog and analog-to-digital converters are also discussed. These converters enable digital computers to be interfaced with analog monitoring equipment in engineering laboratories and manufacturing plants. The development of such **computer work stations** is rapidly expanding in academia and industry, particularly in conjunction with the CAD/CAM activities.

11.2   SEMICONDUCTOR DIODE

There are many materials whose ability to conduct current lies between a conductor and an insulator. By breaking loose electrons from the outermost orbit of the atoms of such materials, n (cathode) and p (anode) regions can be produced. Electrons flow only from the n-type material to the p-type material. When an n region is placed right next to p region, a so-called pn junction or **semiconductor diode** is formed. Figure 11-1 shows the circuit symbol of a diode and the stacking order of the cathode and anode regions.

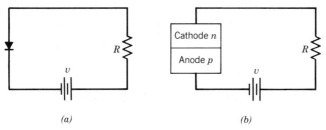

(a)   *(b)*

Figure 11-1   (*a*) Circuit symbol of a diode and (*b*) *pn* junction.

The current leakage i_s of a diode can be related to the current i in a circuit by the equation[1]

$$i_s = \left[\exp\left(\frac{qv}{kT}\right) - 1 \right] \bigg/ i$$

where q is the electron charge equal to 1.6×10^{-19} C, T is the absolute temperature in Kelvin, and k is Boltzmann's constant equal to 8.617×10^{-15} V/°K. A common application of a diode is to convert alternating voltage into an output voltage, which is almost a d.c. voltage. As illustrated in Fig. 11-2, the diode in a rectifier circuit allows only the positive half cycles of the input voltage to pass through. During the negative half cycles, except for leakage, the output voltage is almost zero across the resistor. How to change the pulsating d.c. voltage into smooth and almost constant batterylike voltage is discussed later, when filtering techniques are introduced.

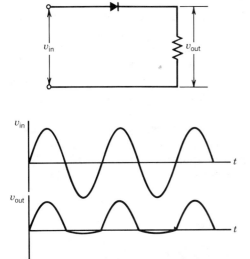

Figure 11-2   Use of a diode in a rectifier.

---

[1] J. K. Fidler and C. Nightingale, *Computer Aided Circuit Design*, Thomas Nelson & Sons Ltd., Hong Kong, 1978.

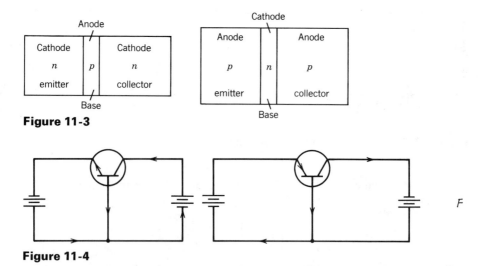

Figure 11-3

Figure 11-4

11.3 BIPOLAR JUNCTION TRANSISTOR

The n and p regions can be combined to form npn and pnp junctions with the middle region being made very narrow. Such structures are the basic forms of the **transistor**. Since they are bipolar, the three sections are called the emitter, base, and collector, as shown in Fig. 11-3. Their respective circuit symbols are illustrated in Fig. 11-4.

The transistor is the basic amplifying device in most electronic equipment. Input signals received at the micro- or millivólt level are often amplified by factors of several thousands in operation of radios, televisions, printers, and scientific instruments.

Integrated circuits are formed by use of a single semiconductor material. Through a diffusion process, transistors, resistors, and capacitors are integrated to form entire circuits.

In a recent computer conference, Donze and co-workers reported the automated design of integrated circuits on a tiny chip. The design specifications, as they described them, can be typed into a computer terminal (Fig. 11-5) to place the required transistors and other circuit elements on the chip and to route the wires that interconnect them.

11.4 AMPLIFIERS

When an npn transistor is used with the emitter and collector arranged as shown in Fig. 11-6a, the variable resistance R_p indicated by the **wiper** W may represent a situation in which a microphone is connected to the circuit. The input to the

Figure 11-5 Computer terminal. (Courtesy of IBM Corporation, Essex Junction, Vermont. R. From R. Donze, J. Sanders, M. Jenkins, and G. Sporzynski, "PHILO—A VLSI Design System," *Proceedings of the 19th Design Automation Conference*, IEEE Publication, 1982, pp. 163–169.)

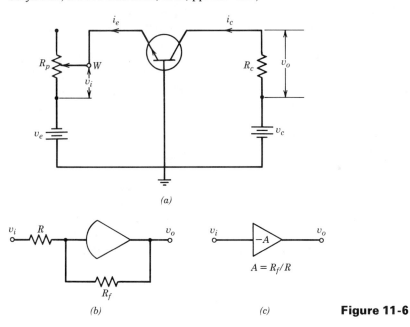

Figure 11-6

microphone is in the form of acoustic waves that vary in time and cause the resistance R_p to change in time. Assuming that the emitter-base junction has an h_{FB} of 0.99, that is, $i_c = h_{FB}i_e = 0.99i_e$, and the collector resistance R_c is equal to 2000 ohms, it is of practical interest to investigate the change in the voltage v_o across the resistor R_c as a consequence of the change in R_p or v_i. Let us first consider the case of $v_e = -0.5$ V and $i_e = 5$ mA, for which we easily obtain $i_c = 4.95$ mA and $v_o = (4.95 \times 10^{-3}) \times (2 \times 10^3) = 9.9$ V. Now, suppose that the microphone input causes the emitter voltage to change from $v_e = -0.5$ V to $v_e = -0.6$ V and the current from $i_e = 5$ mA to $i_e = 10$ mA. These changes cause the collector current to become $i_c = 0.99 \times 10 = 9.9$ mA and the voltage across the resistor R_c to become $v_o = (9.9 \times 10^{-3}) \times (2 \times 10^3) = 19.8$ V. It thus shows that a change of 0.1 V in the emitter-base voltage will cause a change of 9.9 V in v_o. In other words, an amplification of $9.9/0.1 = 99$ times can be achieved by utilizing an npn transistor as illustrated. The design of amplifiers is itself a field of intensive study. There are numerous books[2] devoted exclusively to this topic, and so we shall deal in this chapter only with those subjects pertaining to the CAD applications of analog and hybrid computers.

Figure 11-6b and c shows the commonly used symbols for an amplifier with a gain equal to A. When A is equal to 1, the amplifier is simply called a **sign changer** and the "-1" is omitted inside the triangle in Fig. 11-6c. The amplifier gain A as shown is directly proportional to the feedback resistance R_f and inversely proportional to the input resistance R.

11.5 POTENTIOMETER

In Fig. 11-6a the symbolic form of a wiper indicated by the letter W at the terminal has been used to represent the sliding contact of the arrow with the resistor. It depicts the basic operation of an electric **potentiometer**. The arrow of a potentiometer can be arranged to slide from one end to the other of the length of the resistor. Figure 11-7a gives further detail of the potentiometer operation. By adjusting the location of the arrow, the ratio of $R_2/(R_1 + R_2)$ can be changed. And as a consequence, the output voltage v_o can be scaled down from the level of the input voltage v_i by a factor of $k = R_2/(R_1 + R_2)$. That is, $v_o = kv_i = R_2v_i/(R_1 + R_2)$. The symbolic notation of a potentiometer is shown in Fig. 11-7b.

As will be demonstrated in the ensuing sections, potentiometers are to be used for obtaining decimal numbers that appear as coefficients in most differential and algebraic equations arising in solving engineering problems. The potentiometers available on the market can provide coefficients accurate to at least three decimal places. As most analog computers are equipped with amplifiers having gains of 1 and 10, these amplifiers combined with potentiometers make it possible to generate any real coefficients within the voltage range of analog computers, which are usually limited to ± 100 or ± 10 V. For example, a coefficient 9.054 can be

---

[2] See, for example, R. G. Irvine, *Operational Amplifier Characteristics and Applications*, Prentice-Hall, Englewood Cliffs, New Jersey, 1981.

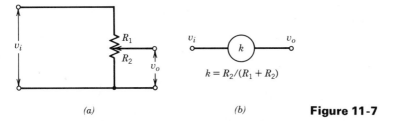

(a)   (b)   **Figure 11-7**

generated from a 1-V source input by use of a potentiometer adjusted to a scaling factor of $k = 0.9054$ and an amplifier of gain 10. Arrangement by connecting patch cords to the input jacks for obtaining such a desired coefficient on an analog computer will be demonstrated later, in the next section on summing integrators.

11.6   OPERATIONAL AMPLIFIERS

In complex electronic systems, various combinations of amplifiers are often used. Operational amplifiers, abbreviated as op amps, are high-gain amplifiers of several stages, most of which are made available on the market in the form of integrated circuits (IC). Simple schematic symbols have been introduced for representation of operational amplifiers. Figure 11-8a is the symbol commonly used for the **differential amplifier**, which has two inputs, v_1 and v_2. The gain for the differential

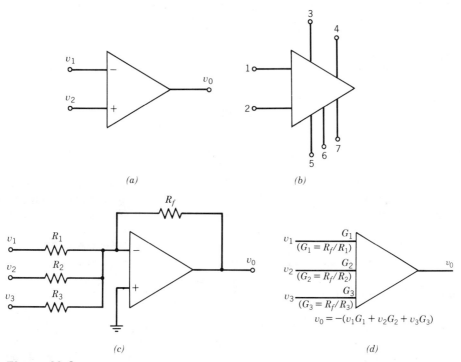

Figure 11-8

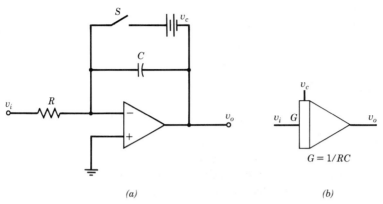

(a)   *(b)*

Figure 11-9

amplifier in this case is to be expressed as $A_d = v_o/(v_2 - v_1)$. Notice that the
subscript d is added to indicate a differential amplifier and that v_2 and v_1 terms
carry signs plus (+) and minus (−), respectively, as specified inside the triangle.
Figure 11-8b illustrates the schematic symbol for an IC differential amplifier, which
has two inputs 1 and 2, two power supplies 3 and 5, two outputs 4 and 7, and a
ground connection 6.

Figure 11-8c and d shows the circuit and schematic symbol for a summing
amplifier. Each of the inputs can be multiplied by a different gain and then added
to result in a desired output.

Op Amp Integrator

An RC circuit can be used together with a differential amplifier to form an integrator. A constant voltage source, v_c, can be added at $t = 0$, as shown in Fig.
11-9a, to account for the integration constant. The output voltage v_o is related
to the input voltage v_i by the equation

$$v_o = -\frac{1}{RC} \int v_i \, dt + v_c$$

Figure 11-9b shows the schematic symbol for an op amp integrator.

Usually the op amp integrators on an analog computer can have several inputs,
each multiplied by different gains. Figure 11-10a is a symbolic representation of
two inputs v_1 and v_2, which are to be multiplied by use of gains 1 and 10,
respectively. V_{IC} is the voltage input representing the initial condition. In actual
practice, a patching board is used for arranging the cords to be properly connected
to the ports on the board. Figure 11-10b indicates how the cords should be connected for obtaining the desired integration of Fig. 11-10a. Notice that the cord
carrying the input voltage v_1 should be plugged into a port marked with a gain
equal to 1, whereas the other input voltage v_2 should be plugged into a 10-gain
port. After all necessary wirings are finished, the patching board is then mounted
onto the analog computer. Figure 11-10c illustrates the implementation of an

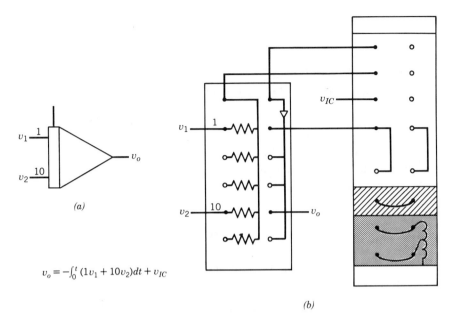

(a)

$$v_o = -\int_0^t (1v_1 + 10v_2)dt + v_{IC}$$

(b)

(c)

Figure 11-10   Patching cords on an op amp integrator. (Courtesy of Electronic Associates, Inc.)

analog simulation. In the following sections the procedure for analog simulation of engineering systems will be delineated.

11.7 ANALOG SIMULATION OF ENGINEERING SYSTEMS

In order to delineate how an engineering system can be simulated by utilizing the various components available in an analog computer that have been discussed in the preceding sections, let us recall the differential equation governing the translational mechanical system shown in Fig. 7-13b. As has already been derived as Eq. (e) in Section 7.6, the differential equation for the displacement $x(t)$ of the vibrating mass is

$$m\frac{d^2x}{dt^2} + c\frac{dx}{dt} + kx = F(t) \tag{a}$$

For use of Eq. (a), the initial conditions of the system are usually prescribed. Let us denote them as

$$x(t = 0) = x_0 \tag{b}$$

$$\left.\frac{dx}{dt}\right|_{t=0} = \dot{x}_0 \tag{c}$$

Hereafter, we shall use $\dot{x}$ and $\ddot{x}$ to denote dx/dt and d^2x/dt^2, respectively.

Analog simulation of Eq. (a) starts with rewriting it, by isolating the highest derivative term, as

$$\ddot{x} = [F(t) - c\dot{x} - kx]/m \tag{d}$$

Based on Eq. (d) and assuming that d^2x/dt^2 can be obtained, we may process by integration to find dx/dt. That is,

$$\dot{x} = \int \ddot{x}\,dt + \dot{x}_0 \tag{e}$$

Similarly, assuming that dx/dt is already available, we can obtain by integration

$$x = \int \dot{x}\,dt + x_0 \tag{f}$$

Equations (e) and (f) can be simulated by use of op amp integrators and sign changers as shown in Fig. 11-11a and b, respectively. Notice that in actual application of analog computers, the sign changers are needed that are in accordance with the sign convention used for an integrator, as depicted in Fig. 11-10a.

Returning to Eq. (d), d^2x/dt^2 can be obtained by use of a summer, gains, and potentiometers. The simulated diagram is presented in Fig. 11-11c. By combining Fig. 11-11a, b, and c, we arrive at the complete analog computer circuit for the translational mechanical system in Fig. 11-11d. However, an important problem remains to be solved. That is, how should the inputs $F(t)$, x_0, and $\dot{x}_0$ be represented in voltages that are within the limits of the analog computer being used?

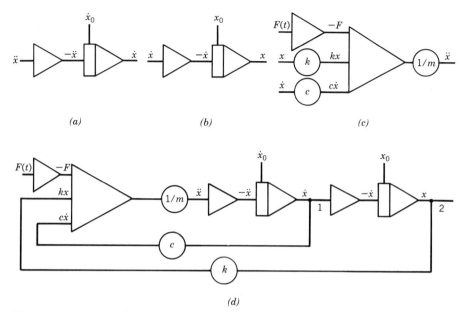

Figure 11-11

Some scaling relationships must be worked out between the computer voltages and the problem variables that may have *any* finite magnitudes. The next section deals with the important task of deciding the scale factors.

11.8 SCALING PROCEDURE

For simplicity in discussion of scaling procedure, let us recall the differential equation governing the translational mechanical system, that is,

$$\ddot{x} + \frac{c}{m}\dot{x} + \frac{k}{m}x = \frac{F(t)}{m} \tag{a}$$

By introducing the damping ratio ζ and natural frequency ω_n such that

$$\omega_n^2 = k/m \tag{b}$$

$$\zeta = c/2(km)^{1/2} \tag{c}$$

Eq. (a) becomes

$$\ddot{x} + 2\zeta\omega_n\dot{x} + \omega_n^2 x = F(t)/m \tag{d}$$

Let s_1 and s_2 be the scale factors to be used for $\dot{x}$ and x, respectively. And let v_m be the maximum voltage level allowed in the adopted analog computer. Referring to Fig. 11-11 and takeoff points 1 and 2, we observe that s_1 and s_2 are to be determined such that

$$s_1\dot{x} = v_1 \leq v_m \qquad s_2 x = v_2 \leq v_m \tag{e, f}$$

From Eq. (d), we may write

$$s_1\dot{x} = -\int \left[s_1 2\zeta\omega_n \dot{x} + s_1\omega_n^2 x - s_1 F(t)/m \right] dt + s_1\dot{x}_0 \tag{g}$$

In the meantime, referring to Fig. 11-10, application of an op amp integrator for obtaining v_1 requires patching the circuit as shown in Fig. 11-12a. The equation for the circuit is

$$s_1\dot{x} = -\int (G_1 K_1 s_1 \dot{x} + G_2 K_2 s_2 x + v_{i1} G_3) \, dt + v_{i2} \tag{h}$$

where G's are the available gains on the analog computer, v_{i1} and v_{i2} are the source inputs, and K's are the potentiometer settings. Since G's can be selected only as equal to 1 or 10, only K's can be adjusted to attain the required values for the coefficients in Eq. (g). In order to find the values for K's, we compare Eq. (g) with Eq. (h) and obtain

$$v_{i1} = -s_1 F(t)/mG_3 \tag{i}$$

$$v_{i2} = s_1\dot{x}_0 \tag{j}$$

$$K_1 = 2\zeta\omega_n/G_1 \tag{k}$$

$$K_2 = s_1\omega_n^2/s_2 G_2 \tag{l}$$

In a similar manner, we integrate $\dot{x}$ to obtain x. Mathematically, that is

$$x = \int \dot{x} \, dt + x_0$$

After multiplying the scale factor s_2 and adding signs, it becomes

$$s_2 x = -\int (-s_2\dot{x}) \, dt + s_2 x_0 \tag{m}$$

Meanwhile, application of an op amp integrator for carrying out the integration requires the analog computer to be patched as shown in Fig. 11-12b; the equation is

$$s_2 x = -\int (-G_4 K_3 s_1 \dot{x}_1) \, dt + v_{i3} \tag{n}$$

Comparison of Eq. (n) with Eq. (m) reveals that

$$v_{i3} = s_2 x_0 \tag{o}$$

$$K_3 = s_2/s_1 G_4 \tag{p}$$

Now let us summarize the observations that we have thus far made. First, v_1, v_2, v_{i1}, v_{i2}, and v_{i3} must not exceed v_m. Equations (e), (f), (i), (j), and (o) describe these conditions. Second, G_1, G_2, G_3, and G_4 are either 1 or 10 to be so selected that K_1, K_2, and K_3 have values between 0 and 1. Equations (k), (l), and (p) are to be used for calculation of the K's.

Often the scale factors required for simulation of an engineering problem on a selected analog computer has to be determined by a trial and error process. In order not to overload the analog computer, the problem variables are usually

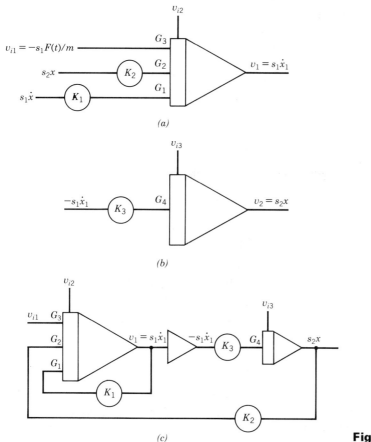

Figure 11-12

overly scaled down in the initial trial. This is followed by a gradual increase of the scaling factors to make the signals monitored on oscilloscopes more distinguishable. In many cases, however, the scaling procedure is quite straightforward. For example, consider the translational mechanical system that is subjected to an initial displacement $x_0 = 0.15$ m only. Then $\dot{x}_0 = 0$ and $F(t) = 0$ for Eq. (a). To be more specific, let us assume that $k = 45$ N/m and $m = 5$ kg. As a result, we obtain from Eq. (b) that the natural frequency is $\omega_m = 3$ rad/sec. Since the system can be expected to undergo sinusoidal motion, the displacement of the mass can be described as $x = A \sin(\omega_n t + \phi)$ and consequently $\dot{x} = A\omega_n \cos(\omega_n t + \phi)$ is the description for the velocity of the mass. Suppose that the maximum allowable voltage on the analog computer is

$$v_m = 10 \text{ V} \tag{q}$$

As the initial displacement $x_0 = 0.15$ m is expected to be damped out eventually, that is,

$$x \le x_{\max} = x_0 = 0.15 \text{ m}$$

Equation (f) hence requires that

$$s_2 x \leq 0.15 s_2 \leq 10$$

or

$$s_2 \leq \frac{10 \text{ V}}{0.15 \text{ m}} = 66.6 \text{ V/m}$$

Or conservatively we can set

$$s_2 = 50 \text{ V/m} \tag{r}$$

Similarly, $\dot{x} \leq \dot{x}_{max} = x_0 \omega_n = 0.45$ m/sec, Eq. (e) can be used to arrive at

$$s_1 \leq 10/0.45 = 22.2 \text{ V-sec/m}$$

For convenience we set

$$s_1 = 20 \text{ V-sec/m} \tag{s}$$

Having determined s_1 and s_2 for scaling of $\dot{x}$ and x, respectively, we can calculate the potentiometer settings according to Eqs. (k), (l), and (p). Since all K's must be less than one and we notice that

$$K_1 = 6\zeta/G_1 \qquad K_2 = 3.6/G_2, \qquad K_3 = 2.5/G_4$$

the gains must therefore have the values of

$$G_2 = G_4 = 10 \tag{t,u}$$

As K_1 and G_1 are dependent on the damping ratio ζ, for ζ ranging from 0.1 to 2.5, the following are examples of K_1 and G_1 values that satisfy the equation $K_1 = 6\zeta/G_1$ and that could be chosen for the simulation studies:

ζ	0.1	0.25	0.5	1.0	2.5
G_1	1	10	10	10	100
K_1	0.6	0.15	0.3	0.6	0.15

Let us recapitulate the problem that we have considered so far, and let us discuss the specific case of $\zeta = 0.5$. Since $F(t) = 0$, $\dot{x}_0 = 0$, $x_0 = 0.15$ m, $m = 5$ kg, and $k = 45$ N/m, Eqs. (b) and (c) lead to $\omega_n = 3$ rad/sec and $c = 15$ kg/sec. The differential equation for this specific problem, from Eq. (d), is

$$\ddot{x} + 3\dot{x} + 9x = 0 \tag{v}$$

Based on the prescribed initial conditions of $\dot{x}_0 = 0$ and $x_0 = 0.15$ m and the maximum allowable voltage for the analog computer $v_m = 10$ V, we have arrived at the gain and potentiometer settings for

$$G_1 = G_2 = G_4 = 10$$

$$K_1 = 0.3, \qquad K_2 = 0.36, \qquad K_3 = 0.25$$

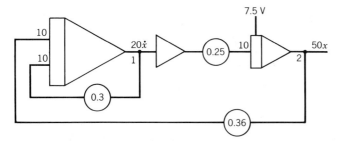

Figure 11-13

And since $F(t) = 0$, the value for G_3 is therefore irrelevant. As the selected scale factors are

$$s_1 = 20 \text{ V-sec/m} \quad \text{and} \quad s_2 = 50 \text{ V/m}$$

consequently, it should be clear that the three input voltages, according to Eqs. (i), (j), and (o), are

$$v_{i1}^{\bullet} = 0, \quad v_{i2} = 0, \quad v_{i3} = 7.5 \text{ V}$$

Figure 11-13 is the final analog diagram for simulation of Eq. (v).

It is worth noting that mathematically the analog diagram of Fig. 11-13 depicts how sine and cosine functions can be generated by use of a constant input voltage. The 7.5 V corresponds to the initial condition, that is, an IC input to an op amp integrator. The voltages at takeoff points 1 and 2 can be monitored on oscilloscopes, which should manifest trigonometric functions of the forms $A \sin \omega t$ and $B \cos \omega t$, respectively. The frequency ω depends on the values of ζ and ω_n involved in the governing equation of the system, Eq. (d). The amplitudes A and B also depend on ζ and ω_n, but in general are time dependent as well.

In the next section we shall demonstrate how analog computers can be employed for generation of simple trigonometric functions, exponential functions, and polynomials.

11.9   FUNCTION GENERATORS

The preceding sections have explained many possible applications of op amp integrators. In this section we shall discuss how op amp integrators can be effectively employed, in some cases together with other electric devices, for generation of commonly used mathematical functions. These function generators are built with the basic knowledge of calculus and on the principle governing how an op amp integrator works. Consider the simple functions of $t^n (n \geq 0)$, $e^{at}(a \neq 0)$, $\sin \omega t$, and $\cos \omega t$, where n, a, and ω are constants. From calculus, we know that

$$\frac{d}{dt} t^n = nt^{n-1} \tag{a}$$

$$\frac{d}{dt} e^{at} = ae^{at} \tag{b}$$

$$\frac{d}{dt} \sin \omega t = \omega \cos \omega t \tag{c}$$

$$\frac{d}{dt} \cos \omega t = -\omega \sin \omega t \tag{d}$$

Suppose that it is desired to generate a polynomial of nth order, $p_n(t) = a_0 + a_1 t + \cdots + a_n t^n$. We observe from Eq. (a) that by n successive differentiations the result of $d^n p_n(t)/dt^n$ is a constant equal to $a_n n!$. Hence, $p_n(t)$ can be obtained by n successive integrations of $d^n p_n(t)/dt^n$ when appropriate initial conditions, IC inputs, are assigned. It suggests that n op amp integrators can be employed for generation of $p_n(t)$. As a simple example, for generation of $p_2(t) = a_0 + a_1 t + a_2 t^2$ we have $dp_2/dt = a_1 + 2a_1 t$ and $d^2 p_2/dt^2 = 2a_1$. An analog diagram can be easily worked out and is presented in Fig. 11-14a. It shows that a battery input of $2a_2$ V, and IC inputs equal to $-a_1$ and a_0 V are required for the two op amp integrators.

In a similar manner, an exponential function of the form $E(t) = Ae^{at}$ can be generated by observing from Eq. (b) that $dE/dt = aAe^{at} = aE$, or $E(t)$ is the solution of the differential equation

$$\frac{d}{dt} E - aE = 0 \tag{e}$$

And since $E(0) = A$ or A is the initial condition of $E(t)$, the analog diagram can thus be easily drawn as the one presented in Fig. 11-14b.

For generation of trigonometric functions $S(t) = A \sin \omega t$ and $C(t) = B \cos \omega t$, we observe, from Eqs. (c) and (d), that $dS/dt = A\omega \cos \omega t$,

$$d^2 S/dt^2 = -A\omega^2 \sin \omega t = -\omega^2 S,$$

and $dC/dt = -B\omega \sin \omega t$, $d^2 C/dt^2 = -B\omega^2 \cos \omega t = -\omega^2 C$. Hence, both S and C satisfy the second-order differential equation

$$\frac{d^2}{dt^2} x(t) + \omega^2 x(t) = 0 \tag{f}$$

But the constant A associated with $S(t)$ is to be obtained by arranging the IC input so that $dS/dt = A\omega$ at $t = 0$, whereas the constant B associated with $C(t)$ is to be obtained by arranging the IC input so that $C = B$ at $t = 0$. Figure 11-14c and d show, respectively, the analog diagrams for generation of $S(t)$ and $C(t)$.

It should be pointed out that all analog diagrams shown in Fig. 11-14 are unscaled computer circuitry. The numerical detail for scaling, as can be obviously seen, depends on the values of a, ω, A, B, and other parameters involved in the specific functions that need to be generated. These parameters have to be replaced by properly selected gains and potentiometer settings in actual analog simulations.

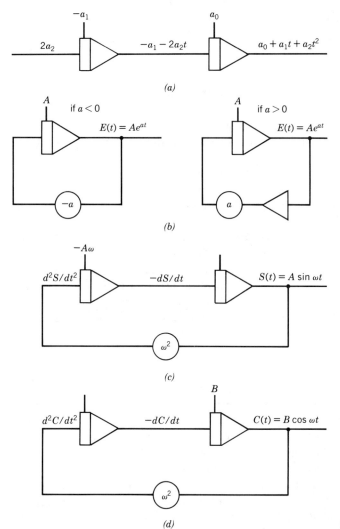

(a)

(b)

(c)

(d)

Figure 11-14

11.10   ANALOG SIMULATION OF DUAL-TANK PROBLEM

In Chapter 7 we discussed the solution of the dual-tank problem by analytical method. It is an ideal example for explaining the analog simulation of two coupled differential equations involving two unknowns. Referring back to Fig. 7-14, the differential equations governing the fluid levels h_1 and h_2 in the two cylindrical tanks of uniform cross-sectional areas A_1 and A_2 are

$$A_1 \frac{dh_1}{dt} = Q_1 - \frac{1}{R_1}(h_1 - h_3) \tag{a}$$

$$A_2 \frac{dh_2}{dt} = Q_2 - \frac{1}{R_2}(h_2 - h_3) \qquad\qquad (b)$$

where

$$h_3 = R_3 q_3 = \frac{1}{R_1}(h_1 - h_3) + \frac{1}{R_2}(h_2 - h_3) \qquad\qquad (c)$$

If h_{10} and h_{20} are the initial levels in tanks 1 and 2 at $t = 0$, respectively, then it is easy to apply op amp integrators for analog simulation of Eqs. (a) and (b). The diagrams are presented in Fig. 11-15a and b. By use of a summer and a sign changer, Eq. (c) can also be simulated by the circuitry in Fig. 11-15c. The reader should be able to combine Fig. 11-15a, b, and c to obtain the final complete circuitry as shown in Fig. 11-15d for analog simulation of the dual-tank problem.

The multipliers r_1, r_2, $1/A_1 R_1$, $1/A_2 R_2$, and $1/R_3$ are to be eventually replaced by gain constants G's and potentiometer settings in the scaling process. The reader is urged to use the numerical data given in Fig. 7-15 to work out the necessary scaling factors for simulation study of the transient study of the dual-tank problem. Also, in Chapter 10 a steady-state problem has been presented and finite-dfference solution is given on page 320. The reader should also use this as an example to practice the determination of scale factors involved in analog simulation of engineering systems.

EXERCISE

A torsional shaft system shown in Fig. 11-16 consists of two disks whose inertias are equal to J_1 and J_2, respectively, a damper of resistance b, and three segments of shafts with spring constants k_1, k_{12}, and k_2. Initially, that is, at $t = 0$, the system is completely at rest. A time-varying external torque $T_a(t)$ is suddenly applied on the disk 2. For monitoring the torques T_1 and T_2, which are transmitted at the two ends of the shaft system, and the motions of the disks, an analog diagram has been partially worked out as shown in Fig. 11-17. In the diagram, it is shown that four op amp integrators numbered 1 through 4, three summers numbered 5 through 7, one differential amplifier numbered 8, and three sign changers numbered 9 through 11 are adopted. The multipliers numbered c_1 through c_6 remain to be specified. Express these multipliers in terms of the physical parameters J_1, J_2, b, k_1, k_{12}, and k_2.

11.11 NONLINEAR ANALOG SIMULATION— MIXING-TANK PROBLEM

The electronic multiplier that generates an output voltage linearly proportional to the product of two input voltages as shown in Fig. 11-18a can be conveniently employed for analog simulation of engineering systems that require nonlinear analysis. Let us again consider the mixing tank that we have discussed in Chapter 10 where a finite-difference solution was presented. Here, we demonstrate how the electronic multiplier and a division circuitry, shown in Fig. 11-18b and c,

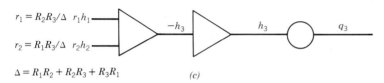

(a)   (b)

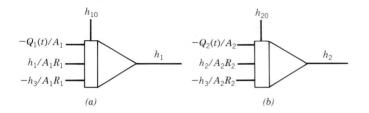

$r_1 = R_2R_3/\Delta$ r_1h_1

$r_2 = R_1R_3/\Delta$ r_2h_2

$\Delta = R_1R_2 + R_2R_3 + R_3R_1$

(c)

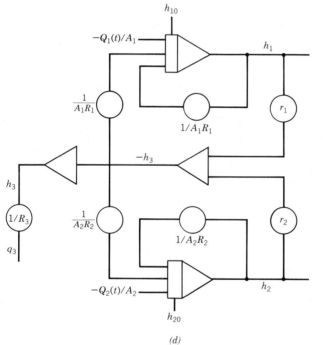

(d)

Figure 11-15

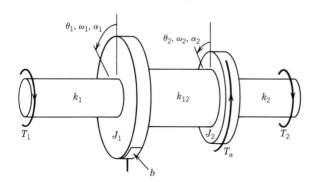

Figure 11-16

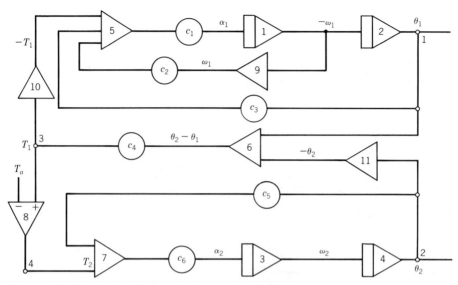

Figure 11-17   Analog diagram for the torsional problem.

which can be constructed by use of an electronic multiplier, can be adopted for analog solution of the problem.

First, let us recall the mixing problem. As shown in Fig. 10-2, the inflows of water, salt, and mixture are denoted as Q_w, Q_s, and Q_m, respectively. It has been shown that the weight W_m and concentration f of the mixture in the tank are governed by the differential equations

$$\frac{dW_m}{dt} = Q_w + Q_s - Q_m \tag{a}$$

and

$$\frac{d(fW_m)}{dt} = Q_s - fQ_m \tag{b}$$

Apparently, the system is coupled in the variables f and W_m, and the product term

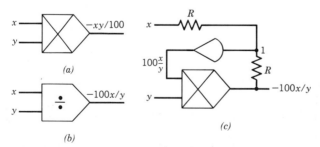

Figure 11-18   (a) Symbol for electronic multiplier, (b) symbol and (c) circuitry for electronic divisor.

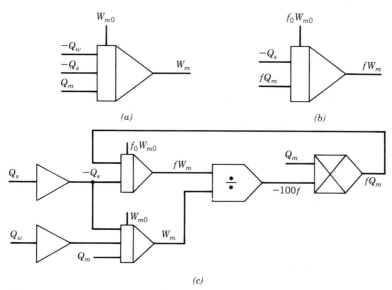

Figure 11-19   Analog diagram for the mixing-tank problem.

fW_m in Eq. (b) clearly indicates that the system is nonlinear and suggests that the electronic multiplier is needed in analog simulation.

Following the previously outlined procedure, Eqs. (a) and (b) can be diagrammed by analog circuitry as shown in Fig. 11-19a and b. f_0 and W_{m0} are the initial concentration and weight, respectively, of the mixture in the tank. It is straightforward to see that the final complete analog diagram shown in Fig. 11-19c is simply the proper combination of Fig. 11-19a and b.

EXERCISE

The electronic multiplier, of which the analog symbol is shown in Fig. 11-18a, can be built in many different ways. Figure 11-20a shows a circuitry that can be employed for generation of an output voltage in square-wave form with unequal positive and negative magnitudes v_a and $-v_b$ as shown in Fig. 11-20c. The values of v_a and $-v_b$ can be adjusted by use of the wipers to obtain variable resistances R_1 and R_2, as shown in Fig. 11-20a. The area per each cycle of the square wave shown in Fig. 11-20c can be easily obtained as

$$A = (v_a - v_b)T/2$$

As an exercise, work out an analog circuit for an electronic multiplier based on this concept.

EXERCISE

If the lower input of the division circuit shown in Fig. 11-18c, namely y, has a value of $\sqrt{x}$, the circuit becomes a square-root circuit. Show that this can be

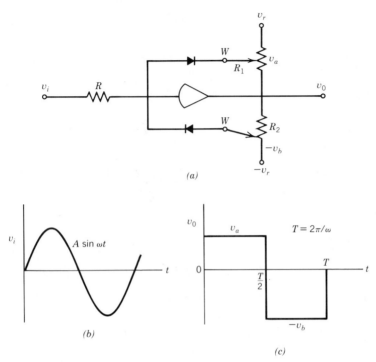

Figure 11-20

accomplished by connecting the output of op amp to both inputs of the electronic multiplier. Derive the output of the square-root circuit by observing that the current at junction 1 shown in Fig. 11-18c is almost equal to zero.

11.12 DIGITAL–ANALOG CONVERTER

In hybrid computer operations, conversions of analog voltages into digital data and vice versa are among the fundamental needs. Relatively speaking, the conversion of digital data into analog voltage is simpler than the conversion of analog voltage into digital data. We shall thus first discuss the former conversion, which in fact is nowadays a prerequisite for achieving the latter conversion in many modern analog-digital converters.

Without loss of generality, a digital number D can be assumed to be in the range of 0 to 1.0 and is represented by n binary digits and a sign bit. The left-most and right-most bits of the n-bit binary number D are referred to as the 0th and n-1st bits, respectively. These bits can carry either "zero" or "one." If the kth bit carries a "one," then $2^{-(k+1)}$ is a part of the value for D. The left-most, 0th bit is most significant because if it carries a "one," it contributes $2^{-1} = 0.5$ to the total value of D. The right-most, n-1st bit is least significant because if it carries a "one," it contributes only 2^{-n}. Consider the case of $D = 0.625$; the 0th and 2nd bits should

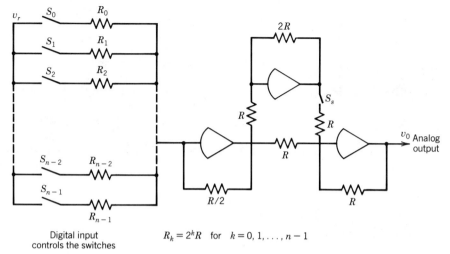

$R_k = 2^k R$ for $k = 0, 1, \ldots, n-1$

Digital input
controls the switches

Figure 11-21 Digital–analog converter.

both contain a "one," whereas all the other bits should contain a "zero" for the reason that $2^{-1} + 2^{-3} = 0.5 + 0.125 = 0.625$.[3]

Figure 11-21 illustrates how in a digital–analog converter the "zero" and "one" bit-structure of digital input data are utilized directly to close and open, respectively, the switches labeled 0 through $n - 1$. Since each of the n bits of the digital input data D carries different *weight*, its contribution to the analog output must thus be accordingly weighted by use of varying resistances in the electric circuit. Since S_0 and R_0 are associated with the 0th bit of D, and S_1 and R_1 are associated with the first bit of D, and so on, and since a "one" in the 0th bit contributes 2^{-1} to D whereas a "one" in the 1st bit contributes 2^{-2} or a half of 2^{-1} and so on, it should be evident that $R_2 = 2R_1$, $R_3 = 2R_2$, and so on. Hence, if we choose $R_0 = R$, R_k should then be equal to $2^k R$ for $k = 0, 1, \ldots, n - 1$.

If $D = 0.625$, then only switches S_0 and S_2 should be closed while the other $n - 2$ switches should be open. The output of amplifier 1 has a magnitude of $-0.625 v_r$, where v_r is the reference voltage. In order to change the sign, the switch S_s, which represents the sign of D, should be open so that amplifier 2 serves as a sign changer. If $D = -0.625$, then switch S_s should be closed. As a consequence, amplifier 2 has two inputs, one from amplifier 1 equal to $-0.625 v_r$ and the other equal to $1.25 v_r$ from amplifier 3. The output of amplifier 2 is thus equal to $-(1.25 - 0.625)v_r = -0.625 v_r$ as it should be.

Obviously, the difficulty in using weighted resistors for conversion of digital data into analog signal is the wide range of resistances required. For a 10-bit digital data and 100-V computer, resistances range from 100 kohm to 512 mohm.

---

[3] In Chapter 3 the binary structure of the "fraction" portion of the floating-point constants stored in a digital computer were discussed. The reader may find it helpful to review the material presented in Sections 3.2 and 3.3.

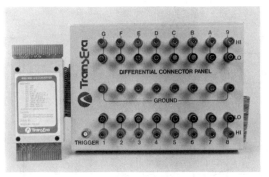

Digital-to-analog (D/A) and analog-to-digital (A/D) converters. (Courtesy of TransEra Corporation, Provo, Utah.)

Accuracy becomes a problem, especially when the input digital value is small and high resistances play dominant roles. This difficulty is partially obviated in practice by use of expensive switches.

11.13 ANALOG–DIGITAL CONVERTER

The principle of converting analog voltage input into digital output can be simply demonstrated by requiring that the output be accurate to only two binary digits. That is to say, the output is to be either 00, 01, 10, or 11. Suppose that the input voltage is v_i. v_i can be compared with three voltage levels $v_r/4$, $v_r/2$, and $3v_r/4$, where v_r is a reference voltage. The output should be either 00, 01, 10, or 11 depending on whether $0 < v_i \le v_r/4$, $v_r/4 < v_i \le v_r/2$, $v_r/2 < v_i \le 3v_r/4$, or $3v_r/4 < v_i$, respectively. In general, if an n-bit binary output is required, then $2^n - 1$ comparators are needed to sort out the result of what the digital output should be. Figure 11-22 shows the schematic of an analog–digital convertor. For example, if $n = 8$, then 255 comparators have to be arranged for the analog–digital conversion. When all comparators are working in parallel, simultaneous comparisons can then be made. It often takes less than 1 μsec to complete the analog–digital conversion. However, only a limited number of comparators are actually employed for most analog–digital converters. Since not all comparisons can be carried out simultaneously, the analog–digital conversion has to be achieved at a reduced rate. The maximum number of conversions per unit time that an analog–digital can perform, or maximum **sampling rate**, is of major concern in hybrid computer operations.

An analog–digital converter is expected to be able to continuously accept input in the form of **transient** voltages (magnitude changing in time) and at selected

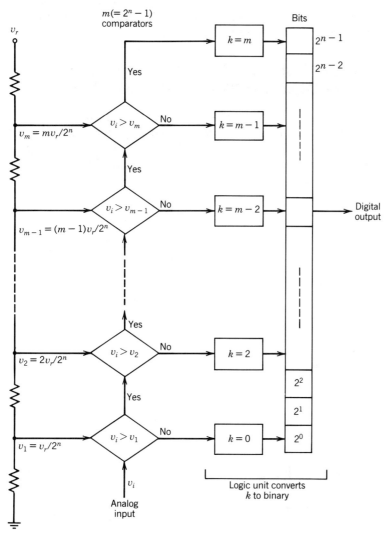

Figure 11-22 Analog–digital converter using $2^n - 1$ comparators for n-bit output.

instants. Usually, the input is provided at a fixed sampling rate, that is, at a specified constant time increment. And the converter is to furnish digit output at the same rate. Because of the multiple input and output situation, the digital–analog converters can be advantageously used together with comparators in analog-digital converters. To explain the concept, let v_i and D_i be the analog input and digital output at $t = t_i$. Suppose that v_1 and D_1 for $t = t_1$ are already available. Because of an insufficient number of comparators, D_1 is not expected to be accurate. Let us convert D_1 back to an analog voltage by use of a digital–analog converter and denote the output as v_1^*. Then, obviously $v_1 - v_1^*$ represents the error. In a way, the digital–analog converter works just like a comparator, and apparently the design of an analog–digital converter by proper combination of comparators and digital–analog converters is itself an attractive CAD topic.

11.14 COMPUTER-AIDED DRAFTING OF ANALOG DIAGRAMS—A. MODULE

As it is evident enough that computer-aided drafting ought to be adopted for drawing the analog diagrams in simulation of engineering systems, a BASIC program called A. MODULE has been developed for this particular purpose. Figure 11-23 is produced by application of A. MODULE. It shows that the 20 user-definable

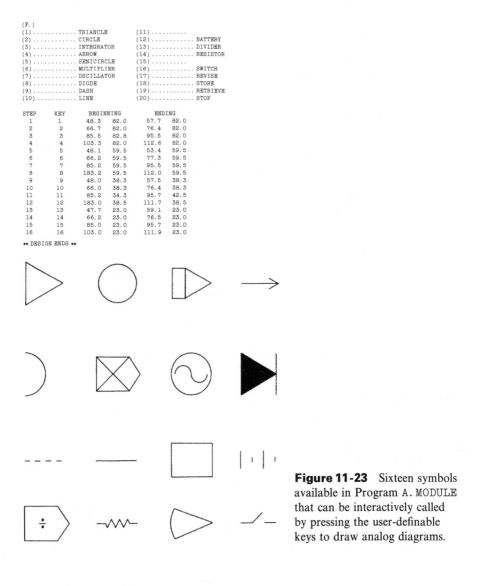

```
(F. )
(1) ............ TRIANGLE      (11) ..........
(2) ............ CIRCLE        (12) ............ BATTERY
(3) ............ INTEGRATOR    (13) ............ DIVIDER
(4) ............ ARROW         (14) ............ RESISTOR
(5) ............ SENICIRCLE    (15) ..........
(6) ............ MULTIPLIER    (16) ............ SWITCH
(7) ............ DSCILLATOR    (17) ............ REVISE
(8) ............ DIODE         (18) ............ STORE
(9) ............ DASH          (19) ............ RETRIEVE
(10) ........... LINE          (20) ............ STOP
```

STEP	KEY	BEGINNING		ENDING	
1	1	48.3	82.0	57.7	82.0
2	2	66.7	82.0	76.4	82.0
3	3	85.5	82.8	95.5	82.0
4	4	103.3	82.0	112.6	82.0
5	5	48.1	59.5	53.4	59.5
6	6	66.2	59.5	77.3	59.5
7	7	85.2	59.5	95.5	59.5
8	8	183.2	59.5	112.0	59.5
9	9	48.0	38.3	57.5	38.3
10	10	66.0	38.3	76.4	38.3
11	11	85.2	34.3	95.7	42.5
12	12	183.0	38.5	111.7	38.5
13	13	47.7	23.0	59.1	23.0
14	14	66.2	23.0	76.5	23.0
15	15	85.0	23.0	95.7	23.0
16	16	103.0	23.0	111.9	23.0

•• DESIGN ENDS ••

Figure 11-23 Sixteen symbols available in Program A. MODULE that can be interactively called by pressing the user-definable keys to draw analog diagrams.

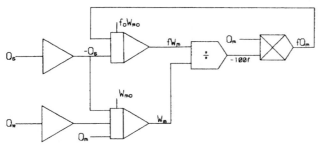

Figure 11-24 Interactively drawn analog diagram by application of Program A. MODULE; hand-drawn version is shown in Fig. 11-19. (Mixing-tank problem.)

keys are utilized for interactive requests by the user through commands entered from the Tektronix 4054 keyboard. It also shows the 16 available symbols built into the program that are most commonly used symbols in construction of analog diagrams. These 16 symbols are to be requested by pressing one of the definable keys numbered 1 through 16. The remaining four keys are arranged for revising, storage, and retrieval of the drawn symbols in a data file and for stopping the program. Detailed explanation of these key assignments is given in the Key Table in Fig. 11-23.

The thumbwheels are used for entering the beginning and ending locations on the screen, based on which the size and orientation of the requested symbol are determined. Flashing messages are shown on the screen to assist the user in maneuvering the thumbwheels and pressing a new key. When key 20 is pressed, all values of the coordinates of the beginning and ending points for drawing the requested symbols are printed on the screen prior to the termination of the program. This is illustrated in Fig. 11-23. A listing of Program A. MODULE is also presented.

Many hand-drawn analog diagrams presented in this chapter can be interactively displayed on the screen first and then hardcopies can be made by application of Program A. MODULE. Figure 11-24 is a machine-made counterpart of Fig. 11-19. The reader should attempt to copy other analog diagrams in this chapter on the screen with A. MODULE.

PROGRAM A. MODULE

```
1 REM * PROGRAM A.MODULE *
2 GO TO 150
3 CALL "WAIT",1000
4 GOSUB 2290
5 GO TO 3
8 GOSUB 3250
9 GO TO 3
12 GOSUB 3175
13 GO TO 3
16 GOSUB 3380
17 GO TO 3
20 GOSUB 3435
```

```
21 GO TO 3
24 GOSUB 370
25 GO TO 3
28 GOSUB 1280
29 GO TO 3
32 GOSUB 3500
33 GO TO 3
36 GOSUB 3290
37 GO TO 3
40 GOSUB 2200
41 GO TO 3
44 GOSUB 2575
45 GO TO 3
48 GOSUB 1785
49 GO TO 3
52 GOSUB 2440
53 GO TO 3
56 GOSUB 1936
57 GO TO 3
60 GOSUB 3570
61 GO TO 3
64 GOSUB 2068
65 GO TO 3
68 GOSUB 2620
69 GO TO 3
72 GOSUB 3635
73 GO TO 3
76 GOSUB 3715
77 GO TO 3
80 GOSUB 2540
89 GO TO 3
150 CHARSIZE 1
151 RINIT
152 PAGE
153 INIT
155 SET KEY
160 GOSUB 200
162 GO TO 318
200 PRINT "KEY TABLE"
203 PRINT "'1'--------TRIANGLE      '11'------RECTANGLE"
205 PRINT "'2'--------CIRCLE        '12'------BATTERY   "
210 PRINT "'3'--------INTEGRATOR    '13'------DIVIDER   "
220 PRINT "'4'--------ARROW         '14'------RESISTOR  "
230 PRINT "'5'-------SEMICIRCLE     '15'------AMPLIFIER"
240 PRINT "'6'--------MULTIPLIER    '16'------SWITCH    "
250 PRINT "'7'--------OSCILLATOR    '17'------REVISE    "
260 PRINT "'8'--------DIODE         '18'------STORE     "
270 PRINT "'9'--------DASH          '19'------RETRIEVE "
280 PRINT "'10'------LINE           '20'------STOP      "
317 RETURN
318 P5=0
319 U7=0
320 U1=0
330 D9=0
340 DIM P9(100),X1(100),Y1(100),X2(100),Y2(100)
350 GOSUB 2900
360 GO TO 3
370 REM * THIS SUBROUTINE PLOTS A MULTIPLIER                    *
380 J=6
400 GOSUB 3815
410 ROTATE T1
415 L=L*0.75
420 MOVE X1(U1),Y1(U1)
430 RDRAW 0,-L/2
440 RDRAW L,0
```

```
450 DRAW X2(U1),Y2(U1)
455 RDRAW -L/0.75*0.25,0.5*L
460 RDRAW -L,0
470 RDRAW 0,-L/2
480 RMOVE 0,L/2
490 RDRAW L,-L
500 RMOVE 0,L
510 RDRAW -L,-L
520 RETURN

1280 REM * THIS SUBROUTINE PLOTS AN OSCILLATOR              *
1300 J=7
1310 GOSUB 3815
1320 MOVE X1(U1),Y1(U1)
1330 ROTATE T1
1340 RMOVE 1*L,0
1350 R7=L*0.5
1360 GOSUB 1700
1370 MOVE X1(U1),Y1(U1)
1380 RMOVE 0.5*L,0
1390 ROTATE 90+T1
1400 R7=4/20*L
1620 GOSUB 1750
1630 MOVE X2(U1),Y2(U1)
1640 ROTATE T1
1650 RMOVE -L*0.5,0
1660 ROTATE 270+T1
1670 R7=4*L/20
1680 GOSUB 1750
1690 RETURN
1700 REM * THIS SUBROUTINE PLOTS A CIRCLE                    *
1710 FOR D5=0 TO 357 STEP 3
1720 RDRAW R7*(COS(D5+3)-COS(D5)),R7*(SIN(D5+3)-SIN(D5))
1730 NEXT D5
1740 RETURN
1750 FOR D5=-90 TO 87 STEP 3
1760 RDRAW R7*(COS(D5+3)-COS(D5)),R7*(SIN(D5+3)-SIN(D5))
1770 NEXT D5
1780 RETURN
1785 REM * THIS SUBROUTINE PLOTS A BATTERY                   *
1795 J=12
1805 GOSUB 3815
1815 MOVE X1(U1),Y1(U1)
1825 ROTATE T1
1835 RMOVE 0,-0.25*L
1845 RDRAW 0,0.5*L
1855 RMOVE L/3,-0.25*L
1865 RMOVE 0,-0.075*L
1875 RDRAW 0,0.15*L
1885 RMOVE L/3,-0.075*L-0.25*L
1895 RDRAW 0,0.5*L
1905 MOVE X2(U1),Y2(U1)
1915 RMOVE 0,-0.075*L
1925 RDRAW 0,0.15*L
1935 RETURN
1936 REM * THIS SUBROUTINE PLOTS A RESISTOR                      *
1940 J=14
1946 GOSUB 3815
1956 MOVE X1(U1),Y1(U1)
1966 ROTATE T1
1976 RDRAW L/5,0
1986 RDRAW L/20,-L/10
1996 RDRAW L/10,L/5

2006 RDRAW L/10,-L/5
2016 RDRAW L/10,L/5
```

```
2026 RDRAW L/10,-L/5
2036 RDRAW L/10,L/5
2046 RDRAW L/20,-L/10
2056 RDRAW L/5,0
2066 RETURN
2068 REM * THIS SUBROUTINE PLOTS A SWITCH          *
2078 J=16
2100 GOSUB 3815
2138 MOVE X1(U1),Y1(U1)
2148 ROTATE T1
2158 RDRAW L/3,0
2168 RDRAW L/3.5,L/3.5
2178 MOVE X2(U1),Y2(U1)
2188 RDRAW -L/3,0
2198 RETURN
2200 REM * THIS SUBROUTINE PLOTS A LINE            *
2210 J=10
2220 GOSUB 3815
2260 MOVE X1(U1),Y1(U1)
2270 DRAW X2(U1),Y2(U1)
2280 RETURN
2290 REM * THIS SUBROUTINE PLOTS A TRIANGLE        *
2300 J=1
2310 GOSUB 3815
2320 MOVE X1(U1),Y1(U1)
2330 ROTATE T1
2340 L5=L/3^0.5
2350 L4=L
2360 GOSUB 2390
2370 RETURN
2380 REM *  SUB TO PLOT A TRIANGLE                 *
2390 RDRAW 0,L5
2400 RDRAW L4,-L5
2410 RDRAW -L4,-L5
2420 RDRAW 0,L5
2430 RETURN
2440 REM *   THIS SUB PLOTS THE DIVIDER             *
2445 J=13
2450 GOSUB 3815
2455 ROTATE T1
2460 L=L*0.75
2465 MOVE X1(U1),Y1(U1)
2470 RDRAW 0,L/2
2475 RDRAW L,0
2476 DRAW X2(U1),Y2(U1)
2480 RDRAW -L/0.75*0.25,-L/2
2485 RDRAW -L,0
2490 RDRAW 0,L/2
2495 RMOVE 0.46*L,0
2500 RDRAW 0.2*L,0
2505 RMOVE -0.08*L,0
2510 RMOVE 0,-0.1*L
2515 R7=L*0.1*0.2
2520 GOSUB 3355
2525 RMOVE 0,0.2*L
2530 GOSUB 3355
2535 RETURN
2540 REM * PROGRAM ENDS                            *
2545 G9=8
2550 RINIT
2555 GOSUB 2765
2560 MOVE 0,10
2565 PRINT '**  DESIGN ENDS  **'
2570 END
2575 REM   *   THIS SUB PLOTS THE RECTANGLE   *
2580 J=11
```

```
2585 GOSUB 3815
2590 MOVE X1(U1),Y1(U1)
2595 DRAW X2(U1),Y1(U1)
2600 DRAW X2(U1),Y2(U1)
2605 DRAW X1(U1),Y2(U1)
2610 DRAW X1(U1),Y1(U1)
2615 RETURN
2620 REM * THIS SUBROUTINE REVISES THE LAST DRAWN ELEMENT        *
2625 RINIT
2630 MOVE 0,80
2635 MOVE 0,20
2640 U2=U1
2645 PRINT "INPUT THE NEW KEY NUMBER USING "
2650 PRINT "  NUMERIC KEYS ; NOT DEFINABLE KEY!!!"
2655 INPUT P9(U2)
2660 PRINT "ADJUST THE THUMBWHEELS TWICE!!!"
2665 PRINT "  TO INPUT THE BEGINNING AND"
2670 PRINT "  ENDING POINTS OF THE ELEMENT"
2675 POINTER X1(U2),Y1(U2),C$
2680 POINTER X2(U2),Y2(U2),C$
2685 D9=1
2690 PAGE
2695 FOR U3=1 TO U1
2700 GOS P9(U3) OF 2290,3250,3175,3380,3435,370,1280,3500,3290,2200,2575
2705 GOSUB P9(U3)-11 OF 1785,2440,1936,3570,2068,2620,3635,3715,2540
2710 NEXT U3
2715 D9=0
2720 MOVE 0,98.3
2725 GOSUB 200
2730 MOVE 0,70
2735 ROPEN 5
2740 PRINT "!!!!!!!!!!!!!!!!!!!!!!!!!!!!!!!!!!!!!!!!"
2745 PRINT "!!  ENTER ANOTHER NEW KEY#         !!"
2750 PRINT "!!!!!!!!!!!!!!!!!!!!!!!!!!!!!!!!!!!!!!!!"
2755 RCLOSE
2760 GO TO 3
2765 REM * THIS SUBROUTINE PRINTS OUT THE ELEMENT COORDIANTES       *
2770 MOVE 0,80
2775 PRINT "STEP# KEY#  BEGINNING       ENDING "
2785 FOR I5=1 TO U1
2790 PRINT USING 2795:I5,P9(I5),X1(I5),Y1(I5),X2(I5),Y2(I5)
2795 IMAGE 2D,4X,2D,2X,4(3D.1D,2X)
2800 NEXT I5
2805 RETURN
2810 REM *           THIS SUB FINDS THE LENGTH AND ROTATION ANGLE    *
2815 SET DEGREES
2820 L=((X1(U1)-X2(U1))^2+(Y1(U1)-Y2(U1))^2)^0.5
2825 IF X1(U1)>X2(U1) AND Y1(U1)=Y2(U1) THEN 2865
2830 IF X1(U1)=X2(U1) AND Y1(U1)>Y2(U1) THEN 2875
2835 IF X1(U1)>X2(U1) AND Y1(U1)>Y2(U1) THEN 2885
2840 IF Y1(U1)=>Y2(U1) THEN 2855
2845 T1=ACS((X2(U1)-X1(U1))/L)
2850 GO TO 2895
2855 T1=ASN((Y2(U1)-Y1(U1))/L)
2860 GO TO 2895
2865 T1=180
2870 GO TO 2895
2875 T1=-90
2880 GO TO 2895
2885 T1=ATN((Y2(U1)-Y1(U1))/(X2(U1)-X1(U1)))
2890 T1=T1+180
2895 RETURN
2900 REM * THIS SUBROUTINE INSTRUCTS THE USER          *
2905 ROPEN 7
2910 IF U1<>0 THEN 2940
2915 MOVE 0,75
```

```
2920 PRINT "******************************************"
2925 PRINT "**        ENTER FIRST KEY#          **"
2930 PRINT "******************************************"
2935 RCLOSE
2940 RETURN
2945 REM * INPUT THE BEGINNING AND ENDING POINTS              *
2950 P5=P5+1
2955 U1=P5+U7
2960 P9(U1)=J
2965 RINIT
2970 MOVE 0,65
2975 ROPEN 1
2980 PRINT "KEY# CHOSEN: ";J
2985 PRINT "ADJUST THE THUMBWHEELS"
2990 PRINT " TO INPUT THE BEGINNING"
2995 PRINT "POINT WHEN READY PRESS"
3000 PRINT " 'RETURN' "

3005 RCLOSE
3010 POINTER X1(U1),Y1(U1),C$
3015 MOVE 0,53
3020 PRINT " "
3025 ROPEN 2
3030 PRINT "COORDINATE OF BEGINNING POINT"
3035 PRINT USING 3065:"X= ",X1(U1)
3040 PRINT USING 3065:"Y= ",Y1(U1)
3045 PRINT "ADJUST THE THUMBWHEELS"
3050 PRINT "TO INPUT THE ENDING"
3055 PRINT "POINT WHEN READY PRESS"
3060 PRINT " 'RETURN' "
3065 IMAGE 4A,5T,3D.1D
3070 RCLOSE
3075 POINTER X2(U1),Y2(U1),C$
3080 MOVE 0,36.5
3085 PRINT " "
3090 ROPEN 3
3095 PRINT "COORDINATE OF ENDING POINT"
3100 PRINT USING 3065:"X= ",X2(U1)
3105 PRINT USING 3065:"Y= ",Y2(U1)
3110 RCLOSE
3115 ROPEN 8
3120 MOVE 0,25
3125 PRINT "!!!!!!!!!!!!!!!!!!!!!!!!!!!!!!!!!!!!"
3130 PRINT "!!!  ENTER ANOTHER NEW KEY#  !!!"
3135 PRINT "!!!!!!!!!!!!!!!!!!!!!!!!!!!!!!!!!!!!"
3140 RCLOSE
3145 RETURN
3150 REM *************************************************************
3155 REM **          U1 IS COUNTER  ;   P9 STORE KEY #          **
3160 REM **          D9 IS INDENTIFICATION FOR WHICH CALLING    **
3165 REM **   G9 IS THE IDENTIFIER TO STORE THE DATA OR NOT     **
3170 REM **        R7 IS THE RADIUS ;                           **
3175 REM * THIS SUBROUTINE PLOTS A INTEGRATOR                  *
3180 J=3
3185 GOSUB 3815
3190 MOVE X1(U1),Y1(U1)
3195 ROTATE T1
3200 L4=L*2/3
3205 L5=L*2/3/3^0.5
3210 RDRAW 0,L5
3215 RDRAW L/3,0
3220 RDRAW 0,-L5*2
3225 RDRAW -L/3,0
3230 RDRAW 0,L5
3235 RMOVE L/3,0
3240 GOSUB 2380
3245 RETURN
3250 REM * THIS SUBROUTINE PLOTS A CIRCLE ELEMENT              *
```

```
3255 J=2
3260 GOSUB 3815
3265 R7=0.5*L
3270 ROTATE T1
3275 MOVE X2(U1),Y2(U1)
3280 GOSUB 3355
3285 RETURN
3290 REM * THIS SUBROUTINE PLOTS A DASH LINE            *
3295 J=9
3300 GOSUB 3815
3305 MOVE X1(U1),Y1(U1)
3310 ROTATE T1
3315 RDRAW L/7,0
3320 RMOVE L/7,0
3325 RDRAW L/7,0
3330 RMOVE L/7,0
3335 RDRAW L/7,0
3340 RMOVE L/7,0
3345 DRAW X2(U1),Y2(U1)
3350 RETURN
3355 REM * THIS SUBROUTINE PLOTS A CIRCLE               *
3360 FOR D5=0 TO 357 STEP 3
3365 RDRAW R7*(COS(D5+3)-COS(D5)),R7*(SIN(D5+3)-SIN(D5))
3370 NEXT D5
3375 RETURN
3380 REM * THIS SUBROUTINE PLOTS A ARROW                *
3385 J=4
3390 GOSUB 3815
3395 MOVE X1(U1),Y1(U1)
3400 DRAW X2(U1),Y2(U1)
3405 ROTATE T1
3410 RMOVE -L/4,L/4/3^0.5
3415 DRAW X2(U1),Y2(U1)
3420 RMOVE -L/4,-L/4/3^0.5
3425 DRAW X2(U1),Y2(U1)
3430 RETURN
3435 REM * THIS SUBROUTINE PLOTS A SEMICIRCLE         *
3440 J=5
3445 GOSUB 3815
3450 ROTATE T1
3455 MOVE X1(U1),Y1(U1)
3460 R7=L
3465 RMOVE 0,-L
3470 GOSUB 3480
3475 RETURN
3480 FOR D5=-90 TO 87 STEP 3
3485 RDRAW R7*(COS(D5+3)-COS(D5)),R7*(SIN(D5+3)-SIN(D5))
3490 NEXT D5
3495 RETURN
3500 REM * THIS SUBROUTINE PLOTS A DIODE                *
3505 J=8
3510 GOSUB 3815
3515 MOVE X1(U1),Y1(U1)
3520 ROTATE T1
3525 RDRAW 0,-L/3^0.5
3530 DRAW X2(U1),Y2(U1)
3535 RDRAW -L,L/3^0.5
3540 DRAW X1(U1),Y1(U1)
3545 MOVE X2(U1),Y2(U1)
3550 RMOVE 0,-L/3^0.5
3555 RDRAW 0,2*L/3^0.5
3560 GOSUB 3850
3565 RETURN
3570 REM     *    THIS SUB PLOTS THE AMPLIFIER          *
3575 J=15
3580 GOSUB 3815
3585 ROTATE T1
3590 MOVE X1(U1),Y1(U1)
```

```
3592 RMOVE L*2/3,0
3595 R7=2*L/3
3600 RMOVE R7*COS(150),R7*SIN(150)
3605 DRAW X2(U1),Y2(U1)
3610 RMOVE -L/3,0
3612 RMOVE R7*COS(150),R7*SIN(150)
3615 FOR D5=150 TO 210 STEP 3
3620 RDRAW R7*(COS(D5+3)-COS(D5)),R7*(SIN(D5+3)-SIN(D5))
3625 NEXT D5
3627 DRAW X2(U1),Y2(U1)
3630 RETURN
3635 REM  *           THIS SUB STORES THE DATA                    *
3640 RINIT
3645 KILL 4
3650 FIND 4
3655 WRITE @33:U1
3660 FOR I5=1 TO U1
3665 WRITE @33:X1(I5),Y1(I5),X2(I5),Y2(I5),P9(I5)
3670 NEXT I5
3675 MOVE 0,60
3680 ROPEN 7
3685 PRINT "!!!!!!!!!!!!!!!!!!!!!!!!!!!!!!!!!!!!!!!!!!!!!!!!!!!!!"
3690 PRINT "!!!      DATA HAS BEEN STORED ON FILE 4      !!!"
3695 PRINT "!!!             ENTER ANOTHOR NEW KEY#        !!!"
3700 PRINT "!!!!!!!!!!!!!!!!!!!!!!!!!!!!!!!!!!!!!!!!!!!!!!!!!!!!!"
3705 RCLOSE
3710 RETURN
3715 REM *          THIS SUB RETRIEVE THE DATA FROM TAPE          *
3720 RINIT
3725 FIND 4
3730 READ @33:U7
3735 FOR I5=1 TO U7
3740 READ @33:X1(I5),Y1(I5),X2(I5),Y2(I5),P9(I5)
3745 NEXT I5
3750 D9=1
3755 FOR U3=1 TO U7
3760 GOS P9(U3) OF 2290,3250,3175,3380,3435,370,1280,3500,3290,2200,2575
3765 GOSUB P9(U3)-11 OF 1785,2440,1936,3570,2068,2620,3635,3715,2540
3770 NEXT U3
3775 D9=0
3780 MOVE 0,60
3785 ROPEN 8
3790 PRINT "!!!!!!!!!!!!!!!!!!!!!!!!!!!!!!!!!!!!!!!!!!!!!!!!!!!!!"
3795 PRINT "!!!         ENTER ANOTHER NEW KEY#      !!!"
3800 PRINT "!!!!!!!!!!!!!!!!!!!!!!!!!!!!!!!!!!!!!!!!!!!!!!!!!!!!!"
3805 RCLOSE
3810 RETURN
3815 REM * THIS SUBROUTINE CHECKS THE CALLING ROUTINE            *
3820 IF D9<>1 THEN 3835
3825 U1=U3
3830 GO TO 3840
3835 GOSUB 2945
3840 GOSUB 2810
3845 RETURN
3850 REM   *   THIS SUB DO THE PAINTING   *
3855 MOVE X1(U1),Y1(U1)
3860 RMOVE 0,L/3^0.5
3865 G8=99
3870 FOR T9=1 TO G8
3875 RMOVE L*T9/G8,-2*L/3^0.5*T9/G8/2
3880 RDRAW 0,-L/3^0.5*2*(1-1/G8*T9)
3885 MOVE X1(U1),Y1(U1)
3890 RMOVE 0,L/3^0.5
3895 NEXT T9
3900 RETURN
```

CHAPTER 12

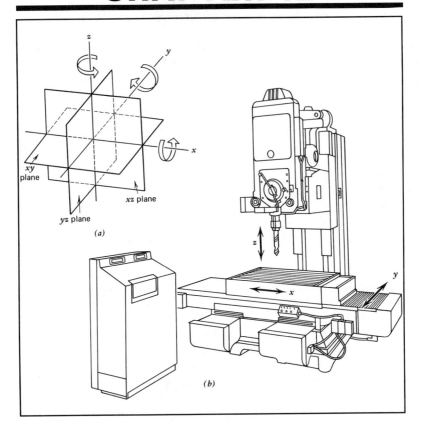

(a)

(b)

ADVANCED AND
MISCELLANEOUS
TOPICS

12.1   INTRODUCTION

The preceding chapters have covered the principal elements with which designers need to be familiar prior to engaging in the CAD activities. In this chapter a number of miscellaneous topics pertaining to CAD/CAM are discussed. First, the effect of changing the values of some of the parameters involved in CAD of an engineering system on the system behavior is studied by use of a few examples. This so-called parametric study is followed by the optimum study, which finds ways to change the parameter values in an optimum combination or combinations so as to make the system perform in the best way possible.

Numerical control, abbreviated as NC, is a well-developed technology in computer-aided manufacturing processes. It is briefly introduced by use of an example, in which a specific cam profile is to be machined. One of the special-purpose programming languages for CAM called APT is utilized in Program CAM PROFILE, which spells out the step-by-step geometric movements of the machine tool in codes necessary for attaining the desired, precise profile of the cam.

The filament-wound axisymmetric composite shell fabricated by continuously winding of fibers around a premade axisymmetric casing serves very well as an example for CAM. It involves the manufacturing of a prescribed geometric shape and requires the manufacturing process to be worked out. In particular, the timing of the latter has to be *programmed* in detailed steps dictating how the fibers should traverse around a rotating casing and what translational and rotational speeds, acceleration, and so forth should be. The so-called winding kinematics is to be discussed later in this chapter.

Following the same vein of the motion study, four-bar linkages, which have been discussed in Chapter 10, are again chosen as a CAD/CAM example for generation of functions that may be required in attaining a desired geometric shape or motion path or manufacturing process sequence. $y = \log x$ is used as an example in the discussion of four-bar linkage syntheses. It could very well be $y = 50 \sin t$ or $y = 3t^2 + 2t + 1$ required as a part of the motion path in a manufacturing process. Mechanical generation of functions is just as important as electrical, analog simulation presented in Chapter 11, as far as CAD and CAM are concerned.

The closing section discusses the follow-up topics beyond the scope of this book that could be pursued by the reader. Principally, it includes the topics in the specialized fields of engineering that are suited for senior or graduate level courses in the engineering curriculum.

12.2   PARAMETRIC STUDY

Inevitably, a number of parameters are involved in every engineering design. How the changes of the parameters will affect the behavior or performance of the designed element or system is of prominent concern, especially in computer-aided design endeavors. Having interactive CAD equipment at hand enables the designer to investigate these effects by a so-called parametric study. Of course, when a

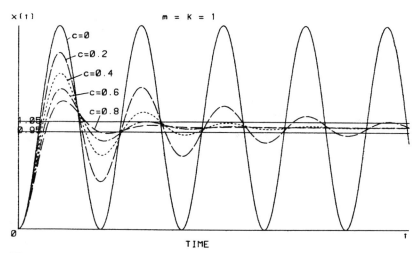

x(t)

m = K = 1

c=0
c=0.2
c=0.4
c=0.6
c=0.8

1.05
0.95

0

TIME

T

Figure 12-1

parameter involved in a design is changed, the designer is most concerned with whether or not it will favorably meet the design specifications.

As an illustration of parametric study, let us recall the translational mechanical system that we used in Chapter 7 for discussion of the design specifications in the time domain. The differential equation governing the system when it is subjected to a unit step excitation is

$$m\frac{d^2x}{dt^2} + c\frac{dx}{dt} + kx = u(t) \tag{a}$$

The steady-state response of the system has been worked out in Section 7.8. To begin the parametric study, let us examine the change in the system response when one of the parameters is changed. Figure 12-1 shows that if the mass m and spring stiffness k are held equal to 1.0 and the damping coefficient c is allowed to change from 0.8 to 0, the responses of the system are significantly different. It indicates that the overshoot and settlement time are increased as c decreases. The delay time and rise time are, however, decreased as c decreases. Figure 12-1 is plotted by use of Program USR. VC.

Suppose that c is held equal to 0.6 and m remains equal to 1.0. Another parametric study can be conducted by varying the spring stiffness k. Figure 12-2 shows the consequent responses of the system. It can be observed from Fig. 12-2 that increase of k results in increase of the overshoot but decrease in rise time and delay time. The settlement time shows very little change. Figure 12-2 is plotted by use of Program USR. VK.

Both Programs USR. VK and USR. VC are listed in Fig. 12-3.

To further demonstrate the frequent need of parametric study in CAD/CAM, let us examine the effect of parametric change on the system response in complying with the design specifications in the frequency domain. For the Bode study of

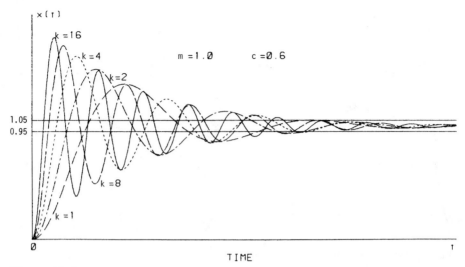

Figure 12-2

```
100 REM * PROGRAM USR.VK *
110 INIT
120 PAGE
130 READ K,M,C,T1,T2,T3
140 DATA 1,1,0.6,0,15,0.05
142 DIM D(5)
144 READ D
146 DATA 3,5,85,2,0
150 A=C/2/M
160 N1=INT((T2-T1)/T3+1.5)
170 DIM T(N1),V(N1)
180 FOR I=1 TO N1
190 T(I)=T1+(I-1)*T3
200 NEXT I
210 VIEWPORT 0,120,20,80
220 WINDOW -0.5,15,0,2
250 MOVE 0,0.95
260 DRAW 30,0.95
264 MOVE 0,1.05
267 DRAW 30,1.05
270 MOVE 0,-5
280 DRAW 0,5
290 MOVE 0,0
300 DRAW 30,0
310 FOR L=1 TO 5
312 DASH D(L)
315 W0=SQR(K/M)
330 W=0.5/M*SQR(4*K*M-C^2)
340 A=C/2/M
```

Figure 12-3

```
350 O=ATN(W/A)
360 FOR I=1 TO N1
370 V(I)=1-WO/W*EXP(-A*T(I))*SIN(W*T(I)+O)
380 NEXT I
390 GOSUB 420
392 K=K*2
400 NEXT L
410 END
420 REM *DRAW TIME FUNCTION V(T)*
430 MOVE 0,0
440 FOR I=1 TO N1
450 DRAW T(I),V(I)
460 NEXT I
470 RETURN

100 REM * PROGRAM USR.VC *
110 INIT
120 PAGE
130 READ K,M,C,T1,T2,T3
140 DATA 1,1,1,0,30,0.1
142 DIM D(5)
144 READ D
146 DATA 3,5,85,2,0
150 WO=SQR(K/M)
160 N1=INT((T2-T1)/T3+1.5)
170 DIM T(N1),V(N1)
180 FOR I=1 TO N1
190 T(I)=T1+(I-1)*T3
200 NEXT I
210 VIEWPORT 0,120,20,80
220 WINDOW -0.5,30.5,0,2
250 MOVE 0,0.95
260 DRAW 30,0.95
264 MOVE 0,1.05
267 DRAW 30,1.05
270 MOVE 0,-5
280 DRAW 0,5
290 MOVE 0,0
300 DRAW 30,0
310 FOR L=1 TO 5
312 DASH D(L)
320 C=C-0.2
330 W=0.5/M*SQR(4*K*M-C^2)
340 A=C/2/M
350 O=ATN(W/A)
360 FOR I=1 TO N1
370 V(I)=1-WO/W*EXP(-A*T(I))*SIN(W*T(I)+O)
380 NEXT I
390 GOSUB 420
400 NEXT L
410 END
420 REM *DRAW TIME FUNCTION V(T)*
430 MOVE 0,0
440 FOR I=1 TO N1
450 DRAW T(I),V(I)
460 NEXT I
470 RETURN
```

Figure 12-3 (*Continued*)

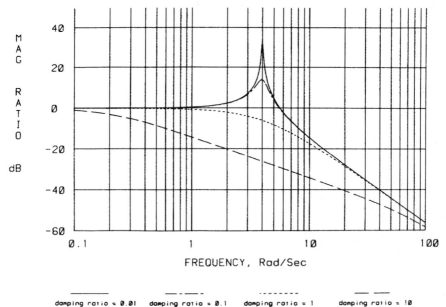

Figure 12-4

Eq. (a), Program BODE. PS has been specifically developed. It is easy to show that the transfer function for the system is

$$G(s) = \frac{1}{m(s^2 + 2\zeta\omega_n s + \omega_n^2)} \tag{b}$$

where

$$\omega_n = (k/m)^{1/2} \tag{c}$$

$$\zeta = c/2m\omega_n \tag{d}$$

Figure 12-4 illustrates the changes in magnitude ratio if the natural frequency ω_n is held equal to 4 rad/sec whereas the damping ratio ζ is varied from 0.01 to 10. Program BODE. PS is listed below.

PROGRAM BODE. PS

```
100 INIT
110 REM * Program BODE.PS *
120 PAGE
130 CHARSIZE 4
140 PRINT "        * BODE ANALYSIS - CHANGING DAMPING RATIO (BODE.PS) *"
150 DIM D3(4)
160 READ D3
170 DATA 0,5,85,3
180 CHARSIZE 2
```

```
190 FOR L=1 TO 4
200 DASH D3(L)
210 MOVE L*25+5,20
220 DRAW L*25+15,20
230 PRINT "_____damping ratio = ";1.0E-3*10^L
240 NEXT L
250 READ C2,O1,B7,F1,C1,D7,P1,Z1
260 DATA 0,0,1,0.1,1,1,0,0
270 I9=2
280 READ X1,X2
290 DATA 1.0E-3,4
300 FOR L=1 TO 4
310 DASH D3(L)
320 X1=X1*10
330 K=1
340 W=F1
350 GOSUB 1410
360 M7=M
370 M1=M
380 W=F1*1000
390 GOSUB 1410
400 M8=M
410 M2=M
420 M1=M1+ABS((M7-M8)*1.1)
430 M2=M2-ABS((M7-M8)*1.3)
440 I1=0.015767*(M1-M2)
450 IF L=1 THEN 470
460 GO TO 930
470 WINDOW -1,3.2,M2-20,M1-20
480 J=1
490 M3=INT(M7+ABS((M7-M8)*0.2))
500 M4=INT(M8-ABS((M7-M8)*0.2))
510 M3=(INT(M3/10)+1)*10
520 M4=(INT(M4/10)-1)*10
530 M3=M3+10
540 FOR I=1 TO 10
550 IF J<>1 THEN 580
560 D1=LGT(I)
570 GO TO 620
580 IF J<>2 THEN 610
590 D1=LGT(I)+1
600 GO TO 620
610 D1=LGT(I)+2
620 MOVE D1,M4
630 DRAW D1,M3
640 NEXT I
650 J=J+1
660 IF J<>4 THEN 540
670 I1=0.015767*(M1-M2)
680 CHARSIZE 4
690 FOR I=M4 TO M3 STEP 20
700 MOVE 0,I
710 DRAW 3,I
720 IF I=0 THEN 800
730 IF I=>-20 THEN 800
740 MOVE -0.3,I-I1
750 PRINT I+20
760 GO TO 820
770 MOVE -0.3,I-I1
780 PRINT 20
790 GO TO 820
```

```
800 MOVE -0.25,I-I1
810 PRINT I+20
820 NEXT I
830 HOME
840 FOR X=0 TO 3 STEP 1
850 IF X=1 THEN 880
860 MOVE X-0.1,M4-3*I1
870 GO TO 890
880 MOVE X-0.08,M4-3*I1
890 PRINT F1*10^X
900 NEXT X
910 GOSUB 1350
920 HOME
930 L9=0
940 T1=F1
950 GOSUB 1190
960 IF L=1 THEN 980
970 GO TO 1040
980 MOVE LGT(4),M4-6*I1
990 CHARSIZE 4
1000 IF I9=2 THEN 1030

1010 PRINT "        FREQUENCY, Hz"
1020 GO TO 1470
1030 PRINT "        FREQUENCY, Rad/Sec"
1040 NEXT L
1100 END
1110 IF I9=2 THEN 1130
1120 W=W/2/PI
1130 RETURN
1140 C=-A6*W^O1
1150 D=A5*W^O1
1160 IF I9=2 THEN 1180
1170 W=W/2/PI
1180 RETURN
1190 REM  * THIS SUB PLOTS THE MAGNITUDE CURVE *
1200 C7=0
1210 FOR W=T1 TO 10*T1 STEP 0.1*T1
1220 GOSUB 1410
1230 D1=LGT(1+0.1*C7)+L9
1240 IF W=F1 THEN 1270
1250 DRAW D1,M-20
1260 GO TO 1280
1270 MOVE D1,M-20
1280 C7=C7+1
1290 NEXT W
1300 T1=T1*10
1310 L9=L9+1
1320 IF L9=3 THEN 1340
1330 GO TO 1200
1340 RETURN
1350 CHARSIZE 4
1360 MOVE -0.5,M3-6*I1
1370 PRINT "M";" A";" G";" ,";" R";" A";" T";" I";" O";
1380 PRINT "     dB."
1390 CHARSIZE 2
1400 RETURN
1410 REM! THIS SUB FINDS THE MAGNITUDE IN DB.
1420 REM! FINAL TERM E+FJ
1430 GOSUB 1490
1440 E=(A*C+B*D)/(C^2+D^2)
```

```
1450 F=(B*C-A*D)/(C^2+D^2)
1460 G=(E^2+F^2)^0.5
1470 M=20*LGT(G)
1480 RETURN
1490 REM! THIS SUB GENERATES THE FUNCTION AS FOLLOWS:
1500 REM! CALCULATE (A+BJ)*(C+DJ) THEN STORES A+BJ
1510 REM!      THEN MULTIPLY (A+BJ)*(E+FJ) AND STORE AS A+BJ
1520 REM!         DOING THIS UNTIL ALL ARE MULTIPLIED
1530 REM!    A1+B1J     ARE      POLES IN DENOMENATER
1540 REM!    A3+B3J     ARE COMPLEX POLES IN DENOMENATER
1550 REM!    A2+B2J     ARE      ZEROS IN NUMERATOR
1560 REM!    A4+B4J     ARE COMPLEX ZEROS IN NUMERATOR
1570 REM!    A+BJ       IS THE FINAL TERM IN NUMERATOR
1580 REM!    C+DJ       IS THE FINAL TERM IN DENOMENATOR.
1590 REM!    B7 IS THE BODE GAIN.
1600 REM!    O1 IS THE ORDER.
1610 IF I9=2 THEN 1620
1620 A2=1
1630 B2=0
1640 A1=1
1650 B1=0
1660 A4=1
1670 B4=0
1680 A3=1-(W/X2)^2
1690 B3=2*X1*W/X2
1700 IF K>C1 THEN 1780
1710 S3=A3
1720 A3=S3*(1-(W/X2)^2)-B3*(2*X1*W/X2)
1730 B3=B3*(1-(W/X2)^2)-S3*(2*X1*W/X2)
1740 K=K+1
1750 GO TO 1700
1760 A3=1
1770 B3=0
1780 A=B7*(A2*A4-B1*B4)
1790 B=B7*(A4*B2+B4*A2)
1800 A5=A1*A3-B1*B3
1810 A6=A1*B3+B1*A3
1820 REM! DEAL WITH ORDER
1830 O2=O1-INT(O1/4)*4
1840 IF O1-INT(O1/2)*2<>0 THEN 1930
1850 C=A5*W^O1
1860 D=A6*W^O1
1870 IF O2=0 THEN 1900
1880 C=-C
1890 D=-D
1900 IF I9=2 THEN 1920
1910 W=W/2/PI
1920 RETURN
1930 C=-A6*W^O1
1940 MOVE L*25,35
1950 IF I9=2 THEN 1970
1960 W=W/2/PI
1970 RETURN
1980 REM  * THIS SUB PLOTS THE MAGNITUDE CURVE *
1990 C7=0
2000 FOR W=T1 TO 10*T1 STEP 0.5*T1

2010 GOSUB 1410
2020 D1=LGT(1+0.5*C7)+L9
2030 IF W=F1 THEN 2060
2040 DRAW D1,M-20
```

```
2050 GO TO 2070
2060 MOVE D1,M-20
2070 C7=C7+1
2080 NEXT W
2090 T1=T1*10
2100 L9=L9+1
2110 IF L9=3 THEN 2130
2120 GO TO 1990
2130 RETURN
```

12.3 OPTIMUM STUDY—ROTATING DISK PROBLEM

Interactive computer graphics provide effective tools for altering designs to make various choices available for final selection. The decision on which design is the best among all the trials will have to be made on the basis of a criterion or a set of criteria. Sometimes, the cost of fabricating a selected design is a dominating factor in decision making. On other occasions, cost may not be a major concern; perhaps keeping the temperature of an operating machine to a minimum is the overriding element. To achieve the best design is the theme of **optimum** studies. A number of specific optimal designs will be discussed here to illustrate how interactive computer graphics can play an essential role in optimal design.

The problem of optimum design of a rotating disk serves very well as an example of how effectively CAD can be applied to examine various disk geometries and to search for the best design. Dr. P. Mahmoodi in a 1969 paper[1] explored this problem and attempted to minimize the disk's hoop stress.

Figure 12-5 is a sketch of a hollow disk that is mounted on a rotating shaft; such a design can be found with a flywheel for machine balancing or a disk pack used in data storage and retrieval operations. The equation for calculation of the tangential stress σ_t in the disk induced during rotation, which has been found to be generally greater than the radial stress, is readily available in the literature.[2] That is,

$$\sigma_t = A - (B/R^2) - \gamma(1 + v)\omega^2 R^2/8g \tag{a}$$

where R is the radial coordinate of a point at which the stress is to be calculated, noting that the stresses are axisymmetric. γ is the density, v is the Poisson's ratio, and ω is the angular speed of the disk. g is the gravitational constant. A and B are constants to be determined from the boundary conditions of the rotating disk.

Mahmoodi derived the expression for the tangential stress, based on Eq. (a), at the innermost rim of the rotating hollow disk, $R = R_3$ to be

$$\sigma_{t3} = \gamma(3 + v)\left(\frac{A}{F} + \frac{B}{C} - \frac{4 + 4v}{3 + v}R_3^2\right)\omega^2 \bigg/ 8g \tag{b}$$

---

[1] P. Mahmoodi, "On the Optimum Design of a Rotating Disk of Nonuniform Thickness," American Society of Mechanical Engineers, Paper No. 69 DE-8, 1969.

[2] W. Flugge, *Handbook of Engineering Mechanics*, McGraw-Hill, New York, 1962, Chapter 37, pp. 15–17.

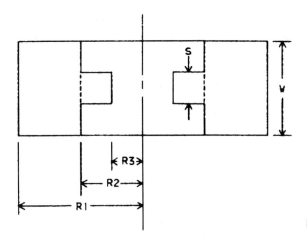

Figure 12-5

where

$$A = \left(\frac{S}{W} - 1\right)R_2^2 - (SR_3^4/WR_2^2) + R_1^2 \tag{c}$$

$$B = (1 - N)R_2^2/R_1^2 + 2N - (1 + N)R_1^2/R_2^2 \tag{d}$$

$$C = FG \tag{e}$$

$$G = (N - 1)/R_2^2 - (1 + N)/R_1^2 \tag{f}$$

$$F = S[1 - (R_3^2/R_2^2)]/2W + [(1/R_1^2) - (1/R_2^2)]/G \tag{g}$$

$$N = v[1 - (S/W)] + (W/S) \tag{h}$$

As can be seen from the above equations, the stress depends on a number of parameters, namely, the geometric parameters S, W, R_1, R_2, and R_3, the material parameters v and γ, and the motion parameter ω. In optimal study, it is desirable in most cases for the problem to be described in **normalized** expressions. The normalized form of Eq. (b) is easily derived as

$$\sigma_{t3}^* = \frac{A^*}{F} + \frac{B}{C^*} - \frac{4 + 4v}{3 + v}(R_{31}^*)^2 \tag{i}$$

where

$$\sigma_{t3}^* = \sigma_{t3} \bigg/ \left[\frac{\gamma(3 + v)}{8g}\omega^2 R_1^2\right] \tag{j}$$

$$A^* = A/R_1^2 = [(S/W) - 1](R_{21}^*)^2 - (S/W)(R_{32}^*/R_{21}^*)^2 + 1 \tag{k}$$

$$C^* = CR_1^2 = F[(N - 1)/(R_{21}^*)^2 - N - 1] \tag{l}$$

$$R_{21}^* = R_2/R_1, \qquad R_{31}^* = R_3/R_1, \qquad R_{32}^* = R_3/R_2 \tag{m,n,o}$$

and for consistency, Eq. (g) can also be written in a normalized form as

$$F = 0.5(S/W)[1 - (R_{32}^*)^2] + [(R_{21}^*)^2 - 1]/[N - 1 - (1 + N)(R_{21}^*)^2] \tag{p}$$

In summary, the normalized tangential stress σ_{t3}^* at the innermost rim of the rotating hollow disk ($R = R_3$) can be studied in terms of the normalized parameters S/W, R_{21}^*, and R_{31}^* (realizing that $R_{32}^* = R_{31}^*/R_{21}^*$ is not an independent parameter). The maximum tangential stress can be minimized for a fixed set of prescribed values of v, γ, and ω by a parametric study of the parameters S/W, R_{21}^*, and R_{31}^*.

Parametric Study

Knowing the material properties of Poisson's ratio v, the density γ of the disk, the rotating speed ω, the normalized maximum tangential stress σ_{t3}^* can be minimized by investigating the first partial derivatives with respect to the normalized parameters S/W, R_{21}^*, and R_{31}^*. However, in view of the complexity of the expressions (i) through (p), it is more practical to conduct parametric studies numerically to investigate the change of σ_{t3}^* relative to these parameters.

A typical parametric study involves selection of the ranges of values in which the particular combination of parameters is expected to provide an optimal solution to the problem being considered. For example, Mahmoodi selected S/W to vary from 0.15 to 0.75, and both R_{21}^* and R_{31}^* to vary from 0.2 to 0.7 for parametric study of the normalized maximum tangential stress σ_{t3}^* of a rotating hollow disk. A number of graphs can then be plotted to provide information about the effects of changing the parameters on σ_{t3}^*. For $v = 0.2$, $\gamma = 0.083$ lb/in^3, and $\omega = 1046.67$ rad/sec, Mahmoodi's graphs show that σ_{t3}^* values decrease as S/W, R_{21}^*, and R_{31}^* are decreased. This conclusion enabled him to proceed to find the optimal designs for various geometric configurations when $R_2 - R_3$ was fixed as equal to 0.25 in.

Table 12-1 lists the values of the circumferential stress at the inner rim of the hollow disk versus R_3 values for R_1 equal to 3, 4, and 5 in. These results were

TABLE 12-1
Results of Parametric Study

R3	Hoop Stress	R3	Hoop Stress	R3	Hoop Stress
0.20	2720.854	0.20	4834.395	0.20	7551.723
0.44	2394.212	0.53	4116.085	0.56	6362.193
0.68	2241.725	0.86	3838.808	0.92	5921.257
0.92	2164.104	1.19	3712.519	1.28	5717.965
1.16	2125.072	1.52	3656.229	1.64	5620.167
1.40	2109.223	1.85	3640.113	2.00	5581.013
1.64	2108.866	2.18	3650.618	2.36	5579.855
1.88	2119.450	2.51	3680.432	2.72	5606.124
2.12	2137.381	2.84	3724.560	3.08	5653.694
2.36	2157.460	3.17	3777.261	3.44	5718.439
2.60	2159.922	3.50	3820.252	3.80	5796.637
($R1 = 3$)		($R1 = 4$)		($R1 = 5$)	

PROGRAM ROTATE.D - MINIMIZIZNG CIRCUMFERENTIAL STRESS

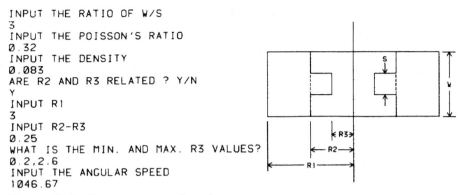

```
INPUT THE RATIO OF W/S
3
INPUT THE POISSON'S RATIO
0.32
INPUT THE DENSITY
0.083
ARE R2 AND R3 RELATED ? Y/N
Y
INPUT R1
3
INPUT R2-R3
0.25
WHAT IS THE MIN. AND MAX. R3 VALUES?
0.2,2.6
INPUT THE ANGULAR SPEED
1046.67
```

Figure 12-6   Program menu and user's response.

generated by Program ROTATE.D, developed with Mahmoodi's method (Fig. 12.6). It shows that the inner radius R_3 has to be increased for larger disks (when R_1 is increased) in order to attain the minimal possible tangential stress at R_3. This trend is manifested by the shifting of the trough of the curves to the right as R_1 is increased from 3 to 5 in.

PROGRAM ROTATE.D

```
100 REM * PROGRAM ROTATE.D *
110 INIT
120 PAGE
130 GOSUB 670
140 HOME
150 GOSUB 270
160 GOSUB 670
170 CHARSIZE 3
180 MOVE 0,77
190 FOR R3=M1 TO M2 STEP 0.1*(M2-M1)
200 GOSUB 1670
210 PRINT USING 220;R3,T
220 IMAGE   3X,3D.2D,5X,6D.3D
230 NEXT R3
240 PRINT " "
250 PRINT "  ** DESIGN END ** "
260 END
270 REM  *   INPUT DATA SECTION   *
280 CHARSIZE 3
290 HOME
300 PRINT "PROGRAM ROTATE.D - MINIMIZIZNG CIRCUMFERENTIAL STRESS "
310 PRINT " INPUT THE RATIO OF W/S"
320 INPUT S1
330 PRINT "INPUT THE POISSON'S RATIO"
340 INPUT V
350 G=32.2*12
360 PRINT "INPUT THE DENSITY"
370 INPUT D
380 PRINT "ARE R2 AND R3 RELATED ? Y/N "
```

```
390 INPUT A$
400 IF A$<>"Y" THEN 460
410 PRINT "INPUT R1"
420 INPUT R1
430 PRINT "INPUT R2-R3"
440 INPUT D1
450 GO TO 480
460 PRINT "INPUT R1,R2"
470 INPUT R1,R2
480 PRINT "WHAT IS THE MIN. AND MAX. R3 VALUES?"
490 INPUT M1,M2
500 PRINT "INPUT THE ANGULAR SPEED "
510 INPUT W
520 CHARSIZE 3
530 PAGE
540 MOVE 0,120
550 PRINT " POISSON'S RATIO = ";V
560 PRINT " DENSITY = ";D
570 IF A$="Y" THEN 600
580 PRINT " INTERMEDIATE RADIUS, R2 = ";R2
590 GO TO 610
600 PRINT " R2-R3= ";D1
610 PRINT " OUTSIDE RADIUS, R1 = ";R1
620 PRINT " W/S = ";S1
630 PRINT " ANGULAR SPEED = ";W
640 PRINT " "
650 PRINT "     R3        HOOP STRESS "
660 RETURN
670 REM  *   THIS SUB PLOTS THE ROTATING DISK
680 MOVE 60,85
690 DRAW 60,70
700 DRAW 100,70
710 DRAW 100,85
720 DRAW 60,85
730 RMOVE 10,0
740 RDRAW 0,-5
750 RDRAW 5,0
760 RDRAW 0,-5
770 RDRAW -5,0
780 RDRAW 0,-5
790 RMOVE 20,0
800 RDRAW 0,5
810 RDRAW -5,0
820 RDRAW 0,5
830 RDRAW 5,0
840 RDRAW 0,5
850 MOVE 80,92
860 DRAW 80,55
870 REM  *    LABEL    *
880 CHARSIZE 2
890 MOVE 60,69
900 DRAW 60,57.5
910 RMOVE 0,1.5
920 SET DEGREES
930 GOSUB 1520
940 RDRAW 8.5,0
950 RMOVE 3,0
960 RDRAW 8.5,0
970 GOSUB 1570
980 MOVE 70,69
```

```
990 DRAW 70,60.5
1000 RMOVE 0,2

1010 GOSUB 1520
1020 RDRAW 3.5,0
1030 RMOVE 3,0
1040 RDRAW 3.5,0
1050 GOSUB 1570
1060 MOVE 75,69
1070 DRAW 75,64.5
1080 RMOVE 0,1.5
1090 GOSUB 1520
1100 RDRAW 1,0
1110 RMOVE 3,0
1120 RDRAW 1,0
1130 GOSUB 1570
1140 MOVE 69.3,58
1150 PRINT "R1"
1160 MOVE 74.3,61.7
1170 PRINT "R2"
1180 MOVE 76.8,65.2
1190 PRINT "R3"
1200 MOVE 101,70
1210 DRAW 104,70
1220 RMOVE -1.5,0
1230 ROTATE 90
1240 GOSUB 1520
1250 ROTATE 0
1260 RDRAW 0,6
1270 RMOVE 0,3
1280 RDRAW 0,6
1290 ROTATE 90
1300 GOSUB 1570
1310 ROTATE 0
1320 MOVE 101,85
1330 DRAW 104,85
1340 MOVE 102.1,76.5
1350 PRINT "W"
1360 MOVE 87.5,72.5
1370 RDRAW 0,2.5
1380 ROTATE 90
1390 GOSUB 1570
1400 MOVE 87.5,82.5
1410 DRAW 87.5,80
1420 GOSUB 1520
1430 MOVE 87.1,82.5
1440 PRINT "S"
1450 ROTATE 0
1460 MOVE 70,75
1470 GOSUB 1620
1480 MOVE 90,75
1490 GOSUB 1620
1500 ROTATE 0
1510 RETURN
1520 RDRAW 1,1*TAN(30)
1530 RMOVE -1,-1*TAN(30)
1540 RDRAW 1,-1*TAN(30)
1550 RMOVE -1,1*TAN(30)
1560 RETURN
1570 RDRAW -1,1*TAN(30)
```

```
1580 RMOVE 1,-1*TAN(30)
1590 RDRAW -1,-1*TAN(30)
1600 RMOVE 1,1*TAN(30)
1610 RETURN
1620 FOR J=1 TO 5
1630 RMOVE 0,0.5/2
1640 RDRAW 0,1.4/2
1650 NEXT J
1660 RETURN
1670 REM  *   THIS IS THE MATH SECTION   *
1680 IF A$()"Y" THEN 1700
1690 R2=R3+D1
1700 B1=D*(3+V)/(G*8)
1710 B=D*(1+3*V)/(8*G)
1720 N=V*(S1-1)+S1
1730 M=(1/S1-1)*R2^2-1/S1*R3^4/R2^2+R1^2
1740 F1=(N-1)/R2^2-(N+1)/R1^2
1750 F2=1/F1*(1/R1^2-1/R2^2)+1/S1/2*(1-(R3/R2)^2)
1760 K=(1-N)*(R2/R1)^2+2*N-(1+N)*(R1/R2)^2
1770 F=F1*F2
1780 T=B1*W^2*(M/F2+K/F-(1+B/B1)*R3^2)
1790 RETURN
```

12.4 OPTIMUM BUS ROUTING

Another notable optimal design problem involves minimizing the milage of school bus routing. Professor P. T. McCoy of the University of Nebraska has developed an operator interactive program to be used on an Apple II microcomputer system to design school bus routes.[3] The program, called SBR/M, contains a number of modules that users can request through conversational question-and-answer dialogue by viewing the messages on the display screen and responding with input from the keyboard of the microcomputer.

As shown in Fig. 12-7, if the locations of stops and the available routes are prescribed, SBR/M has modules for input of real distances, display of the existing routes, evaluation of route length, and database editing. More important, there are modules that optimize rectilinear and real distances and create new stops and routes. There are a total of 23 modules and four data files.

A so-called Lockset Method is used in the routing analysis. There are n stops and the distances between the ith stop and the school are denoted as d_{is} for $i = 1, 2, \ldots, n$. If the bus is to route from school to the ith stop and then return directly to school, denoted as R_{sis}, the route length is

$$L_i = 2d_{is}$$

Suppose that the routing is school–stop$_i$–stop$_j$–school, denoted as R_{sijs}. The route length will then be

$$L_{ij} = d_{is} + d_{ij} + d_{js}$$

---

[3] P. T. McCoy and R. T. Hsueh, "Nebraska School Bus Routing Package, Microcomputer Version I-1 Users Manual," Department of Civil Engineering, University of Nebraska, Lincoln, August 1982.

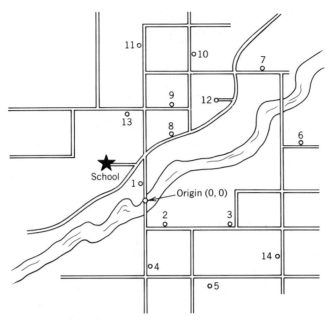

Figure 12-7   Geographic locations of bus stops. (Courtesy of Professor P. T. McCoy, the University of Nebraska—Lincoln.)

with d_{ij} being the distances between stops i and j. The saving in taking R_{sijs} rather than R_{sis} and R_{sjs} can be denoted as

$$S_{ij} = L_i + L_j - L_{ij}$$
$$= 2d_{is} + 2d_{js} - (d_{is} + d_{ij} + d_{js})$$
$$= d_{is} + d_{js} - d_{ij}$$

To find the shortest route is to determine the greatest saving in routing. Mathematically, it is necessary to determine the stopping sequence $i_1 i_2 \cdots i_n$, where i_1 through i_n have the distinct values of 1 through n such that the route $R_{s i_1 i_2 \cdots i_n s}$ will have the greatest saving. That is, $S_{i_1 i_2 \cdots i_n}$ will be maximum among all possible outcomes. For instance, there are a total of 16 possible considerations—S_{1234}, $S_{1243}, S_{1324}, \ldots, S_{4312}, S_{4321}$—when $n = 4$. It should be apparent that computer manipulation aided with operator interactive decision on excluding unrealistic routes is mandatory in such an optimal study.

In actual application the SBR/M program has saved 7 and 20% in school bus routing in the districts in the cities of Blair and Hemingford, Nebraska, respectively.

Optimization study itself is a vast field. Many techniques and methods have been developed, some of which are particularly suited for computer applications. For a comprehensive study on the subject, the book *Foundations of Optimization* by D. J. Wilde and C. S. Beightler (Prentice-Hall, Englewood Cliffs, N.J., 1967) is suggested.

12.5 NUMERICAL CONTROL (NC) OF MANUFACTURING PROCESS

"Numerical control," or "numerically controlled" (both abbreviated as NC), is a field of intensive study in CAM. A numerical control system uses coded numbers to regulate the action of machines. In manufacturing operations NC is being employed in drafting, machining, drilling, rivoting, spot welding, flamecutting, pipe bending, rocket-motor filament winding, electric wiring, electronic assemblying, petroleum refining, and many others. The earliest automatic controls were the music box built around 1650 in Holland and the automatic playing piano by M. Fourneaux built in 1863. The latter was designed by utilizing perforated paper. This concept of punched holes was later employed by M. Falcon in the early 1700s for a knitting machine and certainly is best known in its usage by H. Hollerith in the early 1900s for data storage in U.S. Census studies. The 80-column cards punched with coded characters are still an important part of digital-computer operations, even though remote terminals may eventually completely take over their role in input/output of statements and data.

In manual operations, tolerance accumulates from many sources such as from the manufacturing machine itself, the setup, operator judgment, interpretation of the blueprint drawing, and tooling. An NC system has tolerance from the machine response alone. Generally, NC systems consist of a machine mechanism to be controlled, a control system, and data processing equipment. The attributes of NC systems are accuracy, repeatability, flexibility, and versatility.

Based on the concept of an automatic playing piano, a punched tape may be prepared as a control that spells out the sequence of operations to be performed by an NC machine. It should be easy to perceive that a cassette tape or a floppy

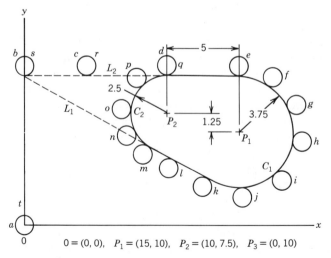

$0 = (0, 0), \quad P_1 = (15, 10), \quad P_2 = (10, 7.5), \quad P_3 = (0, 10)$

All dimensions in cm

Figure 12-8 Machining of a cam profile comprised of parts of C_1, L_1, C_2, and L_2. Cutting path consists of $a \rightarrow b \rightarrow \cdots \rightarrow s \rightarrow t$.

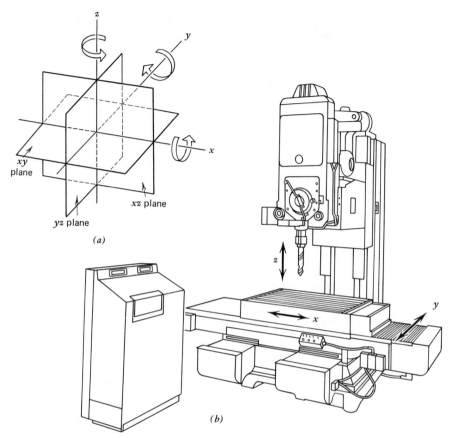

Figure 12-9 Six-degree-of-freedom of three translations and three rotations as shown in (a) for a most versatile machine whereas a vertical-spindle machine shown in (b) can be controlled in three translations and one rotation.

disk will work just as well. There are various special languages such as APT, COMPACT, and CL[4] developed as source languages to expedite the generation of the control tapes needed for every manufacturing process. We shall give an illustrative example using APT for controlled machining of a part detailed in Fig. 12-8.

The abbreviation APT, which stands for automatically programmed tools,[5] is a programming language system facilitating the programmer to write the tool motion commands in simple source language, to compile them, and finally to generate a control tape for NC machines. The next section demonstrates how the part shown in Fig. 12-8 is to be made by use of a thin sheet on a vertical-spindle machine such as the one shown in Fig. 12-9.

---

[4] *Numerical Control*, published by Graphics Technology Corporation, Boulder, Colorado, 1981.

[5] *APT Part Programming*, IIT Research Institute, McGraw-Hill, New York, 1967.

12.6   A SAMPLE NC PROGRAM IN APT LANGUAGE— MACHINING OF CAM PROFILE

A sample APT program is presented in Fig. 12-10 to give a brief introduction to how the cutting path is to be defined by use of APT commands such as GO, GORGT, GOFWD, GOLFT, and GOTO. This program is for implementation of machining the cam profile shown in Fig. 12-8. The statements of the program are to be punched on 80-column IBM cards. As in FORTRAN, columns 73–80 are nonexecutable and to be used for placing the remarks. Here these columns are used to label line numbers.

Line 10 simply uses the command PARTNO to identify the program with a description CAM PROFILE. Line 20 requests that all the tool-end coordinates be printed. Line 30 specifies a cutter of 1.25 cm in diameter be used. Lines 40–120 define points (the numbers after the / being the x, y, z coordinates, respectively), circles (x, y coordinates of the center followed by the radius), lines (starting from the point P_3 and drawn to the right or left and tangent to the circle C_1), and a plane (passing the points P_1, P_2, and P_3). Lines 130–210 delineate how the cutter should be moved: starting from the setup point SETPT, going past (PAST) line $L2$ to the plane XYPLN, then to the right (GORGT) on $L2$ tangent to the circular path $C1$, continuously going forward (GOFWD) tangent to $C1$ until reaching line $L1$, forwarding on $L1$ until tangent to the circular path $C2$, going forward on $C2$ until passing line $L2$, moving to the left (GOLFT) to line $L1$, returning to the setup point, and ending the program (FINI). The cutting path should follow the sequence $a \to b \to \cdots \to s \to t$ as illustrated in Fig. 12-8.

```
                     1          2         3    7       8
Column 12345678901234567890123456789 0...34567890

        PARTNO     CAM PROFILE                        10
                   CLPRNT                             20
                   CUTTER/1.25                        30
        SETPT   =  POINT/0,0,0                         40
        C1      =  CIRCLE/15,10,3.75                   50
        C2      =  CIRCLE/10,7.5,2.5                   60
        P1      =  POINT/15,10,0                        70
        P2      =  POINT/10,7.5,0                       80
        P3      =  POINT/0,10,0                          90
        L1      =  LINE/P3,RIGHT,TANTO,C1              100
        L2      =  LINE/P3,LEFT,TANTO,C1               110
        XYPLN   =  PLANE/P1,P2,P3                      120
                   FROM/SETPT                          130
                   GO/PAST,L2,TO,XYPLN                 140
                   GORGT/L2,TANTO,C1                   150
                   GOFWD/C1,TANTO,L1                   160
                   GOFWD/L1,TANTO,C2                   170
                   GOFWD/C2,PAST,L2                    180
                   GOLFT/L2,TO,L1                      190
                   GOTO/SETPT                          200
                   FINI                                210
```

Figure 12-10   A sample APT program for machining of the cam profile shown in Fig. 12.8.

Figure 12-11   Pocket tool path used in machinging process. (Courtesy of Graphics Technology Corporation.)

The APT program CAM PROFILE demonstrates only the general concept of numerical control programming; it is far from presenting a complete picture of this important field. What has been depicted is merely a preview that brings the readers to the threshold of this interesting field. For example, how should the coolant be applied during machining? Should the spindle be turned clockwise or counterclockwise and at what speed? And what should the translation feed rate be? Apparently, a higher level CAD/CAM course could be devoted to addressing these problems! A practical example of a machining path used in industry is illustrated in Fig. 12-11.

EXERCISE

Use the dynamic graphics capability of CRT to simulate the cutter motion traversing the path $a \rightarrow b \rightarrow \cdots \rightarrow s \rightarrow t$ for machining of the cam profile shown in Fig. 12-8. That is, draw all lines and curves on the screen in **fixed** mode except the rotating circles labeled a through t, which should be drawn in **refreshed** mode by proper use of the VIS, ROPEN, and RCLOSE commands.

12.7 CAD/CAM OF FILAMENT-WOUND AXISYMMETRIC SHELLS

Use of reinforced plastics has increased at a rapid rate since the early 1950s. Filament winding, illustrated in Fig. 12-12, is a reinforced plastic process that continuously wraps resin-impregnated fibers around a rotating mandrel to form an axisymmetric shell of desired geometric shape and dimensions. It was started as a process for making rocket motor cases and launch tubes but nowadays is used for producing automotive leaf strings and other everyday products. The main reason for its popularity is the high strength-to-weight ratio. The fiber-reinforced

Figure 12-12 Filament winding. (Courtesy of McClean Anderson, Inc.)

plastic products can be made lighter than their metallic counterparts but still provide high strength and as good or better electrical, thermal, corrosion, and impact resistances.

Since the filament winding process involves repetitive geometric and motion patterns, it is thus ideal for application of CAD/CAM. Here, a simple case of winding the fibers around an axisymmetric hollow core consisting of a straight cylindrical body and two spherical ends, called a **mandrel**, is adopted for the introduction of the winding kinematics involved in the design of the manufacturing process. As shown in Fig. 12-13a, the three segments are denoted with lengths of L_{AB}, L_{BC}, and L_{CD}. Generally, L_{AB} may not be equal to L_{CD} for the reason that the shell may be required to have openings of unequal sizes at its ends. The core is to be mounted on a rotating shaft by use of proper end fixtures. Many different fiber materials are currently in use, including fiberglass, graphite, boron, and metallic wires. These reinforcement fibers pass through a bath of resin and are mounted on a carriage as shown in Fig. 12-12. The resin can be epoxy, silicone, or one of many newly introduced materials. Usually, the carriage is designed to traverse along the axial, z, direction of the shell and to move along the direction perpendicular to the z axis, a motion called **cross feed**.

To work out the details of the kinematics involved in the process, let us denote

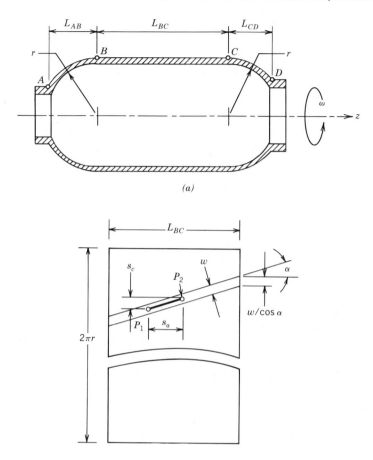

(a)

(b)   **Figure 12-13**

the outside radius of the core as r, the winding angle as α, and the width of the fiber strand as w. As shown in Fig. 12-13b, the total number of strands necessary for complete covering of the surface of the core by one layer of fibers is

$$n = 2\pi r/(w/\cos \alpha) = (2\pi r \cos \alpha)/w \qquad (a)$$

Equation (a) is derived on the basis that each fiber strand covers a portion equal to $w/\cos \alpha$ of the total circumferential length $2\pi r$ of the cylindrical segment of the shell. For simplicity of discussion, consider the case when the fiber winding is to be carried out by a constant angular speed, ω, of the mandrel. During a time increment Δt, let us assume that the point where the fibers touch the shell have moved from P_1 to P_2 as shown in Fig. 12-13b. In order to maintain a desired, constant winding angle α, the distances traversed by the fibers along the circumferential direction, s_c, and along the axial direction, s_a, must be precisely proportioned so that the following equation is satisfied:

$$s_c = s_a \tan \alpha \qquad (b)$$

Let t_{BC} be the time required for the fibers to traverse from one end, B, to the other end, C, of the straight, cylindrical segment of the shell. To find t_{BC} in terms of ω, L_{BC}, and α, we observe that all of the points on the surface of the shell, such as P_1 and P_2, have a circumferential velocity equal to $r\omega$. Hence, for a time increment of Δt,

$$s_c = r\omega \, \Delta t \tag{c}$$

Meanwhile, fibers are mounted on a carriage and pass a device called **eye** and then tightly wrapped around the rotating shell. Let the carriage velocity be denoted as v; the fiberaxial displacement can then be written as

$$s_a = v \, \Delta t \tag{d}$$

Upon combining Eqs. (b), (c), and (d), we arrive at

$$v = s_a/\Delta t = (s_c \cot \alpha)/\Delta t = r\omega \cot \alpha \tag{e}$$

This indicates that if r, ω, and α are constants, the carriage must also traverse with a constant axial velocity.

As a first attempt in the determination of the fiber winding kinematics, we shall schedule the velocity change of the carriage using a constant acceleration/ deceleration approach. According to Eq. (e), for the segment BC of the shell, $v_B = v_C = v$ and the acceleration, a_{BC}, is equal to zero. A velocity reversal has to take place at the ends A and D, which requires $v_A = v_D = 0$. The simplest way to find a feasible acceleration for the spherical segments is to assume it to be constant. For example, let a_{AB} be a constant, which can then be computed by use of the following kinematics equations:[6]

$$v_B = v_A + a_{AB}t_{AB} \tag{f}$$

$$L_{AB} = v_A t_{AB} + \tfrac{1}{2}a_{AB}t_{AB}^2 \tag{g}$$

Since $v_A = 0$ and $v_B = v$, it is easy to obtain from Eqs. (f) and (g) that

$$a_{AB} = v^2/2L_{AB} \tag{h}$$

$$t_{AB} = 2L_{AB}/v \tag{i}$$

Obviously, the same approach for deciding the axial deceleration of the carriage can be followed for the segment CD and furthermore be employed for the returning path $D \rightarrow C \rightarrow B \rightarrow A$ of the fiber winding.

In Table 12-2 a summarized, segment-by-segment detail of the carriage kinematics is presented for a complete cycle of the fiber winding path, namely $A \rightarrow B \rightarrow C \rightarrow D \rightarrow C \rightarrow B \rightarrow A$. Since each turnaround wraps two strands of fibers on the surface of the shell, the total time required for complete covering of the surface of the shell with fibers can be computed with Eqs. (a) and (e), and also with the aid of Table 12-2. It is

---

[6] J. L. Meriam, *Dynamics*, John Wiley & Sons, New York, 1978, pp. 15–34.

TABLE 12-2
Carriage Kinematics (with Constant Angular Speed ω of the Mandrel and $v = r\omega \cot \alpha^a$)

Segments	Velocities		Acceleration	Time Increment
$A \to B$	$v_A = 0$	$v_B = v$	$a_{AB} = v^2/2L_{AB}$	$t_{AB} = 2L_{AB}/v$
$B \to C$	$v_B = v$	$v_C = v$	$a_{BC} = 0$	$t_{BC} = L_{BC}/v$
$C \to D$	$v_C = v$	$v_D = 0$	$a_{CD} = -v^2/2L_{CD}$	$t_{CD} = 2L_{CD}/v$
$D \to C$	$v_D = 0$	$v_C = v$	$a_{DC} = v^2/2L_{CD}$	$t_{DC} = 2L_{CD}/v$
$C \to B$	$v_C = v$	$v_B = v$	$a_{CB} = 0$	$t_{CB} = L_{BC}/v$
$B \to A$	$v_B = v$	$v_A = 0$	$a_{BA} = -v^2/2L_{AB}$	$t_{BA} = 2L_{AB}/v$

^{a} r is the outer radius of the cylindrical shell and α is the fiber winding angle.

$$T = \frac{n}{2} \times 2(t_{AB} + t_{BC} + t_{CD}) = \frac{2\pi r \cdot \cos \alpha}{w}(2L_{AB} + L_{BC} + L_{CD})/v$$

$$= \frac{2\pi r^2 \cdot \cos \alpha}{w}(2L_{AB} + L_{BC} + 2L_{CD})/r\omega \cot \alpha$$

$$= 2\pi r^2(2L_{AB} + L_{BC} + 2L_{CD})\cos^2 \alpha/w\omega \sin \alpha \qquad \text{(j)}$$

It should again be pointed out that Eq. (j) is a result of using a constant angular speed ω for the mandrel rotation and constant axial acceleration/deceleration of the carriage for winding the fibers around the spherical segments of the shell.

Program FIBER. S has been developed for interactive CAD of the manufacturing kinematics of filament-wound axisymmetric shells. A numerical sample case is presented in Table 12-2. See the listing of the program at the end of this section.

Referring again to Fig. 12-13a, when the fibers are wrapped around the cylindrical, BC, segment of the shell, the strands of width w should be arranged to adjoin each other without overlapping. This requires that the number of strands n, defined by Eq. (a), for complete covering of the shell surface with one layer of fibers must be an integer. Hence, the values of r, α, and w must be accurately related in accordance with Eq. (a) in order to fulfill this requirement. In the meantime, it should be obvious that the fiber strands are, in most cases, overlapped in the spherical segments, AB and CD, of the segment. As a consequence, the wall thickness varies from the cylinder–sphere junction (B or C) to the end (A or D) but remains constant in the segment BC of the shell. There are various ways of design[7] for the end segments (spherical or ellipsoidal) of the shell according to different criteria. These specialized topics are, however, beyond the scope of this book and should therefore be left as individual pursuits for the interested reader.

Another noteworthy observation is that the combination of the constant axial acceleration/deceleration of the carriage for fiber winding of the spherical segment

---

[7] Y. C. Pao, "Momentless Design of Composite Structures with Variable Elastic Constants," *Journal of Composite Materials*, Vol. 3, 1969, pp. 604–616.

of the shell, a_{AB}—to be computed by use of Eq. (h)—and the constant velocity for the cylindrical segment of the shell, $a_{BC} = 0$, will cause a **jerk** of the carriage. This jerk motion occurs at the junctions B and C and during both the forwarding ($AB \rightarrow BC$ and $BC \rightarrow CD$) and returning ($DC \rightarrow CB$ and $CB \rightarrow BA$) paths of the carriage. How to alleviate or minimize such jerk motion is also an intensive research topic by itself[8] and worthy of exploration for possible application of CAD/CAM.

PROGRAM FIBER. S

```
100 REM ****** PROGRAM FIBER.S ******
110 INIT
120 PAGE
130 CHARSIZE 4
140 SET DEGREES
150 PRINT "* PROGRAM FIBER.S *"
160 DIM L(3),T3(3),V9(51),A9(51),T4(4),A3(3),S9(51)
170 READ L,R,F,A,N
180 DATA 10,30,8,12,30,45,50
190 CHARSIZE 2
200 PRINT "_LENGTHS OF SHELL SEGMENTS ARE (CM.) :";
210 PRINT "  L(AB) = ";L(1);"   L(BC) = ";L(2);"   L(CD) = ";L(3)
220 PRINT "_OUTER RADIUS OF SHELL IS (CM.) : ";R
230 PRINT "_ANGULAR SPEED OF MANDREL IS (RPM.) : ";F
240 PRINT "_NUMBER OF FIBER STRANDS PER LAYER IS : ";N
250 V=R*F*2*3.1416/60*COS(A)/SIN(A)
260 T3(1)=2*L(1)/V
270 T3(2)=L(2)/V
280 T3(3)=2*L(3)/V
290 A3(1)=V^2/2/L(1)
300 A3(2)=0
310 A3(3)=-(V^2)/2/L(3)
320 T4(1)=0
330 T4(2)=T4(1)+T3(1)
340 T4(3)=T4(2)+T3(2)
350 T4(4)=T4(3)+T3(3)
360 W=2*3.1416*R*COS(A)/N
370 PRINT "_FIBER STRAND WIDTH SHOULD BE EQUAL TO (CM.) : ";W
380 T6=2*(T3(1)+T3(2)+T3(3))
390 PRINT "_TOTAL WINDING TIME PER LAYER IS (SEC.) : ";T6
400 D6=T6/100
410 PRINT "_           TIME            VELOCITY           ACCELERATION";
420 PRINT "   AXIAL DISTANCE"
430 FOR C=1 TO 101 STEP 1
440 T=(C-1)*D6
450 IF C>51 THEN 500
460 GOSUB 540
470 REM A>B>C>D WINDING PATH
480 PRINT T,V9(C),A9(C),S9(C)
490 GO TO 520
```

[8] Y. C. Pao, "A User's Manual of WK2 Computer Program (*W*inding *K*inematics Analysis of CAD/CAM of Axisymmetric Filament Composite Shells, Version *2*)," Company Report, The Brunswick Corporation, August 31, 1981.

```
500 REM D>C>B>A WINDING PATH
510 PRINT T,-V9(102-C),-A9(102-C),S9(102-C)
520 NEXT C
530 END
540 REM *WINDING KINEMATICS*
550 IF T<=T4(2) AND T=>T4(1) THEN 670
560 IF T<=T4(3) AND T=>T4(2) THEN 620
570 REM PATH CD
580 A9(C)=A3(3)
590 V9(C)=V+A3(3)*(T-T4(3))
600 S9(C)=S9(C-1)+V9(C-1)*D6+0.5*A3(3)*D6^2
610 RETURN
620 REM PATH BC
630 A9(C)=0
640 V9(C)=V
650 S9(C)=S9(C-1)+V*D6
660 RETURN
670 REM PATH AB
680 A9(C)=A3(1)
690 V9(C)=A3(1)*T
700 IF C=1 THEN 730
710 S9(C)=S9(C-1)+V9(C-1)*D6+0.5*A3(1)*D6^2
720 GO TO 740
730 S9(1)=0
740 RETURN
```

* PROGRAM FIBER.S *

LENGTHS OF SHELL SEGMENTS ARE (CM.) : L(AB) = 10   L(BC) = 30   L(CD) = 8

OUTER RADIUS OF SHELL IS (CM.) : 12

ANGULAR SPEED OF MANDREL IS (RPM.) : 30

NUMBER OF FIBER STRANDS PER LAYER IS : 50

FIBER STRAND WIDTH SHOULD BE EQUAL TO (CM.) : 1.06620439861

TOTAL WINDING TIME PER LAYER IS (SEC.) : 3.50140056022

TIME	VELOCITY	ACCELERATION	AXIAL DISTANCE
0	0	71.061484032	0
0.0350140056022	2.4881472	71.061484032	0.04356
0.0700280112045	4.9762944	71.061484032	0.17424
0.105042016807	7.4644416	71.061484032	0.39204
0.140056022409	9.9525888	71.061484032	0.69696
0.175070028011	12.440736	71.061484032	1.089
0.210084033613	14.9288832	71.061484032	1.56816
0.245098039216	17.4170304	71.061484032	2.13444
0.280112044818	19.9051776	71.061484032	2.78784
0.31512605042	22.3933248	71.061484032	3.52836
0.350140056022	24.881472	71.061484032	4.356
0.385154061625	27.3696192	71.061484032	5.27076
0.420168067227	29.8577664	71.061484032	6.27264
0.455182072829	32.3459136	71.061484032	7.36164
0.490196078431	34.8340608	71.061484032	8.53776
0.525210084034	37.322208	71.061484032	9.801
0.560224089636	37.6992	0	11.121
0.595238095238	37.6992	0	12.441
0.63025210084	37.6992	0	13.761
0.665266106443	37.6992	0	15.081
0.700280112045	37.6992	0	16.401
0.735294117647	37.6992	0	17.721
0.770308123249	37.6992	0	19.041
0.805322128852	37.6992	0	20.361
0.840336134454	37.6992	0	21.681
0.875350140056	37.6992	0	23.001
0.910364145658	37.6992	0	24.321
0.945378151261	37.6992	0	25.641
0.980392156863	37.6992	0	26.961
1.01540616247	37.6992	0	28.281
1.05042016807	37.6992	0	29.601
1.08543417367	37.6992	0	30.921
1.12044817927	37.6992	0	32.241

1.15546218487	37.6992	0	33.561
1.19047619048	37.6992	0	34.881
1.22549019608	37.6992	0	36.201
1.26050420168	37.6992	0	37.521
1.29551820728	37.6992	0	38.841
1.33053221289	37.322208	-88.82685504	40.10655
1.36554621849	34.212024	-88.82685504	41.3589
1.40056022409	31.10184	-88.82685504	42.50235
1.43557422969	27.991656	-88.82685504	43.5369
1.47058823529	24.881472	-88.82685504	44.46255
1.50560224090	21.771288	-88.82685504	45.2793
1.54061624651	18.661104	-88.82685504	45.98715
1.57563025210	15.55092	-88.82685504	46.5861
1.61064425771	12.440736	-88.82685504	47.07615
1.64565826331	9.330552	-88.82685504	47.4573
1.68067226891	6.220368	-88.82685504	47.72955
1.71568627451	3.110184	-88.82685504	47.8929
1.75070028011	2.046363079E-12	-88.82685504	47.94735
1.78571428571	-3.110184	88.82685504	47.8929
1.82072829132	-6.220368	88.82685504	47.72955
1.85574229692	-9.330552	88.82685504	47.4573
1.89075630252	-12.440736	88.82685504	47.07615
1.92577030812	-15.55092	88.82685504	46.5861
1.96078431373	-18.661104	88.82685504	45.98715
1.99579831933	-21.771288	88.82685504	45.2793
2.03081232493	-24.881472	88.82685504	44.46255
2.06582633053	-27.991656	88.82685504	43.5369
2.10084033613	-31.10184	88.82685504	42.50235
2.13585434174	-34.212024	88.82685504	41.3589
2.17086834734	-37.322208	88.82685504	40.10655
2.20588235204	-37.6992	0	38.841
2.24089635854	-37.6992	0	37.521
2.27591036415	-37.6992	0	36.201
2.31092436975	-37.6992	0	34.881
2.34593837535	-37.6992	0	33.561
2.38095238095	-37.6992	0	32.241
2.41596638655	-37.6992	0	30.921
2.45098039216	-37.6992	0	29.601
2.48599439776	-37.6992	0	28.281
2.52100840336	-37.6992	0	26.961
2.55602240896	-37.6992	0	25.641
2.59103641457	-37.6992	0	24.321
2.62605042017	-37.6992	0	23.001
2.66106442577	-37.6992	0	21.681
2.69607843137	-37.6992	0	20.361
2.73109243697	-37.6992	0	19.041
2.76610644258	-37.6992	0	17.721
2.80112044818	-37.6992	0	16.401
2.83613445378	-37.6992	0	15.081
2.87114845938	-37.6992	0	13.761
2.90616246499	-37.6992	0	12.441
2.94117647059	-37.6992	0	11.121
2.97619047619	-37.322208	-71.061484032	9.801
3.01120448179	-34.8340608	-71.061484032	8.53776
3.0462184874	-32.3459136	-71.061484032	7.36164
3.081232493	-29.8577664	-71.061484032	6.27264
3.1162464986	-27.3696192	-71.061484032	5.27076
3.1512605042	-24.881472	-71.061484032	4.356
3.1862745098	-22.3933248	-71.061484032	3.52836
3.22128851541	-19.9051776	-71.061484032	2.78784
3.25630252101	-17.4170304	-71.061484032	2.13444
3.29131652661	-14.9288832	-71.061484032	1.56816
3.32633053221	-12.440736	-71.061484032	1.089
3.36134453782	-9.9525888	-71.061484032	0.69696
3.39635854342	-7.4644416	-71.061484032	0.39204
3.43137254902	-4.9762944	-71.061484032	0.17424
3.46638655462	-2.4881472	-71.061484032	0.04356
3.50140056022	0	-71.061484032	0

EXERCISE

Modify the Program FIBER.S to incorporate a parabolic acceleration/deceleration approach for a_{AB} so that the acceleration is at a relative maximum at the end of the shell, point A, and is equal to zero at the junction, point B, shown in Fig. 12-13a. The jerk, which is the time derivative of the acceleration, will then be linear in time, maximum at point A, and equal to zero at the point B. *Hint:* Write $a_{AB} = c_1 + c_2 t + c_3 t^2$. Then $da_{AB}/dt = c_2 + 2c_3 t$. The conditions to be satisfied are $a_{AB} = 0$ and $da_{AB}/dt = 0$ at $t = t_B$, and $da_{AB}/dt = j_A$ at $t = t_A = 0$, where j_A is a selected initial value of the carriage jerk.

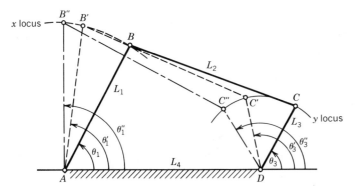

Figure 12-14

12.8   FOUR-BAR LINKAGE SYNTHESIS OF $y = \log x$

In Chapter 11 analog circuits were presented for generation of commonly used functions—sine, cosine, exponential, and polynomial functions. Here, we shall discuss a geometric method for generation of functions using planar four-bar linkages. The function $y = \log x$ is used as an illustration in discussion of the procedure involved. It is adopted for simplicity of discussion; the procedure delineated can be followed for generation of any other functions. Since it is a geometric method, interactive graphics can thus be employed to the best advantage.

Figure 12-14 shows a planar four-bar linkage $ABCD$. The driving crank is AB and the driven crank is CD. For generation of the function $y = \log x$, let the locus of B represents the independent variable x and the locus of C represents the dependent variable y. The problem is then to determine the dimensions, namely, the lengths L_1 through L_4, of the linkage. Suppose that it is required to achieve $y_C = \log x_B$, $y_{C'} = \log x_{B'}$, and $y_{C''} = \log x_{B''}$, where x_B, $x_{B'}$, and $x_{B''}$ are three selected points[9] within the interval of $x_i \le x \le x_l$, for which the linkage is to be designed to generate the logarithmic function. It is easy to observe from Fig. 12-14 the following geometric relationships:

$$L_1 = (x_{B''} - x_B)/(\theta_1'' - \theta_1) = (x_{B'} - x_B)/(\theta_1' - \theta_1) \qquad \text{(a)}$$

$$L_3 = (y_{C''} - y_C)/(\theta_3'' - \theta_3) = (y_{C'} - y_C)/(\theta_3' - \theta_3) \qquad \text{(b)}$$

$$y_C = \log x_B \qquad y_{C'} = \log x_{B'} \qquad y_{C''} = \log x_{B''} \qquad \text{(c,d,e)}$$

The graphical procedure for finding the necessary lengths L_1 through L_4 involves (1) assuming a convenient length A_1D_1 as shown in Fig. 12-15, (2) choosing an appropriate length A_1B_1 along with θ_1 to locate the point B_1, (3) using an adequately chosen θ_3 to locate the poles P' and P'', (4) drawing the lines $P'C_1$ and $P''C_1$ based on the measured angles ϕ and ψ, and (5) measuring A_1B_1, B_1C_1, and

---

[9] These are called accuracy points. For a discussion of how these points should be selected to achieve the least error in generation of functions, see R. S. Hartenberg and J. Denavit, *Kinematic Synthesis of Linkages*, McGraw-Hill, New York, 1964.

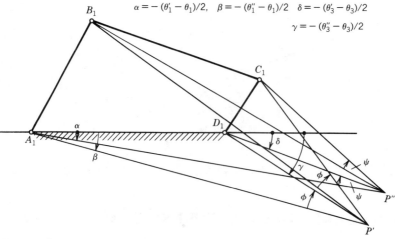

$$\alpha = -(\theta_1' - \theta_1)/2, \quad \beta = -(\theta_1'' - \theta_1)/2 \quad \delta = -(\theta_3' - \theta_3)/2$$

$$\gamma = -(\theta_3'' - \theta_3)/2$$

Figure 12-15

C_1D_1, D_1A_1, which are the lengths of the linkage for generation of $y = \log x$. To locate the poles P' and P'', the selected θ_1, x_B, $x_{B'}$, and $x_{B''}$ are used for calculation of θ_1' and θ_1'' based on Eq. (a) and subsequently α and β shown in Fig. 12-15. The selected θ_3 value together with Eqs. (b to e) enable the values θ_3', θ_3'', δ, and γ to be calculated. P' is obtained as the intercept of the line drawn from A_1 having an angle of α measured from the A_1D_1 line, and the line drawn from D_1 having angle of δ measured from A_1D_1 line. P'' is obtained in a similar manner but using the angles β and γ instead.

The selection of A_1B_1 determines the location of B_1. Two lines B_1P' and B_1P'' can then be drawn and the angles $A_1P'B_1$ and $A_1P''B_1$ can be measured and called ϕ and ψ. Using ϕ and ψ, $P'C_1$ and $P''C_1$ can be drawn to locate the intercept C_1. This constitutes step 4 mentioned above.

It should be evident that the locations of the poles P' and P'' depend on the selection of θ_1, the length of A_1B_1, and θ_3. Sometimes considerable trial-and-error effort is necessary to arrive at a satisfactory solution. Hence, interactive computer graphics can be effectively utilized for such a task; Program LOGX has therefore been developed. It can be easily changed for generation of any function $y = f(x)$ by simply observing the fact that the Eqs. (c), (d), and (e) contain the only change that needs to be made in the program. Figures 12-16 and 12-17 are presented to illustrate a sample application of Program LOGX, for which x_i and x_l have been set equal to 1.0 and 2.0, respectively, for the range of the x. The choice of $\theta_1 = 45°$, $\theta_3 = 10°$, $A_1D_1 = L_4 = 1$ in., and $A_1B_1 = 1$ in. allows us to arrive at the presented results.

PROGRAM LOGX

```
100 INIT
110 CHARSIZE 4
120 PAGE
```

```
* PROGRAM LOGX - 4-BAR LINKAGE SYNTHESIS OF Y=LOG(X) *

RANGE OF X (X1<X<X5). ENTER X1,X5 (1,2 FOR SAMPLE RUN): 1,2

ENTER THE THREE ACCURACY POINTS (X2,X3,X4) DESIRED.
  (1.1, 1.5, AND 1.9 FOR SAMPLE RUN) : 1.1,1.5,1.9


 Y2=  0.0414  Y3=  0.1761  Y4=  0.2788

ENTER DESIRED RANGE OF ROTATION (IN DEGREES) OF INPUT AND
OUTPUT CRANKS AND DISTANCE BETWEEN FIXED PIVOTS.
  (60, 60, 1 FOR SAMPLE RUN): 60,60,1

ENTER THE DESIRED INITIAL POSITIONS (IN DEGREES) OF THE
FIRST ACCURACY POINTS OF THE INPUT AND OUTPUT CRANKS.
  (45 AND 0 FOR SAMPLE RUN): 45,0


 L1= -1.049  L2=  2.900  L3= -2.545  L4=  1.000

TO DISPLAY 4-BAR MECHANISM, ENTER Y:  TO DISPLAY DATA FROM 4-BAR OUTPUT,
 ENTER D. Y
```

Figure 12-16

```
130 PRINT "* PROGRAM LOGX - 4-BAR LINKAGE SYNTHESIS OF Y=LOG(X) *"
260 O1=4
290 GO TO 2390
930 REM *THIS SUBROUTINE DOES THE PLOTTING*
940 WINDOW 0.8*-B7,2*B7,-B7*1.3,B7*1.3
950 GOSUB 1240
960 CHARSIZE 3
970 V2=L4-L1*COS(S1)+L3*COS(T3)
980 V2=V2/L2
990 IF ABS(V2)>1 THEN 1160
1000 T2=ACS(V2)

1010 IF L1*COS(S1)+L2*COS(T2)-L3*COS(T3)<>L4 THEN 1160
1020 MOVE 0,0
1030 DRAW L1*COS(S1),L1*SIN(S1)
1040 MOVE 0,-0.235*ABS(L1)
1050 PRINT "A"
1060 MOVE L4,0
1070 DRAW L3*COS(T3)+L4,L3*SIN(T3)
1080 DRAW L1*COS(S1),L1*SIN(S1)
1090 MOVE L1*COS(S1)*1.09,L1*SIN(S1)*1.09
1100 PRINT "B"
1110 MOVE L4*1.025,-0.235*ABS(L1)
1120 PRINT "D"
1130 MOVE L3*COS(T3)*1.05+L4,L3*SIN(T3)*1.05
1140 PRINT "C"
1150 RETURN
1160 REM    ITERATION FAILED
1190 MOVE 0,B7
1200 PRINT "ITERATION FOR THETA3 FAILED"
1210 MOVE 0,0.8*B7
1220 PRINT "WHEN THETA1 =";S1
1230 END
1240 REM *DRAW THE PINS*
1250 MOVE L4+L3*COS(T3)-0.008*B7,L3*SIN(T3)-0.02*B7
```

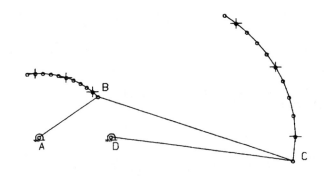

Y=LOG(X) FUNCTION GENERATOR

X-LOCUS OF B   Y-LOCUS OF C

RANGE OF X : 1.0 TO 2.0

+ INDICATES LOCATION OF CRITICAL POINTS.

TO DISPLAY 4-BAR MECHANISM, ENTER "Y"; TO DISPLAY DATA, ENTER "D". FOR GRAPH OF DATA, ENTER "G";
TO RETURN TO START OF PROGRAM, ENTER "S"; TO END PROGRAM, PRESS <RETURN> D

ERROR IN 4-BAR FUNCTION GENERATOR

X	T1,DEG	T3,DEG	Y	Y-MECH	ERROR
1.00	39.0000	-9.1234	0.0000	-0.0044	-0.0044
1.10	45.0000	0.0000	0.0414	0.0414	0.0000
1.20	51.0000	7.7915	0.0792	0.0805	0.0013
1.30	57.0000	14.7064	0.1139	0.1152	0.0012
1.40	63.0000	21.0049	0.1461	0.1468	0.0006
1.50	69.0000	26.8475	0.1761	0.1761	0.0000
1.60	75.0000	32.3386	0.2041	0.2036	-0.0005
1.70	81.0000	37.5487	0.2304	0.2298	-0.0007
1.80	87.0000	42.5273	0.2553	0.2548	-0.0005
1.90	93.0000	47.3098	0.2788	0.2788	0.0000
2.00	99.0000	51.9217	0.3010	0.3019	0.0000

DIMENSIONS OF 4-BAR MECHANISM

LINK	LENGTH
AB	-1.0489
BC	2.9000
CD	-2.5453
DA	1.0000

INPUT DATA

RANGE OF X: 1 TO 2
CRITICAL ACCURACY PTS: X= 1.1 , 1.5 , 1.9
RANGE OF ROTATION OF INPUT AND OUTPUT CRANKS IN DEGREES: D1=60 D3=60
POSITION OF FIRST ACCURACY POINTS FOR INPUT AND OUTPUT CRANKS IN DEGREES: T1=45 T3=0
DISTANCE BETWEEN FIXED SUPPORTS: L4= 1 INCHES

TO DISPLAY 4-BAR MECHANISM, ENTER "Y"; TO END PROGRAM; PRESS <RETURN>.

Figure 12-17

```
1260 CHARSIZE 1
1270 PRINT "o"
1280 MOVE L1*COS(S1)-0.006*B7,L1*SIN(S1)-B7*0.02
1290 PRINT "o"
1300 MOVE -0.008*B7,-B7*0.02
1310 PRINT "o"
1320 MOVE L4-0.008*B7,-B7*0.02
1330 PRINT "o"
1340 CHARSIZE 4
1350 RETURN
1360 REM *DRAW THE SUPPORTS AT A AND D*
1370 WINDOW 0.8*-B7,2*B7,-B7*1.3,B7*1.3
1380 MOVE L4*0.05,-L4*0.05
1390 DRAW L4*0.05,0
1400 FOR C1=1 TO 180 STEP 10
1410 DRAW L4*0.05*COS(C1),L4*0.05*SIN(C1)
1420 NEXT C1
1430 DRAW -L4*0.05,-L4*0.05
1440 MOVE L4*1.05,-0.05*L4
1450 DRAW L4*1.05,0
1460 FOR C4=1 TO 180 STEP 10
1470 DRAW L4*0.05*COS(C4)+L4,L4*SIN(C4)*0.05
1480 NEXT C4
1490 DRAW L4*0.95,-L4*0.05
1500 MOVE -L4*0.08,-0.05*L4
1510 DRAW L4*0.08,-0.05*L4
1520 MOVE L4-L4*0.08,-0.05*L4
1530 DRAW L4*1.08,-0.05*L4
1540 FOR C7=-1 TO 29 STEP 6
1550 MOVE -L4*0.07+L4*C7/500,-0.07*L4
1560 DRAW -L4*0.05+C7/500*L4,-0.05*L4
1570 NEXT C7
1580 FOR C8=-1 TO 29 STEP 6
1590 MOVE L4*0.93+L4*C8/500,-0.07*L4
1600 DRAW L4*0.95+L4*C8/500,-0.05*L4
1610 NEXT C8
1620 HOME
1630 CHARSIZE 4
1640 RETURN
1650 REM *THIS SUBROUTINE DOES THE SCALING*
1660 IF ABS(L1)>ABS(L2) THEN 1690
1670 B7=ABS(L2)
1680 GO TO 1700
1690 B7=ABS(L1)
1700 IF B7>ABS(L3) THEN 1720
1710 B7=ABS(L3)
1720 IF B7>ABS(L4) THEN 1740
1730 B7=ABS(L4)
1740 RETURN
1910 HOME
1920 PRINT "DO YOU WANT TO TRY ANOTHER MECHANISM?"
1930 PRINT "ENTER Y FOR YES, N FOR NO. ";
1940 INPUT B$
1950 IF B$="N" THEN 1990
1960 PAGE
1970 HOME
1980 GO TO 190
1990 PRINT "_END OF PROGRAM"
2000 PRINT

2020 HOME
2030 END
2380 END
```

```
2390 REM   *4-BAR FUNCTION GENERATOR SUBROUTINE*
2400 DIM Y(5),Z(5),X(11),P(11),Q(11),R(11),E(11),F(11)
2540 PRINT "_RANGE OF X (X1<X<X5, ENTER X1,X5 (1,2 FOR SAMPLE RUN): ";
2550 INPUT Z(1),Z(5)
2560 PRINT "_ENTER THE THREE ACCURACY POINTS (X2,X3,X4) DESIRED.  "
2562 PRINT "  (1.1, 1.5, AND 1.9 FOR SAMPLE RUN) : ";
2570 INPUT Z(2),Z(3),Z(4)
2580 FOR I=1 TO 5
2590 X(I)=Z(I)
2600 Y(I)=LGT(X(I))
2610 NEXT I
2620 PRINT USING 2630:"_ Y2=",Y(2)," Y3=",Y(3)," Y4=",Y(4)
2630 IMAGE L,5A,3D.4D,5A,3D.4D,5A,3D.4D
2640 PRINT "_ENTER DESIRED RANGE OF ROTATION (IN DEGREES) OF INPUT AND"
2650 PRINT "OUTPUT CRANKS AND DISTANCE BETWEEN FIXED PIVOTS.  "
2652 PRINT "  (60, 60, 1 FOR SAMPLE RUN): ";
2670 INPUT D1,D3,L4
2680 D1=D1*0.0174533
2690 D3=D3*0.0174533
2700 PRINT "_ENTER THE DESIRED INITIAL POSITIONS (IN DEGREES) OF THE"
2710 PRINT "FIRST ACCURACY POINTS OF THE INPUT AND OUTPUT CRANKS."
2712 PRINT "  (45 AND 0 FOR SAMPLE RUN): ";
2720 INPUT T1,T4
2730 A1=T1*0.0174533
2740 B1=T4*0.0174533
2750 SET RADIANS
2760 A2=(Z(3)-Z(2))/(Z(5)-Z(1))*D1+A1
2770 A3=(Z(4)-Z(2))/(Z(5)-Z(1))*D1+A1
2780 B2=(Y(3)-Y(2))/(Y(5)-Y(1))*D3+B1
2790 B3=(Y(4)-Y(2))/(Y(5)-Y(1))*D3+B1
2800 REM *SOLUTION OF THREE FREUDENSTEIN EQUATIONS*
2810 U1=COS(A1)-COS(A2)
2820 U2=COS(B1)-COS(B2)
2830 U3=COS(A1-B1)-COS(A2-B2)
2840 U4=COS(A1)-COS(A3)
2850 U5=COS(B1)-COS(B3)
2860 U6=COS(A1-B1)-COS(A3-B3)
2870 K1=(U2*U6-U3*U5)/(U2*U4-U1*U5)
2880 K2=(U1*U6-U3*U4)/(U2*U4-U1*U5)
2890 K3=COS(A1-B1)-K1*COS(A1)+K2*COS(B1)
2900 L1=L4/K2
2910 L3=L4/K1
2920 L2=SQR(L1^2+L3^2+L4^2-2*L1*L3*K3)
2930 PRINT USING 2940:"_ L1=",L1," L2=",L2," L3=",L3," L4=",L4
2940 IMAGE 1L,6A,3D.3D,5A,3D.3D,5A,3D.3D,5A,3D.3D
2950 PRI "_TO DISPLAY 4-BAR MECHANISM, ENTER Y:  TO DISPLAY DATA FROM";
2960 PRINT " 4-BAR OUTPUT, ENTER D. ";
2970 INPUT A$
2980 PAGE
2990 D=Z(5)-Z(1)
3000 A4=A1-D1/D*(Z(2)-Z(1))

3010 REM  S1=A4/0.0174533+180
3020 D2=Y(5)-Y(1)
3030 A5=B1-D3/D2*(Y(2)-Y(1))
3040 GOSUB 4610
3050 S1=P(1)+180
3060 T3=Q(1)+180
3070 IF A$="D" THEN 3580
3080 REM  **DRAW 4-BAR ROUTINE**
3090 SET DEGREES
3100 GOSUB 1650
```

```
3110 GOSUB 1360
3120 GOSUB 930
3130 CHARSIZE 1
3140 FOR I=1 TO 11
3150 S1=P(I)+180
3160 T3=Q(I)+180
3170 MOVE L4+L3*COS(T3)-0.008*B7,L3*SIN(T3)-0.02*B7
3180 PRINT "o"
3190 MOVE L1*COS(S1)-0.006*B7,L1*SIN(S1)-B7*0.02
3200 PRINT "o"
3210 IF I=1 THEN 3280
3220 S2=P(I-1)+180
3230 T6=Q(I-1)+180
3240 MOVE L4+L3*COS(T6),L3*SIN(T6)
3250 DRAW L4+L3*COS(T3),L3*SIN(T3)
3260 MOVE L1*COS(S2),L1*SIN(S2)
3270 DRAW L1*COS(S1),L1*SIN(S1)
3280 NEXT I
3290 S1=P(1)+180
3300 T3=Q(1)+180
3310 SET DEGREES
3320 X1=B7*3
3330 MOVE L1*COS(A1/0.0174533+180),L1*SIN(A1/0.0174533+180)
3340 GOSUB 4420
3350 MOVE L1*COS(A2/0.0174533+180),L1*SIN(A2/0.0174533+180)
3360 GOSUB 4420
3370 MOVE L1*COS(A3/0.0174533+180),L1*SIN(A3/0.0174533+180)
3380 GOSUB 4420
3390 MOVE L4+L3*COS(B1/0.0174533+180),L3*SIN(B1/0.0174533+180)
3400 GOSUB 4420
3410 MOVE L4+L3*COS(B2/0.0174533+180),L3*SIN(B2/0.0174533+180)
3420 GOSUB 4420
3430 MOVE L4+L3*COS(B3/0.0174533+180),L3*SIN(B3/0.0174533+180)
3440 GOSUB 4420
3450 MOVE 0,0
3460 CHARSIZE 4
3470 PRINT "____Y=LOG(X) FUNCTION GENERATOR"
3480 CHARSIZE 2
3490 PRINT "___                         X-LOCUS OF B          ";
3500 PRINT "                   Y-LOCUS OF C"
3510 PRINT USING 3520:"__RANGE OF X : ",Z(1)," TO ",Z(5)
3520 IMAGE 43X,15A,2D.1D,4A,2D.1D
3530 PRINT USING 3540:"+ INDICATES LOCATION OF CRITICAL POINTS."
3540 IMAGE 2L,36X,40A
3550 PRINT "___     "
3560 CHARSIZE 2
3570 GO TO 4490
3580 REM  **ERROR TABLE ROUTINE**
3590 CHARSIZE 3
3600 PRINT "_                      ERROR IN 4-BAR FUNCTION GENERATOR_"
3610 CHARSIZE 1
3620 PRINT USING 3640:"X","T1,DEG","T3,DEG","  Y  ","Y-MECH","ERROR"
3630 PRINT
3640 IMAGE 15X,1A,16X,6A,15X,6A,15X,6A,14X,6A,16X,8A
3650 FOR I=1 TO 11
3660 PRINT USING 3670:X(I),P(I),Q(I),R(I),F(I),E(I)
3670 IMAGE 12X,2D.2D,12X,4D.4D,12X,4D.4D,12X,4D.4D,12X,4D.4D,12X,4D.4D
3680 NEXT I
3690 CHARSIZE 4
3700 PRINT "___ DIMENSIONS OF 4-BAR MECHANISM"
3710 CHARSIZE 3
3720 PRINT "_      LINK          LENGTH_"
3730 PRINT USING 3732:"AB",L1
```

```
3732 IMAGE 6X,2A,9X,2D.4D
3734 PRINT USING 3732;"BC",L2
3736 PRINT USING 3732;"CD",L3
3738 PRINT USING 3732;"DA",L4
3740 CHARSIZE 4
3750 PRINT "____ INPUT DATA"
3760 CHARSIZE 2
3770 PRINT "_ RANGE OF X:  ";Z(1);" TO ";Z(5)
3780 PRINT " CRITICAL ACCURACY PTS:  X= ";Z(2);" , ";Z(3);" , ";Z(4)
3790 PRINT " RANGE OF ROTATION OF INPUT AND OUTPUT CRANKS IN DEGREES: ";
3800 PRINT "D1=";D1/0.0174533;"  D3=";D3/0.0174533
3810 PRINT " POSITION OF FIRST ACCURACY POINTS FOR INPUT AND OUTPUT ";
3820 PRINT "CRANKS IN DEGREES:  T1=";T1;"  T3=";T4
3830 PRINT " DISTANCE BETWEEN FIXED SUPPORTS:  L4= ";L4;" INCHES"
3860 PRINT "____ "
3870 CHARSIZE 2
3880 PRINT "____      TO DISPLAY";
3890 PRINT " 4-BAR MECHANISM, ENTER ""Y""; TO END PROGRAM: PRESS";
3900 PRINT " <RETURN>. ";
3910 INPUT A$
3920 PAGE
3930 IF A$="Y" THEN 3090
3940 GO TO 1990
3950 REM   **GRAPH DISPLAY ROUTINE**
3960 CHARSIZE 2
3970 VIEWPORT 30,110,30,90
3980 WINDOW Z(1),Z(5),Y(1),Y(5)
3990 AXIS 0.1,0.1
4000 MOVE X(1),R(1)

4010 DRAW X,R
4020 RMOVE 0.01*(Z(5)-Z(1)),0.01*(Y(5)-Y(1))
4030 PRINT "Y=F(X)"
4040 MOVE X(1),F(1)
4050 DRAW X,F
4060 RMOVE 0.01*(Z(5)-Z(1)),-0.01*(Y(5)-Y(1))
4070 PRINT "4-BAR"
4080 GOSUB 4310
4090 VIEWPORT 0,130,0,100
4100 WINDOW 0,130,0,100
4110 MOVE 68,26
4120 CHARSIZE 3
4130 PRINT "X-AXIS"
4140 MOVE 19,60
4150 PRINT "Y-AXIS"
4160 MOVE 26,27
4170 IMAGE 3D.3D
4180 PRINT USING 4170;Z(1)
4190 MOVE 106,27
4200 PRINT USING 4170;Z(5)
4210 MOVE 18,29
4220 PRINT USING 4170;Y(1)
4230 MOVE 18,89
4240 PRINT USING 4170;Y(5)
4250 MOVE 40,20
4260 CHARSIZE 4
4270 PRINT "Y=F(X) VS. 4-BAR MECHANISM OUTPUT."
4280 MOVE 5,8
4290 CHARSIZE 2
4300 GO TO 4490
4310 REM *PLOT ACCURACY POINTS*
4320 SET DEGREES
```

```
4330 X1=D
4340 MOVE Z(2),Y(2)
4350 GOSUB 4420
4360 MOVE Z(3),Y(3)
4370 GOSUB 4420
4380 MOVE Z(4),Y(4)
4390 GOSUB 4420
4400 SET RADIANS
4410 RETURN
4420 REM *DRAW + ROUTINE*
4430 FOR K=1 TO 4
4440 ROTATE (K-1)*90
4450 RDRAW 0.01*X1,0
4460 RMOVE -0.01*X1,0
4470 NEXT K
4480 RETURN
4490 PRINT "_TO DISPLAY 4-BAR MECHANISM, ENTER ""Y"":  TO DISPLAY DATA";
4500 PRINT ", ENTER ""D"":  FOR GRAPH OF DATA, ENTER ""G"":"
4510 PRINT "TO RETURN TO START OF PROGRAM, ENTER ""S"":  TO END ";
4520 PRINT "PROGRAM, PRESS <RETURN>.";
4530 INPUT A$
4540 PAGE
4550 IF A$="D" THEN 3580
4560 IF A$="Y" THEN 3090
4570 IF A$="G" THEN 3950
4580 CHARSIZE 4
4590 IF A$="S" THEN 100
4600 GO TO 4780
4610 REM  **CALCULATE TEN POSITIONS OF 4-BAR MECHANISM**
4620 SET RADIANS
4630 FOR I=1 TO 11
4640 X(I)=Z(1)+(I-1)*D/10
4650 P(I)=D1/D*(X(I)-Z(1))+A4
4660 A=SIN(P(I))
4670 B=L4/L1+COS(P(I))
4680 C=L4/L3*COS(P(I))+(L1*L1-L2*L2+L3*L3+L4*L4)/(2*L1*L3)
4690 Q(I)=2*ATN((A-SQR(A*A+B*B-C*C))/(B+C))
4700 R(I)=LGT(X(I))
4710 F(I)=D2/D3*(Q(I)-A5)+Y(1)
4720 E(I)=F(I)-R(I)
4730 P(I)=P(I)/0.0174533
4740 Q(I)=Q(I)/0.0174533
4750 NEXT I
4760 SET DEGREES
4770 RETURN
4780 END
```

12.9 CLOSING REMARKS

The common implication of the topics discussed in this chapter is that many different subjects have to be further explored in depth. As it is the intent of this book to present a broad spectrum of the **elements** that are considered essential to the CAD/CAM, the coverage throughout this book has been maintained approximately at the junior level and has encompassed all engineering fields. Depending on his or her specific field, the reader may want to extend the fundamental knowledge acquired from this book in many different areas. Among the possible senior/graduate level subjects, the following undoubtedly are the most prominent:

CAD of integrated circuits

CAD of mechanisms

CAD of process plants

CAD of control systems

CAD of structures

CAM and robotics

As was pointed out at the outset of this book, these specialized subjects should be contained in books written by experts in the field. Only faculty members, practicing engineers, and researchers in the particular field can provide in-depth information and the most up-to-date materials pertaining to the CAD/CAM subject.

In addition to emphasizing of software development for CAD/CAM, the software and hardware available on the market should be surveyed, explained, and utilized in these senior/graduate level books. These packages and equipment should have already incorporated the prevailing methods of analysis, design codes, and necessary databases for implementation of a good many of CAD/CAM tasks. It spares the user from time-consuming software development and enables the user to try different designs and manufacturing processes expeditiously.

APPENDIXES

APPENDIX A

MATRIX ALGEBRA, CRAMER'S RULE, AND GAUSSIAN ELIMINATION

INTRODUCTION

Solving a number of unknowns from a given set of linear equations is a frequent need in computing. It may be a case like the following:

$$2x_1 + 3x_2 + 5x_3 = 23$$
$$4x_1 + x_2 + 2x_3 = 12 \qquad \text{(a)}$$
$$x_1 + 9x_2 + 7x_3 = 40$$

There are three unknowns x_{1-3} to be solved from three linear algebraic equations.

A general system of linear equations has the form

$$a_{11}x_1 + a_{12}x_2 + a_{13}x_3 + \cdots + a_{1n}x_n = c_1$$
$$a_{21}x_1 + a_{22}x_2 + a_{23}x_3 + \cdots + a_{2n}x_n = c_2 \qquad \text{(b)}$$
$$\vdots \quad \vdots \quad \vdots \quad \vdots \quad \vdots \quad \vdots \quad \vdots \quad \vdots$$
$$a_{n1}x_1 + a_{n2}x_2 + a_{n3}x_3 + \cdots + a_{nn}x_n = c_n$$

where a_{ij}'s and c_i's are known constants and x_j's are unknowns. If we introduce the matrix notation, Eq. (b) can be written in a compact form

$$AX = C \qquad \text{(b')}$$

A **matrix** is an ordered rectangular array of elements. For example,

$$A = \begin{bmatrix} a_{11} & a_{12} & a_{13} & \cdots & a_{1n} \\ a_{21} & a_{22} & a_{23} & \cdots & a_{2n} \\ \vdots & \vdots & \vdots & \vdots & \vdots \\ a_{n1} & a_{n2} & a_{n3} & \cdots & a_{nn} \end{bmatrix} = [a_{ij}] \qquad \text{(c)}$$

There are $n \times n$ **elements** in matrix A. We refer to a_{ij} as the element of A at the location of ith **row** and jth **column**.

A matrix that has m rows and n columns is called an $m \times n$ matrix, and it is said that the matrix is of **order** $m \times n$. A matrix whose number of rows is equal to its number of columns is called a **square** matrix. A matrix having all its elements

426

equal to zero is called a **null** or **zero** matrix. A matrix that has only one row is called a **row** matrix, whereas a matrix that has only one column is called a **column** or a **vector** such as in

$$
X = \begin{bmatrix} x_1 \\ x_2 \\ \vdots \\ x_n \end{bmatrix}
\quad \text{and} \quad
C = \begin{bmatrix} c_1 \\ c_2 \\ \vdots \\ c_n \end{bmatrix}
\tag{d,e}
$$

Main diagonal of a matrix refers to the line connecting all the **diagonal elements**, which in the case of matrix A in Eq. (c) are the elements $a_{11}, a_{22}, a_{33}, \ldots a_{nn}$. A **diagonal matrix** is a square matrix having all its elements equal to zero except those along the diagonal. A **unity** or **identity matrix** is a diagonal matrix whose diagonal elements are all equal to unity.

A square matrix $M = [m_{ij}]$ is called a **symmetric matrix** if its elements are such that $m_{ij} = m_{ji}$, an **antisymmetric** or **skew-symmetric matrix** if $m_{ij} = -m_{ji}$.

Transpose of a matrix A, written as A^T, is a new matrix obtained from A by interchanging the rows and columns, so

$$
A^T = \begin{bmatrix}
a_{11} & a_{21} & a_{31} & \cdots & a_{n1} \\
a_{12} & a_{22} & a_{32} & \cdots & a_{n2} \\
\vdots & \vdots & \vdots & \vdots & \vdots \\
a_{1n} & a_{2n} & a_{3n} & \cdots & a_{nn}
\end{bmatrix}
\tag{f}
$$

In the case of Eq. (a), the coefficient matrix and its transpose are

$$
A = \begin{bmatrix} 2 & 3 & 5 \\ 3 & 2 & 2 \\ 1 & 9 & 7 \end{bmatrix}
\quad
A^T = \begin{bmatrix} 2 & 3 & 1 \\ 3 & 2 & 9 \\ 5 & 2 & 7 \end{bmatrix}
$$

ADDITION AND SUBTRACTION OF MATRICES

Two matrices of **same order** can be added or subtracted to obtain a resultant or difference matrix of the same order. The operation is accomplished by adding or subtracting corresponding elements. If $A = [a_{ij}]$, $B = [b_{ij}]$, $C = [c_{ij}]$, $D = [d_{ij}]$, and $C = A + B$ and $D = A - B$ then

$$
c_{ij} = a_{ij} + b_{ij}
\tag{g}
$$

$$
d_{ij} = a_{ij} - b_{ij}
\tag{h}
$$

Example:

$$
A = \begin{bmatrix} 1 & 2 & 3 \\ 2 & 3 & 1 \\ 3 & 1 & 2 \end{bmatrix}
\quad
B = \begin{bmatrix} 4 & 5 & 6 \\ 5 & 6 & 4 \\ 6 & 4 & 5 \end{bmatrix}
$$

$$C = A + B = \begin{bmatrix} 1+4 & 2+5 & 3+6 \\ 2+5 & 3+6 & 1+4 \\ 3+6 & 1+4 & 2+5 \end{bmatrix} = \begin{bmatrix} 5 & 7 & 9 \\ 7 & 9 & 5 \\ 9 & 5 & 7 \end{bmatrix}$$

$$D = A - B = \begin{bmatrix} 1-4 & 2-5 & 3-6 \\ 2-5 & 3-6 & 1-4 \\ 3-6 & 1-4 & 2-5 \end{bmatrix} = \begin{bmatrix} -3 & -3 & -3 \\ -3 & -3 & -3 \\ -3 & -3 & -3 \end{bmatrix}$$

MULTIPLICATION OF MATRICES

Two matrices $A = [a_{ij}]$ and $B = [b_{ij}]$ can be multiplied, written as AB (A post-multiplied by B, or B premultiplied by A) in that order only when the number of columns of matrix A is equal to the number of rows of matrix B. If A is of order $l \times m$ and B of order $m \times n$, $E = AB = [e_{ij}]$ will be a matrix of order $l \times n$. The elements of E are calculated by use of the formula

$$e_{ij} = \sum_{k=1}^{m} a_{ik}b_{kj} \tag{i}$$

Example:

$$A_{3\times3} = \begin{bmatrix} 0 & 1 & 2 \\ 1 & 2 & 0 \\ 2 & 0 & 1 \end{bmatrix} \qquad B_{3\times2} = \begin{bmatrix} 1 & 2 \\ 3 & 1 \\ 2 & 3 \end{bmatrix}$$

$$e_{11} = \sum_{k=1}^{3} a_{1k}b_{k1} = a_{11}b_{11} + a_{12}b_{21} + a_{13}b_{31} = 0 \times 1 + 1 \times 3 + 2 \times 2 = 3 + 4 = 7$$

$$e_{12} = \sum_{k=1}^{3} a_{1k}b_{k2} = a_{11}b_{12} + a_{12}b_{22} + a_{13}b_{32} = 0 \times 2 + 1 \times 1 + 2 \times 3 = 1 + 6 = 7$$

$$\vdots \vdots \vdots \vdots \vdots \vdots \vdots \vdots \vdots \vdots \vdots \vdots \vdots$$

$$E_{3\times2} = \begin{bmatrix} 7 & 7 \\ 7 & 4 \\ 4 & 7 \end{bmatrix}$$

In general $AB \neq BA$. For example,

$$A = \begin{bmatrix} 1 & 0 & 1 \\ 0 & 1 & 0 \\ 0 & 0 & 1 \end{bmatrix} \qquad B = \begin{bmatrix} 1 & 0 & 2 \\ 0 & 1 & 0 \\ 1 & 0 & 1 \end{bmatrix}$$

$$AB = \begin{bmatrix} 2 & 0 & 3 \\ 0 & 1 & 0 \\ 1 & 0 & 1 \end{bmatrix} \qquad BA = \begin{bmatrix} 1 & 0 & 3 \\ 0 & 1 & 0 \\ 1 & 0 & 2 \end{bmatrix}$$

Having established the rule of matrix multiplication, we can now show that Eq. (b') really represents Eq. (b). From Eq. (c) and Eqs. (d) and (e) it is evident that

$$AX = \begin{bmatrix} a_{11} & a_{12} & \cdots & a_{1n} \\ a_{21} & a_{22} & \cdots & a_{2n} \\ \vdots & \vdots & \vdots & \vdots \\ a_{n1} & a_{n2} & \cdots & a_{nn} \end{bmatrix} \begin{bmatrix} x_1 \\ x_2 \\ \vdots \\ x_n \end{bmatrix} = \begin{bmatrix} a_{11}x_1 + a_{12}x_2 + \cdots + a_{1n}x_n \\ a_{21}x_1 + a_{22}x_2 + \cdots + a_{2n}x_n \\ \vdots\vdots\vdots\vdots\vdots\vdots\vdots \\ a_{n1}x + a_{n2}x_2 + \cdots + a_{nn}x_n \end{bmatrix} = \begin{bmatrix} c_1 \\ c_2 \\ \vdots \\ c_n \end{bmatrix} = C$$

The following multiplication formulas are commonly used in matrix operations:

$$kA = k[a_{ij}] = [ka_{ij}] \tag{j}$$

$$AI = IA = A \tag{k}$$

$$(AB)^T = B^T A^T \tag{l}$$

$$L^2 = V^T V = \begin{bmatrix} v_1 & v_2 & \cdots & v_n \end{bmatrix} \begin{bmatrix} v_1 \\ v_2 \\ \vdots \\ v_n \end{bmatrix} = v_1^2 + v_2^2 + \cdots + v_n^2 \tag{m}$$

In the above equations, k is a scaler, I is a unity matrix, V is a vector, and L is the length of V.

INVERSE OF A SQUARE MATRIX

A matrix is called the **inverse** of a square matrix A, denoted as A^{-1}, if

$$AA^{-1} = A^{-1}A = I \tag{n}$$

The **determinant** of a square matrix A of order n ($n \times n$) is

$$D = |A| = \begin{vmatrix} a_{11} & a_{12} & \cdots & a_{1n} \\ a_{21} & a_{22} & \cdots & a_{2n} \\ \vdots\vdots\vdots\vdots\vdots\vdots \\ a_{n1} & a_{n2} & \cdots & a_{nn} \end{vmatrix}$$

$$= \sum (-1)^h a_{1i_1} a_{2i_2} \cdots a_{n-1,\, i_{n-1}} a_{ni_n} \tag{o}$$

where the summation extends over all possible arrangements $(i_1, i_2, i_3, \ldots, i_n)$ of the n second subscripts $(1, 2, \ldots, n)$, and h is the total number of inversions in the sequence of the second subscripts, that is, the total number of times any of the numbers $i_1, i_2, \ldots, i_n$ precedes a smaller number. Example:

$$D = \begin{vmatrix} a_{11} & a_{12} & a_{13} \\ a_{21} & a_{22} & a_{23} \\ a_{31} & a_{32} & a_{33} \end{vmatrix} = \begin{vmatrix} 1 & 2 & 3 \\ 2 & 3 & 1 \\ 3 & 1 & 2 \end{vmatrix}$$

$$= (-1)^0 a_{11}a_{22}a_{33} + (-1)^1 a_{11}a_{23}a_{32} + (-1)^1 a_{12}a_{21}a_{33}$$
$$\quad + (-1)^2 a_{12}a_{23}a_{31} + (-1)^2 a_{13}a_{21}a_{32} + (-1)^3 a_{13}a_{22}a_{31}$$
$$= 6 - 1 - 8 + 6 + 6 - 27 = -18$$

A matrix whose determinant is equal to zero is called a **singular** matrix. A submatrix obtained by striking out the ith row and the jth column of a matrix $A = [a_{ij}]$ is called the **minor** of a_{ij}. **Cofactor** of a_{ij} is equal to $(-1)^{i+j}$ multiplied by the determinant of the minor of a_{ij}, usually denoted as A_{ij}. **Adjoint** of a matrix A is a matrix whose transpose has A_{ij} as its elements, denoted as

$$\text{Adj } A = [A_{ij}]^T \tag{p}$$

Inverse of a matrix A can be calculated by the equation

$$A^{-1} = (\text{Adj } A)/D \tag{q}$$

A convenient way of finding the determinant of a square matrix is the **Laplace expansion**, which states that a determinant is equal to the sum of the product of elements of any single row i or column j multiplied by their cofactors

$$A = \sum_{k=1}^{n} a_{ik} A_{ik}, \quad \text{or} \quad A = \sum_{k=1}^{n} a_{kj} A_{kj} \tag{r}$$

where n is the order of the square matrix.

CRAMER'S RULE

A simple method for solving three simultaneous linear algebraic equations is called **Cramer's rule.**[1] It requires the use of determinants. The method may best be explained by solving a system of equations, such as

$$9x_1 + x_2 + x_3 = 10$$
$$3x_1 + 6x_2 + x_3 = 14 \tag{a}$$
$$x_1 + 2x_2 + 3x_3 = 2$$

Cramer's rule proceeds to calculate the determinants D, D_1, D_2, and D_3, where D is the determinant of the coefficient matrix of the system of equations (a) and D_i for $i = 1, 2, 3$ are the determinants of the matrices when the ith column of the coefficient matrix is replaced by the constants appearing on the right-hand sides of Eq. (a). Explicitly, these determinants are

$$D = \begin{vmatrix} 9 & 1 & 1 \\ 3 & 6 & 1 \\ 1 & 2 & 3 \end{vmatrix} = 136 \qquad D_1 = \begin{vmatrix} 10 & 1 & 1 \\ 14 & 6 & 1 \\ 2 & 2 & 3 \end{vmatrix} = 136$$

$$D_2 = \begin{vmatrix} 9 & 10 & 1 \\ 3 & 14 & 1 \\ 1 & 2 & 3 \end{vmatrix} = 272 \qquad D_3 = \begin{vmatrix} 9 & 1 & 10 \\ 3 & 6 & 14 \\ 1 & 2 & 2 \end{vmatrix} = 136$$

The solutions x_i for $i = 1, 2, 3$ of the system (a) determined by Cramer's rule are

$$x_i = D_i/D \tag{b}$$

---

[1] S. Perlis, *Theory of Matrices*, Addison-Wesley, Reading, Massachusetts, 1958.

Numerically, they are

$$x_1 = 136/136 = 1 \qquad x_2 = 272/136 = 2 \qquad x_3 = 136/136 = 1$$

A subroutine called CRAMER coded in both BASIC and FORTRAN, is presented below. It is utilized frequently in this text.

SUBROUTINE CRAMER(A,B,M,C)

```
C
C       USE CRAMER'S RULE TO SOLVE MATRIX EQUATION AC=B  (ORDER M<=3)
C       NEEDS SUBROUTINE DETRM3
        DIMENSION A(M,M),B(M),C(M),T(3,3)
        GO TO (1,2,3),M
      1 C(1)=B(1)/A(1,1)
        RETURN
      2 D=A(1,1)*A(2,2)-A(1,2)*A(2,1)
        C(1)=(B(1)*A(2,2)-B(2)*A(1,2))/D
        C(2)=(A(1,1)*B(2)-A(2,1)*B(1))/D
        RETURN
      3 CALL DETRM3(A,D)
        DO 15 K=1,3
        DO 10 I=1,3
        DO 10 J=1,3
        IF(J.EQ.K) GO TO 5
        T(I,J)=A(I,J)
        GO TO 10
      5 T(I,J)=B(I)
     10 CONTINUE
        CALL DETRM3(T,C(K))
     15 C(K)=C(K)/D
        RETURN
        END

        SUBROUTINE DETRM3(A,D)
C       CALCULATE THE DETERMINANT OF THE 3*3 MATRIX A
        DIMENSION A(3,3)
      3 D=A(1,1)*A(2,2)*A(3,3)+A(2,1)*A(3,2)*A(1,3)+A(3,1)*A(1,2)
     *   *A(2,3)-A(1,3)*A(2,2)*(3,1)-A(2,3)*A(3,2)*A(1,1)-A(3,3)
     *   *A(1,2)*A(2,1)
        RETURN
        END
    100 REM SUBROUTINE CRAMER - CRAMER'S RULE
    110 REM SOLVES MATRIX EQUATION AC=B (ORDER M<=3)
    120 DIM A(M,M),B(M),C(M),T(3,3)
    130 GO TO M OF 140,160,200
    140 C(1)=B(1)/A(1,1)
    150 RETURN
    160 D=A(1,1)*A(2,2)-A(1,2)*A(2,1)
    170 C(1)=(B(1)*A(2,2)-B(2)*A(1,2))/D
    180 C(2)=(A(1,1)*B(2)-A(2,1)*B(1))/D
    190 RETURN
    200 REM*SUBROUTINE DETERM - CALCULATES THE DETERMINANT OF A 3X3
            MATRIX D
    210 D=A(1,1)*A(2,2)*A(3,3)+A(2,1)*A(3,2)*A(1,3)+A(3,1)*A(1,2)
            *A(2,3)
    220 D=D-A(1,3)*A(2,2)*A(3,1)-A(2,3)*A(3,2)*A(1,1)-A(3,3)*A(1,2)
            *A(2,1)
    230 FOR K=1 TO 3
```

```
240 FOR I=1 TO 3
250 FOR J=1 TO 3
260 IF J=K THEN 290
270 T(I,J)=A(I,J)
280 GO TO 300
290 T(I,J)=B(I)
300 NEXT J
310 NEXT I
320 D=T(1,1)*T(2,2)*T(3,3)+T(2,1)*T(3,2)*T(1,3)+T(3,1)*T(1,2)
      *T(2,3)
330 D=D-T(1,2)*T(2,2)*T(3,1)-T(2,3)*T(3,2)*T(1,1)-T(3,3)*T(1,2)
      *T(2,1)
340 C(K)=D/D
350 NEXT K
360 RETURN
```

SOLUTION OF MATRIX EQUATIONS

There are many methods for the solution of systems of linear algebraic equations.[2] Here, we shall delineate the most commonly used method, called **Gaussian elimination**. Again, let us describe the procedure by use of the system of three equations that we have used for discussion of the Cramer's rule. The system will be referred to as

$$AX = C \tag{a}$$

where

$$A = [a_{ij}] = \begin{bmatrix} 9 & 1 & 1 \\ 3 & 6 & 1 \\ 1 & 2 & 3 \end{bmatrix} \qquad X = [x_i] = \begin{bmatrix} x_1 \\ x_2 \\ x_3 \end{bmatrix} \qquad [C] = [c_i] = \begin{bmatrix} 10 \\ 14 \\ 2 \end{bmatrix}$$

The first step is called the **normalization** of the first equation—that is, we divide the entire first equation by the coefficient of the x_1 term. The general formula for deriving the **new** coefficients from the **old** coefficients is, noting that $n = 3$,

$$(a_{1,j})_{\text{new}} = (a_{1,j})_{\text{old}}/(a_{1,1})_{\text{old}} \qquad \text{for} \qquad j = 1, 2, \ldots, n \tag{b}$$

and

$$(c_1)_{\text{new}} = (c_1)_{\text{old}}/(a_{1,1})_{\text{old}} \tag{c}$$

The new first equation is

$$x_1 + 0.111x_2 + 0.111x_3 = 1.111$$

And the modified coefficient and constant matrices become

$$[a_{ij}] = \begin{bmatrix} 1 & 0.111 & 0.111 \\ 3 & 6 & 1 \\ 1 & 2 & 3 \end{bmatrix} \qquad [c_i] = \begin{bmatrix} 1.111 \\ 14 \\ 2 \end{bmatrix}$$

---

[2] M. L. James, G. M. Smith, and J. C. Wolford, *Applied Numerical Methods for Digital Computation with FORTRAN and CSMP*, Harper & Row, New York, 1977, Chapter 3.

Next, the x_1 term in the second equation is eliminated by multiplying the entire normalized first equation by the coefficient associated with x_1, namely, $a_{2,1}$, then subtracting the resulting equation from the second equation. Specifically, it involves the operation as follows:

$$
\begin{array}{r}
3x_1 + \quad 6x_2 + \quad x_3 = 14 \\
- \; \lfloor 3(1x_1 + 0.111x_2 + 0.111x_3 = 1.111) \\
\hline
0x_1 + 5.667x_2 + 0.667x_3 = 10.667
\end{array}
$$

In fact, the new coefficients of the reduced second equation can be calculated by the **elimination** formulas

$$(a_{2,j})_{\text{new}} = (a_{2,j})_{\text{old}} - (a_{2,1})_{\text{old}}(a_{1,j})_{\text{old}}$$

for $j = 1, 2, \ldots, n$, and

$$(c_2)_{\text{new}} = (c_2)_{\text{old}} - (a_{2,1})_{\text{old}}(c_1)_{\text{old}}$$

Similar elimination of the x_1 term can be carried out for the third (which happens to be the last, nth) equation except $(a_{3,1})_{\text{old}}$ should be multiplied to the first equation. This suggests that the above equations for elimination of the x_1 term from the second through nth equation can be extended to

$$(a_{k,j})_{\text{new}} = (a_{k,j})_{\text{old}} - (a_{k,1})_{\text{old}}(a_{1,j})_{\text{old}} \qquad \text{(d)}$$

and

$$(c_k)_{\text{new}} = (c_k)_{\text{old}} - (a_{k,1})_{\text{old}}(c_1)_{\text{old}} \qquad \text{(e)}$$

for $k = 2, 3, \ldots, n$ and $j = 1, 2, \ldots, n$. The reduced coefficient and constant matrices are

$$
[a_{ij}] = \begin{bmatrix} 1 & 0.111 & 0.111 \\ 0 & 5.667 & 0.667 \\ 0 & 1.889 & 2.889 \end{bmatrix} \qquad [c_i] = \begin{bmatrix} 1.111 \\ 10.667 \\ 0.889 \end{bmatrix}
$$

The normalization is then carried out for the second equation with the formulas

$$(a_{2,j})_{\text{new}} = (a_{2,j})_{\text{old}}/(a_{2,2})_{\text{old}} \qquad \text{for } j = 2, 3, \ldots, n \qquad \text{(f)}$$

and

$$(c_2)_{\text{new}} = (c_2)_{\text{old}}/(a_{2,2})_{\text{old}} \qquad \text{(g)}$$

The coefficient and constant matrices can be easily obtained to be

$$
[a_{ij}] = \begin{bmatrix} 1 & 0.111 & 0.111 \\ 0 & 1 & 0.118 \\ 0 & 1.889 & 2.889 \end{bmatrix} \qquad [c_i] = \begin{bmatrix} 1.111 \\ 1.882 \\ 0.889 \end{bmatrix}
$$

For elimination of the x_2 term in the third equation, the normalized second equation $x_2 + 0.118x_3 = 1.882$ is multiplied by $a_{3,2}$ and the resulting equation is subtracted from the third equation. The new coefficients in the reduced third equation are calculated, following Eqs. (d) and (e), from the formulas

$$(a_{k,j})_{new} = (a_{k,j})_{old} - (a_{k,2})_{old}(a_{2,j})_{old} \qquad \text{(h)}$$

$$(c_k)_{new} = (c_k)_{old} - (a_{k,2})_{old}(c_2)_{old} \qquad \text{(i)}$$

for $k = 3, \ldots, n$ and $j = 1, 2, \ldots, n$. For the specific example that we are currently working, $n = 3$ and Eqs. (h) and (i) need to be applied only once; but for the general case, the x_2 terms need to be eliminated from the third equation through the last, nth equation. The coefficient and constant matrices now have the form

$$[a_{ij}] = \begin{bmatrix} 1 & 0.111 & 0.111 \\ 0 & 1 & 0.118 \\ 0 & 0 & 2.666 \end{bmatrix} \qquad [c_i] = \begin{bmatrix} 1.111 \\ 1.882 \\ -2.666 \end{bmatrix}$$

Apparently, the third unknown x_3 can now be solved by the last normalization step

$$(a_{3,j})_{new} = (a_{3,j})_{old}/(a_{3,3})_{old} \qquad \text{for } j = 3, \ldots, n \qquad \text{(j)}$$

and

$$(c_3)_{new} = (c_3)_{old}/(a_{3,3})_{old} \qquad \text{(k)}$$

The resulting coefficient and constant matrices are

$$[a_{ij}] = \begin{bmatrix} 1 & 0.111 & 0.111 \\ 0 & 1 & 0.118 \\ 0 & 0 & 1 \end{bmatrix} \qquad [c_i] = \begin{bmatrix} 1.111 \\ 1.882 \\ -1 \end{bmatrix} \qquad \text{(l,m)}$$

The last equation becomes simply $x_3 = -1$. The other unknowns can be solved by a **backward substitution** process, namely,

$$x_2 = [c_2 - a_{2,3}x_3]/a_{2,2} = \left[c_2 - \sum_{k=3}^{3} a_{2,k}x_k \right]/a_{2,2}$$

$$= [1.882 - 0.118x(-1)]/1 = 1.882 + 0.118 = 2$$

and

$$x_1 = [c_1 - a_{1,2}x_2 - a_{1,3}x_3]/a_{1,1} = \left[c_1 - \sum_{k=2}^{3} a_{1,k}x_k \right]/a_{1,1}$$

$$= [1.111 - 0.111 \times 2 - 00.111 \times (-1)]/1 = 1$$

   In summary, for the particular problem of a three-unknown system, three normalization steps and two elimination steps need to be implemented in order to arrive at an **upper-triangular matrix** as shown in Eq. (l), of which the elements below the main diagonal are all equal to zero and the diagonal elements are all equal to one. For the general case when there are n unknowns, the normalization formulas can be derived by generalizing Eqs. (b), (c), (f), and (g) to become

$$(a_{i,j})_{\text{new}} = (a_{i,j})_{\text{old}}/(a_{i,j})_{\text{old}} \tag{n}$$

and

$$(c_i)_{\text{new}} = (c_i)_{\text{old}}/(a_{i,i})_{\text{old}} \tag{o}$$

for $i = 1, 2, \ldots, n$ and $j = 1, 2, \ldots, n$. And following each normalization of the ith equation, x_i terms are to be eliminated from the $i + 1$st equation through the last, nth equation. The general ith elimination step is to be carried out by use of the formulas generalized from Eqs. (d), (e), (h), and (i):

$$(a_{k,j})_{\text{new}} = (a_{k,j})_{\text{old}} - (a_{k,i})_{\text{old}}(a_{i,j})_{\text{old}} \tag{p}$$

and

$$(c_k)_{\text{new}} = (c_k)_{\text{old}} - (a_{k,k})_{\text{old}}(c_i)_{\text{old}} \tag{q}$$

for $k = i + 1, i + 2, \ldots, n$ and $j = 1, 2, \ldots, n$. Since all elements below the diagonal, a_{ij} for $i < j$, are equal to zero, Eq. (p) needs to be applied only for $j = i$, $i + 1, \ldots, n$.

After implementing $n - 1$ combinations of normalization and elimination steps and a final nth normalization of the last, nth equation, the coefficient matrix becomes an upper-triangular matrix with elements equal to one along the main diagonal. Then, one proceeds to the backward substitution by use of the following general formula

$$x_i = c_i - \sum_{k=i+1}^{n} a_{i,k}x_k \qquad \text{for } i = n - 1, n - 2, \ldots, 2, 1 \tag{r}$$

Equations (n) through (r) can be readily used to develop a computer program for solution of the matrix equation $AX = C$ of order n. There are numerous computer programs developed for Gaussian elimination, and other numerical methods. Readers should be able to find one from a standard textbook on computer methods, best suited for their particular needs. In this text we present Program BANDEQ, which is specially designed for solution of the stiffness, fluidity, and admittance matrix equations discussed in Chapter 8.

SUBROUTINE BANDEQ

In practical cases, the stiffness, fluidity, and admittance matrix equations introduced in Chapter 8 are of considerable size; that is, the order of the coefficient matrix, n, of the matrix equation $AX = C$ is large. Notice that instead of discussing the matrices $[K]$, $[F]$, and $[Y^*]$ specifically, the notation A is adopted for convenience of deriving new formulas by reducing those obtained for the Gaussian elimination in the preceding section. Although the matrix A is large in size, we are fortunate in dealing with a matrix that is symmetric and contains only nonzero elements in a finite number of diagonal bands. Figure A-1 shows that if a $n \times n$ symmetric matrix A has all of its nonzero elements in the $2b - 1$ diagonal bands,

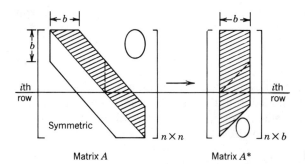

Matrix A Matrix A^* **Figure A-1**

it is advantageous for the sake of saving computer storage space to store these nonzero elements as a rectangular $n \times b$ matrix A^*. The formulas for generating elements of A^*, called a^*_{ij}, are[3]

$$a_{ij} \to a^*_{i, j-i+1} \tag{a}$$

for $i = 1, 2, \ldots, n$ and $i \le j \le l_i$, where the arrow is to indicate "stored as" and

$$l_i = \text{Minimum } (i + b - 1, n) \tag{b}$$

If an element a_{ij} below the main diagonal, that is, for $j < i$, of the square matrix A is needed, it can be retrieved from the rectangular matrix A^* when it is realized that

$$a_{ij} = a_{ji} \to a^*_{j, i-j+1} \tag{c}$$

for $m_i \le j \le i$ with m_i being defined as

$$m_i = \text{Maximum } (1, i - b + 1) \tag{d}$$

Equations (b) and (d) are necessary for controlling where nonzero elements of the ith row of the matrix A end and begin, respectively. For example, for the last row, $i = n$ and $l_i = n$ because a_{nn} is the last element. Also, for the last row, $m_i = n - b + 1$, which indicates that the first nonzero element is $a_{n,n-b+1}$.

Based on the relationships (a) and (c), the Gaussian elimination formulas presented in the preceding section are accordingly modified here. First, the normalization steps use the formulas

$$(a^*_{i, j-i+1})_{\text{new}} = (a^*_{i, j-i+1})_{\text{old}}/(a^*_{i,1})_{\text{old}} \tag{e}$$

$$(c_i)_{\text{new}} = (c_i)_{\text{old}}/(a^*_{i,1})_{\text{old}} \tag{f}$$

for $i = 1, 2, \ldots, n$ and $j = 1, 2, \ldots, l_i$. The elimination formulas become

$$(a^*_{k, j-k+1})_{\text{new}} = (a^*_{k, j-k+1})_{\text{old}} - (a^*_{k, i-k+1})_{\text{old}}(a^*_{i, j-i+1})_{\text{old}} \tag{g}$$

$$(c_k)_{\text{new}} = (c_k)_{\text{old}} - (a^*_{k,1})_{\text{old}}(c_i)_{\text{old}} \tag{h}$$

---

[3] Y. C. Pao, "On Computations Involving Stiffness Matrix Stored in Rectangular Form," *International Journal for Numerical Methods in Engineering*, Vol. 9, 1975, pp. 250–252.

for $k = i + 1, i + 2, \ldots, n$ and $j = 1, 2, \ldots, l_i$. Last, the backward substitution steps follow the formulas

$$x_i = c_i - \sum_{k=i+1}^{n} a^*_{i, k-i+1} x_k \qquad \text{(i)}$$

for $i = n - 1, n - 2, \ldots, 1$, together with $x_n = c_n$.

Programs BANDEQ in both **BASIC** and **FORTRAN** languages have been written using the rectangular-coefficient matrix approach for the Gaussian-elimination solution of $AX = C$. It is employed frequently in this text. Listings of both versions are presented below.

PROGRAM BANDEQ

```
100 REM * POROGRAM BANDEQ - SOLVES THE MATRIX EQUATION KX=D
110 REM WHERE K IS A N3XN3 MATRIX WITH 2*N4-1 DIAGONALS
120 REM STORED AS A N3XN4 RECTANGULAR MATRIX, X EXITS AS D
130 REM BOTH K AND D ARE NOT PRESERVED
140 REM * ELIMINATION THEN NORMALIZATION
150 FOR I3=1 TO N3
160 FOR I4=2 TO N4
170 IF K(I3.I4)=0 THEN 250
180 T=K(I3,I4)/K(I3,I4)
190 FOR J4=I4 TO N4
200 IF K(I3,I4)=0 THEN 220
210 K(I3+I4-1,J4-I4+1)=K(I3+I4-1,J4-I4+1)-T*K(I3,J4)
220 NEXT J4
230 K(I3,I4)=T
240 D(I3+I4-1)=D(I3+I4-1)-T*D(I3)
250 NEXT I4
260 D(I3)=D(13)/K(I3,1)
270 NEXT I3
280 REM * BACKWARD SUBSTITUTION
290 FOR I3=2 TO N3
300 J=N3+1-I3
310 FOR I4=2 TO N4
320 IF K(J,I4)=0 THEN 340
330 D(J)-K(J,I4)*D(J+I4-1)
340 NEXT I4
350 NEXT I3
360 RETURN


      SUBROUTINE BANDEQ(A,N,K,D)
      COMMON/B/NBAND
C
C     SOLVE FOR X FROM THE MATRIX EQUATION AX=D WHERE A IS A NXN
C       MATRIX WITH 2K-1 DIAGONAL BANDS STORED AS AN NXK
          RECTANGULAR MATRIX.
C     X EXITS AS D, SO BOTH A AND D ARE NOT PRESERVED.
C
      DIMENSION A(N,1),D(1)
C
C     ELIMINATION THEN NORMALIZATION.
C
      DO 60 IEQ=1,N
      DO 50 IR=2,K
      IF (A(IEQ,IR).EQ.0.) GO TO 50
```

```
      RATIO=A(IEQ,IR)/A(IEQ,1)
      DO 30 JC=IR,K
      IF (A(IEQ,JC).NE.0.) A(IEQ+IR-1,JC-IR+1)=A(IEQ+IR-1,JC-IR+1)
     *                                          -RATIO*A(IEQ,JC)
   30 CONTINUE
   40 A(IEQ,IR)=RATIO
      D(IEQ+IR-1)=D(IEQ+IR-1)-RATIO*D(IEQ)
   50 CONTINUE
   60 D(IEQ)=D(IEQ)/A(IEQ,1)
C
C     BACKWARD SUBSTITUTION.
C
      DO 70 L=2,N
      I=N+1-L
      DO 70 J=2,K
      IF (A(I,J).NE.0.) D(I)=D(I)-A(I,J)*D(I+J-1)
   70 CONTINUE
      RETURN
      END
```

APPENDIX B

BAIRSTOW'S SOLUTION OF POLYNOMIAL ROOTS

INTRODUCTION

As this text deals mainly with linear problems, there are many occasions where the roots of a polynomial have to be found, most notably when the characteristics equation of an engineering system needs to be studied. Here, we shall discuss an iterative method, named after L. Bairstow,[1] by which all roots, whether real or complex, of an nth order polynomial can be determined by successive factorization of quadratic terms. Let the polynomial be written as

$$p(x) = x^n + a_1 x^{n-1} + a_2 x^{n-2} + \cdots + a_{n-1} x + a_n \qquad \text{(a)}$$

Bairstow suggested that Eq. (a) be factorized as

$$p(x) = (x^2 + c_1 x + c_2)(x^{n-2} + b_1 x^{n-3} + b_2 x^{n-4}$$
$$+ \cdots + b_{n-3} x + b_{n-2}) + r_1 x + r_2 \qquad \text{(b)}$$

The coefficients c_1 and c_2 of the quadratic expression $x^2 + c_1 x + c_2$ are to be iteratively adjusted to make the coefficients r_1 and r_2 of the remainder expression $r_1 x + r_2$ as small as possible. When r_1 and r_2 are small enough within a prescribed tolerance, the final iterated coefficients c_1 and c_2 can then be utilized to find two roots of $p(x) = 0$ by solving the quadratic equation

$$x^2 + c_1 x + c_2 = 0 \qquad \text{(c)}$$

As it is well known that Eq. (c) will have different types of roots depending on the values of c_1 and c_2; the various outcomes are

1. $c_1^2 - 4c_2 > 0$, two distinct real roots are to be calculated by

$$x_{1,2} = [c_1 \pm (c_1^2 - 4c_2)^{1/2}]/2 \qquad \text{(d,e)}$$

2. $c_1^2 - 4c_2 = 0$, two real roots are identical and equal to $c_1/2$.

3. $c_1^2 - 4c_2 < 0$, two complex roots exist and they are a conjugate pair to be calculated by the equations

$$x_{1,2} = (c_1/2) \pm [(4c_2 - c_1^2)^{1/2}/2]j \qquad \text{(f,g)}$$

where $j = (-1)^{1/2}$.

---

[1] L. Bairstow, "Investigations Relating to the Stability of the Aeroplane," *Reports and Memoranda*, No. 154, Advisory Committee of Aeronautics, 1914.

439

The iterative procedure may be continuously employed to extract other quadratic factors from the quotient expression $x^{n-2} + b_1 x^{n-3} + \cdots + b_{n-3} x + b_{n-2}$ in Eq. (b) until the quotient expression becomes of order 2 or 1. In this manner, the roots of the given polynomial $p(x) = 0$ can all be found providing that the iterations converge. Since the iterations proceed by continuously adjusting the quadratic coefficients c_1 and c_2 in attempting to make r_1 and r_2 in Eq. (b) as small as possible, we next explain how the adjustments in c_1 and c_2 are numerically calculated.

ITERATIVE PROCEDURE— NEWTON–RAPHSON METHOD

It should be clear that the remainder terms in Eq. (b), $r_1 x + r_2$, depend on the guessed quadratic factor $x^2 + c_1 x + c_2$. In other words, r_1 and r_2 are functions of the chosen coefficients c_1 and c_2. Let us write them as

$$r_1 = f_1(c_1, c_2) \tag{h}$$

$$r_2 = f_2(c_1, c_2) \tag{i}$$

Suppose that for a pair of guessed values c_1 and c_2, the remainder terms are not equal to zero. In that case, it is then desirable to find an appropriate pair Δc_1 and Δc_2 that when added to c_1 and c_2 will lead to r_1 and r_2 both equal to zero. That is to say, according to Eqs. (h) and (i),

$$f_1(c_1 + \Delta c_1, c_2 + \Delta c_2) = f_2(c_1 + \Delta c_1, c_2 + \Delta c_2) = 0 \tag{j,k}$$

In the meantime, Taylor's theorem[2] for functions of two variables can be applied to f_1 to f_2 to yield the expressions, for $i = 1, 2$

$$\begin{aligned} f_i(c_1 + \Delta c_1, c_2 + \Delta c_2) = {} & f_i(c_1, c_2) + f_{i,1}(c_1, c_2)\Delta c_1 + f_{i,2}(c_1, c_2)\Delta c_2 \\ & + [f_{i,11}(c_1, c_2)(\Delta c_1)^2 + 2f_{i,12}(c_1, c_2)\Delta c_1 \Delta c_2 \\ & + f_{i,22}(c_1, c_2)(\Delta c_2)^2]/2! + \cdots \end{aligned} \tag{l}$$

where the abbreviations for the partial derivatives, $f_{1,1} \equiv \partial f_1/\partial c_1$, $f_{2,2} \equiv \partial f_2/\partial c_2$, $f_{1,12} \equiv \partial^2 f_1/\partial c_1 \partial c_2$, and so on have been adopted. Notice that all partial derivatives and the functions themselves are evaluated at the guessed values (c_1, c_2).

If the iterations have been carried out long enough, the adjustments Δc_1 and Δc_2 that need to be taken to yield Eqs. (j) and (k) should be small enough making the second-order and higher-order terms in Eq. (l) negligible. In so doing, we arrive at two equations for Δc_1 and Δc_2, namely,

$$0 = f_1(c_1, c_2) + f_{1,1}(c_1, c_2)\Delta c_1 + f_{1,2}(c_1, c_2)\Delta c_2$$

$$0 = f_2(c_1, c_2) + f_{2,1}(c_1, c_2)\Delta c_1 + f_{2,2}(c_1, c_2)\Delta c_2$$

---

[2] See, for example, F. B. Hildebrand, *Advanced Calculus for Applications*, Prentice-Hall, Englewood Cliffs, New Jersey, 1962.

In matrix form, the above expressions can be better organized as

$$\begin{bmatrix} f_{1,1} & f_{1,2} \\ f_{2,1} & f_{2,2} \end{bmatrix} \begin{bmatrix} \Delta c_1 \\ \Delta c_2 \end{bmatrix} = \begin{bmatrix} -f_1 \\ -f_2 \end{bmatrix} \tag{m}$$

It should be noted again that the partial derivatives and f_1 and f_2 in Eq. (m) are all to be evaluated at (c_1, c_2).

The matrix equation (m) enables the adjustments of c_1 and c_2, namely, Δc_1 and Δc_2, to be calculated during successive iterations. This is a Newton–Raphson method for two-variable problems.

FORMULAS FOR COMPUTATIONS OF b's AND PARTIAL DERIVATIVES

In order to examine whether or not a guessed pair of (c_1, c_2) values will produce zero remainder coefficients r_1 and r_2, it is necessary to derive formulas for the quotient coefficients b's as well for r_1 and r_2. Let us actually carry out the division of $p(x) = 0$ by $x^2 + c_1 x + c_2$. According to Eq. (b), we can have

$$
\begin{array}{l}
\quad\quad\quad b_1 \quad\quad\quad\quad\quad b_2 \quad\quad\quad\quad\quad b_3 \\
\quad\quad\quad \| \quad\quad\quad\quad\quad \| \quad\quad\quad\quad\quad \| \\
x^{n-2} + (a_1 - c_1)x^{n-3} + (a_2 - b_1 c_1 - c_2)x^{n-4} + (a_3 - b_2 c_1 - b_1 c_2)x^{n-5} + \cdots
\end{array}
$$

$$
\begin{array}{ll}
x^2 + c_1 x + c_2 & \big| x^n \;\; + a_1 x^{n-1} \;\;\;\; + a_2 x^{n-2} \;\;\;\; + a_3 x^{n-3} \;\;\;\; + a_4 x^{n-4} \;\;\; + \cdots \\
& - \big| x^n \;\; + c_1 x^{n-1} \;\;\;\; + c_2 x^{n-2} \\
\hline
& \;\;\;\;\;\; (a_1 - c_1)x^{n-1} + (a_2 - c_2)x^{n-2} \;\;\;\; + a_3 x^{n-3} \;\;\;\; + a_4 x^{n-4} \;\;\; + \cdots \\
& \;\;\;\;\;\; - \big| (a_1 - c_1)x^{n-1} + b_1 c_1 x^{n-2} \;\;\;\; + b_1 c_2 x^{n-3} \\
\hline
& \;\;\;\;\;\;\;\;\;\;\;\;\;\;\;\;\;\; (a_2 - b_1 c_1 - c_2)x^{n-2} + (a_3 - b_1 c_2)x^{n-3} \;\;\;\; + a_4 x^{n-4} \;\;\; + \cdots \\
& \;\;\;\;\;\;\;\;\;\;\;\;\;\;\;\;\;\; - \big| (a_2 - b_1 c_1 - c_2)x^{n-2} + b_2 c_1 x^{n-3} \;\;\;\; + b_2 c_2 x^{n-4} \\
\hline
& \; (a_3 - b_2 c_1 - b_1 c_2)x^{n-3} + (a_4 - b_2 c_2)x^{n-4} + \cdots \\
& \; - \big| (a_3 - b_2 c_1 - b_1 c_2)x^{n-3} + b_3 c_1 x^{n-4} \\
\hline
\end{array}
$$

It can be observed that

$$b_1 = a_1 - c_1 \tag{n}$$

$$b_2 = a_2 - b_1 c_1 - c_2 \tag{o}$$

and

$$b_i = a_i - b_{i-1} c_1 - b_{i-2} c_2 \quad \text{for } i = 3, 4, \ldots, n-2 \tag{p}$$

The remainder terms are simply the extension of Eq. (p), namely,

$$r_1 = a_{n-1} - b_{n-2} c_1 - b_{n-3} c_2 \tag{q}$$

and

$$r_2 = a_n - b_{n-1} c_1 - b_{n-2} c_2 \tag{r}$$

In fact, Eq. (p) can be modified to allow i to extend from 3 to n with the understanding of $r_1 \equiv b_{n-1}$ and $r_2 \equiv b_n$.

For every guessed pair of (c_1, c_2) values, the remainder coefficients r_1 and r_2 can thus be calculated with Eqs. (n) through (r). Returning to Eqs. (h), (i), and (m), we notice that in order to determine the adjustments Δc_1 and Δc_2 in attempt of making r_1 and r_2 equal to zero, formulas for evaluation of the partial derivatives must also be derived. By actually carrying out the partial differentiation of Eqs. (q) and (r), we obtain

$$\frac{\partial f_1}{\partial c_1} = \frac{\partial r_1}{\partial c_1} = \frac{\partial}{\partial c_1}(a_{n-1} - b_{n-2}c_1 - b_{n-3}c_2)$$

$$= -\frac{\partial}{\partial c_1}(b_{n-2}c_1) - \frac{\partial}{\partial c_1}(b_{n-3}c_2)$$

$$= -b_{n-2} - c_1\frac{\partial}{\partial c_1}b_{n-2} - c_2\frac{\partial}{\partial c_1}b_{n-3} \tag{s}$$

$$\frac{\partial f_2}{\partial c_1} = -b_{n-1} - c_1\frac{\partial}{\partial c_1}b_{n-1} - c_2\frac{\partial}{\partial c_1}b_{n-2} \tag{t}$$

$$\frac{\partial f_1}{\partial c_2} = -b_{n-3} - c_1\frac{\partial}{\partial c_2}b_{n-2} - c_2\frac{\partial}{\partial c_2}b_{n-3} \tag{u}$$

$$\frac{\partial f_2}{\partial c_2} = -b_{n-2} - c_1\frac{\partial}{\partial c_2}b_{n-1} - c_2\frac{\partial}{\partial c_2}b_{n-2} \tag{v}$$

Since the b's are related by Eq. (p), the following recurrence formula for partial derivatives of b's with respect to c_1 and c_2 can be obtained

$$\frac{\partial b_i}{\partial c_1} = -b_{i-1} - c_1\frac{\partial b_{i-1}}{\partial c_1} - c_2\frac{\partial b_{i-2}}{\partial c_1} \tag{w}$$

$$\frac{\partial b_i}{\partial c_2} = -b_{i-2} - c_1\frac{\partial b_{i-1}}{\partial c_2} - c_2\frac{\partial b_{i-2}}{\partial c_2} \tag{x}$$

for $i = 3, 4, \ldots, n-2$. Also, from Eqs. (n) and (o), the beginning derivatives are

$$\frac{\partial b_1}{\partial c_1} = -1, \quad \frac{\partial b_1}{\partial c_2} = 0, \quad \frac{\partial b_2}{\partial c_1} = -b_1, \quad \frac{\partial b_2}{\partial c_2} = -1 \tag{z}$$

Indeed, the calculation of partial derivatives of the remainder coefficients r_1 and r_2 with respect to the guessed c_1 and c_2 are tedious. The derived formulas are, however, recursive and thus best suited for machine computation. Program BAIRSTOW has been prepared for this specific purpose. Its listing and a sample application of solving $x^6 + 2x^5 + 3x^4 + 4x^3 + 5x^2 + 6x + 7 = 0$ are presented. The iterations are to be terminated if $|\Delta c_1| + |\Delta c_2|$ is less than a prescribed tolerance. For the sample problem, a tolerance equal to 0.00001 was specified for obtaining the listed six roots, which all happen to be complex.

PROGRAM BAIRSTOW

---

* PROGRAM BAIRSTOW *

SOLVING POLYNOMIAL ROOTS BY QUADRATIC FACTORIZATION.

INPUT C1,C2 OF THE QUADRATIC FACTOR X↑2+C1*X+C2 : 0,0

INPUT THE ORDER OF POLYNOMIAL : 6

INPUT THE ACCURACY TOLERANCE : .00001

INPUT 6 COEFFICIENTS (IN DESCENDING POWER OF X)
2,3,4,5,6,7

	REAL PART	IMAGINARY PART	ITERATIONS
X(6) =	-1.3079	0.5933	9
X(5) =	-1.3079	-0.5933	9
X(4) =	0.7104	1.1068	9
X(3) =	0.7104	-1.1068	9
X(2) =	-0.4025	1.3417	1
X(1) =	-0.4025	-1.3417	1

```
100 REM    ROOTS OF POLYNOMIAL BY BAIRSTOW'S METHOD
110 INIT
120 PAGE
130 U0=1
140 V0=1
150 E=1.0E-6
160 PRINT "INPUT UI,VI,EPSI,N"
170 INPUT U0,V0,E,N
180 DIM A(N),B(N),C(N)
190 PRINT "INPUT ";N;" COEFFICIENTS (IN DESCENDING POWER)"
200 INPUT A
210 PRINT "           REAL PART    IMAGINARY PART    ITERATIONS"
220 IF N<>0 THEN 250
230 PRINT "N=0"
240 STOP
250 IF N>1 THEN 320
260 P=-A(1)
270 Q=0
280 I=1
290 PRINT USING 300:"X(",N,")=",P,Q,I
300 IMAGE 2A,2D,2A,2X,4D.4D,6X,4D.4D,10X,3D
310 GO TO 1010
320 IF N>2 THEN 560
330 U=A(1)
340 V=A(2)
350 I=1
360 P=-U/2
370 R=U^2-4*V
380 IF R>0 THEN 460
390 R=-R
400 Q=SQR(R)/2
410 PRINT USING 300:"X(",N,")=",P,Q,I
420 N=N-1
430 Q=-Q
440 PRINT USING 300:"X(",N,")=",P,Q,I
```

```
450 GO TO 500
460 Q=SQR(R)/2
470 PRINT USING 300:"X(",N,")=",P+Q,0,I
480 N=N-1
490 PRINT USING 300:"X(",N,")=",P-Q,0,I
500 N=N-1
510 IF N<=0 THEN 1010
520 FOR K=1 TO N
530 A(K)=B(K)
540 NEXT K
550 GO TO 220
560 U=U0
570 V=V0
580 REM   MAX OF 200 ITERATIONS
590 FOR I=1 TO 200
600 B(1)=A(1)-U
610 B(2)=A(2)-B(1)*U-V
620 FOR K=3 TO N
630 B(K)=A(K)-B(K-1)*U-B(K-2)*V
640 NEXT K
650 C(1)=B(1)-U
660 C(2)=B(2)-C(1)*U-V
670 FOR K=3 TO N-1
680 C(K)=B(K)-C(K-1)*U-C(K-2)*V
690 NEXT K
700 IF N>3 THEN 780
710 D0=C(N-1)-C(N-2)^2
720 IF D0<>0 THEN 750
730 PRINT "     DENOMINATOR IS ZERO"
740 STOP
750 D1=(B(N)-B(N-1)*C(N-2))/D0
760 D2=(C(N-1)*B(N-1)-C(N-2)*B(N))/D0
770 GO TO 820
780 D0=C(N-1)*C(N-3)-C(N-2)^2
790 IF D0=0 THEN 730
800 D1=(B(N)*C(N-3)-B(N-1)*C(N-2))/D0
810 D2=(C(N-1)*B(N-1)-C(N-2)*B(N))/D0
820 U=U+D1
830 V=V+D2
840 S=ABS(D1)+ABS(D2)
850 IF S<E THEN 360
860 IF I>1 THEN 890
870 S0=S
880 GO TO 970
890 IF I<>50 OR S<S0 THEN 940
900 PRINT "DIVERGENCE OCCURRING"
910 PRINT USING 920:"U=",U,"V=",V,"DELU=",D1,"DELV=",D2
920 IMAGE 2A,FE,3X,2A,FE,3X,4A,FE,3X,4A,FE
930 STOP
940 IF I<>100 THEN 970
950 PRINT "  CONVERGENCE IS SLOW"
960 PRINT USING 920:"U=",U,"V=",V,"DELU=",D1,"DELV=",D2
970 NEXT I
980 PRINT "  ITERATING STOPPED AFTER 200 ITERATIONS"
990 PRINT USING 920:"U=",U,"V=",V,"DELU=",D1,"DELV=",D2
1000 STOP
1010 RETURN
```

APPENDIX C
LAPLACE TRANSFORM

INTRODUCTION

The Laplace transform has been a powerful tool for solution of differential equations arising from the engineering analysis of control and feedback systems, electric circuit networks, and steady-state and transient vibration problems. It converts the differential equations in the time domain into algebraic equations and thereby expedites the solution. However, difficulty may result in the inversion of the algebraic solution back to the time domain. For practical problems, analytical Laplace inversion is either not possible or too complex and tedious. Numerical inversion and computer application become more realistic and in many cases the only workable approach.

The Laplace transform of a function $f(t)$, usually denoted as $L[f(t)]$, is defined by the integral

$$L[f(t)] = \int_0^\infty f(t)e^{-st}\,dt = F(s) \tag{a}$$

where s is a complex number. In Eq. (a), the resulting integral, which is dependent on the complex number s, is designated as $F(s)$. The function $f(t)$ is referred to as the inverse Laplace transform of $F(s)$. In terms of $F(s)$, $f(t)$ can be obtained by use of the formula[1]

$$L^{-1}[F(s)] = f(t) = \frac{1}{2\pi j}\int_{a-j\infty}^{a+j\infty} F(s)e^{st}\,ds \tag{b}$$

where the constant a is arbitrary but must be greater than zero and be so chosen that the derivative of $F(s)$ exists at all points to the right of a, and $j = (-1)^{1/2}$.

A number of frequently used formulas of Laplace transform pairs are listed in Table C-1. It would be helpful to demonstrate a few examples of how these results are derived. Referring to Fig. C-1, we first demonstrate the derivation of the Laplace transform of a unit step function $u(t - t_i)$. According to Eq. (a),

$$L[u(t - t_i)] = \int_0^\infty e^{-st}u(t - t_i)\,dt$$

$$= \int_{t_i}^\infty e^{-st}\,dt = -\frac{1}{s}e^{-st}\Big|_{t_i}^\infty = \frac{1}{s}e^{-st_i} \tag{c}$$

In particular, for $t_i = 0$ we have

---

[1] M. G. Smith, *Laplace Transform Theory*, Van Nostrand, London, 1966.

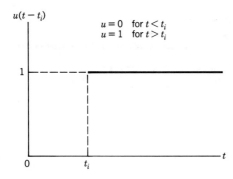

$u = 0$ for $t < t_i$
$u = 1$ for $t > t_i$

Figure C-1   Unit step function.

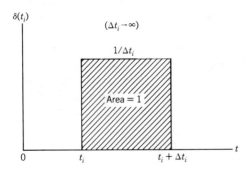

$(\Delta t_i \to \infty)$

$1/\Delta t_i$

Area = 1

Figure C-2   Unit impulse function.

$$L[u(t)] = \frac{1}{s} \tag{d}$$

Next, we consider the unit impulse function $\delta(t - t_i)$. To derive its Laplace transform, we follow a limiting approach by treating $\delta(t - t_i)$ as the limit as $\Delta t_i \to 0$ of the unit area shown in Fig. C-2. That is,

$$L[\delta(t - t_i)] = L\left[\lim_{\Delta t_i \to 0} \{u(t - t_i) - u(t - t_i - \Delta t_i)\}/\Delta t_i\right]$$

According to Eq. (c), this yields

$$L[\delta(t - t_i)] = \lim_{\Delta t_i \to 0} e^{-st}(1 - e^{-s\Delta t_i})/s\,\Delta t_i$$

By applying L'Hospital's rule

$$L[\delta(t - t_i)] = \lim_{\Delta t_i \to 0} se^{-s(t_i + \Delta t_i)}/s = e^{-st_i} \tag{e}$$

In particular, for $t_i = 0$ the result is

$$L[\delta(t)] = 1 \tag{f}$$

For deriving the Laplace transform of trigonometric functions, it is convenient to use Euler's formulas, which relate the Cartesian and polar forms of complex numbers. The formulas are

$$e^{\pm j\theta} = \cos \theta \pm j \sin \theta \tag{g,h}$$

and

TABLE C-1
Commonly Used Time Functions and Their Laplace Transforms

Formula No.	$f(t) = L^{-1}[F(s)]$	$F(s) = L[f(t)]$
1	$A\delta(t)$	A
2	$A\delta(t - t_i)$	Ae^{-st_i}
3	A, or, $Au(t)$	A/s
4	$Au(t - t_i)$	Ae^{-st_i}/s
5	At^n	$An!/s^{n+1}$
6	$A \sin \omega t$	$A\omega/(s^2 + \omega^2)$
7	$A \cos \omega t$	$As/(s^2 + \omega^2)$
8	$Ae^{\sigma t}$	$A/(s - \sigma)$
9	$Af(t) + Bg(t)$	$AF(s) + BG(s)$
10	$df(t)/dt$	$sF(s) - f(0)$
11	$d^n f(t)/dt^n$	$s^n F(s) - \sum_{k=0}^{n-1} s^{n-1-k} \dfrac{d^k f}{dt^k}\Big|_{t=0}$
12	$\int_0^t f(x)\,dx$	$F(s)/s$
13	$tf(t)$	$-dF(s)/ds$
14	$f(t)/t$	$\int_s^\infty F(x)\,dx$
15	$e^{\sigma t}f(t)$	$F(s - \sigma)$
	Example: $L[e^{\sigma t} \cos \omega t] = \dfrac{s - \sigma}{(s - \sigma)^2 + \omega^2}$	
16	$f(t - t_i)u(t - t_i)$	$e^{-st_i}F(s)$
	$L[f(t)u(t - t_i)] = e^{-st_i}L[f(t + t_i)]$	

$$\cos \theta = (e^{j\theta} + e^{-j\theta})/2 \qquad\qquad\qquad (i)$$
$$\sin \theta = (e^{j\theta} - e^{-j\theta})/2j \qquad\qquad\qquad (j)$$

With Eqs. (i) and (j), formulas 6 and 7 in Table C-1 can readily be obtained. For example,

$$L[A \cos \omega t] = \int_0^\infty A \cos \omega t e^{-st}\,dt$$

$$= \frac{A}{2} \int_0^\infty \left[e^{(-s+j\omega)t} + e^{(-s-j\omega)t} \right] dt$$

$$= \frac{A}{2} \left[\frac{e^{(-s+j\omega)t}}{-s+j\omega} + \frac{e^{(-s-j\omega)t}}{-s-j\omega} \right]_0^\infty$$

$$= \frac{A}{2}\left(\frac{-1}{-s+j\omega} + \frac{-1}{-s-j\omega}\right) = \frac{As}{s^2+\omega^2} \tag{k}$$

The result of $e^{(-s+j\omega)t}$ equal to 0 as $t \to \infty$ can be so argued that s can be chosen to make $e^{-st} \to 0$ more rapidly than $e^{j\omega t} \to \infty$.

TRANSFORM DIFFERENTIAL EQUATIONS INTO ALGEBRAIC EQUATIONS

Consider the case of a series LCR circuit discussed in Chapter 7. The differential equation governing the current $i(t)$ in the circuit is, with $v(t)$ being the driving voltage,

$$L\frac{d^2i}{dt^2} + R\frac{di}{dt} + \frac{1}{C}i = \frac{dv(t)}{dt} \tag{a}$$

To apply the Laplace transform to Eq. (a) requires formulas 10 and 11 in Table C-1. The derivation of these formulas needs the method of integration by parts, that is,

$$\int_a^b u\,dv = uv\Big|_a^b - \int_a^b v\,du \tag{b}$$

The Laplace transform of the first derivative of a time function $f(t)$ can now be derived as follows:

$$L[df(t)/dt] = \int_0^\infty e^{-st}\frac{df(t)}{dt}\,dt$$

$$= e^{-st}f(t)\Big|_0^\infty - \int_0^\infty f(t)\,de^{-st}$$

$$= -f(t=0) + s\int_0^\infty e^{-st}f(t)\,dt$$

$$= sF(s) - f(t=0) \tag{c}$$

Again, in arriving at Eq. (c), the argument that $e^{-st}f(t) \to 0$ as $t \to \infty$ has been employed with appropriately chosen s values.

Formula 11 in Table C-1 is the extension of the above-shown result. For both formulas 10 and 11 in Table C-1, the **initial values** of the function $f(t)$ and its derivatives have to be prescribed in order to arrive at the required Laplace transforms.

By applying the Laplace transform to Eq. (a), the following algebraic equation in terms of the $I(s)$, which is $L[i(t)]$, and $V(s)$, which is $L[v(t)]$, can be derived:

$$L\left[s^2I(s) - si(t=0) - \frac{di(t)}{dt}\Big|_{t=0}\right] + R[sI(s) - i(t=0)] + \frac{1}{C}I(s)$$

$$= sV(s) - v(t=0)$$

After simplification, the algebraic equation of $I(s)$ is

$$\left(Ls^2 + Rs + \frac{1}{C}\right)I(s) = (s+1)i_0 + i_{,t0} + sV(s) + v_0 \tag{d}$$

where $i_0 \equiv i(t = 0)$, $v_0 \equiv v(t = 0)$, and i_{t0} is the value of di/dt at $t = 0$.

For a coupled system of differential equations, the Laplace transform technique can also be applied to convert it into a matrix equation. Take the case of dual-tank problem discussed in Chapter 7. Let us consider the numerical case of

$$2000 \frac{d}{dt} h_1 + 91.7 h_1 - 83.3 h_2 = 0$$

$$-83.3 h_1 + 2000 \frac{d}{dt} h_2 + 166.7 h_2 = 0$$

with the initial conditions of $h_1 = 30$ and $h_2 = 0$. The resulting Laplace transform equations are

$$2000[sH_1(s) - 30] + 91.7 H_1(s) - 83.3 H_2(s) = 0$$

$$-83.3 H_1(s) + 2000[sH_2(s) - 0] + 166.7 H_2(s) = 0$$

Upon manipulation and expressing the equations into a matrix form, the results become

$$\begin{bmatrix} 2000s + 91.7 & -83.3 \\ -83.3 & 2000s + 166.7 \end{bmatrix} \begin{bmatrix} H_1(s) \\ H_2(s) \end{bmatrix} = \begin{bmatrix} 60,000 \\ 0 \end{bmatrix} \tag{e}$$

The $H_1(s)$ and $H_2(s)$ can easily be solved by Cramer's rule:

$$H_1(s) = [60,000(2000s + 166.7) - 83.3^2]/C(s)$$
$$= (120s + 9.995) \times 10^6/C(s) \tag{f}$$

$$H_2(s) = -60,000 \times 83.3/C(s) = -4.998 \times 10^6/C(s) \tag{g}$$

where $C(s) = 0$ is the characteristics equation of the system derived by formation of the determinant of the coefficient matrix of Eq. (e), namely,

$$C(s) = (2000s + 91.7)(2000s + 166.7) - 83.3^2$$
$$= 4 \times 10^6 \times (s^2 + 0.129s - 0.00208) = 0 \tag{h}$$

Of course, the ultimate purpose is to find $h_1(t)$ and $h_2(t)$ as far as the dual-tank problem is concerned. We have so far determined their respective Laplace transforms, $H_1(s)$ and $H_2(s)$ described by Eqs. (f) and (g). It remains to be explained how $H_1(s)$ and $H_2(s)$ should be inverted to obtain $h_1(t)$ and $h_2(t)$, respectively.

INVERSION OF LAPLACE TRANSFORM

Instead of using Eq. (b) in the Introduction section to invert $H_1(s)$ by complex integration, $h_1(t)$ can be obtained by the method of **partial fractions** and utilization of Table C-1. It is easy to show that the roots[2] of $C(s)$ are -0.0189 and -0.1103. Consequently, Eq. (f) can be written as

---

[2] In general cases, the characteristics equation can be solved by application of the Bairstow iterative method discussed in Appendix B.

$$H_1(s) = \frac{A}{s + 0.0189} + \frac{B}{s + 0.1103}$$

$$= [A(s + 0.1103) + B(s + 0.0189)]/C(s)$$

By comparison of the coefficients of s and constant terms in the above expression with Eq. (f), the constants A and B must satisfy the equations

$$A + B = 30 \quad \text{and} \quad 0.1103A + 0.0189B = 2.499$$

Again, by application of Cramer's rule, we obtain $A = 21.15$ and $B = 8.85$. The resulting expression of $H_1(s)$ in partial fractions is

$$H_1(s) = \frac{21.15}{s + 0.0189} + \frac{8.85}{s + 0.1103}$$

for which, according to Table C-1, the inverse function can be easily observed to be

$$h_1(t) = 21.15e^{-0.0189t} + 8.85e^{-0.1103t}$$

Likewise, $h_2(t)$ can also be determined by application of the partial-fraction method for $H_2(s)$.

The Laplace transform technique is introduced here solely for the purpose of explaining how the transfer function of a system can be derived so it can be applied for CAD in Chapter 8. It is not the objective of this text to extensively discuss the topic of Laplace transforms and their inversion. For an in-depth study, such as on the ramification of the partial-fraction method related to the multiple real roots and complex conjugate roots of the characteristics equations and on numerical inversion of Laplace transform, advanced texts and publications on Laplace transform should be consulted.[3,4]

---

[3] On analytical Laplace inversion, see, for example, C. R. Wylie, Jr., *Advanced Engineering Mathematics*, McGraw-Hill, New York, 1960.

[4] On numerical inversion of Laplace transform, see, for example, C. M. Lin, *Theory and Application of Transform Methods for Vibration Analysis*, M. S. Thesis, Department of Engineering Mechanics, the University of Nebraska–Lincoln, 1982, Y. C. Pao, Advisor.

APPENDIX D

FORTRAN VERSION AND APPLE, IBM, AND TRS BASIC VERSIONS OF THE DEVELOPED PROGRAMS

FORTRAN VERSION

1. Programs CIRCLES and R. ARRAY

In Chapter 4, we briefly discussed the FORTRAN versions of the BASIC programs developed in this text. Here, more examples are given to familiarize the readers with the conversion of BASIC language into FORTRAN language. Furthermore, the use of PLOT 10 software[1] is demonstrated as it is nowadays a popular means in FORTRAN plotting and graphics programs.

First, let us list the plot FORTRAN Programs CIRCLES and R. ARRAY. These are the two programs for producing the four concentric circles of radii equal to 10, 20, 30, and 40, and the 7 × 7 rectangles shown in Figs. 4-3 and 4-4, respectively. A VAX 11-780 computer system[2] with Tektronix 4010 graphics terminals were used in running these programs.

PROGRAM CIRCLES

```
      CALL GRSTRT(4010,1)
      CALL DEGREE
      X1=65
      Y1=50
      DO 200 I=1,4
      R=10.*I
      CALL MOVE(X1,Y1)
      CALL ARC(R,0.,360.)
  200 CONTINUE
      CALL GRSTOP
      STOP
      END
```

---

[1] *PLOT 10, 4010C01, Interactive Graphics Library, User's Manual*, No. 061-1956–00, Tektronix, Inc., 1978.

[2] Digital Equipment Corporation, Maynard, Massachusetts.

PROGRAM R. ARRAY

```
      CALL GRSTRT(4010,1)
      D1=8.0
      D2=5.0
      DO 200 I=50,110,10
      X1=FLOAT(I)
      X2=X1+D1
      DO 200 J=90,30,-10
      Y1=FLOAT(J)
      Y2=Y1-D2
      CALL REC(X1,X2,Y1,Y2)
  200 CONTINUE
      CALL GRSTOP
      STOP
      END

      SUBROUTINE REC(X1,X2,Y1,Y2)

      CALL MOVE(X1,Y1)
      CALL DRAW(X1,Y2)
      CALL DRAW(X2,Y2)
      CALL DRAW(X2,Y1)
      CALL DRAW(X1,Y1)
      RETURN
      END
```

The subroutine GRSTRT initiates and the subroutine GRSTOP terminates the graphics terminal 1040. The second argument of CALL GRSTRT statement specifies what type of option. MOVE and DRAW both move the cursor on the screen to the point whose coordinates are given by the two arguments inside the parentheses; but the DRAW statement draws a line from the current position to the destination, whereas the MOVE statement does not. The function FLOAT converts an integer into a floating-point number. CALL DEGREE requests that from here on all angles are in units of degrees. The subroutine ARC draws a whole or a part of a circle having a radius specified by the first argument, as well as the starting and ending angles specified by the second and third arguments, respectively. Subroutine REC draws a rectangle having $(X1, Y1)$ and $(X2, Y2)$ as the coordinates of the end points of one of its two diagonals.

2. Program GEAR

The BASIC program GEAR presented in Chapter 4 for drawing the complete tooth profile of a gear shown in Fig. 4-9 can be best utilized for demonstration of the repeated application of an available graph by rotating and placing it at different locations. Here, the PLOT 10 subroutines NEWPAG, SKIP, and POLY together with the newly developed subroutines CIRCLE, CENTLN, INVOLU, and ZROT are employed for obtaining the desired display of a gear.

NEWPAG erases the display screen. SKIP and POLY facilitate the use of subscripted arrays for plotting graphs, in place of MOVE and DRAW. SKIP causes the cursor to move to the first point whose coordinates are $[X(I), Y(I)]$, and the CALL POLY $[N, X(I), Y(I)]$ statement continues to draw the straight-line sides of a *poly*gon by connecting the points P_j and P_{j+1} whose coordinates are $[X(J), Y(J)]$

and for $j = 1, 2, \ldots, N - 1$. If $I = 1$, then the second and third arguments of the CALL POLY statement, $X(I)$ and $Y(I)$, can be simply written as X and Y.

The newly developed subroutines CIRCLE, CENTLN, INVOLU, and ZROT draw a circle, center lines, and involute curve and perform the rotation about the z axis, respectively. The arguments involved are quite self-explanatory and have been described in Chapter 4 when their BASIC counterparts were presented.

PROGRAM GEAR

```
C       * PROGRAM GEAR *
        DIMENSION O1(22),O2(22),T1(22),T2(22)
        DATA C1/96./,C1/60./,R1/20./,A1/95./,A2/60./
        CALL GRSTRT(4010,1)
        CALL DEGREE
        CALL NEWPAG
        CALL CIRCLE(C1,C2,R1)
        CALL CENTLN(C1,C2,R1)
        CALL INVOLU(C1,C2,R1,A1,A2,T1,T2)
        CALL SKIP
        CALL POLY(22,T1,T2)
        DO 100 C3=20,340,20
        DO 200 I=1,22
        X1=T1(I)-C1
        Y1=T2(I)-C2
        CALL ZROT(X1,Y1,C3,X2,Y2)
        O1(I)=X2+C1
        O2(I)=Y2+C2
200     CONTINUE
        CALL SKIP
        CALL POLY(22,O1,O2)
100     CONTINUE
        CALL GRSTOP
        STOP
        END

C       * PLOT CENTER LINE *
        SUBROUTINE CENTLN(C1,C2,R1)
        CALL MOVE(-1.25*R1+C1,C2)
        CALL DRAW(-0.1*R1+C1,C2)
        CALL MOVE(-0.05*R1+C1,C2
        CALL DRAW(0.05*R1+C1,C2
        CALL MOVE(0.1*R1+C1,C2)
        CALL DRAW(1.25*R1+C1,C2)
        CALL MOVE(C1,-1.25*R1+C2)
        CALL DRAW(C1,-0.1*R1+C2)
        CALL MOVE(C1,-0.05*R1+C2)
        CALL DRAW(C1,0.05*R1+C2)
        CALL MOVE(C1,0.1*R1+C2)
        CALL DRAW(C1,1.5*R1+C2)
        RETURN
        END

C       * PLOT CIRCLE *
        SUBROUTINE CIRCLE(C1,C2,R1)
        CALL MOVE(C1,C2)
        CALL ARC(R1,0.,360.)
        RETURN
        END
```

```
C        * PLOTS GEAR TOOTH INVOLUTE *
         SUBROUTINE INVOLU(C1,C2,R1,A1,A2,T1,T2)
         DIMENSION T1(22),T2(22)
         N=11
         A3=(A1-A2)/10.
         DO 300 J=I,N
         A4=A1-(J-1)*A3
         TL=R1*0.017453*(J-1)*A3
         T1(J)=R1*COS(A4*3.14159/180.)+C1-TL*COS((A4-90.)*3.14159/180.)
         T2(J)=R1*SIN(A4*3.14159/180.)+C2-TL*SIN((A4-90.)*3.14159/180.)
         T1(23-J)=-T1(J)+2,*C1
         T2(23-J)=T2(J)
300      CONTINUE
         RETURN
         END

C        * Z ROTATE *
         SUBROUTINE ZROT(X1,Y1,C3,X2,Y2)
         X2=COS(C3*3.14159/180.)*X1-SIN(C3*3.14159/180.)*Y1
         Y2=SIN(C3*3.14159/180.)*X1+COS(C3*3.14159/180.)*Y1
         RETURN
         END
```

3. Program PERSPECT

PLOT 10 has subroutines WINDOW and VWPORT for mapping the user's data points onto a selected viewport on the display screen. They both have four arguments defining the ranges of the data values. Program PERSPECT presented in Chapter 4 in BASIC language is here converted into FORTRAN to demonstrate the use of WINDOW and VWPORT and some other subroutines of PLOT 10.

Lines 0035 and 0036 in Program PERSPECT illustrate that the user's data fall within the range $-5 \leq x \leq 5$ and $-5 \leq y \leq 5$ and that they are to be displayed in a viewport $10 \leq x \leq 90$ and $10 \leq y \leq 90$ both in graphics display units (GDU). CALL TXAM prepares the program for entering a literal string with up to 80 characters (AM, multiple A format) from the keyboard of 4010. The subroutine GETURN captures as many real numbers as specified by the third argument and stores them in the array specified by the fourth argument. The total number of real data actually entered from the keyboard is kept in the variable specified as the last argument. The prompting message is entered as the second argument while the first argument counts the total number of characters used in the prompting literal string.

Subroutine ROTATE has two arguments specifying the rotation in degrees about the x and y axes, respectively. Positive argument indicates that a counterclockwise rotation is desired. Subroutines VECABS and VECREL can be called to request that the subsequent displacements of the cursor refer to a defined origin or to the current position of the cursor, respectively.

PROGRAM PERSPECT

```
0001    C        * PROGRAM PERSPECT *
0002             DIMENSION X0(24),Y0(24),Z0(24),X(24),Y(24),Z(24)
0003    C        * FIRST 8 FOR BOX, THEN 16 FOR LETTERS F, T, AND R.
0004             DATA X0/0.,4.,4.,0.,0.,4.,4.,0.,
0005       *         3.4,3.4,3.4,3.8,3.7,0.2,0.4,0.6,0.4,4.,4.,4.,4.,
0006       *         4.,4.,4./
```

```
0007            DATA Y0/0.,0.,2.,2.,0.,0.,2.,2.,
0008      *         1.4,1.6,1.8,1.8,1.6,2.,2.,2.,1.4,1.6,1.8,1.8,
0009      *         1.6,1.6.1.4/
0010            DATA Z0/0.,0.,0.,0.,3.,3.,3.,3.,
0011      *         3.,3.,3.,3.,3.,0.2 0.2,0.2,0.6,0.6,0.6,0.6,0.2,
0012      *         0.2,0.4,0.2/
0013            DATA T1/-2./,T2/-1./,T3/-1.5/
0014            CALL GRSTRT(4010,1)
0015            CALL NEWPAG
0016            N=24
0017      C     * TRANSLATE SO THAT X-Y PLANE BISECTS
0018      C            THE R AND T PLANES OF THE BRICK
0019            DO 300 I=I,N
0020            X(I)=X0(I)+T1
0021            Y(I)=Y0(I)+T2
0022            Z(I)=Z0(I)+T3
0023      300   CONTINUE
0024      C     * PERSPECTIVE VIEWING *
0025            CALL TXAM
0026            CALL GETURN(33,'INPUT AXIAL COORDINATE OF SCREEN '
0027      *              ,1,Z1,IGOT)
0028            CALL GETURN(30,'INPUT AXIAL COORDINATE OF EYE '
0029      *              ,1,Z2,IGOT)
0030            DO 400 I=1,N
0031            X(I)=X(I)*(Z2-Z1)/(Z2-Z(I))
0032            Y(I)=Y(I)*(Z2-Z1)/(Z2-Z(I))
0033      400   CONTINUE
0034      C     * COMMENCE PLOTTING *
0035            CALL WINDOW(-5.,5.,-5.,5.)
0036            CALL VWPORT(10.,90.,10.,90.)
0037      C     * DRAW X-Y PROJECTION OF 3-D BOX *
0038            CALL MOVE(X(5),Y(5))
0039            CALL DRAW(X(6),Y(6))
0040            CALL DRAW(X(2),Y(2))
0041            CALL DRAW(X(1),Y(1))
0042            CALL DRAW(X(5),Y(5))
0043            CALL DRAW(X(8),Y(8))
0044            CALL DRAW(X(7),Y(7))
0045            CALL DRAW(X(3),Y(3))
0046            CALL DRAW(X(4),Y(4))
0047            CALL DRAW(X(8),Y(8))
0048            CALL MOVE(X(6),Y(6))
0049            CALL DRAW(X(7),Y(7))
0050            CALL MOVE(X(3),Y(3))
0051            CALL DRAW(X(2),Y(2))
0052            CALL MOVE(X(1),Y(1))
0053            CALL DRAW(X(4),Y(4))
0054      C      * DRAW "F"
0055            CALL MOVE(X(9),Y(9))
0056            CALL DRAW(X(11),Y(11))
0057            CALL DRAW(X(12),Y(12))
0058            CALL MOVE(X(10),Y(10))
0059            CALL DRAW(X(13),Y(13))
0060      C     * DRAW "T"
0061            CALL MOVE(X(14),Y(14))
0062            CALL DRAW(X(16),Y(16))
0063            CALL MOVE(X(17),Y(17))
0064            CALL DRAW(X(15),Y(15))
0065      C     * DRAW "R"
0066            CALL MOVE(X(18),Y(18))
0067            CALL DRAW(X(20),Y(20))
```

```
0068            CALL DRAW(X(21),Y(21))
0069            CALL DRAW(X(22),Y(22))
0070            CALL DRAW(X(19),Y(19))
0071            CALL MOVE(X(23),Y(23))
0072            CALL DRAW(X(24),Y(24))
0073            CALL GRSTOP
0074            STOP
0075            END
```

4. Program RESISTOR

To illustrate the application of PLOT 10 in conjunction with the interactive use of the thumbwheels available on the Tektronix 4010 terminal, we extract from the BASIC Program E.MODULE a portion of the statements connected with the drawing of a resistor. This new program written in FORTRAN language is called RESISTOR. The reader, after having become familiarized with the procedure, should be able to extract other information from the BASIC Program E.MODULE for incorporation of more electric elements with interactive graphics arrangements.

The coordinates of the beginning and ending points are entered by maneuvering of the thumbwheels. By calling the subroutine INPUT, these four data are stored in $XP1$, $YP1$, $XP2$, and $YP2$. CALL TEXT results in displays of an alphanumeric string specified by the second argument, while the first argument indicates the number of characters in the string. The argument in the CALL TXFCUR specifies where the cursor should be positioned after the display of a text. The cursor will be one line below the initial cursor position, one character to the right of the last character of the displayed text, and returned to the initial position when the argument is set equal to 1, 2, and 3, respectively.

Subroutine LOCATE can be called to store the coordinates of the points on the display screen selected by use of the cross-hair cursor. The second and third arguments in the calling sequence specify the variables where the x and y coordinates, respectively, are to be stored. They can be subscripted arrays of size defined by the first argument. The fourth and fifth arguments store the characters entered accompanying the points and the number of points selected, respectively.

Subroutine RNUMBR enables the value of a real variable specified by the first argument to be printed according to the formats described by the second and third arguments (decimal places and maximum number of characters), respectively. For example, CALL RNUMBR (XP1, 2, 5) results in printing on the screen the value of $XP1$ by use of five characters and with two decimal places.

RESIST, INPUT, and LENANG have been specifically prepared for drawing the resistor, specifying its beginning and ending locations ($XP1$, $YP1$, $XP2$, $YP2$), and calculating the length RLEN and angle ANG, respectively to aid the PLOT 10 programs.

PROGRAM RESISTOR

```
0001     C      MAIN PROGRAM
0002     C      PLOTTING THE COMPONENTS BY USING 4010
0003     C
0004            CALL GRSTRT(4010,1)
0005            CALL WINDOW(0.,130.,0.,100.)
```

```
0006          CALL NEWPAG
0007          CALL INPUT(XP1,XP2,YP1,YP2)
0008          CALL RESIST(XP1,XP2,YP1,YP2)
0009          CALL GRSTOP
0010          STOP
0011          END


0001   C
0002          SUBROUTINE RESIST(XP1,XP2,YP1,YP2)
0003          CALL LENANG(RL,RANG,XP1,XP2,YP1,YP2)
0004          CALL MOVE(XP1,YP1)
0005          CALL ROTATE(RANG,RANG)
0006          CALL VECREL
0007          CALL DRAW(RL/5.,0)
0008          CALL DRAW(RL/20.,-RL/10.)
0009          CALL DRAW(RL/10.,RL/5.)
0010          CALL DRAW(RL/10.,-RL/5.)
0011          CALL DRAW(RL/10.,RL/5.)
0012          CALL DRAW(RL/10.,-RL/5.)
0013          CALL DRAW(RL/10.,RL/5.)
0014          CALL DRAW(RL/20.,-RL/10.)
0015          CALL DRAW(RL/5.,0.)
0016          CALL TXFCUR(1)
0017          CALL VECABS
0018          CALL MOVE(0.,10.)
0019          CALL TEXT(18.'*DRAWING FINISHED*')
0020          RETURN
0021          END
       C
              SUBROUTINE INPUT(XP1,XP2,YP1,YP2)
              CALL TXAM
              CALL TXFCUR(1)
              CALL MOVE(0.,85.)
              CALL TEXT(22,'ADJUST THE THUMBWHEELS')
              CALL TEXT(28,'TWICE TO INPUT THE BEGINNING')
              CALL TEXT(29,'AND ENDING POINTS, WHEN READY')
              CALL TEXT (27,'PRESS ANY ALPHANUMERICAL KEY')
              CALL TEXT(12,'AND "RETURN"')
              CALL LOCATE(1,XP1,YP1,IDAT,IGOT)
              CALL LOCATE(1,XP2,YP2,IDAT,IGOT)
              CALL TXFCUR(2)
              CALL MOVE(0.,20.)
              CALL TEXT(17,'BEGINNING POINT X= ')
              CALL RNUMBR(XP1,2,5)
              CALL TEXT(4,' Y= ')
              CALL RNUMBR(YP1,2,5)
              CALL MOVE(0.,15.)
              CALL TEXT(17,'ENDING POINT X= ')
              CALL RNUMBR(XP2,2,5)
              CALL TEXT(4,' Y= ')
              CALL RNUMBR(YP2,2,5)
              RETURN
              END


       C      THIS SUBROUTINE FINDS THE LENGTH AND ANGLE
       C
              SUBROUTINE LENANG(RLEN,ANG,XP1,XP2,YP1,YP2)
              RLEN=((XP1-XP2)**2+(YP1-YP2)**2)**0.5
              IF (XP1.EQ.XP2) GOTO 2001
```

```
        ANG=ATAN((YP2-YP1)/(XP2-XP1))*180/3.14159
        IF ((YP2.GT.YP1).AND.(ANG.LT.0)) ANG=ANG+180
        IF ((YP2.LT.YP1).AND.(ANG.GT.0)) ANG=ANG+180
        IF ((YP2.EQ.YP1).AND.(XP2.LT.XP1)) ANG=ANG+180
        GOTO 2002
2001    IF (YP2.LE.YP2) ANG=90.
        IF (YP1.GE.YP2) ANG=-90.
2002    RETURN
        END
```

BASIC SUBROUTINE VIEW.WIN

In Chapter 4, the graphics commands VIEWPORT and WINDOW available on Tektronix 4054 graphics system were discussed. Basically, they enable the user's data points to be displayed on the 4054 screen, which has a resolution of 4096 × 3025 pixels and covers 130 graphics display units (GDU) horizontally and 100 GDU vertically.

For microcomputers, on which no VIEWPORT and WINDOW commands are available, a need then arises to develop a BASIC subroutine that can implement the tasks performed by these two commands. This is an essential need because VIEWPORT and WINDOW appear in almost all of the developed BASIC programs throughout this text. Toward that end, we first observe that the coordinate system adopted on microcomputers is different from that on the Tektronix 4054 system. As shown in Fig. D-1b, the origin of the $x \sim y$ coordinate system on a microcomputer is in most cases located at the upper left corner of the display screen with the vertical, y axis directed downward. And returning to Fig. 4-14 in Chapter 4, the Tektronix 4054 display screen has the origin of the $x \sim y$ coordinate system placed at the lower left corner with the vertical, y axis pointing upward. A transformation procedure is hence needed for these two different coordinate systems.

Since the user's data are usually defined in reference to the conventional $x \sim y$ coordinate system as shown in Fig. D-1a, which is in agreement with the Tektronix 4054 system, it is therefore a matter of transforming the data points in Fig. D-1a into those in Fig. D-1b. It may appear that all that is needed is to rotate about the x axis by 180° and to use proper scales for the two axes; but that will result in an upside-down look of the displayed image, namely, points P_3 and P_4 will be positioned above points P_1 and P_2 in Fig. D-1b. In order to preserve the right-side-up displayed image, special transformation equations must be derived.

Let the user's data points (x_i, y_i) be denoted with superscript (w) (WINDOW) and those on the screen with superscript (v) (VIEWPORT). Let their respective ranges be $W_1 \leq x_i^{(w)} \leq W_2$, $W_3 \leq y_i^{(w)} \leq W_4$, $V_1 \leq x_i^{(v)} \leq V_2$, and $V_3 \leq y_i^{(v)} \leq V_4$. The mappings that maintain the right-side-up image should be worked out in such a way that the data point $P_i^{(w)}$ is transformed to $P_i^{(v)}$ on the screen for $i = 1, 2, \ldots, N$ with N being the total number of points for producing a desired image. By utilizing the relationships among the four vertices of the two rectangles shown in Fig. D-1a and b, it is easy to arrive at the following two transformation equations:

$$x_i^{(v)} = V_1 + (x_i^{(w)} - W_1)(V_2 - V_1)/(W_2 - W_1) \tag{a}$$

$$y_i^{(v)} = V_4 - (y_i^{(w)} - W_3)(V_4 - V_3)/(W_4 - W_3) \tag{b}$$

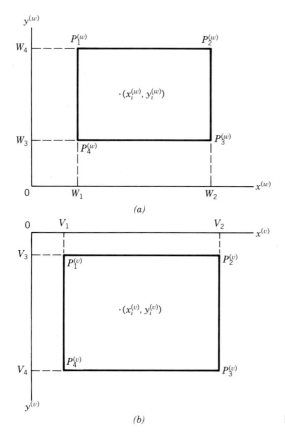

(a)

(b)   **Figure D-1**

The above two equations can be readily used for modifying the data points (x_i, y_i) for $i = 1, 2, \ldots, N$ whenever the statements

$$\text{VIEWPORT V1, V2, V3, V4}$$
$$\text{WINDOW W1, W2, W3, W4}$$

appear in a BASIC program presented in this text. It should be pointed out that the ranges of the available viewport on a microcomputer are often limited, depending on the graphics resolution, that is, $0 \leq V_1 < V_2 \leq x_{max}$, $0 \leq V_3 < V_4 \leq y_{max}$. These limits, (x_{max}, y_{max}), also depend on the level of the microcomputer's supporting graphics software.

A subroutine called VIEW. WIN has been developed based on Eqs. (a) and (b). Its listing is provided below. In the ensuing pages, there are a number of examples showing the applications of this subroutine.

```
100 REM * SUBROUTINE VIEW.WIN *
110 REM    TRANSFORMS N (X,Y) POINTS
120 REM    FROM WINDOW W1,W2,W3,W4
130 REM       TO VIEWPORT V1,V2,V3,V4
140 FOR I=1 TO N
150 X(I)=V1+(X(I)-W1)*(V2-V1)/(W2-W1)
```

```
160 Y(I)=V4-(Y(I)-W3)*(V4-V3)/(W4-W3)
170 NEXT I
180 RETURN
```

The pair of commands GET and PUT for the popular home microcomputers perform almost the same as the set of VIEWPORT and WINDOW for the Tektronix system. The program GETPUT listed below has been prepared to illustrate how Fig. 4-16 can be displayed by use of the GET and PUT commands. GET (X1,Y1)-(X2, Y2), T saves in the array T the graph shown in the rectangular area, which has $(X1, Y1)$ and $(X2, Y2)$ defined as the left lower and upper right corners. The graph can subsequently be modified and then displayed in a specified rectangular area by a PUT (X3,Y3)-(X4,Y4), T statement. To help explain the other statements in Program GETPUT, we present in the next three sections some BASIC programs for the APPLE, IBM, and Radio Shack microcomputers.

```
10 REM * PROGRAM GETPUT*
20 REM *   MICROCOMPUTER VERSION OF
30 REM *   VIEWPORT and WINDOW
40 PCLS : PMODE 3,1 : SCREEN 1,1
50 DIM T(20,20)
60 DATA 0,0,16,10
70 READ X1,Y1,X2,Y2
80 GOSUB 220
90 FOR H=1 TO 7
100 C1=X1+(H-1)*32
110 FOR V=1 TO 7
120 C2=Y2+(V-1)*20
130 PUT (C1,C2)-(C1+20,C2+20),T
140 NEXT V : NEXT H
150 GOTO 150

220 LINE (X1,Y1)-(X1,Y1),PSET
230 LINE -(X1,Y2),PSET
240 LINE -(X2,Y2),PSET
250 LINE -(X2,Y1),PSET
260 LINE -(X1,Y1),PSET
270 RETURN
```

APPLE VERSION OF BASIC PROGRAMS

1. Programs CIRCLES, R. ARRAY, and GEAR

To demonstrate how the Tektronix BASIC programs developed in this text can be easily converted into the versions for the APPLE microcomputer, we again give a number of examples beginning with the Programs CIRCLES, R. ARRAY, and GEAR. By simple comparison of the APPLE version with the Tektronix version in Chapter 4, it is evident that (1) the HPLOT command replaces the MOVE command and (a) the HPLOT TO command replaces the DRAW command.

Also, it can be observed from the listings of new versions that the COS and SIN functions on APPLE microcomputers adopt an argument that must have units in radians, unlike the Tektronix 4054, which can be either in radians or in degrees simply by including a SET DEGREE or SET RADIAN statement in the program. Program GEAR also shows that colons can be used as separators to allow several statements to appear in a same line.

PROGRAM CIRCLES

```
100 D1=8
110 D2=5
120 FORX=50 TO 110 STEP 10
130 X1=X
140 X2=X1+D1
150 FORY=90 TO 30 STEP -10
160 Y1=Y
170 Y2=Y1-D2
180 GOSUB 220
190 NEXT Y
200 NEXT X
210 END
220 HPLOT X1,Y1
230 HPLOT TO X1,Y2
240 HPLOT TO X2,Y2
250 HPLOT TO X2,Y1
260 HPLOT TO X1,Y1
270 RETURN
```

PROGRAM R. ARRAY

```
110 X1=90
120 Y1=60
130 C1=20
140 FOR R=40 TO 10 STEP -10
150 HPLOT X1+R,Y1
160 FOR C=C1 TO 360 STEP C1
170 X=X1+R*COS(C*3.1416/180)
180 Y=Y1+R*SIN(C*3.1416/180)
190 HPLOT TO X,Y
200 NEXT C
210 NEXT R
220 END
```

PROGRAM GEAR AND SUBROUTINES CENTLN, CIRCLE, INVOLU, AND Z.ROTATE

```
100REM   * PROGRAM GEAR *
120 DIM D1(22),D2(22)
130 DATA 96,60,20,95,60
140 READ C1,C2,R1,A1,A2
150 GOSUB 450
160 GOSUB 510
170 GOSUB 320
180 HPLOT T1(1),T2(1)
190 FOR I=2 TO 22
191 HPLOT TO T1(I),T2(I)
192 NEXT I
200 FOR C3=20 TO 340 STEP 20
210 FOR I=1 TO 22
220 X1=T1(I)-C1
230 Y1=T2(I)-C2
240 GOSUB 650
250 D1(I)=X2+C1
260 D2(I)=Y2+C2
270 NEXT I
280 HPLOT D1(I),D2(I)
```

```
290 FOR I=2 TO 22: HPLOT TO D1(I),D2(I): NEXT I
300 NEXT C3
310 END

320 REM * SUBROUTINE INVOLU - PLOTS GEAR TOOTH INVOLUE *
330 N=11
340 A3=(A1-A2)/10
350 DIM T1(22),T2(22)
360 FOR J=1 TO N
370 A4=A1-(J-1)*A3
380 L=R1*0.017453*(J-1)*A3
390 T1(J)=R1*COS(A4*3.14159/180)+C1-L*COS((A4-90)*3.14159/180)
400 T2(J)=R1*SIN(A4*3.14159/180)+C2-L*SIN((A4-90)*3.14159/180)
410 T1(23-J)=-T1(J)+2*C1
420 T2(23-J)=T2(J)
430 NEXT J
440 RETURN
450 REM  * SUBROUTINE CIRCLE
460 HPLOT C1+R1,C2
470 FOR J=5 TO 360 STEP 5
480 HPLOT TO C1+R1*COS(J*3.14159/180),C2+R1*SIN(J*3.14159/180)
490 NEXT J
500 RETURN


510 REM * SUBROUTINE CENTLN
520 HPLOT -1.25*R1+C1,C2
530 HPLOT TO -0.1*R1+C1,C2
540 HPLOT -0.05*R1+C1,C2
550 HPLOT TO 0.05*R1+C1,C2
560 HPLOT 0.1*R1+C1,C2
570 HPLOT TO 1.25*R1+C1,C2
580 HPLOT C1,-1.25*R1+C2
590 HPLOT TO C1,-0.1*R1+C2
600 HPLOT C1,-0.05*R1+C2
610 HPLOT TO C1,0.05*R1+C2
620 HPLOT C1,0.1*R1+C2
630 HPLOT TO C1,1.5*R1+C2
640 RETURN


650 REM * SUBROUTINE Z.ROTATE
660 X2=COS(C3*3.14159/180)*X1-SIN(C3*3.14159/180)*Y1
670 Y2=SIN(C3*3.14159/180)*X1+COS(C3*3.14159/180)*Y1
680 RETURN
```

2. Program PERSPECT

We will demonstrate the use of subscripted arrays by looking at the APPLE
BASIC version of the Program PERSPECT, which draws the perspective view of a
$2 \times 3 \times 4$ brick. Unlike the Tektronix 4054 BASIC, the APPLE BASIC cannot
read or print the entire array. Lines 160, 190, and 220 in Program PERSPECT are
examples. The CLEAR command in APPLE BASIC works the same way as
INITIAL in Tektronix BASIC. The commands HCOLOR and HGR in line 415 of

Program PERSPECT are to set up the color in a color APPLE microcomputer and to request the high-graphics-resolution software support.

Notice that the APPLE graphics also uses the PRINT command, for example, line 350, to display **message** on the screen, and the INPUT command, for example, line 360, to enter data from the keyboard. Lines 6000 through 6080 give the details on how the WINDOW and VIEWPORT statements of the Tektronix-version programs are replaced in accordance with the derivation presented in the preceding section.

PROGRAM PERSPECT

```
100 REM * PROGRAM PERSPECT *
110 CLEAR
120 HOME
130 N=24
140 DIM X0(N),Y0(N),Z0(N),Z(N),Y(N),Z(N)
150 REM * FIRST 8 FOR BOX, THEN 16 FOR LETTERS F, T, AND R.
160 FOR I=1 TO N: READ X0(I): NEXT I
170 DATA 0,4,4,0,0,4,4,0
180 DATA 3.4,3.4,3.4,3.8,3.7,0.2,0.4,0.6,0.4,4,4,4,4,4,4,4
190 FOR I=1 TO N: READ Y0(I): NEXT I
200 DATA 0,0,2,2,0,0,2,2
210 DATA 1.4,2.6,1.8,1.8,1.6,2,2,2,2,1.4,1.6,1.8.1.8.1.6,1.6,1.4
220 FOR I=1 TO N: READ Z0(I): NEXT I
230 DATA 0,0,0,0,3,3,3,3
240 DATA 3,3,3,3,3,0.2,0.2,0.2,0.6,0.6,0.6,0.6,0.2,0.2,0.4,0.2
250 REM * TRANSLATE SO THAT X-Y PLANE BISECTS *
260 REM THE R AND T PLANES OF THE BRICK
270 READ T1,T2,T3
280 DATA -2,-1,-1.5
290 FOR I=1 TO N
300 X(I)=X0(I)+T1
310 Y(I)=Y0(I)+T2
320 Z(I)=Z0(I)+T3
330 NEXT I
340 REM * PERSPECTIVE VIEWING *
350 PRINT 'INPUT AXIAL COORDINATE OF SCREEN AND EYE ";
360 INPUT Z1,Z2
370 FOR I=1 TO N
380 X(I)=X(I)*(Z2-Z1)/(Z2-Z(I))
390 Y(I)=Y(I)*(Z2-Z1)/(Z2-Z(I))
400 NEXT I
410 REM * COMMENCE PLOTTING
415 HCOLOR=3: HGR
420 REM * WINDOW -5,5,-5,5
430 REM * VIEWPORT 10,90,10,90
435 GOSUB 6000
440 REM 8 DRAW X-Y PROJECTION OF 3-D BOX
450 HPLOT X(5),Y(5)
460 HPLOT TO X(6),Y(6)
470 HPLOT TO X(2),Y(2)
480 HPLOT TO X(1),Y(1)
490 HPLOT TO X(5),Y(5)
500 HPLOT TO X(8),Y(8)
510 HPLOT TO X(7),Y(7)
520 HPLOT TO X(3),Y(3)
530 HPLOT TO X(4),Y(4)
```

```
540 HPLOT TO X(8),Y(8)
550 HPLOT X(6),Y(6)
560 HPLOT TO X(7),Y(7)
570 HPLOT X(3),Y(3)
580 HPLOT TO X(2),Y(2)
590 HPLOT X(1),Y(1)
600 HPLOT TO X(4),Y(4)
610 REM DRAW "F"
620 HPLOT X(9),Y(9)
630 HPLOT TO X(11),Y(11)
640 HPLOT TO X(12),Y(12)
650 HPLOT X(10),Y(10)
660 HPLOT TO X(13),Y(13)
670 REM DRAW "T"
680 HPLOT X(14),Y(14)
690 HPLOT TO X(16),Y(16)
700 HPLOT X(17),Y(17)
710 HPLOT TO X(15),Y(15)
720 REM DRAW "R"
730 HPLOT X(18),Y(18)
740 HPLOT TO X(20),Y(20)
750 HPLOT TO X(21),Y(21)
760 HPLOT TO X(22),Y(22)
770 HPLOT TO X(19),Y(19)
780 HPLOT X(23),Y(23)
790 HPLOT TO X(24),Y(24)
800 HOME
810 END


6000 REM * SUBROUTINE VIEWPORT
6010 REM * ORIGINAL VIEWPORT 10,90,10,90 *
6020 REM * ORIGINAL WINDOW -5,5,-5,5 *
6030 REM * THE X,Y ARRAY TO BE CHANGED *
6040 FOR I=1 TO N
6050 X(I)=((X(I)-(-5))/(5-(-5))*(90-10)/130+10/130)*259
6060 Y(I)=((5-Y(I))/(5-(-5))*(90-10)/100+(100-90)/100)*159
6070 NEXT I
6080 RETURN
```

IBM VERSION OF BASIC PROGRAMS

1. Programs CIRCLES, R. ARRAY, and GEAR

The converted versions of Programs CIRCLES, R. ARRAY, and GEAR show that the IBM Personal Computer uses CLS, PSET, and LINE in place of PAGE, MOVE, and DRAW commands in the Tektronix 4054 version. However, it should be pointed out that PSET (X, Y) results in a dot being displayed at the location (X, Y).

Line 110 SCREEN 2 requests a high-resolution graphics software to be made available. In the case of Program GEAR, the high-resolution graphics makes it possible to utilize 500 and 200 GDU in the horizontal and vertical directions, respectively. Notice also that adding this (aspect) ratio of 500/200 to a number of statements, lines 175, 187, and others, helped in removing the problem of distorting a circle into an ellipse on the display screen.

PROGRAM CIRCLES

Program CIRCLES

```
90 CLS
100 SCREEN  2
110 X1=120
120 Y1=100
130 C1=20
140 FOR R=80 TO 20 STEP -20
145 X=(X1+R)*500/200
150 PSET (X,Y1)
160 FOR C=C1 TO 360 STEP C1
165 C2=C*3.141593/180
170 X=(X1+R*COS(C2))*500/200
180 Y=Y1+R*SIN(C2)
190 LINE -(X,Y)
200 NEXT C
210 NEXT R
220 END
```

PROGRAM R.ARRAY

Program R.ARRAY

```
80 CLS
90 SCREEN 2
100 D1=40
110 D2=10
120 FOR X=150 TO 450 STEP 50
130 X1=X
140 X2=X1+D1
150 FOR Y=140 TO 40 STEP -20
160 Y1=Y
170 Y2=Y1-D2
180 GOSUB 220
190 NEXT Y
200 NEXT X
210 END
220 PSET (X1,Y1)
230 LINE -(X1,Y2)
240 LINE -(X2,Y2)
250 LINE -(X2,Y1)
260 LINE -(X1,Y1)
270 RETURN
```

PROGRAM GEAR AND SUBROUTINES CENTLN, CIRCLE, INVOLU, AND Z.ROTATE

Program GEAR and Subroutines CENTLN, CIRCLE, INVOLU and Z.ROTATE

```
100 REM   *PROGRAM GEAR*
105 P1=3.141593/180 :CLS
110  SCREEN 2
120  DIM Q1(22),Q2(22)
125 PRINT
130 DATA 300,100,50,95,60
140 READ C1,C2,R1,A1,A2
150 GOSUB 450
160 GOSUB 510
170 GOSUB 320
175 H=(T1(1)-C1)*500/200+C1
180 PSET (H,T2(1))
185 FOR K=1 TO 22
187 H=(T1(K)-C1)*500/200+C1
190 LINE -(H,T2(K))
195 NEXT K
200 FOR C3=20 TO 340 STEP 20
210 FOR I=1 TO 22
220 X1=T1(I)-C1
230 Y1=T2(I)-C2
240 GOSUB 650
250 Q1(I)=X2*500/200+C1
260 Q2(I)=Y2+C2
270 NEXT I
280 PSET (Q1(1),Q2(1))
285 FOR K=1 TO 22
290 LINE -(Q1(K),Q2(K))
295 NEXT K
300 NEXT C3
310 END
```

```
450 REM  * SUBROUTINE CIRCLE *
460 PSET (C1+R1*500/200,C2)
470 FOR J=5 TO 360 STEP 5
475 X3=C1+(R1*COS(J*P1))*500/200
476 Y3=C2+R1*SIN(J*P1)
480 LINE -(X3,Y3)
490 NEXT J
500 RETURN

510 REM * SUBROUTINE CENTLN *
520 PSET (-1.25*R1*500/200+C1,C2)
530 LINE -(-.1*R1*500/200+C1,C2)
540 PSET (-.05*R1*500/200+C1,C2)
550 LINE -(.05*R1*500/200+C1,C2)
560 PSET (.1*R1*500/200+C1,C2)
570 LINE -(1.25*R1*500/200+C1,C2)
580 PSET (C1,-1.25*R1+C2)
590 LINE -(C1,-.1*R1+C2)
600 PSET (C1,-.05*R1+C2)
610 LINE -(C1,.05*R1+C2)
620 PSET (C1,.1*R1+C2)
630 LINE -(C1,1.5*R1+C2)
640 RETURN

650 REM * SUBROUTINE Z.ROTATE *
660 X2=COS(C3*P1)*X1-SIN(C3*P1)*Y1
670 Y2=SIN(C3*P1)*X1+COS(C3*P1)*Y1
680 RETURN
```

```
320 REM  *SUBROUTINE INVOLU - PLOTS GEAR TOOTH INVOLUTE*
330 N=11
340 A3=(A1-A2)/10
350 DIM T1(22),T2(22)
360 FOR J=1 TO N
370 A4=A1-(J-1)*A3
380 L=R1*.017453*(J-1)*A3
390 T1(J)=R1*COS(A4*P1)+C1-L*COS((A4-90)*P1)
400 T2(J)=R1*SIN(A4*P1)+C2-L*SIN((A4-90)*P1)
410 T1(23-J)=-T1(J)+2*C1
420 T2(23-J)=T2(J)
430 NEXT J
440 RETURN
```

2. Program PERSPECT

The IBM BASIC also cannot read an entire subscripted array by merely mentioning of the variable name alone. Lines 155 through 225 in Program PERSPECT verify that the elements of subscripted arrays must be read one by one. Like the APPLE microcomputer, the IBM Personal Computer allows several statements to be separated by colons and listed in the same line. The window–viewport subroutine is listed in lines 890 through 940.

PROGRAM PERSPECT

```
100 REM  * PROGRAM PERSPECT *
110 CLEAR: CLS
120 SCREEN 2
130 N=24
140 DIM XO(N),YO(N),ZO(N),X(N),Y(N),Z(N)
150 REM  *FIRST 8 FOR BOX, THEN 16 FOR LETTERS F, T, AND ·R
155 FOR K=1 TO N
160 READ XO(K)
165 NEXT K
170 DATA 0,4,4,0,0,4,4,0
180 DATA 3.4,3.4,3.4,3.8,3.7,0.2,0.4,0.6,0.4,4,4,4,4,4,4,4
185 FOR K=1 TO N
190 READ YO(K)
195 NEXT K
200 DATA 0,0,2,2,0,0,2,2
210 DATA  1.4,1.6,1.8,1.8,1.6,2,2,2,2,1.4,1.6,1.8,1.8,1.6,1.6,1.4
215 FOR K=1 TO N
220 READ ZO(K)
225 NEXT K
230 DATA 0,0,0,0,3,3,3,3
240 DATA 3,3,3,3,3,0.2,0.2,0.2,0.6,0.6,0.6,0.6,0.2,0.2,0.4,0.2
250 REM  * TRANSLATE SO THAT X-Y PLANE BISECTS
260 REM        THE R AND T PLANES OF THE BRICK *
270 READ T1,T2,T3
280 DATA  -2,-1,-1.5
290 FOR I=1 TO N
300 X(I)=XO(I)+T1
310 Y(I)=YO(I)+T2
320 Z(I)=ZO(I)+T3
330 NEXT I
340 REM   * PERSPECTIVE VIEWING *
350 PRINT "INPUT AXIAL COORDINATES OF SCREEN AND EYE ";
360 INPUT Z1,Z2
370 FOR I=1 TO N
```

```
380 X(I)=X(I)*(Z2-Z1)/(Z2-Z(I))
390 Y(I)=Y(I)*(Z2-Z1)/(Z2-Z(I))
400 NEXT I
410 REM  * COMMENCE PLOTTING *
420 W1=-5: W2=5: W3=-5: W4=5
430 V1=140: V2=500: V3=10: V4=190
435 GOSUB 900
440 REM  *DRAW X-Y PROJECTION OF 3-D BOX *
450 PSET (X(5),Y(5))
460 LINE -(X(6),Y(6))
470 LINE -(X(2),Y(2))
480 LINE - (X(1),Y(1))
490 LINE -(X(5),Y(5))                890 REM  * WINDOW-VIEWPORT SUBROUTINE *
500 LINE -(X(8),Y(8))                900 FOR I=1 TO N
510 LINE -(X(7),Y(7))                910 X(I)=V1+(X(I)-W1)*(V2-V1)/(W2-W1)
520 LINE -(X(3),Y(3))                920 Y(I)=V4-(Y(I)-W3)*(V4-V3)/(W4-W3)
530 LINE -(X(4),Y(4))                930 NEXT I
540 LINE -(X(8),Y(8))                940 RETURN
550 PSET (X(6),Y(6))
560 LINE -(X(7),Y(7))
570 PSET (X(3),Y(3))
580 LINE -(X(2),Y(2))
590 PSET (X(1),Y(1))
600 LINE -(X(4),Y(4))
610 REM    * DRAW "F"
620 PSET (X(9),Y(9))
630 LINE -(X(11),Y(11))
640 LINE -(X(12),Y(12))
650 PSET  (X(10),Y(10))
660 LINE -(X(13),Y(13))
670 REM  * DRAW "T"
680 PSET (X(14),Y(14))
690 LINE -(X(16),Y(16))
700 PSET (X(17),Y(17))
710 LINE -(X(15),Y(15))
720 REM    * DRAW "R"
730 PSET (X(18),Y(18))
740 LINE -(X(20),Y(20))
750 LINE -(X(21),Y(21))
760 LINE -(X(22),Y(22))
770 LINE -(X(19),Y(19))
780 PSET (X(23),Y(23))
790 LINE -(X(24),Y(24))
800 END
```

3. Program A. C. I

As an additional illustration, Program A. C. I presented in Chapter 6 for calculations of cross-sectional area, position of centroid, and moments of inertia has been converted as well into IBM BASIC language. In this new version, a number of new statements need to be discussed. Line 310 exemplifies a decimal format for printing out floating-point numbers. The logical IF statement shown in line 590 is similar to that in the Tektronix version; but those in lines 1380 and 1390 are more FORTRAN-like than BASIC. Line 280 also demonstrates a logical IF statement involving a string of characters and the use of quotes.

LOCATE commands appear in lines 880, 900, and others. They cause the cursor to be positioned at a specified location on the display screen, and often are followed by a PRINT statement so that a string of characters can start to be

printed there. It may seem that both MOVE X, Y in Tektronix BASIC and
LOCATE X, Y in IBM BASIC carry out the same function; but the former X, Y
are in graphics display units (GDU), whereas the latter X, Y are in "block" units
when the screen is being used in **text** mode. Section 3.5 discussed the distinction
between text and graphics modes. Tektronix 4054 graphics terminal uses GDU
for both modes. Microcomputers connected to appropriate graphics terminals
usually allow only a finite number of text lines, each line with limited number
of characters. For example, if only 24 lines of characters are allowed, the X and
Y in LOCATE statements must have values within the ranges of $0 \leq X \leq 23$ and
$0 \leq Y \leq 79$.

PRESET is different from PSET in that it results in a background-colored,
instead of foreground-colored, dot, which is hence invisible.

As stated by line 390, the IBM version of BASIC does not have an AXIS
command. Consequently, the drawing of the axes shown in Fig. 6-7 has to be
omitted. It could, however, be a good exercise for the readers to add necessary
statements to Program A. C. I to fulfill this frequent need.

PROGRAM A. C. I

```
100 CLEAR
110 SCREEN 2
120 CLS
130 PRINT "PROGRAM   A.C.I."
140 PRINT: PRINT "   CALCULATES CROSS-SECTIONAL AREA."
150 PRINT "   DETERMINES LOCATION OF CENTROID"
160 PRINT "   AND COMPUTES MOMENTS OF INERTIA."
170 PRINT: PRINT "INPUT :   COORDINATES OF VERTICES "
180 PRINT "          DESCRIBING THE CROSS-SECTIONAL"
190 PRINT "          SHAPE.  MUST ALL BE POSITIVE."
200 PRINT: PRINT "WANT TO SEE A DEMONSTRATION.  Y/N?   ";
210 INPUT A$
220 IF A$="Y" THEN 1260
230 PRINT "ENTER THE NUMBER OF VERTICES. ";
240 INPUT N1
250 GOTO 1120
260 PRINT "WANT TO LIST THE COORDINATES, Y/N?   ";
270 INPUT P$
280 IF P$="N" THEN  360
290 PRINT "NO. OF VERTICES = ";N1
300 FOR K=1 TO N1
310 PRINT USING "###.###"; X(K),Y(K)
330 NEXT K
340 PRINT "PUSH KEY 'Y' TO CONTINUE ! ";
350 INPUT A$
360 V1=280: V2=600: V3=10: V4=170
370 REM  * REPLACE SCALE STATEMENT WITH WINDOW COORDINATES. *
380 W1=0: W2=W: W3=0: W4=W
390 REM   * SKIP 400 TO 480 DUE TO LACK OF IBM AXIS STATEMENT *
485 GOSUB 1500
490 PRESET (X3(N1),Y3(N1))
495 FOR I=1 TO N1
500 LINE -(X3(I),Y3(I))
505 NEXT I                      520 X0(K)=0
507 FOR K=1 TO N1               530 Y0(K)=0
510 A(K)=0                      540 I1(K)=0
```

```
550 I2(K)=0                          600 K1=1
560 NEXT K                           610 H=Y(K1)-Y(K)
570 FOR K=1 TO N1                     620 B=X(K1)-X(K)
580 K1=K+1                            630 X1=X(K)+2*B/3
590 IF K<N1 THEN 610                  640 Y1=Y(K)+H/3
650 X2=X(K)+B/2
660 Y2=Y(K)/2
670 A1=B*H/2
680 A2=2*Y2*B
690 A(K)=A1+A2
700 IF B=0 THEN 750
710 X0(K)=(X1*A1+X2*A2)/A(K)
720 Y0(K)=(Y1*A1+Y2*A2)/A(K)
730 I1(K)=B*H^3/36+A1*(Y1-Y0(K))^2+B*(2*Y2)^3/12+A2*(Y2-Y0(K))^2
740 I2(K)=H*B^3/36+A1*(X1-X0(K))^2+2*Y2*B^3/12+A2*(X2-X0(K))^2
750 NEXT K
760 C1=0
770 C2=0
780 FOR K=1 TO N1
790 C1=C1+A(K)*X0(K)
800 C2=C2+A(K)*Y0(K)
810 NEXT K
814 A0=0
815 FOR K=1 TO N1
820 A0=A0+A(K)
825 NEXT K
830 C1=C1/A0
840 C2=C2/A0
850 REM
860 REM
870 REM
880 LOCATE 19,35
890 PRINT "OUTPUT: "
900 LOCATE 21,35
910 PRINT "AREA = ";A0
920 LOCATE 22,35
930 PRINT "(X)CENTROID = ";C1
940 LOCATE 23,35
950 PRINT "(Y)CENTROID = ";C2
955 I4=0: I5=0
960 FOR K=1 TO N1
970 I1(K)=I1(K)+A(K)*(Y0(K)-C2)^2
980 I2(K)=I2(K)+A(K)*(X0(K)-C1)^2
982 I4=I4+I1(K)
984 I5=I5+I2(K)
990 NEXT K
1000 LOCATE 21,60
1010 PRINT "Ix = ";I4
1020 LOCATE 22,60
1030 PRINT "Iy = ";I5
1040 GOSUB 1560
1050 LOCATE (C4+8)/8, (C5+8)/8
1060 PRINT " X "
1065 PRESET (C5,C4)
1070 LINE -(C3,C4)
1080 LINE -(C3,C6)
1085 LOCATE C6/8,C3/8
1090 PRINT " Y "
1095 LOCATE 23,1
1100 END
```

```
1120 DIM X(N1),Y(N1),A(N1),XO(N1),YO(N1),X3(N1),Y3(N1)
1130 DIM I1(N1),I2(N1)
1140 PRINT "ENTER COORDINATES:   X1,Y1/X2,Y2/...etc"
1150 PRINT "   (ENTER TWO NUMBERS THEN PUSH 'RETURN')"
1160 PRINT "FOR POSITIVE AREA, ENTER VERTICES IN "
1170 PRINT "   CLOCKWISE ORDER."
1180 FOR L1=1 TO N1
1190 INPUT X(L1),Y(L1)
1200 NEXT L1
1210 GOTO 1340
1220 INPUT X
1230 PRINT "ENTER THE ";N1;" Y VALUES ";
1240 INPUT Y
1250 GOTO 1340
1260 N1=8
1270 DIM X(N1),Y(N1),A(N1),XO(N1),YO(N1),X3(N1),Y3(N1)
1280 DIM I1(N1),I2(N1)
1285 FOR I=1 TO N1
1290 READ X(I)
1295 NEXT I
1300 DATA 3,3,1,1,6,6,4,4
1305 FOR I=1 TO N1
1310 READ Y(I)
1315 NEXT I
1320 DATA 1,4,4,5,5,4,4,1
1330 P$="Y"
1340 REM * SCALING *
1350 X2=X(1)
1360 Y2=Y(1)
1370 FOR K=1 TO N1
1380 IF X2 < X(K) THEN X2=X(K)
1390 IF Y2 < Y(K) THEN Y2=Y(K)
1400 NEXT K
1410 IF X2>Y2 THEN 1440
1420 W=1.05*Y2
1430 GOTO 260
1440 W=1.05*X2
1450 GOTO 260
1500 REM    * WINDOW-VIEWPORT SUBROUTINES *
1510 FOR K=1 TO N1
1520 X3(K)=V1+(X(K)-W1)*(V2-V1)/(W2-W1)
1530 Y3(K)=V4-(Y(K)-W3)*(V4-V3)/(W4-W3)
1540 NEXT K
1550 RETURN
1560 C3=V1+(C1-W1)*(V2-V1)/(W2-W1)
1570 C4=V4-(C2-W3)*(V4-V3)/(W4-W3)
1580 C5=V1+((C1+.3*W)-W1)*(V2-V1)/(W2-W1)
1590 C6=V4-((C2+.3*W)-W3)*(V4-V3)/(W4-W3)
1600 RETURN
```

RADIO SHACK TRS-80 BASIC PROGRAMS

1. Programs CIRCLES and R.ARRAY

TRS-80 color microcomputers have also been employed for development of
BASIC programs CIRCLES and R.ARRAY. In these BASIC versions, the statements PMODE 0,1 and SCREEN 1,1 enable a graphics display resolution of
256×192 to be made available for generation of graphs. The command PCLS
clears the screen. The statement LINE (X,Y)-(X,Y),PSET, for example, line 150

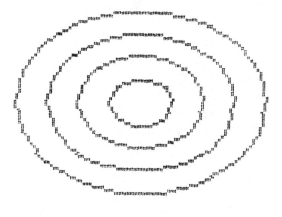

Figure D-2

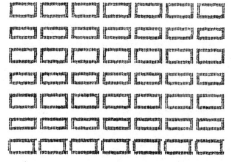

Figure D-3

in Program CIRCLES and line 220 in Program R. ARRAY, replaces the MOVE statement in the Tektronix BASIC but results in displaying a dot at position (X, Y) on the screen. The statement LINE-(X, Y), PSET draws a line from the current position of the cursor to the point (X, Y).

The (aspect) ratio of 192/256 helped in removing the distortion of a circle into an ellipse and in correcting the actual 16 × 10 (width by height) shape of the rectangles displayed on the screen when Programs CIRCLES and R. ARRAY are run, respectively. However, it should be particularly pointed out that when these displayed images, though corrected on the screen, were copied on a printer, they were distorted again, as manifested by Figs. D-2 and D-3. This is due to the fact that the printer has a different horizontal-to-vertical resolution from that of the display screen.

PROGRAM CIRCLES

```
10 REM " PROGRAM CIRCLES *
20 PMODE 0,1
30 PCLS
40 SCREEN 1,1
50 S=192/256
100 PI=3.1416
110 X1=125
120 Y1=95
130 C1=20
```

```
10 REM * PROGRAM R.ARRAY *
20 PMODE 0,1
30 PCLS
40 SCREEN 1,1
50 S=192/256
100 D1=16
110 D2=10
120 FOR X=50 TO 170 STEP 20
130 X1=X
```

```
140 FOR R=80 TO 20 STEP -20
150 LINE (X1+R,Y1*S)-(X1+R,Y1*S),PSET
160 FOR C=C1 TO 360 STEP C1
170 X=X1+R*COS(C/180*PI)
180 Y=Y1+R*SIN(C/180*PI)
190 LINE -(X,Y*S),PSET
200 NEXT C
210 NEXT R
220 GOTO 220
```

```
140 X2=X1+D1
150 FOR Y=150 TO 30 STEP -20
160 Y1=Y
170 Y2=Y1-D2
180 GOSUB 220
190 NEXT Y
200 NEXT X
210 GOTO 210
220 LINE (X1,Y1*S)-(X1,Y1*S),PSET
230 LINE -(X1,Y2*S),PSET
240 LINE -(X2,Y2*S),PSET
250 LINE -(X2,Y1*S),PSET
260 LINE -(X1,Y1*S),PSET
270 RETURN
```

2. Program ROTATE

In Chapter 4, Program ROTATE was used for showing the rotation of a $2 \times 3 \times 4$ brick with "*F*," "*T*," and "*R*" labeled on its front, top, and right sides, respectively. A TRS-80 BASIC version of this program has been developed and its listing is presented below. Again, from the listing it can be observed that the TRS-80 BASIC like APPLE and IBM BASICs has to read the elements of subscripted arrays by referring to them one by one. This is illustrated by lines 170 through 234. Aside from the new statement LINE (X1, Y1)-(X2, Y2), PSET, which replaces MOVE or DRAW, the other statements are almost identical to those in Tektronix 4054 BASIC language. This fact is verified by the subroutines TRANSF and MAX. MIN presented here and those counterparts in Chapter 4.

Subroutine VIEW. WIN makes it possible to replace VIEWPORT and WINDOW statements in the Tektronix 4054 BASIC programs. This subroutine is listed in lines 1410 through 1460.

It should be repeatedly pointed out that the ratio 256/192 used in a number of statements in Program ROTATE enables undistorted images to be generated on the display screen although the printed copy will still show distorted shapes, as demonstrated in Fig. D-4, because of different resolutions adopted by the printer and the display screen.

PROGRAM ROTATE

```
10 REM * PROGRAM ROTATE *
20 PMODE 0,1
30 PCLS
40 SCREEN 1,1
50 PI=3.1416
130 REM * 8 POINTS FOR BRICK ;  16 POINTS FOR LETTERS
140 N=24
150 DIM X0(N),Y0(N),Z0(N), X(N),Y(N),Z(N)
170 FOR I=1 TO 24
172 READ X0(I)
174 NEXT I
180 DATA 0,4,4,0,0,4,4,0
190 DATA 3.4,3.4,3.4,3.8,3.7,.2, .4,.6,.4,4,4,4,4,4,4,4
200 FOR I=1 TO 24
202 READ Y0(I)
204 NEXT I
210 DATA 0,0,2,2,0,0,2,2
```

Figure D-4

```
220 DATA 1.4,1.6,1.8,1.8,1.6,2,2,2,2,1.4,1.6,1.8,1.8, 1.6,1.6
    ,1.4
230 FOR I=1 TO 24
232 READ Z0( I )
234 NEXT I
240 DATA 0,0,0,0,3,3,3,3
250 DATA 3,3,3,3,.2,.2,.2,.6,.6,.6,.6,.2,.2,.4,.2
270 REM C-ROTATION ; S-SCALE ;  T-TRANSLATION
275 REM SCALING OF 192/256=.75  FOR ADJUSTING Y/X
    RESOLUTION OF TRS.
280 FOR L=1 TO 3
290 READ S1,S2,S3,T1,T2,T3,C1,C2,C3
295 REM * GENERATE "F" VIEW *
300 DATA 1,.75,.75,0,0,0,0,0,0
305 REM * SET THE VIEWPORT *
310 READ H0,V0,S
320 DATA 20,95,90
330 V1=H0
332 V2=H0+S
334 V3=V0
336 V4=V0+S
340 REM * TRANSFORMATION *
350 GOSUB 480
355 REM * FIND MAX & MIN *
360 GOSUB 1210
370 IF L>1 THEN 390
380 W=X2-X1
382 IF Y2-Y1>W THEN 388
384 GOTO 390
388 W=Y2-Y1
390 W1=X1
392 W2=X1+W
394 W3=Y1
396 W4=Y1+W
400 REM * WINDOW MAPPING *
405 GOSUB 1410
407 REM * PLOTTING *
410 GOSUB 760
```

```
415 REM * GENERATE "T" VIEW *
420 DATA 1,.75,.75,0,0,0,1.5708,0,0
425 REM * SET THE VIEWPORT *
430 DATA 20,0,90
435 REM * GENERATE "R" VIEW *
440 DATA 1,.75,1.0,0,0,0,0,-1.5708,0
445 REM * SET THE VIEWPORT *
450 DATA 150,95,90
460 NEXT L
470 GOTO 470



480 REM * SUBROUTINE TRANSF *  TRANSLATE, ROTATE & SCALE
490 A=COS(C3)*COS(C2)
500 B=SIN(C3)*COS(C2)
510 C=-SIN(C2)
520 D=-SIN(C3)*COS(C1)  +COS(C3)*SIN(C2)*SIN(C1)
530 E=COS(C3)*COS(C1)   +SIN(C2)*SIN(C3)*SIN(C1)
540 F=COS(C2)*SIN(C1)
550 G=SIN(C3)*SIN(C1)   +COS(C3)*SIN(C2)*COS(C1)
560 H=-COS(C3)*SIN(C1)  +SIN(C3)*SIN(C2)*COS(C1)
570 P=COS(C2)*COS(C1)
580 FOR I=1 TO N
590 X(I)=A*S1*(X0(I)+T1)+D*S2*(Y 0(I)+T2)+G*S3*(Z0(I)+T3)
600 Y(I)=B*S1*(X0(I)+T1)+E*S2*(Y 0(I)+T2)+H*S3*(Z0(I)+T3)
610 Z(I)=C*S1*(X0(I)+T1)+F*S2*(Y 0(I)+T2)+P*S3*(Z0(I)+T3)
620 NEXT I
630 RETURN
760 REM * SUBROUTINE PLOT *           PLOT A 2X3X4 BRICK
770 LINE (X(5),Y(5))-(X(5),Y(5))          ,PSET
780 LINE -(X(6),Y(6)),PSET
790 LINE -(X(2),Y(2)),PSET
800 LINE -(X(1),Y(1)),PSET
810 LINE -(X(5),Y(5)),PSET
820 LINE -(X(8),Y(8)),PSET
830 LINE -(X(7),Y(7)),PSET
840 LINE -(X(3),Y(3)),PSET
850 LINE -(X(4),Y(4)),PSET
860 LINE -(X(8),Y(8)),PSET
870 LINE (X(6),Y(6))-(X(6),Y(6)) ,PSET
880 LINE -(X(7),Y(7)),PSET
890 LINE (X(3),Y(3))-(X(3),Y(3)) ,PSET
900 LINE -(X(2),Y(2)),PSET
910 LINE (X(1),Y(1))-(X(1),Y(1)) ,PSET
920 LINE -(X(4),Y(4)),PSET
930 REM * DRAW "F"
940 LINE (X(9),Y(9))-(X(9),Y(9)) ,PSET
950 LINE -(X(11),Y(11)),PSET
960 LINE -(X(12),Y(12)),PSET
970 LINE (X(10),Y(10))-(X(10),Y( 10)),PSET
980 LINE -(X(13),Y(13)),PSET
990 REM * DRAW "T"
1000 LINE (X(14),Y(14))-(X(14),Y (14)),PSET
1010 LINE -(X(16),Y(16)),PSET
1020 LINE (X(17),Y(17))-(X(17),Y (17)),PSET
1030 LINE -(X(15),Y(15)),PSET
```

```
1040 REM * DRAW "R"
1050 LINE (X(18),Y(18))-(X(18),Y (18)),PSET
1060 LINE -(X(20),Y(20)),PSET
1070 LINE -(X(21),Y(21)),PSET
1080 LINE -(X(22),Y(22)),PSET
1090 LINE -(X(19),Y(19)),PSET
1100 LINE (X(23),Y(23))-(X(23),Y (23)),PSET
1110 LINE -(X(24),Y(24)),PSET
1120 RETURN
1210 REM * SUBROUTINE MAX.MIN *
1220 REM FIND MAXIMA & MINIMA OF  X & Y ARRAYS OF LENGTH N & STORE THEM
IN X1, Y1 (MIN)      & X2, Y2 (MAX).
1230 X1=X(1)
1240 X2=X(1)
1250 Y1=Y(1)
1260 Y2=Y(1)
1270 FOR I=2 TO N
1280 IF X(I)>X1 THEN 1300
1290 X1=X(I)
1300 IF X(I)<X2 THEN 1320
1310 X2=X(I)
1320 IF Y(I)>Y1 THEN 1340
1330 Y1=Y(I)
1340 IF Y(I)<Y2 THEN 1360
1350 Y2=Y(I)
1360 NEXT I
1370 RETURN




1410 REM * SUBROUTINE VIEW.WIN *     MAPS W1, W2, W3 & W4 OF  WINDOW TO V
1, V2, V3 & V4      OF VIEWPORT. TRANSFORMS X & Y ARRAYS OF
     LENGTH N.
1420 FOR I=1 TO N
1430 X(I)=V1+(X(I)-W1)*(V2-V1)/        (W2-W1)
1440 Y(I)=V4-(Y(I)-W3)*(V4-V3)/        (W4-W3)
1450 NEXT I
1460 RETURN
```

AUTHOR INDEX

AUTHOR INDEX

SUBJECT INDEX

SUBJECT INDEX

FORTRAN, BASIC AND COMPUTER GRAPHICS COMMANDS

FORTRAN, BASIC AND COMPUTER GRAPHICS COMMANDS

493

INTERACTIVE GRAPHICS AND CAD/CAM PROGRAMS INDEX

INTERACTIVE GRAPHICS AND CAD/CAM PROGRAMS INDEX

497